Hairy Nanoparticles

Hairy Nanoparticles

From Synthesis to Applications

Edited by Zhiqun Lin and Yijiang Liu

WILEY-VCH

Editors

Dr. Zhiqun Lin
Department of Chemical and
biomolecular Engineering
National University of Singapore
Singapore 117585
Singapore

Prof. Yijiang Liu
Xiangtan University
College of Chemistry
XiJiaoYangGuTang
Yuhu District
411105 Xiangtan
China

Cover Image: © Westend61/Getty Images

Library of Congress Card No.: applied for

British Library Cataloguing-in-Publication Data
A catalogue record for this book is available from the British Library.

Bibliographic information published by the Deutsche Nationalbibliothek
The Deutsche Nationalbibliothek lists this publication in the Deutsche Nationalbibliografie; detailed bibliographic data are available on the Internet at <http://dnb.d-nb.de>.

Print ISBN: 978-3-527-35005-6
ePDF ISBN: 978-3-527-83585-0
ePub ISBN: 978-3-527-83586-7
oBook ISBN: 978-3-527-83587-4

Typesetting Straive, Chennai, India
Printing and Binding CPI Group (UK) Ltd, Croydon, CR0 4YY

C117641_130623

Contents

Preface

Hairy nanoparticles (NPs), also known as polymer-grafted NPs, represent an important class of hybrid NPs, composed of a layer of polymer shell tethered to an NP core. The past several decades have witnessed rapid advances in synthesis, self-assembly, and application of a rich variety of hairy NPs, including zero-dimensional (0D) plain, hollow and core–shell NPs, one-dimensional (1D) plain, hollow and core–shell nanorods and nanowires, and two-dimensional (2D) nanosheets. Owing to their intriguing characteristics enabled by judiciously tailoring dimensions, compositions, architectures, and surface chemistry, these hairy nanomaterials find applications in optics, optoelectronics, magnetic devices, catalysis, sensors, and biotechnology, among other areas. Despite the impressive developments noted above, a book centered on a comprehensive discussion of hairy NPs herein referred to as hairy nanocrystals of different shapes (e.g. sphere-, rod-, star-, sheet-, prism-, and dumbbell-like) regarding their synthesis, self-assembly, interfacial properties, functionalities, and applications is yet to be available.

This motivates us to edit this book that offers the current field of knowledge from synthesis to self-assembly, property, and application of hairy NPs. The synthesis method for hairy NPs represents the primary focus of this book, as manifested from Chapters 1 to 4. Chapter 1 extensively surveys recent advances in the synthesis of hairy NPs via surface-initiated controlled radical polymerization (CRP) and their characterization. The preparation of hairy NPs via bulk microphase separation and solution self-assembly of linear block copolymers are discussed in Chapter 2. It is crucial to precisely control the topology, architecture, composition, size, surface chemistry, and self-assembly of hairy NPs, as these characteristics in turn greatly affect their properties and applications. As such, unimolecular block copolymer micelles are employed as nanoreactors to direct the synthesis of monodisperse hairy NPs in Chapter 3. Chapter 4 features the synthesis and environmentally responsive behavior of silica (nano)particles with binary mixed homopolymer brushes and thermoresponsive polymer brushes.

Self-assembly renders a facile, scalable, and cost-effective route to NP ensembles with well-defined structures and functionalities and thus readily tailorable properties. Notably, hairy NPs stand out as the appealing building blocks for self-assembly due to the highly tunable functionality of grafted polymers. Chapter 5 describes the recent progress in the self-assembly of hairy NPs and their representative applications, highlighting the interactions that govern the assembly and methods that enable the construction of NP assemblies. In Chapter 6, interfacial behavior of hairy NPs, including interfacial assembly, interfacial entropy consideration, and interfacial jamming, as well as single-chain NPs, is examined.

It is notable that significant progress in the design, synthesis, and applications of hairy hollow NPs with different morphologies and diverse internal, shell, and hairy structures is detailed in Chapter 7. The self-assembly of binary mixed homopolymer brush-grafted silica NPs (BMNPs) is presented in Chapter 8, including computer simulations and various factors influencing the self-assembled morphology of BMNPs. Chapter 9 investigates the energy transfer and plasmonic coupling scenarios of hairy plasmonic NPs via expediently integrating plasmonic NPs with hairy ligands.

In addition to the applications discussed in Chapters 1–9, the utility of hairy NPs in catalysis and oil lubrication is also summarized in this book. Particularly, Chapter 10 assesses how polymer ligands impact the catalytic efficiency of hairy metal NPs. Chapter 11 outlines the implementation of oil-soluble polymer brush-grafted NPs, synthesized via surface-initiated reversible deactivation radical polymerization, as environmentally friendly lubricant additives for friction and wear reduction.

This book aims to be an introductory resource for scientists and engineers in the fields of chemistry, materials science and engineering, polymer science and engineering, nanobiotechnology, and biomedicine in both academia and industry.

Finally, we greatly appreciate all of the authors who contributed to this book. In the course of editing this book, we received great support from Wiley, and we would like to particularly acknowledge the assistance of Alice Qian and Katrina Maceda.

Yijiang Liu, Xiangtan University
Zhiqun Lin, National University of Singapore

1

Synthesis of Hairy Nanoparticles

Zongyu Wang[1], *Jiajun Yan*[1,2], *Michael R. Bockstaller*[1], *and Krzysztof Matyjaszewski*[1]

[1] *Carnegie Mellon University, Department of Chemistry, 4400 Fifth Avenue, Pittsburgh, PA 15213, USA*
[2] *ShanghaiTech University, School of Physical Science and Technology, 393 Middle Huaxia Road, Shanghai 201210, China*

1.1 Introduction to Grafting Chemistry

Recent progress in the field of macromolecular science affords economically feasible materials with mechanical robustness, light weight, and advanced optical, thermal, or electronic transport performance to impact fields, such as energy storage, transportation, electronics, and bioengineering. Hybrid materials that derive novel and enhanced properties from the synergisms between distinct organic and inorganic (or biological) constituents play a particularly important role. Research in this area is often inspired by "nature" that employs multifunctional hybrid materials (such as "mollusk shells") in which novel properties arise due to the hierarchical arrangement of constituents. An important theme is the role of interfaces in mediating the interactions between the constituents. Surface functionalization via anchored polymer chains has become a ubiquitously applied method to tune the physiochemical properties of the surface, leading to significant improvements in interface chemistry and engineering [1–11]. Advances in surface-initiated polymerization have enabled the synthesis of brush (or "hairy") nanoparticles, a novel class of hybrid material "building blocks" that can be assembled into functional material architectures or that can be applied as fillers to augment the performance of polymer materials. The incorporation of nanoparticles allows for enhancing the performance of the polymer host without sacrificing the superior processability features of the host matrix. The ability to construct polymeric materials with defined or desired thermal, optical, catalytic, electronic, and mechanical performance rendered the so-called polymer nanocomposites one of the most active fields in modern macromolecular chemistry and engineering. Applications of brush particles to realize stimuli-responsive polymer hybrids [2, 8, 12, 13], antifouling paints [5, 14–16], colloidal stabilizers [17], adhesives [18], catalytic systems [19], electronic devices [20], and biosensors [21] have been demonstrated. Moreover, hairy nanoparticles prepared via surface-initiated polymerization have

Hairy Nanoparticles: From Synthesis to Applications, First Edition. Edited by Zhiqun Lin and Yijiang Liu.

found application for the functionalization of various novel substrates, including nanofibers, mesoporous constituents, nanotubes, graphene, living cells, and protein nanocomposites [22–25]. This contribution summarizes recent advances in the field of surface-initiated polymerization, in particular, based on reversible-deactivation polymerization methods, that have been fundamental to the advances in polymer hybrid materials [10, 26, 27].

1.2 Surface Functionalization of Nanoparticles

The preparation of hairy nanoparticles starts with the introduction of functional groups onto nanoparticle surfaces to enable the subsequent coupling of polymer chains ("grafting-onto" approach) or initiating groups for surface-initiated polymerization ("grafting-from" approach). The generation of strong bonding between the polymer chains and inorganic substrates surface relies on the precise and proper selection of desired anchoring groups.

1.2.1 Surface Modification by Chemical Treatment

A large number of anchoring reagents with distinct functionalities were explored to fulfill the surface modification of the respective desired substrates. Table 1.1 includes some general functional groups and their suitable surfaces. Examples of anchoring reagents and the applicable functional groups are summarized and listed below. One of the most commonly used reagents for surface functionalization is silane-based coupling agents, as many commercially available functional silanes, including trimethoxy(vinyl)silane, (3-aminopropyl)triethoxysilane, (3-mercaptopropyl)trimethoxysilane, and (3-chloropropyl)triethoxysilane, are available at a low price from the industrial sources. The introduced amino moieties are subsequently converted to different functionalities, including atom transfer radical polymerization (ATRP) initiating sites. Many other functionalities can also be incorporated through hydrosilylation reactions. The silanol group on the surface can react with the functional groups. Silane-based coupling agents can consume up to three halide or alkoxy functional groups. Hence, the anchoring reagent with multiple functionalities can form a stronger covalent bond to the surface of nanoparticles. However, the multifunctional silane coupling agents tend to self-polymerize and form a multilayered microstructure [60]. The construction of the multilayered structure empowers the coupling agents with functional groups to enclose the surface, therefore making them widely applicable to a broad range of substrates.

Pyrocatechol, also known as catechol, and its derivatives have attracted great attention in the past years [34, 61]. Pyrocatechol, especially dopamine or polydopamine, was originally recognized to assemble stable adhesion to the surfaces of metal oxide substrates by forming a chelate attachment [62]. Nevertheless, the breakthrough of dopamine-like peptides in the byssus of mussels expanded the application of (poly)dopamine as a universal surface modifier. Generally, dopamine self-polymerizes into a multilayered polydopamine covering the substrate [34]. The

Table 1.1 Surface anchoring groups and applicable surfaces.

Anchoring group	Surface	Examples	References
Trihalo/alkoxysilane 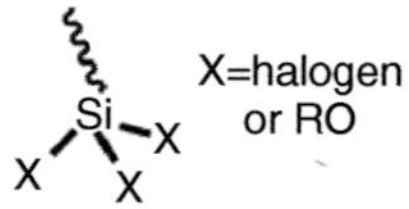	Silica, metal oxide, metal, etc.	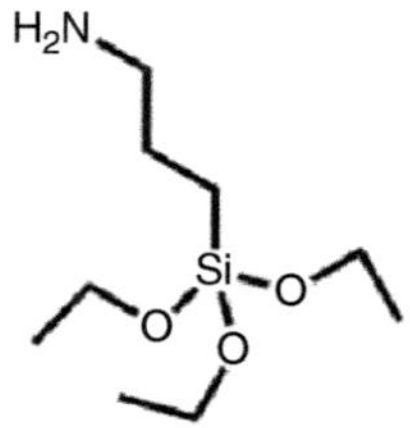 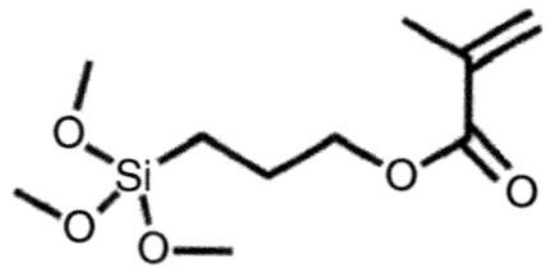	[28–31]
Halo/alkoxydimethylsilane 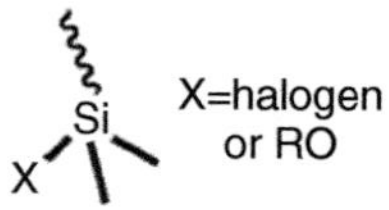	Silica	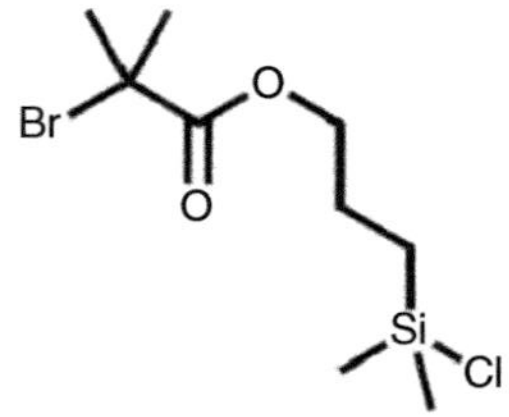	[32, 33]
Catechol 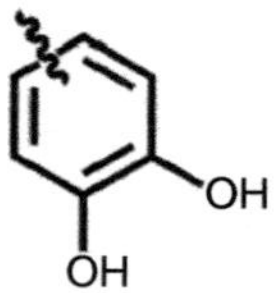	Metal, metal oxide, carbon, etc.	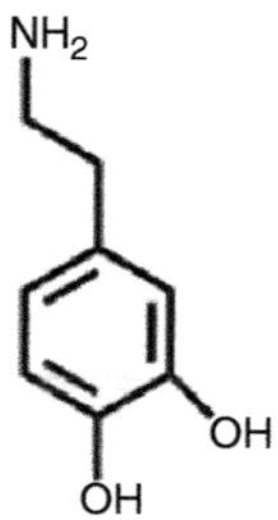	[34–37]

Table 1.1 (Continued)

Anchoring group	Surface	Examples	References
Amino 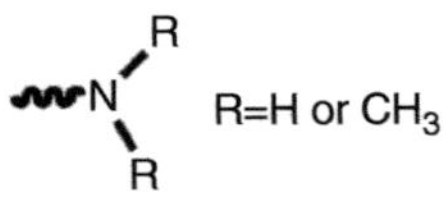	Metal, metal oxide, carboxylic acid-functionalized surface	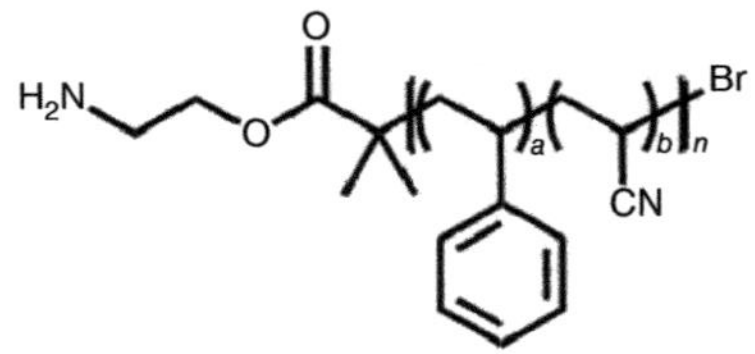	[38]
Carboxylic acid 	Metal, metal oxide, amino-functionalized surface	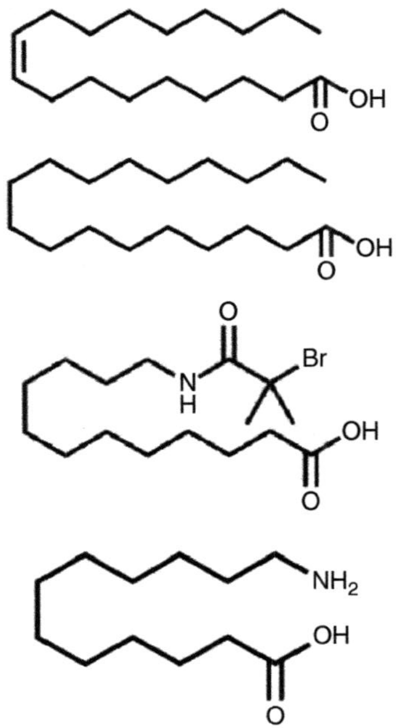	[39–42]

Activated acyl 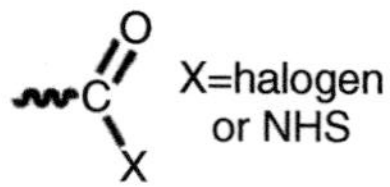	Hydroxy/amino-functionalized surface, cellulose	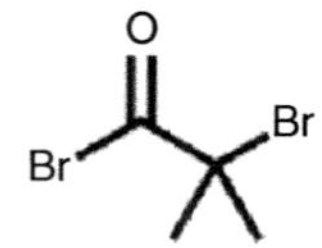	[43–45]
Phosphonic acid/ phosphates 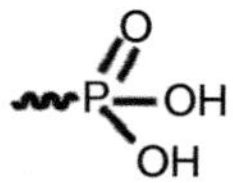	Metal oxide	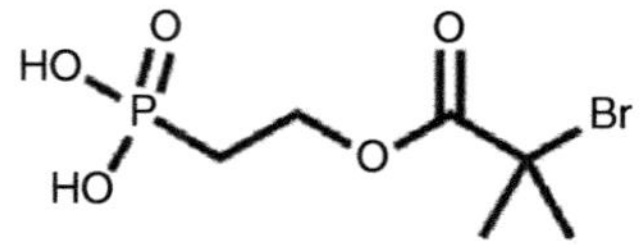	[46, 47]
Radical	Metal, metal oxide, carbon	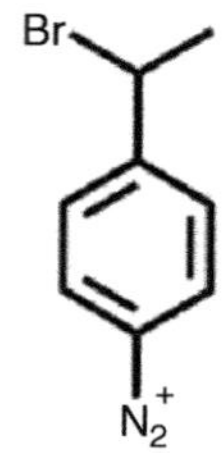 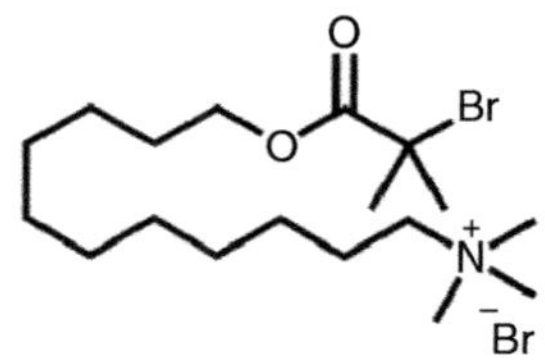	[48–50]

Table 1.1 (Continued)

Anchoring group	Surface	Examples	References
Thiol/disulfide 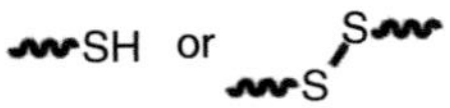	Metal	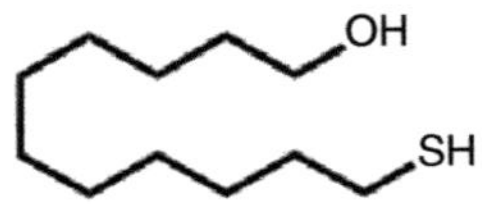 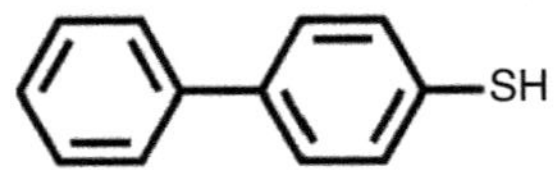	[51–56]
Allyl 	Hydrogen-treated silicon	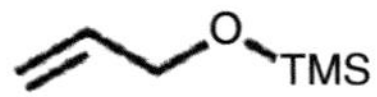	[32]
Phosphine oxide 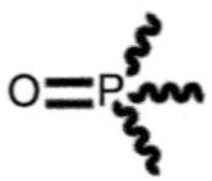	Quantum dots	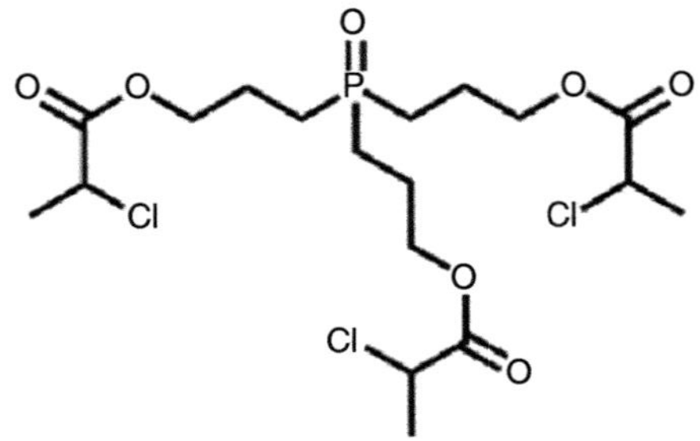	[57, 58]

Source: Yan et al. [59], table 1 (p. 198)/Reproduced with permission from Elsevier.

formation of polydopamine affords a large variety of surface functional groups, such as amino, hydroxyl, aromatics, and conjugated carbonyl [61, 63, 64]. Besides dopamine, other natural and synthetic compounds based on phenol/catechol were employed as anchoring reagents, including catechol-bearing peptides and tannic acid [65, 66].

The application of aliphatic acids to the functionalization of metal or metal oxide surfaces has been well explored in the field of inorganic surface modification [67]. Later, aliphatic acids, or amines, were utilized to stabilize inorganic colloid nanocrystals in organic dispersion [68–70]. Either carboxyl or amino groups can generate coordinative interaction with the metal atoms on the surfaces [70–72]. Nevertheless, low-cost aliphatic acids or amines, including oleic acid, stearic acid, octylamine, or dodecylamine, initially served simply for compatibilization with no reactive functional groups incorporated. Recent research proved the introduction of ATRP initiating sites or even polymer chains with carboxylic acid or amino chain ends onto the nanofiller surfaces [38, 39]. A carboxylate-based anchor showed high efficiency to modify a large variety of inorganic substrates. The incorporation of ATRP initiating moieties allowed the application of these anchoring reagents for the synthesis of hybrid polymer nanocomposite [39, 73]. Besides aliphatic acids or amines, another alternative approach to modify the surface of the inorganic substrate is the deposition of aniline or pyrrole [74]. However, their control over the surface functionality as well as the anchoring efficiency is not as facile as silane or dopamine coupling agents [75]. The hydroxyl- or amino-modified surfaces can directly react with the coupling agents based on derivatives of carboxylic acids, including acyl halide or active esters [43, 44, 76]. Compared to carboxylic acids, as one alternative surface anchoring agent, functional phosphonates and phosphates were used to graft polymer chains onto/from salts and metal oxides as they can build strong bond with the surface of metal oxide substrates [46, 77, 78]. Phosphine oxides functionalized the surface of selective quantum dots, including cadmium selenide or zinc sulfide [57, 58]. Additionally, radical species generated through heat, light, or redox reaction from the surface or precursors introduced functional moieties and formed covalent carbon–carbon/metal bonds on the substrate surfaces. For instance, thermal initiators attached to the surface of carbon substrates upon heating [79]; aryl radicals were produced through the electrochemical reduction of diazonium compounds [48, 80, 81]; and radical coupling with alkenes yields photochemically active surfaces [82, 83]. Similar to the typical process of radical coupling, hydrosilylation can connect allyl-based anchoring reagents and pretreated silicon substrates through either a thermal/photoinduced radical path or an organometallic catalytic path [32]. Organosulfur compounds can form a stable linkage with a wide range of metals, including gold, silver, mercury, iron, or copper. In these systems, thiols and disulfides were efficiently anchored to the metal surface, providing proper coupling functionalities. For instance, thiols are commonly employed to tune the size of gold nanoparticles [84]. Disulfides were prone to split into thiolates upon chemisorption on metal surfaces [85, 86]. Initiating sites or polymer ligands were incorporated onto surfaces of gold via coordination with compounds containing mercapto groups or thiol-terminated polymer ligands [51–53]. The essence of the sulfur–metal interaction is still under investigation [86, 87].

1.2.2 Surface Modification by Plasma Treatment

Plasma is a moderately ionized vapor of free electrons, ions, and radicals. It can be defined as a quasi-neutral particle system in the form of a gaseous or fluid-like blend [88–91]. During plasma treatment, functional groups are immobilized onto the surfaces or free radicals are produced. The radicals can react with oxygen from the atmosphere and be used for subsequent coupling or grafting reactions [92]. Gases such as argon, helium, oxygen, nitrogen, ammonia, and carbon tetrafluoride are particularly widely applied. The anchored moieties can further be used to bind other molecules or polymers to the surface to afford the targeted properties. For example, oxygen plasma treatment induced oxygen-containing functionalities including hydroxyl groups, peroxide groups, and carboxyl groups. Carboxyl and hydroxyl groups may also be incorporated via carbon dioxide or CO-plasma treatment [93]. The carbon dioxide plasma treatment creates ketones, aldehydes, and esters [94]. Nitrogen, ammonia, and nitrogen/hydrogen plasmas produce primary, secondary, and tertiary amines, and amides, which can be used to initiate polymerization in the postirradiation grafting procedure [95]. The pulsed plasma treatment can entail the deposition of halogen-containing initiator films on the surface of the substrate (Scheme 1.1) [96–98].

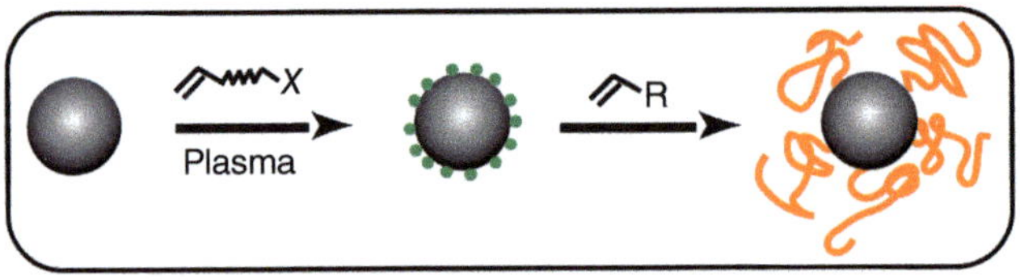

Scheme 1.1 Surface-initiated polymerization from pulsed plasma deposited halogen-containing initiator layers.

1.2.3 Synthesis of Functionalized Nanoparticles Through Initiator-Containing Precursors

Instead of tethering the initiators onto the nanoparticle surfaces, a one-step process to prepare uniform 3 nm initiator-containing organo-silica hybrid nanoparticles was reported [99], which relied on the polycondensation of brominated organosilane precursors, 3-(triethoxysilyl)alkyl α-bromoisobutyrate (TES-ABMP). The utilization of an initiator-modified precursor enabled one to spare the surface modification steps prior to polymerization, hence avoiding the generation of additional silica layers from coupling reactions. In Scheme 1.2, "green Br" refers to the bromine initiating sites that are essential for synthesizing the hairy nanoparticles, and "*x*" indicates the number of $-CH_2-$ units, which are attributed to preparing the corresponding organosilane precursor. The hybrid nanoparticles can readily be polymer-tethered through surface-initiated atom transfer radical polymerization (SI-ATRP) without additional post-functionalization [6].

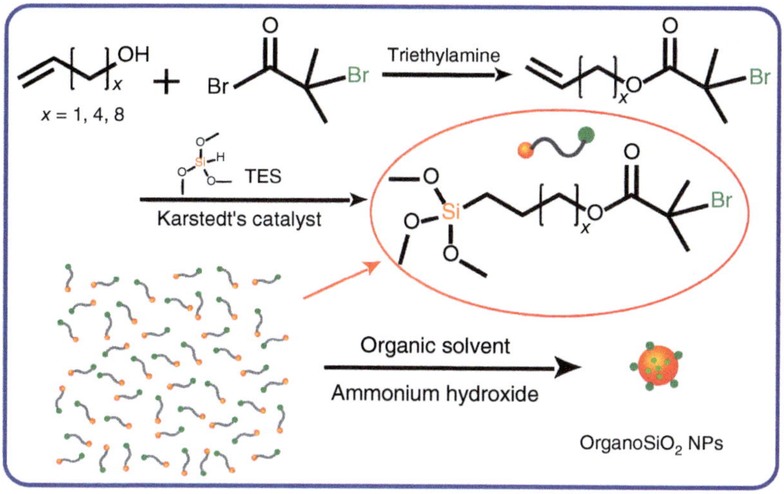

Scheme 1.2 Synthesis of $oSiO_2$ nanoparticles. Source: Han et al. [99], scheme 1 (p. 1219)/Reproduced with permission from the American Chemical Society.

1.3 Synthesis of Hairy Nanoparticles

In the past decades, numerous methodologies and techniques have been explored for polymer–inorganic hybrid material synthesis. Modification of inorganic substrates with tethered polymer ligands optimally integrates the properties of both ingredients [100]. Polymer ligands could be synthesized either by "grafting-from" or "grafting-onto" approaches [101–103]. The "grafting-onto" method exploits the benefits of coupling reactions between the surface functionalities and the complementary anchoring blocks or (chain ends) of polymer ligands to be attached to the substrate surfaces [104], which is experimentally straightforward. On the other hand, the "grafting-from" modification is often preferred as it enables higher grafting densities and polymer shell thicknesses. Both the "grafting-onto" and "grafting-from" approaches involve reactions at a solid surface. In an alternative approach, hairy nanoparticles were synthesized through a "polymer-first" approach, for instance, by applying block copolymers as a template [105]. This procedure was further advanced to prepare covalently bonded hairy nanoparticles with more complicated morphologies, including nano-capsules [106], molecular bottlebrushes [107–109], and star polymers [110], as polymeric templates to prepare precisely controlled polymer–inorganic nanocomposites.

1.3.1 Surface-Initiated Polymerization/The "Grafting-from" Approach

In the "grafting-from" method (Scheme 1.3), tethered polymer ligands grow directly from the modified surfaces, enabling higher grafting density, which is one of the most significant advantages of this approach. The grafting density (unit: chains nm^{-2}) of hairy nanoparticles is defined as the average number of polymer chains per unit surface area (unit: nm^2) and represents a crucial parameter for tuning both the chemical and physical properties of brush-like composites.

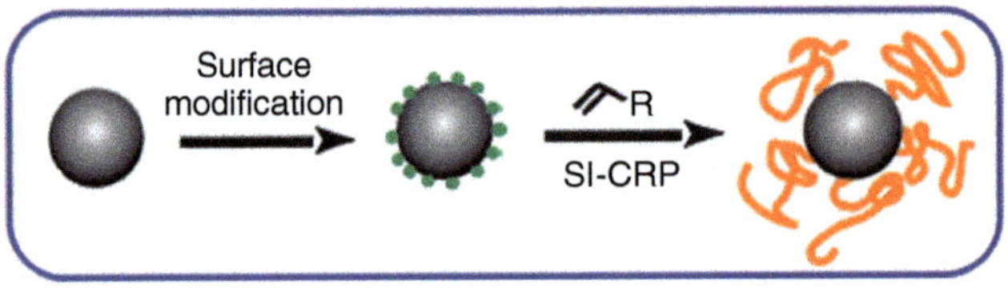

Scheme 1.3 The "grafting-from" approach.

1.3.1.1 SI-Free Radical Polymerization

Conventional (free) radical polymerization (FRP) is the most industrially utilized polymerization technique [111]. It has a long history of grafting polymer brushes from inorganic particles [112, 113]. To perform SI-FRP from nanoparticles, a radical-generating moiety needs to be immobilized. Radicals may be generated via conventional azo initiators [113, 114], photoinitiators [115, 116], or ionizing irradiations [117, 118]. Similar to FRP in solution, surface-generated radicals undergo radical addition to vinyl monomers and the reaction proceeds via a chain-growth mechanism. The multifunctional nature of the nanoparticle "macroinitiators" allows the growth of multiple chains simultaneously on a single nanoparticle.

FRP is one of the least expensive polymerization techniques while it is compatible with the widest variety of vinyl monomers. It is also tolerant to many impurities, such as protic solvents (e.g. water or alcohol), coordinating/chelating agents, and electrophiles, and various polymerization conditions, including bulk, solution, suspension, and (mini/micro)emulsion polymerization [119]. SI-FRP inherits all these features.

Despite such advantages, its intrinsic limitations render SI-FRP a nonideal choice in the preparation of hairy nanoparticles. Due to the high frequency of diffusion-controlled random termination reactions, broad molecular weight distribution is virtually guaranteed for FRP. It may not be a problem for homogeneous systems, but with several hundred growing chains on each nanoparticle, such random distribution leads to a large particle-to-particle difference, and hence a large batch-to-batch difference, as well as gelation due to radical termination between particle brushes.

1.3.1.2 SI-ATRP

In FRP it is challenging to simultaneously optimize each parameter of the targeted materials, particularly when the polymerization of the tethered chains is carried out from surfaces. The advances in surface-initiated controlled radical polymerization (SI-CRP), also known as surface-initiated reversible deactivation radical polymerization (SI-RDRP), allow precise control over chain length, composition, brush shell thickness, and eventually polymer architecture at the same time. They afford an approach to modify various substrates with polymeric shells of different thicknesses meanwhile maintaining the robustness and versatility of the living polymerization technique. The high tolerance of SI-CRP as a synthetic method toward a wide variety of functionalities and extensive applicability enabled it to be broadly applied for the fabrication of hairy nanoparticles.

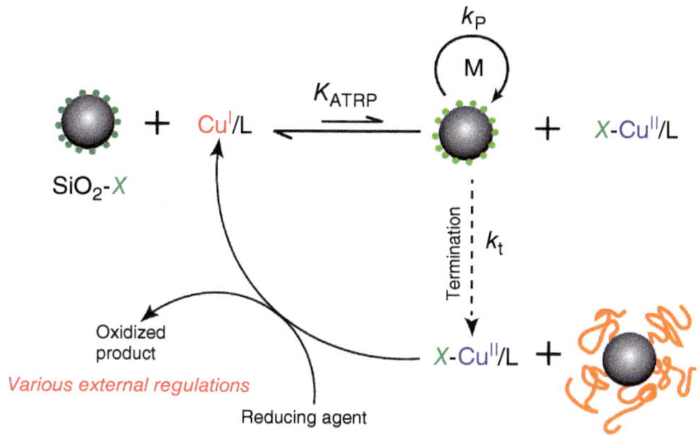

Scheme 1.4 General scheme for SI-ATRP.

Because of the high accessibility of alkyl halide functional groups on the surface and its high tolerance toward various reaction environments, process requirements, and impurities, SI-ATRP amounts to a majority of all "grafting-from" approaches (Scheme 1.4). SI-ATRP as well as its derivative techniques have been recognized as the most common controlled radical polymerization (CRP) approach for growing polymer ligands from substrate surfaces [4, 6, 10, 120], substantially augmenting the toolkit of radical polymerization. SI-ATRP with well-preserved chain-end fidelity was employed for the preparation of precisely controlled, densely tethered polymer ligands from colloidal nanocrystals [120–122]. Based on the dynamic equilibrium between propagating radicals and dormant species, a typical ATRP procedure is tempered by a redox pair of transition metal complex catalysts, especially copper complexes (Cu^I/L, Cu^{II}/L), Figure 1.1 [123–126]. A conventional ATRP process typically includes initiation, propagation, activation/deactivation, and termination steps, similar to SI-ATRP. However, the heterogeneous system presents some special characteristics.

Determined by the diverse morphology of nanoparticles and the degree of surface functionalization, the density of initiating sites could vary in a wide range, up to a couple thousand per particle, resulting in hairy nanoparticles with a very high grafting density. To achieve good control throughout the process, the overall rate of the polymerization should be well-tuned to maintain a sufficient diffusion rate of monomers to the chain-end radicals.

Additionally, when the overall number of initiating sites or deactivators is not high enough, then reversible deactivation becomes too slow, and extra sacrificial initiators [127] or deactivators [110] are added to the reaction to maintain a sufficiently fast reversible deactivation to enable a controlled process. Besides, based on the general gelation theory, for functionalized particles containing a thousand initiating sites on the surface, just about 0.1% of interparticle couplings could result in macroscopic gelation [128]. Reagents containing α-bromoisobutyrate functional groups are commonly used to introduce initiating sites on the surface of nanoparticles for SI-ATRP. Recent work reported the development of a tetherable ATRP initiator, 12-(α-bromoisobutyramido)dodecanoic acid (BiBADA), which contains a long

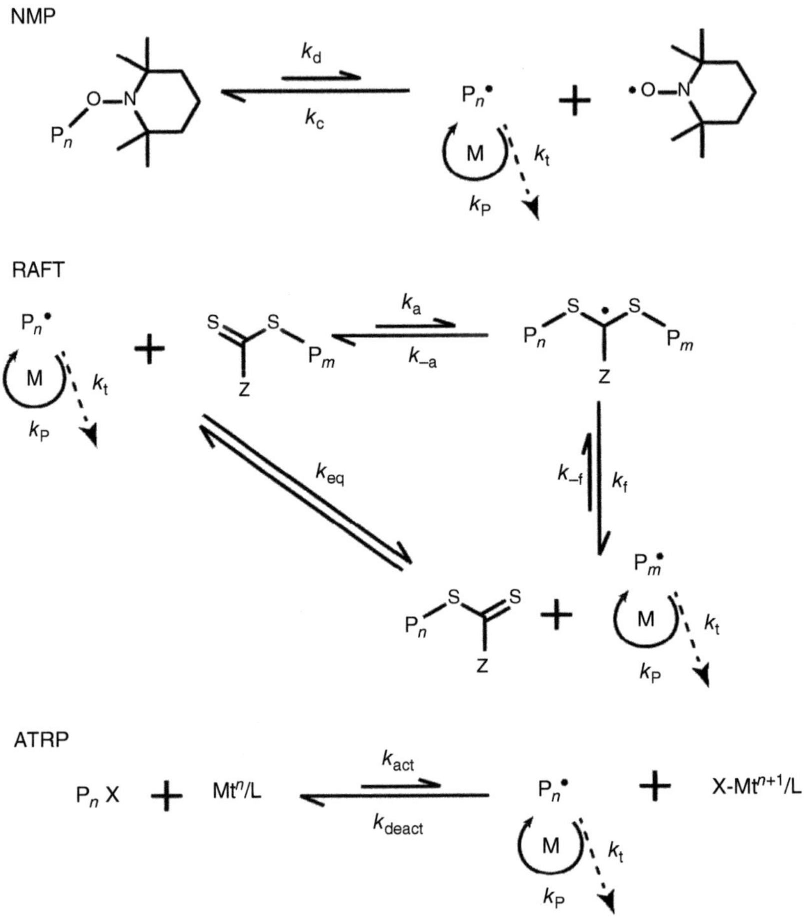

Figure 1.1 Illustration of equilibrium of typical RDRP techniques.

aliphatic spacer between a carboxyl group and an α-bromoisobutyramido chain end. Due to its versatility, BiBADA was used as a universal anchor for the surface modification of metal oxide nanoparticles (Table 1.2, Figure 1.2).

For some applications, the residual catalysts from SI-ATRP should be separated from the ultimate product. In the past decades, numerous methodologies were exploited to afford a precisely controlled polymerization with only ppm levels of copper complex catalyst. Reducing agents were employed to restore the activator of copper complexes with high reactivity, such as Cu/Me_6TREN or Cu/TPMA [129]. There are examples utilizing chemical reducing agents, including activator regenerated by electron transfer (ARGET) ATRP [130, 131], supplemental activator and reducing agent (SARA) ATRP [132, 133], and initiator for continuous activator regeneration (ICAR) ATRP [134]. Later, external stimuli [135], for instance, electrochemical method [136, 137], photo-irradiation [138, 139], and ultrasound agitation [140–143], were applied to produce the reducing environments (Figure 1.3). These methods not only enabled ppm levels of copper complex catalyst

Table 1.2 Summary of polymer-grafted metal oxide nanoparticles synthesized by SI-ATRP using BiBADA.[a)]

	Entry	Particle	Size (nm)	Monomer	M_n[b)]	M_w/M_n[b)]	σ (nm^{-2})[c)]	D_h (nm)[d)]
Alkaline earth	1	MgO	20	MMA	1.32×10^5	1.60	0.08	1600 ± 200
Transition metal	2	TiO_2	15	MMA	7.24×10^4	1.25	0.03	403 ± 5
	3	Co_3O_4	10–30	MMA	1.03×10^5	1.83	0.14	4800 ± 100
	4[e)]	NiO	10–20	MMA	7.69×10^4	1.28	0.14	236 ± 3
	5	ZnO	18	MMA	8.77×10^4	1.33	0.17	282 ± 1
	6	Y_2O_3	10	MMA	1.66×10^5	1.72	0.24	650 ± 10
	7	ZrO_2	40	MMA	5.56×10^4	1.52	0.15	236 ± 1
	8[e)]	La_2O_3	10–100	MMA	6.35×10^4	1.23	0.48	317 ± 2
	9	CeO_2	10	MMA	6.88×10^4	1.27	0.13	244 ± 1
	10	WO_3	60	MMA	2.36×10^5	1.98	0.28	762 ± 5
Post-transition	11	α-Al_2O_3	30	MMA	2.37×10^5	2.10	0.06	501 ± 4
	12	α-Al_2O_3	30	BA	2.42×10^4	1.24	0.06	385 ± 1
	13	In_2O_3	20–70	MMA	1.40×10^5	1.49	0.20	377 ± 9
	14	ITO	20–70	MMA	1.23×10^5	1.92	0.11	396 ± 3
	15[e)]	SnO_2	35–55	MMA	1.64×10^5	2.24	0.22	377 ± 1
Metalloid	16	Sb_2O_3	80–200	MMA	3.66×10^5	1.93	0.14	870 ± 20
Metallate	17[f)]	BTO	200	MMA	1.85×10	2.38	0.43	715 ± 4

a) Typical reaction conditions: [MO_x-Br, assuming 1 Br $nm^{-2}]_0/[M]_0/[CuBr_2]_0/[Me_6TREN]_0$ = 1/1000/0.2/0.5, 50 vol% anisole, 1.0 mm × 1 cm copper wire, room temperature.
b) Determined by SEC.
c) Determined by molar mass and inorganic contents.
d) *Z*-Averaged hydrodynamic size in THF determined by DLS.
e) Nanoparticles functionalized with BiBADA.
f) [BTO-Br, assuming 1 Br $nm^{-2}]_0/[M]_0$ = 1/3000.
Source: Reproduced with permission of Yan et al. [39], Copyright 2017, American Chemical Society.

but also allowed for spatial and/or temporal control over the process. Besides the Cu complex, other transition metal compounds, including Fe [144–146], Ru [147, 148], or Ir [149], could also regulate an ATRP equilibrium. The latest advancement of metal-free ATRP solved the dilemma of transition metal impurities in the polymer product. However, it is still challenging to reach a level of high versatility as well as good control over the reaction that is similar to Cu complexes [150–152]. The ATRP process was lately programmed by applying a DNA synthesizer, further expanding the versatility of this technique and promoting its efficiency [153]. Due to its potential to pattern the substrates with polymer ligands, ATRP procedures with external stimuli have drawn considerable attention to systems involving macroscopic substrates [154–159].

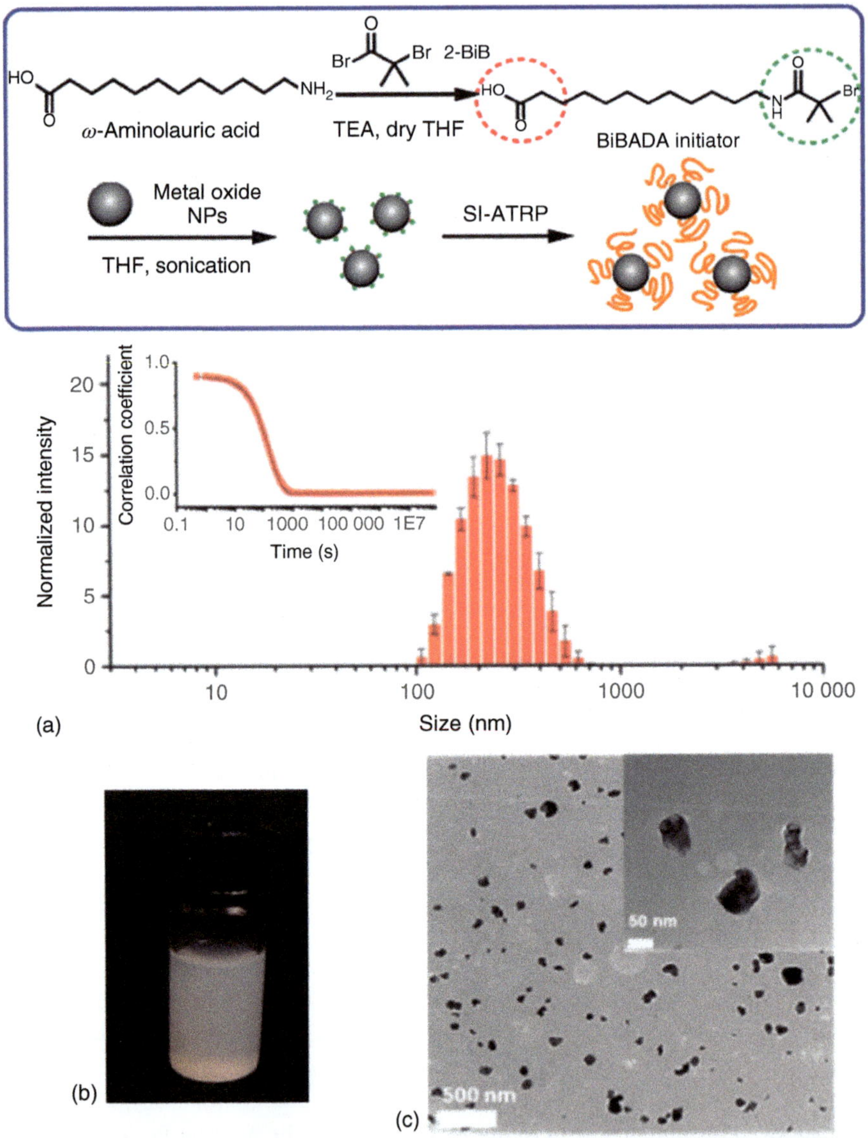

Figure 1.2 Top scheme: synthesis of BiBADA and surface functionalization of metal oxide nanoparticles with polymer ligands. Characterization of ZrO_2-*g*-PMMA nanoparticles: (a) Intensity-weighted hydrodynamic size distributions of ZrO_2-*g*-PMMA as an example. (b) Photograph of a uniform dispersion of ZrO_2-*g*-PMMA in THF. (c) TEM images of ZrO_2-g-PMMA. Source: Yan et al. [39], Reproduced with permission of American Chemical Society.

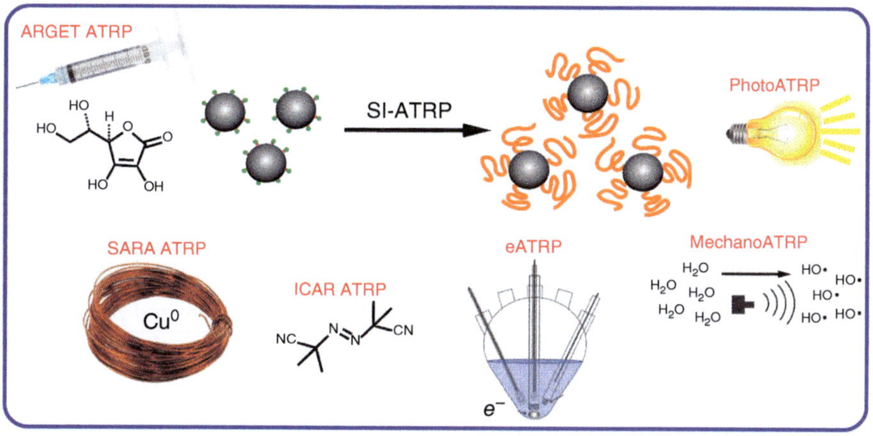

Figure 1.3 External control for various ATRP techniques. Source: Dmitriy Kazitsyn/Adobe Stock.

The emergence of ARGET ATRP not only enabled the dramatic reduction of the concentration of copper complex catalyst to a ppm level but also facilitated the polymerization reaction to be tolerant to limited amounts of air [130, 131]. ARGET ATRP can be recognized as a "green" approach, which consumes ppm amount of the catalyst incorporated with the proper reducing agents including tin(II) 2-ethylhexanoate ($Sn(EH)_2$) [130], ascorbic acids [160], phenol [161], hydrazine and phenyl hydrazine [134], excess inexpensive ligands [162], amines, or nitrogen-containing monomers [163]. ARGET ATRP confirmed that SI-ATRP is readily applicable to large-scale manufacturing on macroscopic substrates [28, 157, 164], even under a certain level of oxygen exposure [165–169]. The repeated redox cycle between the transition metal complex and excess reducing agents consumed all oxygen in the reaction vessel [164]. Another important benefit of ARGET ATRP is that the transition metal complex triggered side reactions are significantly reduced [170]. This helps to further push an ATRP reaction to completion (full conversion) and synthesize copolymers with larger molar mass while preserving chain-end functionality [171, 172], which was confirmed by efficient chain extensions [173].

In ICAR ATRP, an addition of standard free radical initiators is employed to continuously regenerate the extremely low levels of Cu/L catalyst concentration (5–50 ppm). The use of initiators in the continuous activator regeneration procedure could be considered as a "reverse" ARGET ATRP. At this very low concentration of copper activator, in some applications, removing or recycling the transition metal catalyst residues is no longer necessary. The polymerization is promoted to high conversion with low concentrations of a source of organic free-radical initiators [134]. The polymerization rate in ICAR ATRP is determined by the rate of decomposition of the added initiator, while the rate of deactivation and the molecular weight distribution are governed by K_{ATRP} [174, 175].

Cu wire (Cu^0) can act as a reducing agent and induce a Cu^{II} deactivator comproportionation to produce the Cu^I species [176]. Cu^0 can also play the role of a supplemental activator, where it directly reacts with alkyl halides and generates a propagating radical, even though a majority of the activation of alkyl halides is triggered by the Cu^I activator. Hence, this procedure is known as SARA ATRP [177]. The use of other transition metals, such as metallic Zn, Mg, Fe, and Ag, was explored to lower the deactivator concentration in ATRP [177, 178].

The use of chemical reducing agents generated oxidized residues in the polymer product. Therefore, it is important to develop a procedure of reduction via nonchemical means. Electrochemical reductions provide various easily tunable parameters to tune polymerization rates by pursuing the preferred concentration of the redox-active transition metal complexes. For example, a desired percentage of the $Cu^{II}Br_2/Me_6TREN$ deactivator species can be electrochemically reduced to Cu^IBr/Me_6TREN activators to initiate a controlled ATRP reaction. The employed potential determines the activator/deactivator ratio ($[Cu^I/L]:[Cu^{II}/L]$), thus the rate of polymerization [136]. Temporal control of the reaction has become particularly valuable in SI-ATRP, as it offers the possibility to "pause/restart" the polymerization to check and monitor the status of reactions [179]. This procedure also enables temporal control over the polymerization, simply by switching on/off the current. The molar mass of polymer chains formed in the ***e***ATRP process grew linearly with monomer conversion and a low dispersity was achieved. The concentration of catalytic complex as low as 50 ppm was sufficient to retain a controlled polymerization showing first-order kinetics and narrow molecular weight distribution. Cu can be electrodeposited on the electrode and stripped, affording efficient catalytic complex regeneration [180]. ***e***ATRP was also employed to synthesize gradient copolymer grafted hairy nanoparticles where the thickness of polymer shell was governed by tuning space/location of the supporting substrates from the electrode [158, 181].

Due to the simple set-up, insignificant usage of additives, and a possible choice of employing daylight, the PhotoATRP procedure attracted considerable attention [138, 156, 182–190]. PhotoATRP was expanded from copper to iron as the metal catalytic complex [146]. PhotoATRP was successfully performed with ppm amounts of copper catalysts [183, 184, 191]. PhotoATRP in either organic solvents or aqueous solutions was carried out. Precise and well-defined control over polymerization in PhotoATRP enabled efficient chain extension as well as the preparation of block copolymers. The polymerization can be paused and restarted simply by switching on/off the photo-irradiation source. The excited copper catalytic complexes (Cu^{II}/L) were reduced in the presence of electron donors [182]. PhotoATRP from inorganic substrates was later extended to SI-PhotoATRP facilitated by an organic photoredox catalytic complex, generating precisely controlled polymer hybrid nanocomposites without metal residues [33, 179, 192].

The metal-free ATRP was mediated by photo-irradiation with multiple organic photoredox catalysts, including phenothiazines, phenazines, and phenoxazines [150, 152, 193, 194]. The metal-free ATRP showed excellent versatility for a broad range of different methacrylate monomers. Successful chain extension and block copolymer synthesis were combined with other CRP procedures, resulting in synthetic and morphological versatility. Furthermore, phenothiazine derivatives were utilized as novel metal-free photoredox catalytic complexes for the PhotoATRP of polyacrylonitrile (PAN) with desired molar mass and narrow molecular weight distribution. The well-preserved halogen chain-end fidelity of the synthesized PAN was confirmed either by the ^{1}H NMR spectrum or the successful chain extension reaction [151].

A robust mechanically controlled ATRP of methyl acrylate was performed in an ultrasound bath with a ppm level concentration of copper catalyst by means of ultrasonication as the external stimulus and piezoelectric nanoparticles $BaTiO_3$ (barium titanate) as the mechanoelectrical transducing materials in DMSO solution, using a frequency of 40 kHz [141, 195]. It was recently demonstrated that zinc oxide nanoparticles (ZnO) are even more efficient than $BaTiO_3$ as piezoelectric material and could be applied in c. 100 times lower amount than $BaTiO_3$ [140].

1.3.1.3 SI-RAFT

RAFT polymerization is another well-explored CRP technique. The dynamic exchange in a RAFT process is based on the reversible addition-fragmentation of initiating/propagating radicals to chain transfer agents (CTAs), including dithioester or trithiocarbonate, Figure 1.1 [196, 197]. The most important characteristic of RAFT polymerization is that it is mainly based on the classic FRP setups with the added RAFT agents. Free radical initiating species are applied to form the propagating species and maintain the polymerization rate; meanwhile, the concentration of RAFT agents defines the targeted molecular weight. RAFT polymerization involves degenerative chain transfer, which assures all the polymer chains propagate at the same rate, leading to polymers with narrow molecular weight distribution [198]. Compared to ATRP, the RAFT process can be used to polymerize some less reactive or functional monomers [199]. In addition, the colored reactive RAFT chain ends are usually needed to be removed to purify the harmful residues from the polymer product [200]. Moreover, external stimuli were also employed in RAFT polymerizations, such as photoinduced electron/energy transfer (PET)-RAFT polymerization [186, 201]. Unlike the classic RAFT polymerization, the propagating species in PET-RAFT is formed by the PET-excited photoredox catalytic complex from the RAFT agent. Thus, as the process does not require any free radical initiators, it exhibited a superior oxygen tolerance and excellent temporal/spatial control performance [186]. Other external controls, including ultrasound agitation [202] and electrochemical method [203], were also thoroughly investigated for the RAFT polymerization. Occasionally, oxygen can act as an external trigger for RAFT polymerization [204].

SI-RAFT polymerizations were also exploited [205]. The first reported SI-RAFT reaction was accomplished via the mechanistic transformation of an SI-ATRP polymerization [206]. An alternative approach to carry out a RAFT reaction was to trigger the reaction from tethered azo initiators on the surface and control the polymerization through free RAFT agents in the dispersion [207]. Polymerizations were also performed by directly immobilizing CTAs on the surface with initiating species in the reaction mixture [51].

SI-RAFT polymerization could be performed either via surface-tethered free radical initiating sites or surface-functionalized RAFT CTA agents. Generally, there

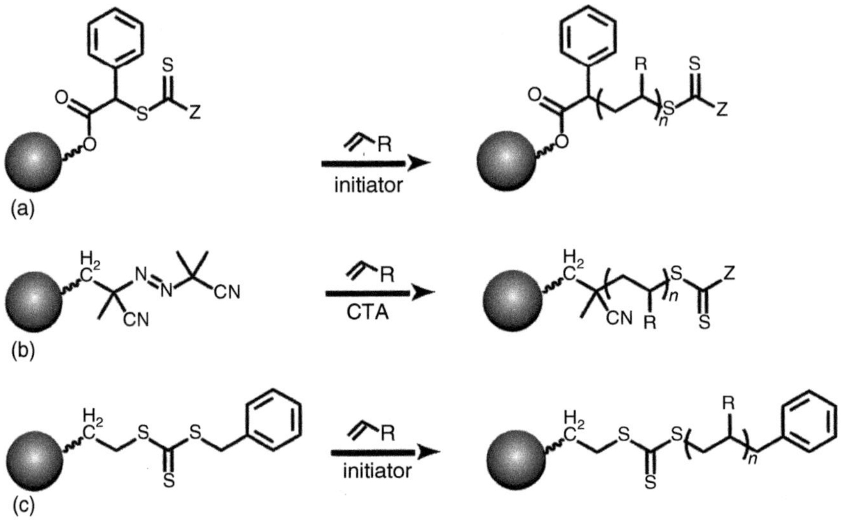

Figure 1.4 Three main approaches to surface RAFT polymerization: (a) Using a R-Group Anchored CTA, (b) Using a Surface Immobilized Initiator, and (c) Using a Z-Group Anchored CTA.

are three different ways to carry out SI-RAFT polymerization from the surfaces, based on how the CTA is immobilized, Figure 1.4. With free radical initiating sites immobilized on the surface and untethered CTAs in solution, SI-RAFT exhibited the same kinetic features as SI-FRP [207]. In other approaches, either the R-group [206, 208] or the Z-group [209, 210] of the CTA was attached to the surface, a process termed SI-RAFT polymerization as the initiation step occurs in the reaction solution [4]. In the R-group anchoring case, the anchored CTA is typically synthesized from an ATRP initiator or its derivatives [208, 211]. The broad range of various functional CTAs, for instance, 4-(4-cyanopentanoic acid) dithiobenzoate, affords an alternative option for immobilizing the CTA functionalities in a direct/indirect manner [212]. The SI-RAFT polymerization performs similarly to SI-ATRP; however, as the system contains the same number of growing radical chain ends and untethered CTAs, the deactivation process is typically less efficient, resulting in broader molecular weight distribution. Furthermore, SI-RAFT polymerization applying Z-group anchored CTA encountered the same dilemma as the "grafting-onto" method, as the tethered CTA on the surface progressively became inaccessible to the active chain ends. Despite the anchoring method applied, untethered free polymers are generally unavoidable in SI-RAFT polymerization. Nevertheless, SI-RAFT polymerization has some advantages, especially the ability to polymerize functional monomers or monomers with low reactivities [213]. Recently, PET-RAFT from surfaces was also reported [214, 215].

One approach to carry out SI-RAFT polymerization is to immobilize an azo initiator to the surface and perform the reaction in the solution containing untethered CTAs. This method has been employed to polymerize a range of monomers, such as (vinylbenzyl)trimethylammonium chloride [216], glucose-based monomers

[217], methacrylate [218], benzyl methacrylate [219], 2-(dimethylamino)ethyl methacrylate [218], *N*-isopropylacrylamine [219–221], *N*-acetylmuramic acid [221], styrene [222], acrylic acids [223, 224], and acrylamide [224]. A more conventional method to perform SI-RAFT polymerization is to immobilize the CTA onto the surface and add the radical initiating species in the solution. This method has been employed to generate polymer chains from multiple substrates, including carbon nanotubes [225, 226], quantum dots [227], gold surfaces [228–230], iron oxide nanoparticles [37, 211, 231–234], graphene [235], SiO_2 nanoparticles [211, 236–253], $BaTiO_3$ [254], ZnO/ZnS [255], and TiO_2 nanoparticles [256]. Although there are two options to immobilize the RAFT agents in the solution-initiated SI-RAFT polymerization, either through the reinitiating R-group of the CTAs or the stable Z-group, there is a preference for tethering the RAFT agent through the R-group.

There are mainly two approaches to functionalize surfaces with RAFT agents. In the first method, the CTA is prepared with an active functional moiety that can be covalently linked to the pristine surface. The second method is based on attaching the CTA to a pre-functionalized substrate to preserve anchoring moieties, corresponding to the functionality of the RAFT agent. The first approach generally is more challenging, but it has been employed for a broad range of substrates. The other approach has its benefits, as it simplified the purification step for the RAFT agents and helped to diminish some general secondary reactions, for instance, the condensation of CTA-bound alkoxysilane groups. Under the same principle, click chemistry was employed to tether CTAs to SiO_2 nanoparticles [257]. Additionally, halide functionalized substrate surfaces can be applied to anchor suitable reactive RAFT agents and their derivatives. Last but not least, pre-synthesized RAFT agents can also be attached to electroconductive substrates through the electrochemical method, such as gold [228, 229].

1.3.1.4 Other Polymerization Techniques

In addition to SI-ATRP and SI-RAFT, other controlled polymerization approaches, such as SI-NMP, living anionic polymerization (LAP), living cationic polymerization (LCP), and ring-opening metathesis polymerization (ROMP), are practical techniques to prepare hair nanoparticles.

NMP was one of the first-reported RDRP methods [258]. In this approach, the dormant species (for example, alkoxyamine) involves homolysis of the carbon–oxygen covalent bond at relatively high temperatures to generate an active propagating radical and a persistent nitroxide species [259–264], Figure 1.1. Later, SI-NMP was reported as another approach to graft polymer chains from the substrates. Only monomers and alkoxyamine-functionalized substrates are required for this technique. Alkoxyamine-functionalized surfaces can be synthesized either via modification of anchoring alkoxyamine [265–267] or by reacting reactive surface radicals with nitroxide radical species [268, 269]. However, SI-NMP presents the limitations of NMP, as it shows the difficulty to polymerize non-styrenic monomers, especially acrylic and methacrylic monomers. Recently developed anchoring derivatives of alkoxyamines with high activity were used in SI-NMP, accomplishing higher efficiency and allowing the polymerization of challenging acrylic monomers [266, 268–270].

In addition to radical polymerizations, ionic polymerizations have been demonstrated as feasible techniques for grafting polymer brushes from surfaces. Both LCP and LAP have a longer history than RDRP [271–273]. Living ionic polymerization was discovered earlier because the ionic intermediates do not undergo bimolecular termination as in radical polymerization. However, cationic polymerization is typically less "living" than anionic polymerization because of proton transfer to monomer reactions, which leads to termination/chain transfer.

Surface-initiated ionic polymerizations are far less studied than SI-RDRP because of the highly reactive carbocation or carbanion intermediates, requiring extremely delicate experimental setups, especially for nanoparticles with high surface areas. In addition, ionic polymerizations are compatible with a narrower selection of monomers than radical polymerization. Nonetheless, ionic polymerizations are employable to monomers that do not undergo radical homopolymerization.

To introduce carbanion to a surface, a similar technique as homogeneous anionic polymerization was used, i.e. the reaction between immobilized 1,1-diphenylethylene and butyl lithium [274]. Carbocation, on the other hand, was introduced in two different ways. It can be generated by Lewis acid-induced cleavage of immobilized ether [275]. Otherwise, in a way similar to surface RAFT polymerization with Z-group-anchored CTAs, surface-initiated cationic polymerization can be initiated by reacting the surface silanol group of silica with p-methoxybenzyl alcohol in the presence of sulfuric acid to generate p-methoxybenzylium cations stabilized by surface adsorbed hydrogen sulfate anions, and therefore proceed with the polymerization with the nanoparticle as a "macro-counterion" [276].

Ring-opening polymerization (ROP) is capable of polymerizing a specific class of cyclic monomers. Therefore, many hairy nanoparticles inaccessible from the polymerization of vinyl polymers can be prepared via SI-ROP. Although the same carbanion or carbocation initiators used for ionic polymerization can also initiate ROP, less reactive initiators are often used because of the easier experimental handling. For example, in anionic ROP, weaker bases such as alkoxides or amines may initiate the polymerization of a wide range of cyclic monomers, such as caprolactone, lactide, or *N*-carboxyanhydride [277–282]. Polymerization of *N*-carboxyanhydride is especially intriguing because it yields peptide brushes [277]. Similarly, cationic ROP can be initiated with alkylation agents such as tosylates. Biocompatible poly(2-oxazoline)-based hairy nanoparticles were prepared this way [283, 284].

ROMP polymerizes cyclic olefin monomers with metallic carbene species as initiators [285–288]. With the rational design of the catalyst, such as Grubbs catalyst third generation, ROMP exhibits excellent control of polymerization and good tolerance to the ambient atmosphere, impurities, and monomer functionalities [289, 290]. Typical ROMP monomers include cyclooctene, norbornene, macrocyclic olefins, and their derivatives [291, 292]. Norbornene derivatives allow an especially rich variety of functionalities, comparable to acrylic monomers of radical polymerization. To initiate ROMP from a surface, a cyclic olefin has to be first immobilized, and a one-step olefin metathesis transfers the metallic carbene to the surface [293, 294]. Subsequently, ROMP can be performed using these immobilized metallic carbenes as initiators. Several intrinsic challenges limit the broad application of SI-ROMP in the preparation of hairy nanoparticles. Both ROMP monomers and catalysts, such as ruthenium, tungsten, and molybdenum complexes, are more expensive than those for RDRP. Such metallic catalysts/initiators in ROMP, become unfavorable impurities in the final product. While ATRP can be performed with a ppm catalyst loading, ROMP requires one metal atom per chain, resulting in higher costs and more metallic residues. Although metal-free ROMP was recently reported, there are still many challenges and no application for surface-initiated systems [295]. Another issue with ROMP is backbiting [286]. As the reactive carbon–carbon double bonds are also present in the polymer backbone, active chain-end attacks on the backbone can lead to undesired loops, free polymers, and/or broadened molecular weight distribution in the preparation of hairy nanoparticles. This also limits the available choice of monomers and catalysts for SI-ROMP when backbiting must be inhibited.

1.3.2 The "Grafting-onto" Approach

Compared to the "grafting-from" approach, the "grafting-onto" approach is generally recognized as a more simple technique. Minimal surface modification is needed as a suitable anchoring functionality can usually be exposed for the pristine surface functional groups. However, due to the strong steric hindrance among the already-grafted polymer chains, it is challenging to obtain high grafting densities through the "grafting-onto" approach [296]. Typically, a more diluted graft layer with a grafting density well below 1.0 chains nm^{-2} was typically obtained through the "grafting-onto" approach, which resulted in a collapsed mushroom-like topology [297].

1.3.2.1 Conventional "Grafting-onto" Approach

Polymer chains can be attached to inorganic substrates when one functional chain end/block is reactive, which can promote connection to the substrate. Generally, such anchoring groups can be the functional end groups of polymer ligands [113, 298], comonomers incorporated through copolymerization or a chain-end functionality tethered by transfer agents [299]. Nevertheless, reacting the active functional end groups through the polymerization requires strict conditions. In the case of copolymer synthesis, when the comonomers are added along with the polymerization, a governing of the sequence of the functional comonomers in the backbone is needed to diminish the generation of unrestrained loop structure on the surface. On the other hand, polymers like poly(ethylene oxide), poly(propylene oxide), and poly(dimethylsiloxane) have intrinsic functional chain ends, which could be directly applied for anchoring [300, 301]. Despite its advantages, this approach is only applicable to a limited range of polymers, as many of them are multifunctional and can lead to the formation of a mixture of hairy nanoparticles and loops. Moreover, the gelation by interparticle coupling could likely occur during the procedure of grafting-onto (Scheme 1.5).

CRP allows precise control over the grafting performance of the polymer ligands on the surface and enables the synthesis of polymers with well-tuned chain length, composition, chain-end fidelity, and chain architectures. For instance, in RAFT polymerization, the sulfur-terminated end group is suitable to be used to

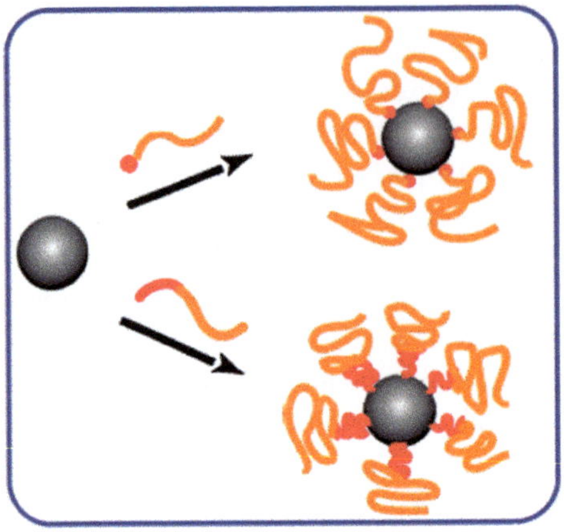

Scheme 1.5 The "grafting-onto" approach.

graft polymer ligands onto noble metal surfaces. Another example was adding sodium borohydride to reduce the dithioester functionality of the thiol group, tethering gold nanoparticles [302, 303]. Polymerizations initiated by functional RAFT agents yield functional group-terminated polymer ligands used in the "grafting-onto" approach. For instance, esterification of propargyl alcohol with 4-cyanopentanoic acid dithiobenzoate produces an alkyne-terminated CTA, resulting in alkyne-terminated polymers. The chains can further be attached to a surface through a copper-catalyzed azide-alkyne cycloaddition (CuAAC) process [304]. A similar process has been applied to the ATRP procedure as well [104]. Meanwhile, as the same copper catalytic complex is involved among all CuAAC, ATRP, and Glaser coupling steps, some secondary side reactions could occur [305]. Another approach is to prepare azide-terminated functional groups by halogen substitution of the chain-end functionality with an azide [306–308]. Additional end functional groups, including phosphonate and thiol, for selectively applying to iron oxide and gold substrates, can be incorporated into polymers through ATRP reaction from pre-functionalized initiators as well [43, 309, 310]. Moreover, Y-shaped polymer ligands grafted substrates were prepared when the functional groups were introduced in the middle of the backbone instead of the chain ends [311].

An alternative option for the "grafting-onto" approach is to use functional block copolymers containing blocks with distinct surface affinities. Normally, just a weak attraction between the anchoring block with surface affinity and the target surface is enough to link the block copolymer ligands to the substrate, the non-tetherable block functions as the polymer "free" chain. If there are multiple anchoring groups present in the block copolymers, the formation of a loop structure can be achieved by the non-affinity segments between the anchoring functionalities [312]. In the diblock copolymers, the blocks without anchoring groups, instead of always fully collapsing, tend to form a partially stretched conformation, as the chain grows [313]. Grafting block copolymers with different anchoring abilities onto surfaces can be applied to disperse and compatibilized inorganic nanofillers

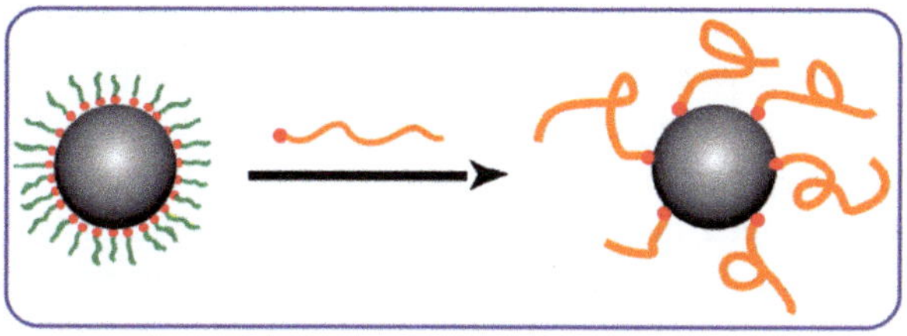

Scheme 1.6 The "ligand exchange" approach.

in different media, including aqueous/organic solutions, or polymer matrices [103, 314, 315]. The different affinities between the two blocks were essential to avoid the agglomeration of nanofillers. For instance, compared to poly(acrylic acid)-*block*-poly(styrene-*co*-acrylonitrile) (PAA-*b*-PSAN) ligands, PAA-*b*-PMMA ligands give better performance as the dispersant for separating ZnO nanoparticles in organic solutions, as the higher polarity acrylonitrile segments might compete with the PAA blocks and cause coupling of ZnO nanoparticles [103]. Besides, amphiphilic triblock copolymers poly(methacrylic acid)-*block*-poly(methyl methacrylate)-*block*-poly(styrene sulfonate) (PMAA-*b*-PMMA-*b*-PSS) were adsorbed to the surfaces of Fe^0/Fe_3O_4 nanoparticles as physisorbed layers to improve the dispersion stability of the nano-iron suspensions in aqueous system and facilitate their adsorption at the water/oil interface [316–318].

1.3.2.2 Ligand Exchange

The "ligand exchange" approach involves the substitution of the small-molecule ligands on the surface of the inorganic substrates with different functionalized polymer ligands, which is a special case of the "grafting-onto" approach. To fulfill the ligand exchange's prerequisite conditions, either the polymer ligands present a stronger affinity to the inorganic surface than the pristine ligands or the small-molecule ligands can be removed from the system during the reaction to promote the ligand substitution process (Scheme 1.6).

The first-reported ligand substitution reaction was carried out between the pristine phosphine oxide ligands and the poly(2-(dimethylamino)ethyl methacrylate) (PDMAEMA) ligands on the CdSe/ZnS quantum dots. In the organic medium, the replacement from labile small-molecule ligands to bulky polymer ligands gave stable and uniform coverage over the quantum dots [319]. However, the polymer-stabilized quantum dots synthesized through this technique exhibited some significant limitations, as the reaction strongly relied on the dynamic equilibrium between the two ligands. The polymer ligands containing carbodithioate functional groups can easily replace the pristine trioctylphosphine oxide ligands due to their substantially larger affinity to the surfaces of CdSe/ZnS quantum dots [320]. In a similar manner, a monofunctional thiol group is generally efficient enough to substitute less stable ligands on noble metal substrates [321].

The surface affinity or the reactivity of the anchoring chain-end groups in polymer ligands will significantly decrease as the chain grows, resulting in an incomplete ligand exchange reaction even with an excessive addition of the desired ligands [322].

This can be resolved in an improved method by using a non-solvent for the small ligands stabilizing nanocrystals as well as the polymer ligands. The precipitation of the nanocrystals with polymer ligands as aggregates from the solutions can locally exclude the fully soluble small pristine ligands and increase the concentration of polymer ligands, the polymer-capped nanocrystals are non-dispersible in the solvent, driving the equilibrium of the ligand exchange reaction further forward to completion.

Compared to the conventional "grafting-onto" approach, a high grafting density can be afforded by the ligand exchange method through gradual substitution of the densely capped nanoparticles with polymer ligands, and a grafting density of up to 1.2 chains nm^{-2} was achieved. For a ZnO-based hairy nanoparticles synthesis, a relatively low boiling point (175 °C) ligand – octylamine was selected as the removable pristine ligands [38]. By heating the reaction mixture above the boiling point of octylamine, the volatile ligands were continuously removed from the system, thus shifting the equilibrium of the ligand exchange process. With the addition of NH_2-terminated PSAN ligand, PSAN-NH_2, which was synthesized through ARGET ATRP, the obtained PSAN-capped ZnO nanoparticles were used as high-refractive index nanofillers in transparent acrylic glasses [323], as well as the precursors of carbon-ZnO hybrids for photocatalytic and electrochemical applications [324, 325]. The achieved grafting density of PSAN-capped nanoparticles through this approach was 0.9–2.5 chains nm^{-2}, depending on the ratio between ZnO nanoparticles and the desired polymer ligands.

1.3.3 Template Synthesis

The "grafting-from" and "grafting-onto" modifications discussed above can be considered as "inorganic-first" approaches. On the other hand, the synthesis of hairy nanoparticles through polymeric templates can be categorized as a "polymer-first" approach. In this approach, inorganic fillers are produced within the inner template, and the remaining outer polymeric shells serve to stabilize the nano-objects. The molar mass, molecular weight distribution, chain composition, and architecture of the pre-synthesized polymer templates can be well-tuned to achieve complex morphologies, which can be also encoded in the hairy nanoparticles.

1.3.3.1 Block Copolymer and Its Derivative Templates

Linear polystyrene (PS)-based block copolymer templates, such as polystyrene-*b*-poly(2-vinyl pyridine) (PS-*b*-P2VP), polystyrene-*b*-poly(ethylene oxide) (PS-*b*-PEO), or polystyrene-*b*-poly(acrylic acid) (PS-*b*-PAA) templates, loaded with different metallic precursors were fabricated into metal or metal oxide nanoparticles/polymer nanocomposite films through reduction reactions [105, 326–328]. Surfactants can prevent aggregation, define the size, and shape of the nanoparticles, as well as their compatibilization with the medium, thus playing a critical role in inorganic nanoparticle preparation [329]. In the presence of surfactants as templates, the precursors in the micelles transform into inorganic nanofillers. Instead of the conventional organic surfactants, the block copolymer micelles were reported as a liquid-phase template [105, 326, 330]. The poly(methyl methacrylate)-*b*-poly(ethylene adipate)-*b*-polyvinylpyrrolidone (PMMA-*b*-PEA-*b*-PVP) random copolymers were used to govern the decomposition of $Co_2(CO)_8$ and induce the formation of well-defined Co nanoparticles [331, 332]. The initial templated approach focused on the use of block copolymers to facilitate the formation of micelles in solution, providing limited rational design beyond the formulation adjustments [333, 334]. The amphiphilic nature of the block copolymers endorses the use of water/oil emulsions. Inorganic/polymer hybrids were synthesized when the inorganic fillers were formed inside the emulsion droplets. On the other hand, porous inorganic materials were generated when the continuous phase was made of inorganic precursors. Later, the polymeric template-assisted synthesis of metal/metal oxide nanoparticles [333, 335–339] was further investigated with polystyrene-*b*-poly(4-vinyl pyridine) (PS-*b*-P4VP) or PS-*b*-P2VP as copolymer templates. Au [336, 338–340], Pd [336, 341], Pt [336], Rh, Co, and CdS [337] colloidal nanoparticles were synthesized from $HAuCl_4$, $Pd(OAc)_2/Na_2PdCl_4$, $K[Pt(C_2H_4)Cl_3]\cdot H_2O$, $[Rh(CO)_2Cl]_2$, $Co_2(CO)_8/CoCl_2$, and $Cd(OAc)_2$, respectively. The catalytic efficiency [338], and magnetic and optical properties of the obtained hybrid nanocomposites were characterized [337, 341]. Moreover, the recent development of polymerization-induced self-assembly (PISA) furthermore facilitated template-assisted synthesis by enabling more complex morphological control, such as the worm-to-vesicle transitions [342–346]. Such a control using PISA templates can allow the synthesis of inorganic nanoparticles of various sizes and shapes [106, 347]. For example, photo-crosslinkable and amine-modified poly(2-hydroxypropyl methacrylate)-poly(2-((3-(4-(diethylamino)phenyl)acryloyl)oxy)ethyl methacrylate) (PHPMA–PDEMA) block copolymer nanoparticles were synthesized via RAFT-PISA, and the crosslinked nanoparticles were applied as templates for polymer/gold hybrid nanoparticles (Figure 1.5) [348].

1.3.3.2 Star/Bottlebrush Polymer Templates

Star and bottlebrush block copolymers can be generally recognized as stable unimolecular templates. Recent publications introduced a universal synthetic route for the preparation of polymer–inorganic hybrid nanoparticles/nanorods using star/bottlebrush polymers with block copolymer arms as nanoreactors Scheme 1.7 [103, 110, 350]. The detailed discussion of the "polymer template approach" is covered in Chapter 3.

1.4 The Role of "Architecture" in Hairy Nanoparticles

The grafted polymer chains on the surface of nanoparticles are generally used to stabilize nanofiller dispersion within the polymeric host matrix and therefore play a critical role in the binary hybrid nanocomposites. Recent advances in hairy nanoparticles synthesis inspired various novel applications in the fields of gas separation [351–354], lubrication [355–361], antifouling [47, 312, 362–364],

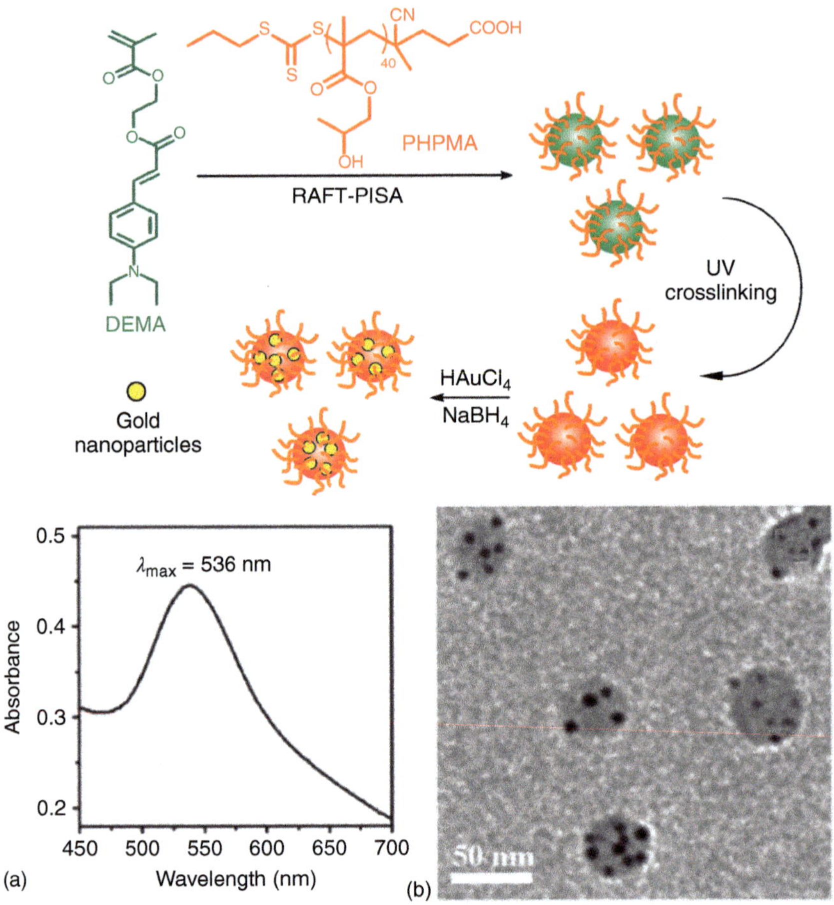

Figure 1.5 Reaction scheme for the synthesis of photo-crosslinked stable PHPMA–PDEMA nanoparticles and preparation of gold/polymer nanoparticle composite. (a) The spectrum of solution after reduction of chloroauric acid using $NaBH_4$, (b) TEM image of the gold/polymer hybrid nanoparticles (molar feed ratio of $HAuCl_4$/tertiary amine is 1/4). Source: (b) Huang et al. [348], Reproduced with permission of John Wiley & Sons Inc.

smart-responsive materials [365–370], so on and so forth, Scheme 1.8. The properties of hairy nanoparticles are strongly associated with the chain architecture as well as the molecular characteristics of the tethered polymer ligands. The advances in synthetic methods to precisely modulate the architecture of polymer ligands, hence give a valuable research opportunity in the field. Major tools of macromolecular engineering include molar mass, dispersity, ligand composition, topology, and functionality. They also include multiple types of copolymers as well as bimodal ligands, miktoarm (binary) ligands that can be separated and assembled into Janus particles.

1.4.1 Conformation of Hairy Nanoparticles

It is important to study the structure of hairy nanoparticles as it affects the physicochemical properties of hairy nanoparticles both in solid and solution states. The

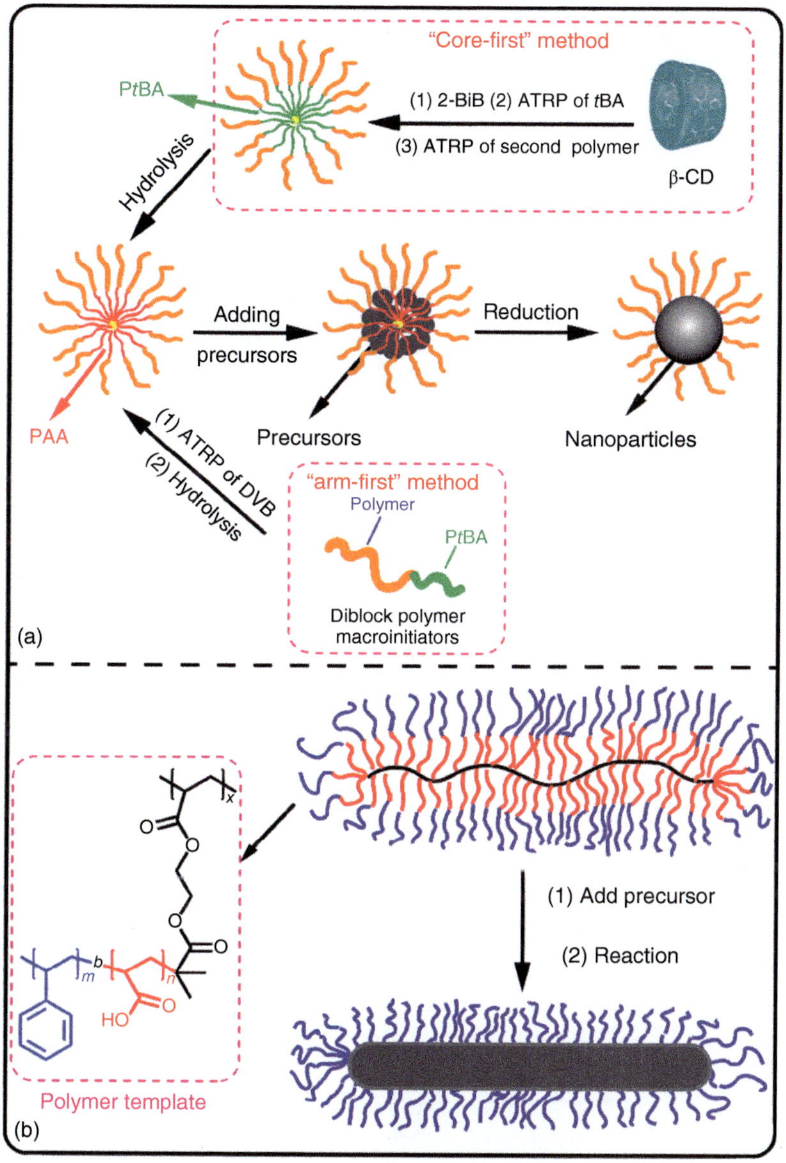

Scheme 1.7 Synthesis of hairy nanoparticles/nanorods through (a) star polymer templates, (b) bottlebrush polymer templates. Source: Reproduced from Wang et al. [349] with permission of American Chemical Society, Copyright 2021.

main features of both the dynamic performance and interactions in hairy nanoparticles can be described using a scaling model that was first established by Daoud and Cotton (DC) to evaluate the structure of star-like polymers, which were later extended to hairy nanoparticle systems [371, 372]. According to the DC model, the structure of hairy nanoparticles can be divided into two regimes: the concentrated particle brush (CPB) regime and the semi-dilute particle brush (SDPB) regime.

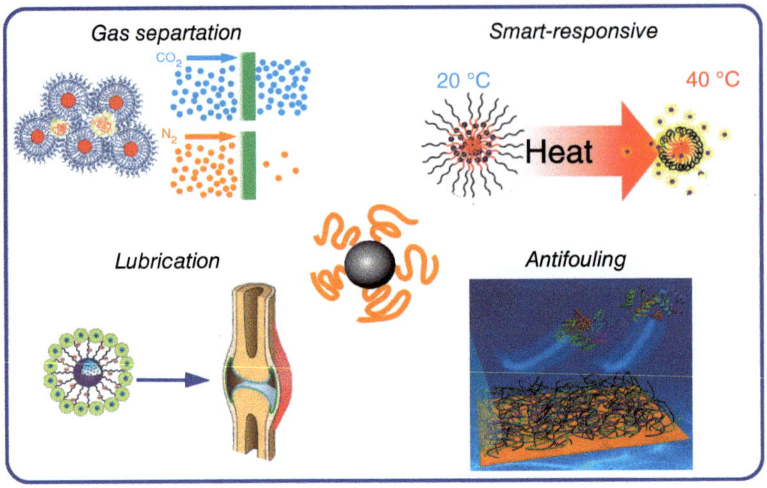

Scheme 1.8 Schematic representation of the advanced application of hairy nanoparticles. Source: Adapted from Refs. [352, 355, 362, 369].

Within the CPB regime, due to the strong excluded volume interactions, chains maintain a stretched conformation, while in the SDPB regime, polymer chains preserve a relaxed mushroom-like conformation (Figure 1.6a). The critical distance $r_c = r_0(\sigma^*)^{1/2}(v^*)^{-1}$ (where r_0 refers to the particle radius, $\sigma^* = \sigma a^2$ is the reduced grafting density, a is the length of a repeat unit, and $v^* = v/(4\pi)^{1/2}$ is the effectively excluded volume parameter) [371] is given to defining the transition threshold between the two regimes: for a total effective hairy nanoparticles size $r_0 + h < r_c$ (where h refers to the thickness of the polymer shell), they are presumed to be in the CPB regime. On the other hand, for a total effective hairy nanoparticle size $r_0 + h > r_c$, the SDPB regime is expected (Figure 1.6b). In the ideal scaling model, in the CPB regime, the thickness of the polymer shell scales $h \sim N^x$ $(1 > x > 3/5)$,

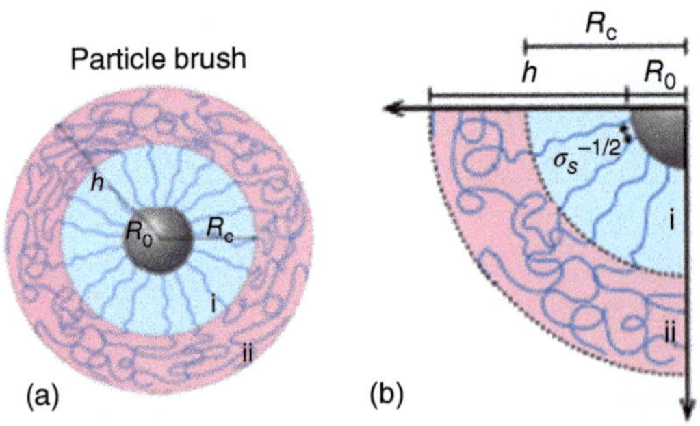

Figure 1.6 Illustration of the transition from concentrated particle brush to semi-dilute particle brush regime. (a) A hairy nanoparticle with radius R_0 and grafting density σ_s shows the concentrated particle brush and semi-dilute particle brush regimes with stretched and relaxed chain conformations. (b) The predicted variation in scaling of the thickness of the polymer shell with the degree of polymerization. Source: Adapted from Choi et al. [373].

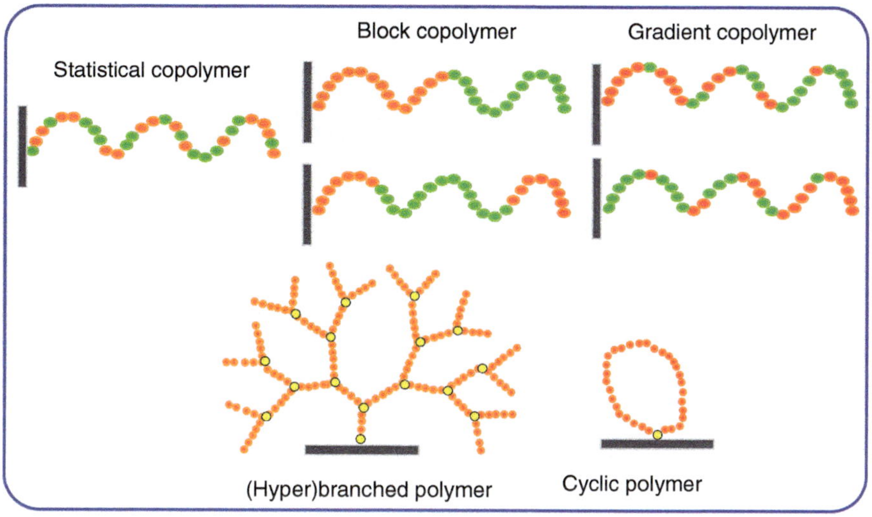

Scheme 1.9 Schematic representation of the architecture of the grafted chains.

where N refers to the degree of polymerization, while in the SDPB regime, the scaling follows $h \sim N^y$ ($y = 3/5$ in good solvents). Besides, the performance of hairy nanoparticles is substantially affected by the chain architecture of the grafted polymer chains, Scheme 1.9. As one of the most versatile methods, SI-CRP has been applied to grow statistical/random [374], block [375, 376], gradient copolymers [377, 378], as well as hyperbranched [379, 380] and cyclic polymers [381–383] from various inorganic substrates.

Tuning the grafting density of the hairy nanoparticles can significantly impact their overall physicochemical properties [384–386]. Based on the modified DC theory, the critical degree of polymerization (N_c) is defined where the grafted polymer ligands enter the SDPB regime from the CPB regime. The N_c value was determined by the surface curvature of the particle core as well as the grafting density of the hairy nanoparticles. As the polymer chains start to relax after the transition from CPB to SDPB regime, the properties of interparticle ligand entanglement and chain penetration can only occur in the SDPB regime [387]. Generally, the final grafting density of the hairy nanoparticle is determined by the density of initiating sites on the surface of nanoparticles, which is controlled during the surface functionalization step. A facile synthetic route to tune the concentration of ATRP initiating sites on the surface of silica nanoparticles is illustrated in Scheme 1.10. The mixture with different ratios of active and "dummy" anchoring ATRP initiators is applied to modulate the concentration of initiating sites and further the grafting density of the hairy nanoparticles [267, 389]. Additionally, in an alternative method, ATRP initiators can be partially removed through high-energy irradiation [390–392]. Hairy nanoparticles with densely grafted and intermediately grafted polymer shells exhibited uniform microstructures, while the sparsely grafted systems showed a string-like superstructure, Figure 1.7 [393]. One way to distinguish whether the hairy nanoparticles form uniform or string-like microstructures is to check

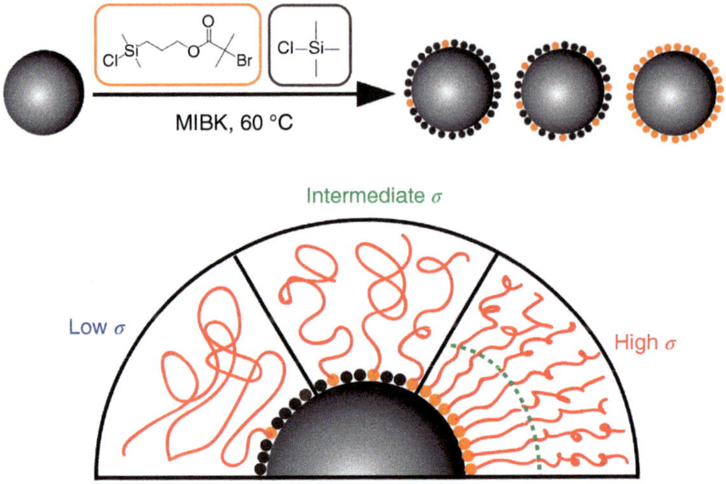

Scheme 1.10 Synthesis of hairy nanoparticles with different grafting densities. Source: Reproduced from Wang et al. [388] with permission of American Chemical Society, Copyright 2020.

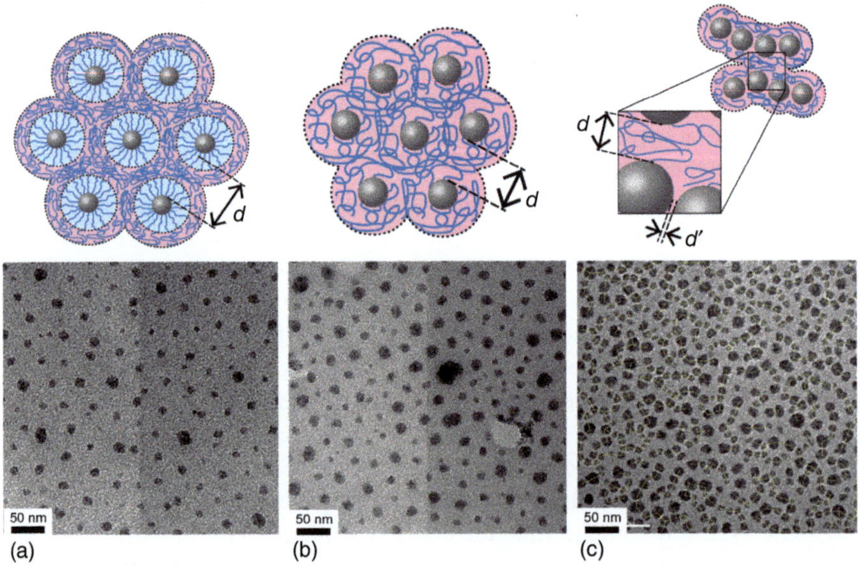

Figure 1.7 Representative bright-field transmission electron micrographs for (a) dense (SiO_2-*d*-S365), (b) intermediate (SiO_2-*i*-S328), and (c) sparse (SiO_2-*s*-S432) PS-brush systems with similar degrees of polymerization. Also shown are schematic illustrations of the corresponding microstructures. Source: Lee et al. [393], Reproduced with permission of American Chemical Society.

their interparticle distance distributions through the image analysis performed on the TEM images. If the distance distribution is monomodal, the structure is considered uniform, which is attributed to the stronger hard-sphere-type attraction in densely/intermediately grafted hairy nanoparticle systems that promote the formation of highly ordered nanostructures, while a bimodal distribution suggests the formation of a string-like superstructure [373, 394–399].

1.4.2 Bimodal Hairy Nanoparticles

A primary issue restraining the application of hairy nanoparticle solids is that short polymer ligands cannot afford sufficient chain entanglements for a tough material with good processibility, while high-molecular-weight ligands with the same grafting density significantly reduce the inorganic content and further the targeted improvements introduced by the inorganic nanofillers. Even though the inorganic fraction can maintain a decent value when applying sparsely grafted high molar mass polymer chains, the nontethered bare surface could result in severe agglomeration, which usually has detrimental effects on the performance of the material [400]. A possible solution to this dilemma was demonstrated by hairy nanoparticles with bimodal molecular weight distribution (Scheme 1.11) [401]. In an ideal case, the short polymer ligands densely cover the pristine nanoparticles without sacrificing the inorganic fraction to prevent the aggregation of nanoparticles. Meanwhile, as the transition from CPB to SDPB regime occurs at the chain ends of the short ligands, the introduction of sparsely grafted high molar mass ligands affords the chain entanglements for enhanced mechanical performance and processibility Figure 1.8.

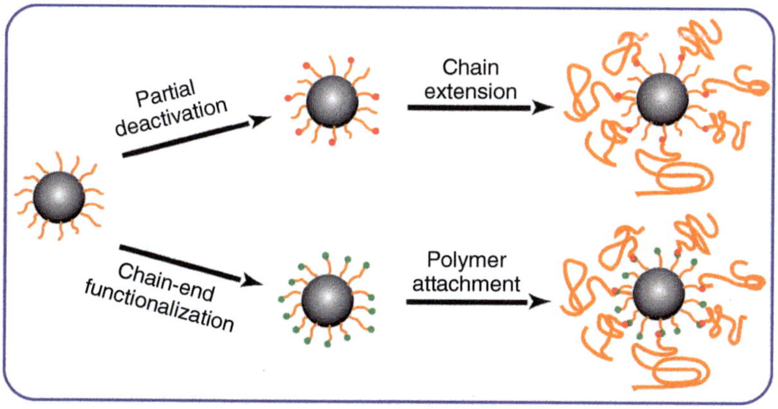

Scheme 1.11 Interactions of hairy nanoparticles with bimodal molecular weight distribution. Polymers with bimodal molecular weight distribution on hairy nanoparticles via "extending-from" (partial deactivation) strategy and "attaching-onto" (polymer attachment) strategy. Source: Reproduced from Yan et al. [401] with permission of American Chemical Society, Copyright 2015.

Bimodal block copolymer hairy nanoparticles with tunable assembling behavior were prepared using functionalized nanoparticles with different concentrations of

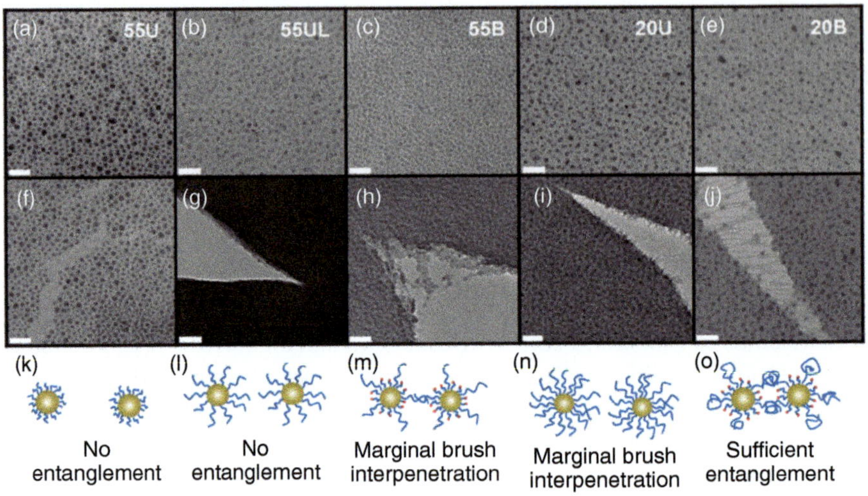

Figure 1.8 TEM images of monolayers (a–e), crack formation (f–j), and illustrations of cracks (k–o) of the five SiO_2-*g*-PS hairy nanoparticles. Unimodal sample **55U** (SiO_2-*g*-PS_{80}, a, f, k): $N < N_e \sim 160$, extensive crack propagation. Unimodal sample **55UL** (SiO_2-*g*-PS_{170}, b, g, l): above entanglement limit but in CPB regime (DP < 250), sharp crack formation. Bimodal sample **55B** (SiO_2-*g*-*bi*-$PS_{13,170}$, c, h, m): long brushes above entanglement limit and slightly beyond CPB–SDPB transition (grafting density ~ 0.11 chains nm^{-2}), plastic deformation. Unimodal sample **20U** (SiO_2-*g*-PS_{250}, d, i, n): $N > N_e$ and slightly beyond CPB–SDPB transition, stent-like undulation formation. Bimodal sample **20B** (SiO_2-*g*-*bi*-$PS_{69,790}$, e, j, o): long brushes in SDPB regime and far above entanglement limit, craze formation. Scale bars = 100 nm. Source: Yan et al. [401], Reproduced with permission of American Chemical Society.

initiating sites. The primary PMMA blocks were extended with PS as the outer shell (Figure 1.9) [388]. The chain extension efficiency and bimodality of the grafting ligands were modulated by the concentration of hairy nanoparticle macroinitiators (SiO_2-g-PMMA-Br). Three different bimodal block copolymer hairy nanoparticles with densely/intermediately/sparsely grafted pristine PMMA blocks were synthesized. Due to their low extension efficiency, the three bimodal SiO_2-g-PMMA-*b*-PS hairy nanoparticles exhibited macroscopically uniform but microscopic string-like features, connected rings, and continuous cluster network morphologies, respectively. These observed different phase-separated structures were attributed to the segregation of PMMA- and PMMA-*b*-PS-grafted chains. This development offers a new path toward designing hierarchically ordered quasi-one-component materials.

1.5 Conclusion

The incorporation of inorganic nanofillers with polymers can achieve composites with overall enhanced performance and novel properties that are derived from the complex superposition of dynamics and interactions. Within the past decade, advancements in synthetic methodologies have enabled the preparation of hairy

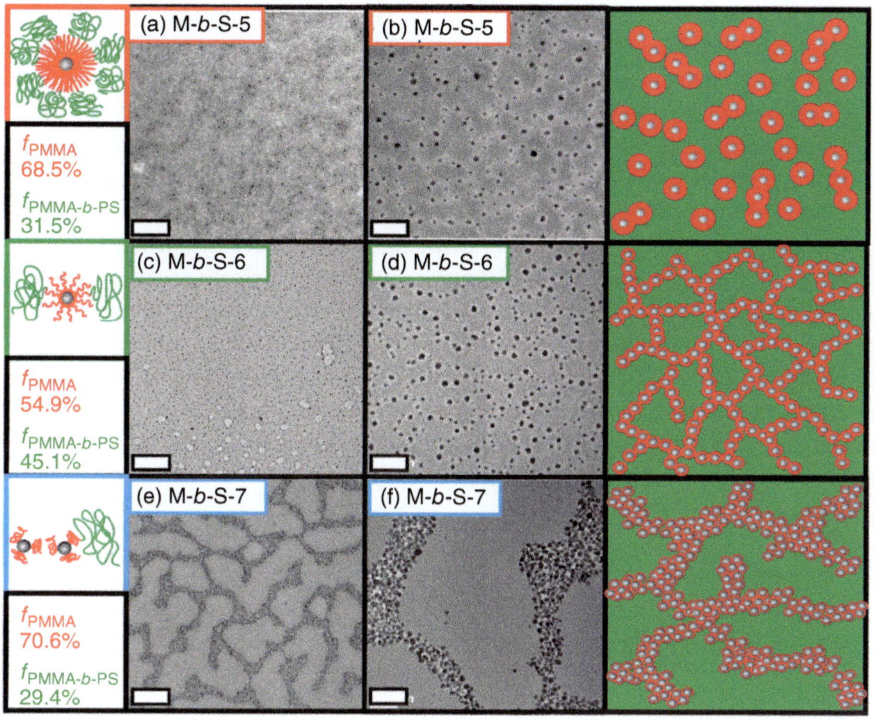

Figure 1.9 TEM images of bimodal SiO_2-*g*-PMMA-*b*-PS hairy nanoparticles. (a, b) densely grafted, (c, d) mediumly grafted, (e, f) sparsely grafted. Scale bar: (a), (c), (e), 500 nm; (b), (d), (f), 100 nm. Source: Wang et al. [388], Reproduced with permission of American Chemical Society.

nanoparticles with precisely engineered structures and compositions. Among the various approaches, SI-CRP is a robust technique that allows the synthesis of hairy nanoparticles with precise control over molar mass, dispersity, grafting density, the microstructure, as well as the architecture of the tethered chains. Although some recent reports present RAFT as a promising methodology to generate surface-grafted polymer chains, SI-ATRP is still the predominant technique in this field, mostly due to facile surface functionalization with ATRP initiators. It is possible to reduce the ATRP catalyst concentration to ppm level and reach a high conversion within a short reaction period with preserved chain-end fidelity. Besides, ATRP offers facile temporal and spatial control over the reaction through external stimuli, including photo-irradiation, electrical current, and ultrasound agitation. The major approaches to tether polymer ligands to inorganic substrates discussed in this chapter are "grafting-from" and "grafting-onto" methods. Generally, the "grafting-from" approach affords grafted polymer ligands with higher and tunable grafting densities; however, in this approach, a step of surface functionalization is needed before polymerization to introduce the initiating sites on the surface of nanoparticles. The "grafting-onto" approach does not always require the pre-treatment step, but due to the steric hindrance among the tether ligands, it is difficult to achieve high grafting densities with this method. Subsequently, "ligand exchange" was developed to obtain high grafting densities with limited selections of interchangeable ligands.

The area of synthesis of hairy nanoparticles is among the most rapidly growing fields, as it provides access to previously unavailable novel materials for high-value potential applications, such as separation science, biosensors, non-fouling coatings, and organic electronics. Vast opportunities are predicted from the basic investigations toward large-scale, low-cost manufacturing of soft materials. Further research will focus on the expansion of the library of monomer selections applicable to SI-RDRP and the further advancement of chemical methods to understand the structure–property correlations and control higher-order chain characteristics, such as the sequence and the spatial distribution of repeat units along with the multicomponent graft systems. The opportunities for developing innovative material systems with better control of the microstructure could have a substantial impact on a broad range of soft matter technologies.

Acknowledgment

This work is funded by the U.S. Department of Energy (DOE) Basic Energy Sciences (DE-SC00 18784).

References

1 Ayres, N. (2010). *Polym. Chem.* 1 (6): 769–777.
2 Chen, T., Ferris, R., Zhang, J. et al. (2010). *Prog. Polym. Sci.* 35 (1): 94–112.
3 Galvin, C.J. and Genzer, J. (2012). *Prog. Polym. Sci.* 37 (7): 871–906.
4 Zoppe, J.O., Ataman, N.C., Mocny, P. et al. (2017). *Chem. Rev.* 117 (3): 1105–1318.
5 Krishnamoorthy, M., Hakobyan, S., Ramstedt, M., and Gautrot, J.E. (2014). *Chem. Rev.* 114 (21): 10976–11026.
6 Hui, C.M., Pietrasik, J., Schmitt, M. et al. (2014). *Chem. Mater.* 26 (1): 745–762.
7 Jiang, H. and Xu, F.-J. (2013). *Chem. Soc. Rev.* 42 (8): 3394–3426.
8 Zhao, B. and Zhu, L. (2009). *Macromolecules* 42 (24): 9369–9383.
9 Chen, T., Amin, I., and Jordan, R. (2012). *Chem. Soc. Rev.* 41 (8): 3280–3296.
10 Barbey, R., Lavanant, L., Paripovic, D. et al. (2009). *Chem. Rev.* 109 (11): 5437–5527.
11 Cayre, O.J., Chagneux, N., and Biggs, S. (2011). *Soft Matter* 7 (6): 2211–2234.
12 Bünsow, J., Kelby, T.S., and Huck, W.T.S. (2010). *Acc. Chem. Res.* 43 (3): 466–474.
13 Olivier, A., Meyer, F., Raquez, J.-M. et al. (2012). *Prog. Polym. Sci.* 37 (1): 157–181.
14 Xu, F.J., Neoh, K.G., and Kang, E.T. (2009). *Prog. Polym. Sci.* 34 (8): 719–761.
15 Hucknall, A., Rangarajan, S., and Chilkoti, A. (2009). *Adv. Mater.* 21 (23): 2441–2446.

16 Chen, H., Zhao, C., Zhang, M. et al. (2016). *Langmuir* 32 (14): 3315–3330.

17 Choudhury, S., Agrawal, A., Wei, S. et al. (2016). *Chem. Mater.* 28 (7): 2147–2157.

18 Sui, T., Song, B., Wen, Y.-h., and Zhang, F. (2016). *Sci. Rep.* 6 (1): 22696.

19 Kirillova, A., Schliebe, C., Stoychev, G. et al. (2015). *ACS Appl. Mater. Interfaces* 7 (38): 21218–21225.

20 Zhang, G., Allahyarov, E., and Zhu, L. (2018). Polymer nanodielectrics: current accomplishments and future challenges for electric energy storage. In: *Nano/Micro-Structured Materials for Energy and Biomedical Applications: Latest Developments, Challenges and Future Directions* (ed. B. Li and T. Jiao), 1–48. Singapore: Springer.

21 Miranda, O.R., Li, X., Garcia-Gonzalez, L. et al. (2011). *J. Am. Chem. Soc.* 133 (25): 9650–9653.

22 Ji, L., Lin, Z., Li, Y. et al. (2010). *Polymer* 51 (19): 4368–4374.

23 Kim, J.Y., Lee, B.S., Choi, J. et al. (2016). *Angew. Chem. Int. Ed.* 55 (49): 15306–15309.

24 Ma, H., Li, D., Sheng, X. et al. (2006). *Langmuir* 22 (8): 3751–3756.

25 Wang, D., Ye, G., Wang, X., and Wang, X. (2011). *Adv. Mater.* 23 (9): 1122–1125.

26 Hansoge, N.K. and Keten, S. (2019). *ACS Macro Lett.* 8 (10): 1209–1215.

27 Staszewski, T. (2020). *J. Phys. Chem. C* 124 (49): 27118–27129.

28 Dunderdale, G.J., Urata, C., Miranda, D.F., and Hozumi, A. (2014). *ACS Appl. Mater. Interfaces* 6 (15): 11864–11868.

29 Esparza-González, S.C., Sánchez-Valdés, S., Ramírez-Barrón, S.N. et al. (2016). *Toxicol. in Vitro* 37: 134–141.

30 Arfat, Y.A., Ahmed, J., Al Hazza, A. et al. (2017). *Int. J. Biol. Macromol.* 101: 1041–1050.

31 Matsuda, T. and Ohya, S. (2005). *Langmuir* 21 (21): 9660–9665.

32 Pietrasik, J., Bombalski, L., Cusick, B. et al. (2005). Controlling polymer chain topology and architecture by ATRP from flat surfaces. In: *Stimuli-Responsive Polymeric Films and Coatings*, ACS Symposium Series, vol. 912, 28–42. Washington: American Chemical Society.

33 Yan, J., Pan, X., Wang, Z. et al. (2016). *Macromolecules* 49 (23): 9283–9286.

34 Lee, H., Dellatore, S.M., Miller, W.M., and Messersmith, P.B. (2007). *Science* 318 (5849): 426–430.

35 Song, Y., Ye, G., Lu, Y. et al. (2016). *ACS Macro Lett.* 5 (3): 382–386.

36 Zeng, Z., Wen, M., Ye, G. et al. (2017). *Chem. Mater.* 29 (23): 10212–10219.

37 Zhou, L., Zheng, L., Yuan, J., and Wu, S. (2012). *Mater. Lett.* 78: 166–169.

38 Wang, Z., Mahoney, C., Yan, J. et al. (2016). *Langmuir* 32 (49): 13207–13213.

39 Yan, J., Pan, X., Wang, Z. et al. (2017). *Chem. Mater.* 29 (11): 4963–4969.

40 Zhang, J., Gu, P., Xu, J. et al. (2016). *Nanoscale* 8 (44): 18578–18595.

41 Wu, L., Zhang, Y., Yang, G. et al. (2016). *RSC Adv.* 6 (74): 69836–69844.

42 Leo, C.P., Cathie Lee, W.P., Ahmad, A.L., and Mohammad, A.W. (2012). *Sep. Purif. Technol.* 89: 51–56.

43 Khabibullin, A., Bhangaonkar, K., Mahoney, C. et al. (2016). *ACS Appl. Mater. Interfaces* 8 (8): 5458–5465.

44 Zeng, G., Liu, M., Shi, K. et al. (2016). *Appl. Surf. Sci.* 390: 710–717.
45 Zaborniak, I., Chmielarz, P., and Matyjaszewski, K. (2019). *Eur. Polym. J.* 120: 109253.
46 Basuki, J.S., Esser, L., Zetterlund, P.B. et al. (2013). *Macromolecules* 46 (15): 6038–6047.
47 Dong, H., Huang, J., Koepsel, R.R. et al. (2011). *Biomacromolecules* 12 (4): 1305–1311.
48 Matrab, T., Chehimi, M.M., Perruchot, C. et al. (2005). *Langmuir* 21 (10): 4686–4694.
49 Liu, T., Jia, S., Kowalewski, T. et al. (2003). *Langmuir* 19 (16): 6342–6345.
50 Böttcher, H., Hallensleben, M.L., Nuß, S. et al. (2002). *J. Mater. Chem.* 12 (5): 1351–1354.
51 Raula, J., Shan, J., Nuopponen, M. et al. (2003). *Langmuir* 19 (8): 3499–3504.
52 Dong, H., Zhu, M., Yoon, J.A. et al. (2008). *J. Am. Chem. Soc.* 130 (39): 12852–12853.
53 Kim, J.-B., Bruening, M.L., and Baker, G.L. (2000). *J. Am. Chem. Soc.* 122 (31): 7616–7617.
54 Mandal, T.K., Fleming, M.S., and Walt, D.R. (2002). *Nano Lett.* 2 (1): 3–7.
55 He, Q., Küller, A., Grunze, M., and Li, J. (2007). *Langmuir* 23 (7): 3981–3987.
56 Ohno, K., Koh, K.-m., Tsujii, Y., and Fukuda, T. (2002). *Macromolecules* 35 (24): 8989–8993.
57 Esteves, A.C.C., Bombalski, L., Trindade, T. et al. (2007). *Small* 3 (7): 1230–1236.
58 Kim, S.-W., Kim, S., Tracy, J.B. et al. (2005). *J. Am. Chem. Soc.* 127 (13): 4556–4557.
59 Yan, J., Bockstaller, M.R., and Matyjaszewski, K. (2020). *Prog. Polym. Sci.* 100: 101180–101230.
60 Acres, R.G., Ellis, A.V., Alvino, J. et al. (2012). *J. Phys. Chem. C* 116 (10): 6289–6297.
61 Ryu, J.H., Messersmith, P.B., and Lee, H. (2018). *ACS Appl. Mater. Interfaces* 10 (9): 7523–7540.
62 Connor, P.A., Dobson, K.D., and McQuillan, A.J. (1995). *Langmuir* 11 (11): 4193–4195.
63 Liebscher, J., Mrówczyński, R., Scheidt, H.A. et al. (2013). *Langmuir* 29 (33): 10539–10548.
64 Lynge, M.E., van der Westen, R., Postma, A., and Städler, B. (2011). *Nanoscale* 3 (12): 4916–4928.
65 Pranantyo, D., Xu, L.Q., Neoh, K.-G. et al. (2015). *Biomacromolecules* 16 (3): 723–732.
66 Kuang, J. and Messersmith, P.B. (2012). *Langmuir* 28 (18): 7258–7266.
67 Allara, D.L. and Nuzzo, R.G. (1985). *Langmuir* 1 (1): 45–52.
68 Zhang, L., He, R., and Gu, H.-C. (2006). *Appl. Surf. Sci.* 253 (5): 2611–2617.
69 Sun, S., Murray, C.B., Weller, D. et al. (2000). *Science* 287 (5460): 1989–1992.
70 Leff, D.V., Brandt, L., and Heath, J.R. (1996). *Langmuir* 12 (20): 4723–4730.

71 Kumar, A., Mandal, S., Selvakannan, P.R. et al. (2003). *Langmuir* 19 (15): 6277–6282.

72 Wu, N., Fu, L., Su, M. et al. (2004). *Nano Lett.* 4 (2): 383–386.

73 Zhang, J., Lee, J., Wang, Z. et al. (2017). *Polymer* 126: 126–132.

74 Armes, S.P., Gottesfeld, S., Beery, J.G. et al. (1991). *Polymer* 32 (13): 2325–2330.

75 Maeda, S., Corradi, R., and Armes, S.P. (1995). *Macromolecules* 28 (8): 2905–2911.

76 Jung, N., Kwon, S., Lee, D. et al. (2013). *Adv. Mater.* 25 (47): 6854–6858.

77 Sahoo, Y., Pizem, H., Fried, T. et al. (2001). *Langmuir* 17 (25): 7907–7911.

78 Burkett, S.L., Ko, N., Stern, N.D. et al. (2006). *Chem. Mater.* 18 (21): 5137–5143.

79 Tsubokawa, N. (1992). *Prog. Polym. Sci.* 17 (3): 417–470.

80 Delamar, M., Hitmi, R., Pinson, J., and Saveant, J.M. (1992). *J. Am. Chem. Soc.* 114 (14): 5883–5884.

81 Iruthayaraj, J., Chernyy, S., Lillethorup, M. et al. (2011). *Langmuir* 27 (3): 1070–1078.

82 Li, Y. and Zuilhof, H. (2012). *Langmuir* 28 (12): 5350–5359.

83 Colavita, P.E., Sun, B., Tse, K.-Y., and Hamers, R.J. (2007). *J. Am. Chem. Soc.* 129 (44): 13554–13565.

84 Leff, D.V., Ohara, P.C., Heath, J.R., and Gelbart, W.M. (1995). *J. Phys. Chem.* 99 (18): 7036–7041.

85 Templeton, A.C., Wuelfing, W.P., and Murray, R.W. (2000). *Acc. Chem. Res.* 33 (1): 27–36.

86 Grönbeck, H., Curioni, A., and Andreoni, W. (2000). *J. Am. Chem. Soc.* 122 (16): 3839–3842.

87 Xue, Y.-Q., Di, J.-M., Luo, Y. et al. (2014). *Oxid. Med. Cell. Longevity* 2014: 765832.

88 Desmet, T., Morent, R., De Geyter, N. et al. (2009). *Biomacromolecules* 10 (9): 2351–2378.

89 Chu, P.K., Chen, J.Y., Wang, L.P., and Huang, N. (2002). *Mater. Sci. Eng. R* 36 (5): 143–206.

90 Hu, S., Ren, X., Bachman, M. et al. (2002). *Anal. Chem.* 74 (16): 4117–4123.

91 Makamba, H., Kim, J.H., Lim, K. et al. (2003). *Electrophoresis* 24 (21): 3607–3619.

92 France, R.M. and Short, R.D. (1997). *J. Chem. Soc., Faraday Trans.* 93 (17): 3173–3178.

93 Wood, T.J. and Badyal, J.P.S. (2012). *ACS Appl. Mater. Interfaces* 4 (3): 1675–1682.

94 Siow, K.S., Britcher, L., Kumar, S., and Griesser, H.J. (2006). *Plasma Processes Polym.* 3 (6, 7): 392–418.

95 Khelifa, F., Ershov, S., Habibi, Y. et al. (2016). *Chem. Rev.* 116 (6): 3975–4005.

96 Teare, D.O.H., Barwick, D.C., Schofield, W.C.E. et al. (2005). *Langmuir* 21 (24): 11425–11430.

97 Kumar, R., Welle, A., Becker, F. et al. (2018). *ACS Appl. Mater. Interfaces* 10 (38): 31965–31976.

98 Coad, B.R., Styan, K.E., and Meagher, L. (2014). *ACS Appl. Mater. Interfaces* 6 (10): 7782–7789.

99 Han, J., Zhai, Y., Wang, Z. et al. (2020). *ACS Macro Lett.* 9 (9): 1218–1223.

100 Kumar, S.K., Benicewicz, B.C., Vaia, R.A., and Winey, K.I. (2017). *Macromolecules* 50 (3): 714–731.

101 Zhao, B. and Brittain, W.J. (2000). *Prog. Polym. Sci.* 25 (5): 677–710.

102 Edmondson, S., Osborne, V.L., and Huck, W.T.S. (2004). *Chem. Soc. Rev.* 33 (1): 14–22.

103 Ding, H., Yan, J., Wang, Z. et al. (2016). *Polymer* 107: 492–502.

104 Gao, H. and Matyjaszewski, K. (2007). *J. Am. Chem. Soc.* 129 (20): 6633–6639.

105 Möller, M., Künstle, H., and Kunz, M. (1991). *Synth. Met.* 41 (3): 1159–1162.

106 Zhang, Y., Wang, Z., Matyjaszewski, K., and Pietrasik, J. (2019). *Eur. Polym. J.* 110: 49–55.

107 Yuan, J., Xu, Y., Walther, A. et al. (2008). *Nat. Mater.* 7 (9): 718–722.

108 Müllner, M., Yuan, J., Weiss, S. et al. (2010). *J. Am. Chem. Soc.* 132 (46): 16587–16592.

109 Djalali, R., Li, S.-Y., and Schmidt, M. (2002). *Macromolecules* 35 (11): 4282–4288.

110 Pang, X., Zhao, L., Han, W. et al. (2013). *Nat. Nanotechnol.* 8 (6): 426–431.

111 Binder, W.H. (2019). *Macromol. Rapid Commun.* 40 (1): 1800610.

112 Laible, R. and Hamann, K. (1975). *Angew. Makromol. Chem.* 48 (1): 97–133.

113 Laible, R. and Hamann, K. (1980). *Adv. Colloid Interface Sci.* 13 (1): 65–99.

114 Prucker, O. and Rühe, J. (1998). *Macromolecules* 31 (3): 592–601.

115 M. Karahasanp lu; A. Onen; I. Serhatm. *Prog. Org. Coat.* 2014, 77, 1079-1084. M. Karahasanoğlu, A. Önen, İ.E. Serhatlı

116 Ahin, M., Krawczyk, K., Roszkowski, P. et al. (2018). *Eur. Polym. J.* 98: 430–438.

117 Lee, H., Ryu, J., Kim, M. et al. (2016). *Radiat. Phys. Chem.* 125: 160–164.

118 Kang, Y., Choi, S.-H., Gopalan, A. et al. (2006). *J. Non-Cryst. Solids* 352: 463–468.

119 Woodworth, B.E., Metzner, Z., and Matyjaszewski, K. (1998). *Macromolecules* 31 (23): 7999–8004.

120 Pyun, J. and Matyjaszewski, K. (2001). *Chem. Mater.* 13 (10): 3436–3448.

121 Ohno, K., Morinaga, T., Koh, K. et al. (2005). *Macromolecules* 38 (6): 2137–2142.

122 Matyjaszewski, K., Miller, P.J., Pyun, J. et al. (1999). *Macromolecules* 32 (20): 6526–6535.

123 Matyjaszewski, K. (2012). *Macromolecules* 45 (10): 4015–4039.

124 Matyjaszewski, K. and Xia, J. (2001). *Chem. Rev.* 101 (9): 2921–2990.

125 Matyjaszewski, K. and Tsarevsky, N.V. (2014). *J. Am. Chem. Soc.* 136 (18): 6513–6533.

126 Wang, J.-S. and Matyjaszewski, K. (1995). *Macromolecules* 28 (23): 7901–7910.

127 Ejaz, M., Yamamoto, S., Ohno, K. et al. (1998). *Macromolecules* 31 (17): 5934–5936.

128 Bombalski, L., Min, K., Dong, H. et al. (2007). *Macromolecules* 40 (21): 7429–7432.

129 Ribelli, T.G., Lorandi, F., Fantin, M., and Matyjaszewski, K. (2019). *Macromol. Rapid Commun.* 40 (1): 1800616.

130 Jakubowski, W. and Matyjaszewski, K. (2006). *Angew. Chem. Int. Ed.* 45 (27): 4482–4486.

131 Jakubowski, W., Min, K., and Matyjaszewski, K. (2006). *Macromolecules* 39 (1): 39–45.

132 Konkolewicz, D., Wang, Y., Krys, P. et al. (2014). *Polym. Chem.* 5 (15): 4396–4417.

133 Konkolewicz, D., Wang, Y., Zhong, M. et al. (2013). *Macromolecules* 46 (22): 8749–8772.

134 Matyjaszewski, K., Jakubowski, W., Min, K. et al. (2006). *Proc. Natl. Acad. Sci. U.S.A.* 103 (42): 15309–15314.

135 Pan, X., Fantin, M., Yuan, F., and Matyjaszewski, K. (2018). *Chem. Soc. Rev.* 47 (14): 5457–5490.

136 Magenau, A.J.D., Strandwitz, N.C., Gennaro, A., and Matyjaszewski, K. (2011). *Science* 332 (6025): 81–84.

137 Chmielarz, P., Fantin, M., Park, S. et al. (2017). *Prog. Polym. Sci.* 69: 47–78.

138 Kwak, Y. and Matyjaszewski, K. (2010). *Macromolecules* 43 (12): 5180–5183.

139 Pan, X., Tasdelen, M.A., Laun, J. et al. (2016). *Prog. Polym. Sci.* 62: 73–125.

140 Wang, Z., Pan, X., Li, L. et al. (2017). *Macromolecules* 50 (20): 7940–7948.

141 Wang, Z., Pan, X., Yan, J. et al. (2017). *ACS Macro Lett.* 6 (5): 546–549.

142 Wang, Z., Wang, Z., Pan, X. et al. (2018). *ACS Macro Lett.* 7 (3): 275–280.

143 Wang, Z., Lorandi, F., Fantin, M. et al. (2019). *ACS Macro Lett.* 8 (2): 161–165.

144 Matyjaszewski, K., Wei, M., Xia, J., and McDermott, N.E. (1997). *Macromolecules* 30 (26): 8161–8164.

145 Dadashi-Silab, S., Pan, X., and Matyjaszewski, K. (2017). *Macromolecules* 50 (20): 7967–7977.

146 Pan, X., Malhotra, N., Dadashi-Silab, S., and Matyjaszewski, K. (2017). *Macromol. Rapid Commun.* 38 (13): 1600651.

147 Kato, M., Kamigaito, M., Sawamoto, M., and Higashimura, T. (1995). *Macromolecules* 28 (5): 1721–1723.

148 Ando, T., Kato, M., Kamigaito, M., and Sawamoto, M. (1996). *Macromolecules* 29 (3): 1070–1072.

149 Fors, B.P. and Hawker, C.J. (2012). *Angew. Chem. Int. Ed.* 51 (35): 8850–8853.

150 Treat, N.J., Sprafke, H., Kramer, J.W. et al. (2014). *J. Am. Chem. Soc.* 136 (45): 16096–16101.

151 Pan, X., Lamson, M., Yan, J., and Matyjaszewski, K. (2015). *ACS Macro Lett.* 4 (2): 192–196.

152 Pan, X., Fang, C., Fantin, M. et al. (2016). *J. Am. Chem. Soc.* 138 (7): 2411–2425.

153 Pan, X., Lathwal, S., Mack, S. et al. (2017). *Angew. Chem. Int. Ed.* 56 (10): 2740–2743.

154 Yan, J., Li, B., Zhou, F., and Liu, W. (2013). *ACS Macro Lett.* 2 (7): 592–596.

155 Li, B., Yu, B., Huck, W.T.S. et al. (2013). *J. Am. Chem. Soc.* 135 (5): 1708–1710.

156 Treat, N.J., Fors, B.P., Kramer, J.W. et al. (2014). *ACS Macro Lett.* 3 (6): 580–584.

157 Sato, T., Dunderdale, G.J., Urata, C., and Hozumi, A. (2018). *Macromolecules* 51 (24): 10065–10073.

158 Li, B., Yu, B., Huck, W.T.S. et al. (2012). *Angew. Chem. Int. Ed.* 51 (21): 5092–5095.

159 Poelma, J.E., Fors, B.P., Meyers, G.F. et al. (2013). *Angew. Chem. Int. Ed.* 52 (27): 6844–6848.

160 Min, K., Gao, H., and Matyjaszewski, K. (2007). *Macromolecules* 40 (6): 1789–1791.

161 Gnanou, Y. and Hizal, G. (2004). *J. Polym. Sci., Part A: Polym. Chem.* 42 (2): 351–359.

162 Kwak, Y., Magenau, A.J.D., and Matyjaszewski, K. (2011). *Macromolecules* 44 (4): 811–819.

163 Dong, H. and Matyjaszewski, K. (2008). *Macromolecules* 41 (19): 6868–6870.

164 Matyjaszewski, K., Dong, H., Jakubowski, W. et al. (2007). *Langmuir* 23 (8): 4528–4531.

165 Yan, W., Dadashi-Silab, S., Matyjaszewski, K. et al. (2020). *Macromolecules* 53 (8): 2801–2810.

166 Yan, W., Fantin, M., Ramakrishna, S. et al. (2019). *ACS Appl. Mater. Interfaces* 11 (30): 27470–27477.

167 Yan, W., Fantin, M., Spencer, N.D. et al. (2019). *ACS Macro Lett.* 8 (7): 865–870.

168 Fantin, M., Ramakrishna, S.N., Yan, J. et al. (2018). *Macromolecules* 51 (17): 6825–6835.

169 Dehghani, E.S., Du, Y., Zhang, T. et al. (2017). *Macromolecules* 50 (6): 2436–2446.

170 Wang, Y., Soerensen, N., Zhong, M. et al. (2013). *Macromolecules* 46 (3): 683–691.

171 Pietrasik, J., Dong, H., and Matyjaszewski, K. (2006). *Macromolecules* 39 (19): 6384–6390.

172 Dong, H., Tang, W., and Matyjaszewski, K. (2007). *Macromolecules* 40 (9): 2974–2977.

173 Mueller, L., Jakubowski, W., Tang, W., and Matyjaszewski, K. (2007). *Macromolecules* 40 (18): 6464–6472.

174 Tang, W., Tsarevsky, N.V., and Matyjaszewski, K. (2006). *J. Am. Chem. Soc.* 128 (5): 1598–1604.

175 Krys, P. and Matyjaszewski, K. (2017). *Eur. Polym. J.* 89: 482–523.

176 Matyjaszewski, K., Tsarevsky, N.V., Braunecker, W.A. et al. (2007). *Macromolecules* 40 (22): 7795–7806.

177 Zhang, Y., Wang, Y., and Matyjaszewski, K. (2011). *Macromolecules* 44 (4): 683–685.

178 Williams, V.A., Ribelli, T.G., Chmielarz, P. et al. (2015). *J. Am. Chem. Soc.* 137 (4): 1428–1431.

179 Yan, J., Pan, X., Schmitt, M. et al. (2016). *ACS Macro Lett.* 5 (6): 661–665.

180 Magenau, A.J.D., Bortolamei, N., Frick, E. et al. (2013). *Macromolecules* 46 (11): 4346–4353.

181 Chmielarz, P., Yan, J., Krys, P. et al. (2017). *Macromolecules* 50 (11): 4151–4159.

182 Ribelli, T.G., Konkolewicz, D., Bernhard, S., and Matyjaszewski, K. (2014). *J. Am. Chem. Soc.* 136 (38): 13303–13312.
183 Mosnáček, J. and Ilčíková, M. (2012). *Macromolecules* 45 (15): 5859–5865.
184 Konkolewicz, D., Schröder, K., Buback, J. et al. (2012). *ACS Macro Lett.* 1 (10): 1219–1223.
185 Anastasaki, A., Nikolaou, V., Zhang, Q. et al. (2014). *J. Am. Chem. Soc.* 136 (3): 1141–1149.
186 Xu, J., Jung, K., Atme, A. et al. (2014). *J. Am. Chem. Soc.* 136 (14): 5508–5519.
187 Ohtsuki, A., Goto, A., and Kaji, H. (2013). *Macromolecules* 46 (1): 96–102.
188 Anastasaki, A., Nikolaou, V., Simula, A. et al. (2014). *Macromolecules* 47 (12): 3852–3859.
189 Alfredo, N.V., Jalapa, N.E., Morales, S.L. et al. (2012). *Macromolecules* 45 (20): 8135–8146.
190 Xu, J., Jung, K., and Boyer, C. (2014). *Macromolecules* 47 (13): 4217–4229.
191 Dadashi-Silab, S., Atilla Tasdelen, M., and Yagci, Y. (2014). *J. Polym. Sci., Part A: Polym. Chem.* 52 (20): 2878–2888.
192 Discekici, E.H., Pester, C.W., Treat, N.J. et al. (2016). *ACS Macro Lett.* 5 (2): 258–262.
193 Theriot, J.C., Lim, C.-H., Yang, H. et al. (2016). *Science* 352 (6289): 1082–1086.
194 Ottou, W.N., Sardon, H., Mecerreyes, D. et al. (2016). *Prog. Polym. Sci.* 56: 64–115.
195 Mohapatra, H., Kleiman, M., and Esser-Kahn, A.P. (2017). *Nat. Chem.* 9 (2): 135–139.
196 Chiefari, J., Chong, Y.K., Ercole, F. et al. (1998). *Macromolecules* 31 (16): 5559–5562.
197 Perrier, S. and Takolpuckdee, P. (2005). *J. Polym. Sci., Part A: Polym. Chem.* 43 (22): 5347–5393.
198 Moad, G., Rizzardo, E., and Thang, S.H. (2005). *Aust. J. Chem.* 58 (6): 379–410.
199 Loiseau, J., Doërr, N., Suau, J.M. et al. (2003). *Macromolecules* 36 (9): 3066–3077.
200 Moad, G., Chong, Y.K., Postma, A. et al. (2005). *Polymer* 46 (19): 8458–8468.
201 Xu, J., Jung, K., Corrigan, N.A., and Boyer, C. (2014). *Chem. Sci.* 5 (9): 3568–3575.
202 McKenzie, T.G., Colombo, E., Fu, Q. et al. (2017). *Angew. Chem. Int. Ed.* 56 (40): 12302–12306.
203 Wang, Y., Fantin, M., Park, S. et al. (2017). *Macromolecules* 50 (20): 7872–7879.
204 Lv, C., He, C., and Pan, X. (2018). *Angew. Chem. Int. Ed.* 57 (30): 9430–9433.
205 Lee, H.J., Nakayama, Y., and Matsuda, T. (1999). *Macromolecules* 32 (21): 6989–6995.
206 Tsujii, Y., Ejaz, M., Sato, K. et al. (2001). *Macromolecules* 34 (26): 8872–8878.
207 Baum, M. and Brittain, W.J. (2002). *Macromolecules* 35 (3): 610–615.
208 Li, C. and Benicewicz, B.C. (2005). *Macromolecules* 38 (14): 5929–5936.
209 Peng, Q., Lai, D.M.Y., Kang, E.T., and Neoh, K.G. (2006). *Macromolecules* 39 (16): 5577–5582.
210 Zhao and Perrier, S. (2006). *Macromolecules* 39 (25): 8603–8608.

211 Ohno, K., Ma, Y., Huang, Y. et al. (2011). *Macromolecules* 44 (22): 8944–8953.

212 Li, C., Han, J., Ryu, C.Y., and Benicewicz, B.C. (2006). *Macromolecules* 39 (9): 3175–3183.

213 Khani, M.M., Abbas, Z.M., and Benicewicz, B.C. (2017). *J. Polym. Sci., Part A: Polym. Chem.* 55 (9): 1493–1501.

214 Niu, J., Lunn, D.J., Pusuluri, A. et al. (2017). *Nat. Chem.* 9 (6): 537–545.

215 Bagheri, A., Arandiyan, H., Adnan, N.N.M. et al. (2017). *Macromolecules* 50 (18): 7137–7147.

216 Demirci, S. and Caykara, T. (2012). *React. Funct. Polym.* 72 (9): 588–595.

217 Kitano, H., Liu, Y., Tokuwa, K.-i. et al. (2012). *Eur. Polym. J.* 48 (11): 1875–1882.

218 Kondo, T., Gemmei-Ide, M., Kitano, H. et al. (2012). *Colloids Surf., B* 91: 215–218.

219 Matsuzaka, N., Takahashi, H., Nakayama, M. et al. (2012). *J. Biomater. Sci., Polym. Ed.* 23 (10): 1301–1314.

220 Takahashi, H., Matsuzaka, N., Nakayama, M. et al. (2012). *Biomacromolecules* 13 (1): 253–260.

221 Takahashi, H., Nakayama, M., Itoga, K. et al. (2011). *Biomacromolecules* 12 (5): 1414–1418.

222 Bain, E.D., Dawes, K., Özçam, A.E. et al. (2012). *Macromolecules* 45 (9): 3802–3815.

223 Yu, H.-Y., Li, W., Zhou, J. et al. (2009). *J. Membr. Sci.* 343 (1): 82–89.

224 Zhou, J., Li, W., Gu, J.-S. et al. (2010). *Sep. Purif. Technol.* 71 (2): 233–240.

225 Macdonald, T., Gibson, C.T., Constantopoulos, K. et al. (2012). *Appl. Surf. Sci.* 258 (7): 2836–2843.

226 Islam, M.R., Bach, L.G., Vo, T.-S. et al. (2013). *Appl. Surf. Sci.* 286: 31–39.

227 Dilag, J., Kobus, H., and Ellis, A.V. (2013). *Forensic Sci. Dermatol. Int.* 228 (1): 105–114.

228 Grande, C.D., Tria, M.C., Jiang, G. et al. (2011). *Macromolecules* 44 (4): 966–975.

229 Tria, M.C.R., Grande, C.D.T., Ponnapati, R.R., and Advincula, R.C. (2010). *Biomacromolecules* 11 (12): 3422–3431.

230 Zengin, A. and Caykara, T. (2012). *J. Polym. Sci., Part A: Polym. Chem.* 50 (21): 4443–4450.

231 Jiao, Y. and Akcora, P. (2014). *J. Polym. Sci., Part A: Polym. Chem.* 52 (12): 1700–1705.

232 Wilczewska, A.Z. and Markiewicz, K.H. (2014). *Macromol. Chem. Phys.* 215 (2): 190–197.

233 Tian, J., Zheng, F., and Zhao, H. (2011). *J. Phys. Chem. C* 115 (8): 3304–3312.

234 Zengin, A., Yildirim, E., Tamer, U., and Caykara, T. (2013). *Analyst* 138 (23): 7238–7245.

235 Ye, Y.-S., Chen, Y.-N., Wang, J.-S. et al. (2012). *Chem. Mater.* 24 (15): 2987–2997.

236 Choi, J., Schattling, P., Jochum, F.D. et al. (2012). *J. Polym. Sci., Part A: Polym. Chem.* 50 (19): 4010–4018.

237 Di Carlo, G., Damin, F., Armelao, L. et al. (2012). *Appl. Surf. Sci.* 258 (8): 3750–3756.

238 Duan, Z., Qu, Z., Hu, F. et al. (2014). *Appl. Surf. Sci.* 300: 104–110.

239 Rotzoll, R. and Vana, P. (2008). *J. Polym. Sci., Part A: Polym. Chem.* 46 (23): 7656–7666.

240 Günay, K.A., Schüwer, N., and Klok, H.-A. (2012). *Polym. Chem.* 3 (8): 2186–2192.

241 Moraes, J., Ohno, K., Maschmeyer, T., and Perrier, S. (2013). *Chem. Mater.* 25 (17): 3522–3527.

242 Gurbuz, N., Demirci, S., Yavuz, S., and Caykara, T. (2011). *J. Polym. Sci., Part A: Polym. Chem.* 49 (2): 423–431.

243 Hua, D., Tang, J., Jiang, J. et al. (2009). *Mater. Chem. Phys.* 114 (1): 402–406.

244 Zengin, A., Karakose, G., and Caykara, T. (2013). *Eur. Polym. J.* 49 (10): 3350–3358.

245 Zhu, L.-J., Zhu, L.-P., Jiang, J.-H. et al. (2014). *J. Membr. Sci.* 451: 157–168.

246 Matsuzaka, N., Nakayama, M., Takahashi, H. et al. (2013). *Biomacromolecules* 14 (9): 3164–3171.

247 Hong, C.-Y., Li, X., and Pan, C.-Y. (2009). *J. Mater. Chem.* 19 (29): 5155–5160.

248 Dukes, D., Li, Y., Lewis, S. et al. (2010). *Macromolecules* 43 (3): 1564–1570.

249 Maillard, D., Kumar, S.K., Rungta, A. et al. (2011). *Nano Lett.* 11 (11): 4569–4573.

250 Huang, X., Appelhans, D., Formanek, P. et al. (2012). *ACS Nano* 6 (11): 9718–9726.

251 Huang, X., Appelhans, D., Formanek, P. et al. (2011). *Macromolecules* 44 (21): 8351–8360.

252 Moll, J.F., Akcora, P., Rungta, A. et al. (2011). *Macromolecules* 44 (18): 7473–7477.

253 Wang, L. and Benicewicz, B.C. (2013). *ACS Macro Lett.* 2 (2): 173–176.

254 Yang, K., Huang, X., Huang, Y. et al. (2013). *Chem. Mater.* 25 (11): 2327–2338.

255 Xu, L., Pan, J., Xia, Q. et al. (2012). *J. Phys. Chem. C* 116 (48): 25309–25318.

256 Xiong, L., Wang, R., Liang, H. et al. (2010). *J. Macromol. Sci. A* 47 (9): 903–908.

257 Demirci, S. and Caykara, T. (2012). *J. Polym. Sci., Part A: Polym. Chem.* 50 (15): 2999–3007.

258 Moad, G. and Rizzardo, E. (1995). *Macromolecules* 28 (26): 8722–8728.

259 Rizzardo, E., Serelis, A., and Solomon, D. (1982). *Aust. J. Chem.* 35 (10): 2013–2024.

260 Hawker, C.J., Barclay, G.G., Orellana, A. et al. (1996). *Macromolecules* 29 (16): 5245–5254.

261 Hawker, C.J. (1997). *Acc. Chem. Res.* 30 (9): 373–382.

262 Hawker, C.J., Bosman, A.W., and Harth, E. (2001). *Chem. Rev.* 101 (12): 3661–3688.

263 Studer, A. and Schulte, T. (2005). *Chem. Rec.* 5 (1): 27–35.

264 Ghannam, L., Parvole, J., Laruelle, G. et al. (2006). *Polym. Int.* 55 (10): 1199–1207.

265 Bartholome, C., Beyou, E., Bourgeat-Lami, E. et al. (2003). *Macromolecules* 36 (21): 7946–7952.

266 Parvole, J., Laruelle, G., Guimon, C. et al. (2003). *Macromol. Rapid Commun.* 24 (18): 1074–1078.

267 Husseman, M., Malmström, E.E., McNamara, M. et al. (1999). *Macromolecules* 32 (5): 1424–1431.

268 Parvole, J., Billon, L., and Montfort, J.P. (2002). *Polym. Int.* 51 (10): 1111–1116.

269 Ghannam, L., Bacou, M., Garay, H. et al. (2004). *Polymer* 45 (21): 7035–7045.

270 Blas, H., Save, M., Boissière, C. et al. (2011). *Macromolecules* 44 (8): 2577–2588.

271 Szwarc, M., Levy, M., and Milkovich, R. (1956). *J. Am. Chem. Soc.* 78 (11): 2656–2657.

272 Higashimura, T., Mitsuhashi, M., and Sawamoto, M. (1979). *Macromolecules* 12 (2): 178–182.

273 Matyjaszewski, K. (1996). *Cationic Polymerizations : Mechanisms, Synthesis, and Applications*. New York: Marcel Dekker.

274 Advincula, R., Zhou, Q., Park, M. et al. (2002). *Langmuir* 18 (22): 8672–8684.

275 Zhao, B. and Brittain, W.J. (2000). *Macromolecules* 33 (2): 342–348.

276 Spange, S., Simon, F., Schutz, H. et al. (1992). *J. Macromol. Sci. A* 29 (11): 997–1006.

277 Whitesell, J.K. and Chang, H.K. (1993). *Science* 261 (5117): 73–76.

278 Schmidt, A.M. (2005). *Macromol. Rapid Commun.* 26 (2): 93–97.

279 Carrot, G., Rutot-Houzé, D., Pottier, A. et al. (2002). *Macromolecules* 35 (22): 8400–8404.

280 Choi, I.S. and Langer, R. (2001). *Macromolecules* 34 (16): 5361–5363.

281 Schneider, M., Fetsch, C., Amin, I. et al. (2013). *Langmuir* 29 (23): 6983–6988.

282 Tsubokawa, N., Funaki, A., Hada, Y., and Sone, Y. (1982). *J. Poylm. Sci: Polym. Chem.* 20 (12): 3297–3304.

283 Jordan, R. and Ulman, A. (1998). *J. Am. Chem. Soc.* 120 (2): 243–247.

284 Jordan, R., West, N., Ulman, A. et al. (2001). *Macromolecules* 34 (6): 1606–1611.

285 Calderon, N. (1972). *J. Macromol. Sci. C* 7 (1): 105–159.

286 Bielawski, C.W. and Grubbs, R.H. (2007). *Prog. Polym. Sci.* 32 (1): 1–29.

287 Schrock, R.R. (1990). *Acc. Chem. Res.* 23 (5): 158–165.

288 Grubbs, R. and Tumas, W. (1989). *Science* 243 (4893): 907–915.

289 Leitgeb, A., Wappel, J., and Slugovc, C. (2010). *Polymer* 51 (14): 2927–2946.

290 Slugovc, C. (2004). *Macromol. Rapid Commun.* 25 (14): 1283–1297.

291 Lynde, B.E., Maust, R.L., Li, P. et al. (2020). *Mater. Chem. Front.* 4 (1): 252–256.

292 Tallon, M.A. (2016). Ring-opening metathesis polymerization (ROMP) using maleic anhydride-based monomers. In: *Handbook of Maleic Anhydride Based Materials: Syntheses, Properties and Applications* (ed. O.M. Musa), 311–398. Berlin: Springer International Publishing.

293 LaNasa, J.A. and Hickey, R.J. (2020). *Macromolecules* 53 (19): 8216–8232.

294 Kim, N.Y., Jeon, N.L., Choi, I.S. et al. (2000). *Macromolecules* 33 (8): 2793–2795.

295 Ogawa, K.A., Goetz, A.E., and Boydston, A.J. (2015). *J. Am. Chem. Soc.* 137 (4): 1400–1403.

296 Mastan, E., Xi, L., and Zhu, S. (2016). *Macromol. Theory Simul.* 25 (3): 220–228.

297 Chevigny, C., Dalmas, F., Di Cola, E. et al. (2011). *Macromolecules* 44 (1): 122–133.

298 Hayashi, S., Handa, S., Oshibe, Y. et al. (1995). *Polym. J.* 27 (6): 623–630.
299 Yakushiji, T., Sakai, K., Kikuchi, A. et al. (1998). *Langmuir* 14 (16): 4657–4662.
300 Tsubokawa, N. and Kogure, A. (1991). *J. Polym. Sci., Part A: Polym. Chem.* 29 (5): 697–702.
301 Hosoya, H., Kumazaki, H., Chida, K. et al. (1990). *Pure Appl. Chem.* 62 (3): 445–450.
302 Lowe, A.B., Sumerlin, B.S., Donovan, M.S., and McCormick, C.L. (2002). *J. Am. Chem. Soc.* 124 (39): 11562–11563.
303 Rossner, C. and Vana, P. (2014). *Angew. Chem. Int. Ed.* 53 (46): 12639–12642.
304 Tao, P., Viswanath, A., Schadler, L.S. et al. (2011). *ACS Appl. Mater. Interfaces* 3 (9): 3638–3645.
305 Leophairatana, P., Samanta, S., De Silva, C.C., and Koberstein, J.T. (2017). *J. Am. Chem. Soc.* 139 (10): 3756–3766.
306 Golas, P.L. and Matyjaszewski, K. (2010). *Chem. Soc. Rev.* 39 (4): 1338–1354.
307 Lutz, J.-F., Börner, H.G., and Weichenhan, K. (2006). *Macromolecules* 39 (19): 6376–6383.
308 Golas, P.L. and Matyjaszewski, K. (2007). *QSAR Comb. Sci.* 26 (11, 12): 1116–1134.
309 Yoshimoto, K., Hirase, T., Madsen, J. et al. (2009). *Macromol. Rapid Commun.* 30 (24): 2136–2140.
310 Yavuz, M.S., Cheng, Y., Chen, J. et al. (2009). *Nat. Mater.* 8 (12): 935–939.
311 Julthongpiput, D., Lin, Y.-H., Teng, J. et al. (2003). *Langmuir* 19 (19): 7832–7836.
312 Huang, J., Koepsel, R.R., Murata, H. et al. (2008). *Langmuir* 24 (13): 6785–6795.
313 Costa, A.C., Geoghegan, M., Vlček, P., and Composto, R.J. (2003). *Macromolecules* 36 (26): 9897–9904.
314 Auschra, C., Eckstein, E., Mühlebach, A. et al. (2002). *Prog. Org. Coat.* 45 (2): 83–93.
315 Bulychev, N.A., Arutunov, I.A., Zubov, V.P. et al. (2004). *Macromol. Chem. Phys.* 205 (18): 2457–2463.
316 Saleh, N., Sirk, K., Liu, Y.Q. et al. (2007). *Environ. Eng. Sci.* 24 (1): 45–57.
317 Saleh, N., Phenrat, T., Sirk, K. et al. (2005). *Nano Lett.* 5 (12): 2489–2494.
318 Saleh, N., Sarbu, T., Sirk, K. et al. (2005). *Langmuir* 21 (22): 9873–9878.
319 Wang, X.-S., Dykstra, T.E., Salvador, M.R. et al. (2004). *J. Am. Chem. Soc.* 126 (25): 7784–7785.
320 Dubois, F., Mahler, B., Dubertret, B. et al. (2007). *J. Am. Chem. Soc.* 129 (3): 482–483.
321 Sardar, R., Park, J.-W., and Shumaker-Parry, J.S. (2007). *Langmuir* 23 (23): 11883–11889.
322 Ehlert, S., Taheri, S.M., Pirner, D. et al. (2014). *ACS Nano* 8 (6): 6114–6122.
323 Wang, Z., Lu, Z., Mahoney, C. et al. (2017). *ACS Appl. Mater. Interfaces* 9 (8): 7515–7522.
324 Ding, H., Yan, J., Wang, Z. et al. (2017). *Polymer* 123: 392.
325 Zhao, Y., Wang, Z., Yuan, R. et al. (2018). *Polymer* 137: 370–377.

326 Moffitt, M., McMahon, L., Pessel, V., and Eisenberg, A. (1995). *Chem. Mater.* 7 (6): 1185–1192.
327 Chan, Y.N.C., Craig, G.S.W., Schrock, R.R., and Cohen, R.E. (1992). *Chem. Mater.* 4 (4): 885–894.
328 Ng Cheong Chan, Y., Schrock, R.R., and Cohen, R.E. (1992). *J. Am. Chem. Soc.* 114 (18): 7295–7296.
329 Park, J., Joo, J., Kwon, S.G. et al. (2007). *Angew. Chem. Int. Ed.* 46 (25): 4630–4660.
330 Möller, M., Spatz, J.P., and Roescher, A. (1996). *Adv. Mater.* 8 (4): 337–340.
331 Hess, P.H. and Parker, P.H. Jr., (1966). *J. Appl. Polym. Sci.* 10 (12): 1915–1927.
332 Thomas, J.R. (1966). *J. Appl. Phys.* 37 (7): 2914–2915.
333 Antonietti, M., Wenz, E., Bronstein, L., and Seregina, M. (1995). *Adv. Mater.* 7 (12): 1000–1005.
334 Burkett, S.L., Sims, S.D., and Mann, S. (1996). *Chem. Commun.* 11: 1367–1368.
335 Platonova, O.A., Bronstein, L.M., Solodovnikov, S.P. et al. (1997). *Colloid. Polym. Sci.* 275 (5): 426–431.
336 Seregina, M.V., Bronstein, L.M., Platonova, O.A. et al. (1997). *Chem. Mater.* 9 (4): 923–931.
337 Zhao, H., Douglas, E.P., Harrison, B.S., and Schanze, K.S. (2001). *Langmuir* 17 (26): 8428–8433.
338 Jaramillo, T.F., Baeck, S.-H., Cuenya, B.R., and McFarland, E.W. (2003). *J. Am. Chem. Soc.* 125 (24): 7148–7149.
339 Spatz, J.P., Mößmer, S., and Möller, M. (1996). *Chem. Eur. J.* 2 (12): 1552–1555.
340 Möller, M., Spatz, J.P., Roescher, A. et al. (1997). *Macromol. Symp.* 117 (1): 207–218.
341 Klingelhöfer, S., Heitz, W., Greiner, A. et al. (1997). *J. Am. Chem. Soc.* 119 (42): 10116–10120.
342 Warren, N.J. and Armes, S.P. (2014). *J. Am. Chem. Soc.* 136 (29): 10174–10185.
343 Charleux, B., Delaittre, G., Rieger, J., and D'Agosto, F. (2012). *Macromolecules* 45 (17): 6753–6765.
344 Wang, G., Schmitt, M., Wang, Z. et al. (2016). *Macromolecules* 49 (22): 8605–8615.
345 Wang, G., Wang, Z., Lee, B. et al. (2017). *Polymer* 129: 57–67.
346 Zhang, Y.-Y., Chen, C.-H., Xu, G.-C., and Xin, W.-B. (2019). *J. Mol. Struct.* 1176: 1–6.
347 Zhang, Y., Filipczak, P., He, G. et al. (2017). *Polymer* 129: 144–150.
348 Huang, J., Li, D., Liang, H., and Lu, J. (2017). *Macromol. Rapid Commun.* 38 (15): 1700202.
349 Wang, Z., Bockstaller, M.R., and Matyjaszewski, K. (2021). *ACS Mater. Lett.* 3 (5): 599–621.
350 Xie, G., Ding, H., Daniel, W.F.M. et al. (2016). *Polymer* 98: 481–486.
351 Adhikari, S., Nikoubashman, A., Leibler, L. et al. (2021). *ACS Polym. Au* 1 (1): 39–46.
352 Bilchak, C.R., Jhalaria, M., Huang, Y. et al. (2020). *ACS Nano* 14 (12): 17174–17183.

353 Wang, Z., Chen, H., Wang, Y. et al. (2021). *ACS Appl. Mater. Interfaces* 13 (23): 27411–27418.

354 Wang, Z., Wang, Y., Chen, J. et al. (2021). *Langmuir* 37 (36): 10875–10881.

355 Adibnia, V., Mirbagheri, M., Faivre, J. et al. (2020). *Prog. Polym. Sci.* 110: 101298.

356 Faivre, J., Montembault, A., Sudre, G. et al. (2019). *Biomacromolecules* 20 (1): 326–335.

357 Seymour, B.T., Wright, R.A.E., Parrott, A.C. et al. (2017). *ACS Appl. Mater. Interfaces* 9 (29): 25038–25048.

358 Seymour, B.T., Fu, W., Wright, R.A.E. et al. (2018). *ACS Appl. Mater. Interfaces* 10 (17): 15129–15139.

359 Wright, R.A.E., Wang, K., Qu, J., and Zhao, B. (2016). *Angew. Chem. Int. Ed.* 55 (30): 8656–8660.

360 Bapat, A.P., Erck, R., Seymour, B.T. et al. (2018). *Eur. Polym. J.* 108: 38–47.

361 Fu, W., Bai, W., Jiang, S. et al. (2018). *Macromolecules* 51 (5): 1674–1680.

362 Xia, Y., Adibnia, V., Shan, C. et al. (2019). *Langmuir* 35 (48): 15535–15542.

363 Xia, Y., Adibnia, V., Huang, R. et al. (2019). *Angew. Chem. Int. Ed.* 58 (5): 1308–1314.

364 Jeong, W., Kang, H., Kim, E. et al. (2019). *Langmuir* 35 (41): 13268–13274.

365 Iqbal, D., Yan, J., Matyjaszewski, K., and Tilton, R.D. (2020). *Colloid. Polym. Sci.* 298 (1): 35–49.

366 Mai, W., Sun, B., Chen, L. et al. (2015). *J. Am. Chem. Soc.* 137 (41): 13256–13259.

367 Raj, W., Jerczynski, K., Rahimi, M. et al. (2021). *Mater. Sci. Eng. C* 130: 112439.

368 Reyes-Ortega, F., Parra-Ruiz, F.J., Averick, S.E. et al. (2013). *Polym. Chem.* 4 (9): 2800–2814.

369 Zhang, Y., Yan, J., Avellan, A. et al. (2020). *ACS Nano* 14 (9): 10954–10965.

370 Maynard, H.D., Heredia, K.L., Li, R.C. et al. (2007). *J. Mater. Chem.* 17 (38): 4015–4017.

371 Ohno, K., Morinaga, T., Takeno, S. et al. (2007). *Macromolecules* 40 (25): 9143–9150.

372 Daoud, M. and Cotton, J.P. (1982). *J. Phys. France* 43 (3): 531–538.

373 Choi, J., Hui, C.M., Schmitt, M. et al. (2013). *Langmuir* 29 (21): 6452–6459.

374 Ignatova, M., Voccia, S., Gilbert, B. et al. (2006). *Langmuir* 22 (1): 255–262.

375 Matyjaszewski, K., Miller, P.J., Shukla, N. et al. (1999). *Macromolecules* 32 (26): 8716–8724.

376 Pyun, J., Jia, S., Kowalewski, T. et al. (2003). *Macromolecules* 36 (14): 5094–5104.

377 Xu, C., Wu, T., Mei, Y. et al. (2005). *Langmuir* 21 (24): 11136–11140.

378 Wang, Z., Liu, T., Zhao, Y. et al. (2019). *Macromolecules* 52 (24): 9466–9475.

379 Mori, H., Seng, D.C., Zhang, M., and Müller, A.H.E. (2002). *Langmuir* 18 (9): 3682–3693.

380 Liu, P., Lu, W., Wang, W.-J. et al. (2012). Synthesis and characterization of PE-b-POEGMA copolymers prepared by linear/hyperbranched telechelic polyethylene-initiated ATRP of oligo(ethylene glycol) methacrylates. In: *Progress* *in Controlled Radical Polymerization: Materials and Applications*, ACS Symposium Series, vol. 1101, 39–64. Washington: American Chemical Society.

381 Morgese, G., Trachsel, L., Romio, M. et al. (2016). *Angew. Chem. Int. Ed.* 55 (50): 15583–15588.

382 Morgese, G., Shirmardi Shaghasemi, B., Causin, V. et al. (2017). *Angew. Chem. Int. Ed.* 56 (16): 4507–4511.

383 Morgese, G., Cavalli, E., Rosenboom, J.-G. et al. (2018). *Angew. Chem. Int. Ed.* 57 (6): 1621–1626.

384 Saigal, T., Dong, H., Matyjaszewski, K., and Tilton, R.D. (2010). *Langmuir* 26 (19): 15200–15209.

385 Choi, J., Hui, C.M., Pietrasik, J. et al. (2012). *Soft Matter* 8 (15): 4072–4082.

386 Ojha, S., Beppler, B., Dong, H. et al. (2010). *Langmuir* 26 (16): 13210–13215.

387 Choi, J., Dong, H., Matyjaszewski, K., and Bockstaller, M.R. (2010). *J. Am. Chem. Soc.* 132 (36): 12537–12539.

388 Wang, Z., Lee, J., Wang, Z. et al. (2020). *ACS Macro Lett.* 9 (6): 806–812.

389 Tugulu, S. and Klok, H.-A. (2008). *Biomacromolecules* 9 (3): 906–912.

390 Yamamoto, S., Ejaz, M., Tsujii, Y., and Fukuda, T. (2000). *Macromolecules* 33 (15): 5608–5612.

391 Lin, C.Y., Marque, S.R.A., Matyjaszewski, K., and Coote, M.L. (2011). *Macromolecules* 44 (19): 7568–7583.

392 Tang, W., Kwak, Y., Braunecker, W. et al. (2008). *J. Am. Chem. Soc.* 130 (32): 10702–10713.

393 Lee, J., Wang, Z., Zhang, J. et al. (2020). *Macromolecules* 53 (4): 1502–1513.

394 Ohno, K., Morinaga, T., Takeno, S. et al. (2006). *Macromolecules* 39 (3): 1245–1249.

395 Chen, J., Fasoli, A., Cushen, J.D. et al. (2017). *Macromolecules* 50 (24): 9636–9646.

396 Voudouris, P., Choi, J., Dong, H. et al. (2009). *Macromolecules* 42 (7): 2721–2728.

397 Morinaga, T., Ohno, K., Tsujii, Y., and Fukuda, T. (2008). *Macromolecules* 41 (10): 3620–3626.

398 Ohno, K., Koh, K., Tsujii, Y., and Fukuda, T. (2003). *Angew. Chem. Int. Ed.* 42 (24): 2751–2754.

399 Nakanishi, Y., Ishige, R., Ogawa, H. et al. (2018). *J. Chem. Phys.* 148 (12): 124902.

400 Natarajan, B., Neely, T., Rungta, A. et al. (2013). *Macromolecules* 46 (12): 4909–4918.

401 Yan, J., Kristufek, T., Schmitt, M. et al. (2015). *Macromolecules* 48 (22): 8208–8218.

2

Hairy Nanoparticles via Self-assembled Linear Block Copolymers

Zhen Zhang, Yi Shi, and Yongming Chen

Sun Yat-Sen University, School of Materials Science and Engineering, No. 135, Xingang Xi Road, Guangzhou 510006, China

2.1 Introduction

Hairy polymer nanoparticles (NPs) with persistent geometric shape and densely tethered polymer hairs represent an important soft nanomaterial that can be used in a variety of applications, ranging from antibacterial agents [1, 2], templating [3, 4], separation [5], catalysis [6], and drug release [7].

There are two strategies to prepare hairy polymer NPs, namely, surface grafting chemistry and self-assembly of block copolymer. Surface grafting chemistry usually requires the introduction of active functional groups or initiators on the surface of nanomaterial particles, and further chemical reactions to covalently connect the polymer chain to the surface of the particles. The preparation process of this method is relatively cumbersome, and the distribution of active functional groups or initiators on the surface of particles is usually poorly controlled, resulting in uneven distribution of polymer chains and low graft density, which is not conducive to subsequent application. The subsequently developed block copolymer self-assembly method can make up for the shortcomings of the former, which is based on the block copolymer self-assembly to prepare hairy NPs with different shapes and compositions. Block copolymer is a kind of polymer that is composed of two or more polymer chains with different properties connected by covalent bonds. Due to different polymers composed of thermodynamic incompatibility, often in selective solvents, namely the solvents, which can dissolve one section of the polymer but not another, the insoluble polymer has not been dissolved and depends on the molecular structure and composition of block copolymers and self-assembly conditions. Hairy NPs ranging from spherical micelles, worm-like micelles, vesicles, composite micelles, and multicellular vesicles, as well as other types that could be prepared for specific functions [8]. This chapter mainly introduces the preparation and functionalization of hairy NPs with different shapes by bulk microphase separation and solution self-assembly of linear block copolymers and also, the challenges and limitations that hairy NPs face and the potential solutions in this field.

Hairy Nanoparticles: From Synthesis to Applications, First Edition. Edited by Zhiqun Lin and Yijiang Liu.

2.2 Hairy NPs via Bulk Microphase Separation of Block Copolymers

Under certain conditions, macroscopic phase separation generally occurs between the polymers when two or more homopolymers with different properties are mixed together, since the thermodynamic incompatibility between the different blocks drives the different components to repel each other. On the other hand, the polymer chains are connected by covalent bonds, so that the blocks are closely linked together. The result of the mutual check and balance between the repulsive force and the attractive force is that the block copolymers cannot undergo macroscopic phase separation in the usual sense, but only form a phase separation structure in the nanometer to micrometer scale. This phase separation connected by chemical bonds is called microphase separation. Block copolymers can form regular periodic structures through microphase separation, and the differences in the number of blocks and component contents in block copolymers form a thermodynamic equilibrium state with very rich structures. From the 1960s to the present, great progress has been made in theoretical and experimental research on the bulk microphase separation of block copolymers. A series of thermodynamically stable nanoperiodic structures with scales ranging from 10 to 100 nm can be formed by bulk microphase separation of block copolymers. These ordered nanostructures can be used in nanoordered arrays [9, 10] and photonic crystals [11, 12], which have potential applications in the fields of information transmission, heterogeneous catalysis, and integrated circuits. The bulk microphase separation systems of linear diblock copolymers and triblock copolymers are currently the most common studies for the preparation of hairy polymer NPs.

2.2.1 Bulk Microphase Separation of Diblock Copolymers

AB diblock copolymers are the simplest type of block copolymers in molecular structure, and their microphase separation has been most studied. For AB diblock copolymers, there is only dispersive force between the A–A, B–B, and A–B polymer chains. Due to the difference in cohesive energy density between the A and B blocks, the two blocks will repel each other, and the magnitude of the repulsive force is described by the Flory–Huggins thermodynamic interaction parameter (χ). When the system temperature is lower than the critical eutectic temperature, the AB diblock copolymer will undergo microphase separation, and when the molecular weight distribution is narrow, a highly ordered thermodynamic equilibrium phase can be formed. Researchers have systematically studied the bulk microphase separation of diblock copolymers from both theoretical and experimental aspects. Figure 2.1 demonstrated four thermodynamically stable and periodically ordered phase morphologies, namely lamellar phase (L), columnar phase (H), bicontinuous phase (Q^{230}), and spherical phase (CPS/Q^{229}) with the changes of the volume fraction of block A (f_A) [13].

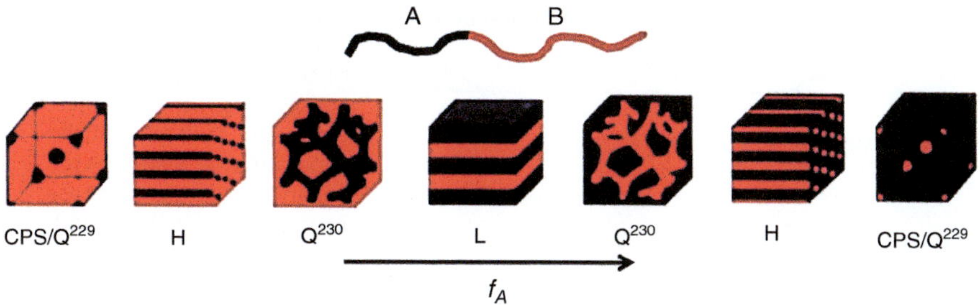

Figure 2.1 Schematic diagram of various bulk microphase separation morphologies of AB diblock copolymer with different volume fractions of each block (CPS/Q^{229}, spherical phase; H, columnar phase; Q^{230}, bicontinuous phase; L, lamellar phase). Source: Orilall and Wiesner [13] Copyright 2010, Royal Society of Chemistry. Reproduced with permission.

2.2.1.1 Theoretical Research

In terms of theoretical research, early researchers proposed the strong segregation theory of block copolymers [14–16] and the weak segregation theory [17, 18]. Using the combination parameter χN of the block copolymer (the product of χ and the total polymerization degree [N] of the copolymer, which reflects the strength of the microphase separation tendency of the block copolymer) and the critical order (heterophase)-the disordered (homogeneous) phase transition parameter $(\chi N)_{\text{ODT}}$ to divide the phase separation of AB diblock copolymers into three regions: weak phase separation region, strong phase separation region, and the mesophase separation region, which is between strong and weak phase separation region. When the value of χN is slightly larger than $(\chi N)_{\text{ODT}}$, the system is in the weak phase separation region; at this time, the phase separation between the A and B blocks is not complete, the interface formed is relatively wide, and the conformation of the polymer chain is close to a random line (Figure 2.2a). When the value of χN is much larger than $(\chi N)_{\text{ODT}}$, the system is in the region of strong phase separation, and the two blocks show strong incompatibility, which makes the phase separation very easy to occur. The two phases are clearly demarcated, the interface phase is narrow, and the chain conformations of the A and B blocks are significantly stretched along the direction perpendicular to the interface (Figure 2.2b). The above theory introduces some unnecessary approximations on the basis of the mean field, which reduces the accuracy of the theoretical prediction and limits the scope of application. Subsequently, Matsen and Bates selected polystyrene-*b*-polyisoprene (PS-*b*-PI) as a model polymer to obtain the thermodynamic phase diagram of microphase separation of diblock copolymers through high-precision self-consistent field theory calculations [19]. The results show that the microphase separation structure is related to parameters, such as χN and f. Then, Cochran and Phillip et al. made a simple correction to the results reported by Matsen [20, 21]. Their results are shown in Figure 2.3a, when the block copolymer combination parameter χN value is greater than 10, the microphase separation of the block copolymer can occur: with the change of the copolymer block ratio, the curvature of the two-phase interface changes [22], and the phase separation

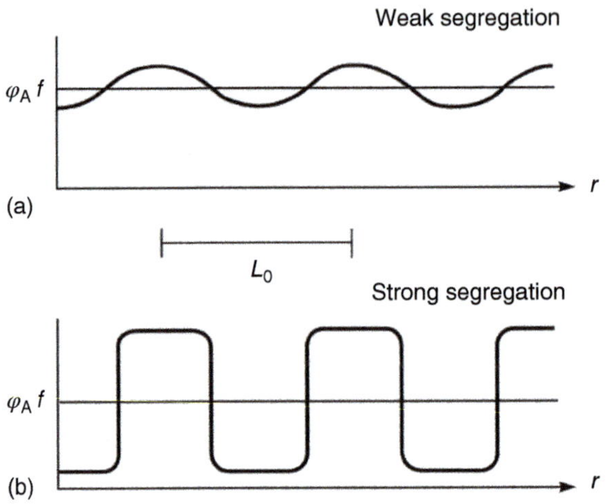

Figure 2.2 Schematic diagram of the phase interface distribution of diblock copolymers with different degrees of bulk microphase separation: (a) weak phase separation; (b) strong phase separation. L_0 is the period length of the phase structure. Source: Bates et al. [16] Copyright 2014, American Chemical Society. Reproduced with permission.

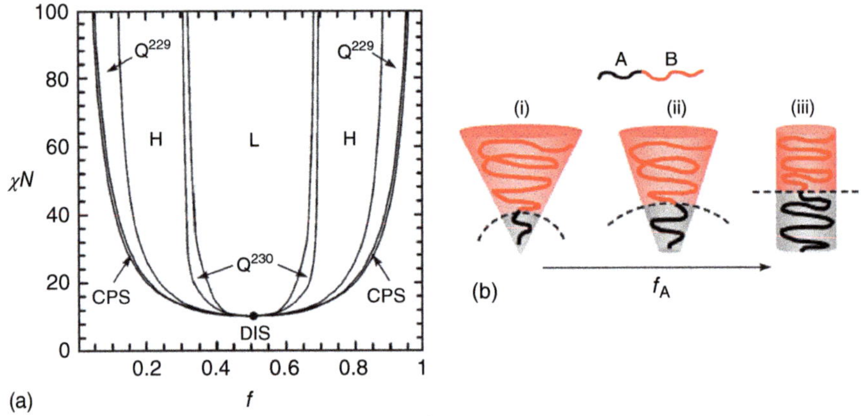

Figure 2.3 (a) Using PS-*b*-PI diblock copolymer as the model polymer, the thermodynamic phase separation diagram is obtained by self-consistent field theory calculation (DIS, disorder phase). Source: Cochran et al. [20] Copyright 2006, American Chemical Society. Reproduced with permission. (b) Schematic diagram of the polymer chains arrangement in the microphase-separated structure of AB diblock copolymers: (i) spherical phase; (ii) columnar phase; (iii) lamellar phase, when the volumes of the two blocks are almost equal, the lamellar phase is formed. Source: Mai and Eisenberg [22] Copyright 2012, Royal Society of Chemistry. Reproduced with permission.

structure also changes accordingly, as shown in Figure 2.3b. The three-phase structures of lamellar phase (L), columnar phase (H), and spherical phase (Q^{229}) occupy most of the area in the phase diagram, indicating that these three-phase structures are relatively easier to be obtained.

2.2.1.2 Experimental Study

In terms of experimental study, the researchers obtained a series of periodically ordered nanostructures by using the bulk microphase separation of the diblock copolymer, which verified the accuracy of the theoretical prediction. Bates et al.

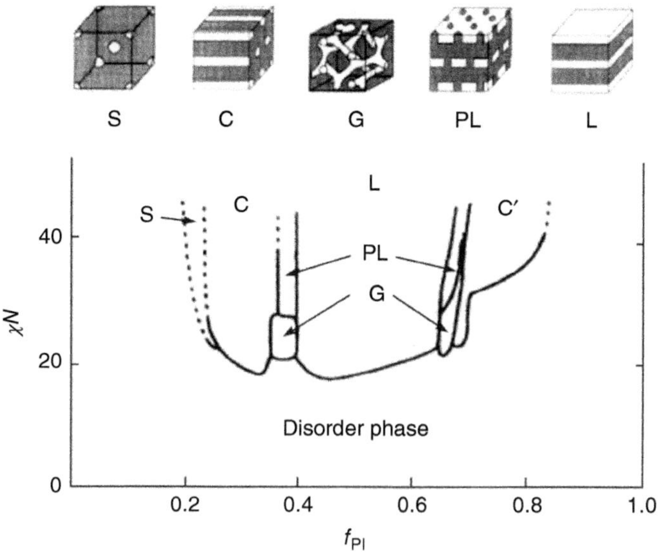

Figure 2.4 Thermodynamic phase diagrams drawn from the experimental results of microphase separation of PS-*b*-PI copolymers with different chain lengths and compositions (S: spherical phase, C: columnar phase, G: bicontinuous phase, PL: porous lamellar phase, L: lamellar phase). Source: Bates and Fredrickson [23] Copyright 1999, AIP Publishing LLC. Reproduced with permission.

studied the microphase separation behavior of PS-*b*-PI diblock copolymers with different chain lengths and compositions [23, 24], and drew a thermodynamic phase diagram as shown in Figure 2.4 (the white area is PI and the gray area is PS in the phase structure schematic diagram). The diagram shows the change in the microphase separation morphology with different values of f and χN. When the value of χN is greater than about 20, strong microphase separation occurs to form an ordered phase structure. The experimental structure is basically consistent with the theoretical calculation results, and in addition to the four thermodynamically stable structures obtained in the experiment, the only difference is that there is a transition state between the bicontinuous phase and the lamellar phase structure, namely porous-layered phase structure, which occupies a small proportion in the phase diagram and is not easy to obtain.

2.2.1.3 Effect Factors

The microphase separation structure of AB diblock copolymers mainly depends on three factors, namely the volume fraction of each block (f_A, f_B), the total polymer degree of the copolymer ($N = N_A + N_B$), and the Flory–Huggins interaction parameter (χ_{AB}). χ_{AB} is essentially used to measure the degree of repulsion between A and B blocks, χN value determines the strength of microphase separation, and f determines the structural morphology formed by microphase separation. In addition, the regularity of the microphase separation structure is also affected by the molecular weight distribution (PDI) of the block copolymer. The narrower PDI the

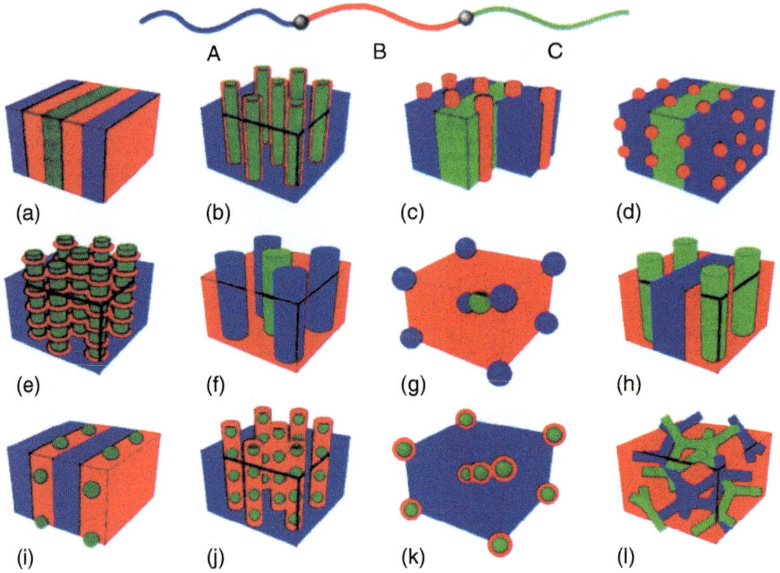

Figure 2.5 The microphase separation structure of ABC triblock copolymer predicted by theoretical calculation, different colors represent different polymer components. Source: Zheng and Wang [30] Copyright 1995, American Chemical Society. Reproduced with permission.

block copolymer has, the more regular the microphase separation structure and the clearer the phase interface they will form.

2.2.2 Bulk Microphase Separation of Triblock Copolymers

Triblock copolymers include two types of structure, namely ABA and ABC. The structure of an ABA copolymer is similar to that of the aggregates of two AB diblock copolymers (A–BB–A). The experimental results also prove that its microphase separation phase behavior is similar to that of AB diblock copolymer [25–29]. For the ABC copolymer, due to the complexity of its own structure, its microphase separation morphology is not only affected by the total polymerization degree of the copolymer ($N = N_A + N_B + N_C$), the volume fraction of each block (f_A, f_B, and f_C), and three Flory–Huggins interaction parameters (χ_{AB}, χ_{BC}, and χ_{AC}), but also affected by factors such as the sequence of the molecular chain (A–B–C, B–C–A, or C–A–B). The change in the block sequence essentially changes the χ between different blocks, that is, changing the mutual repulsion between the blocks will affect the phase separation morphology. Therefore, the microphase separation of ABC triblock copolymers could obtain more complex and abundant phase structures compared with AB diblock copolymers. Zheng and Wang predicted a series of microphase separation structures of ABC triblock copolymers through theoretical calculations (Figure 2.5), and these structures had already been verified in experiments [30].

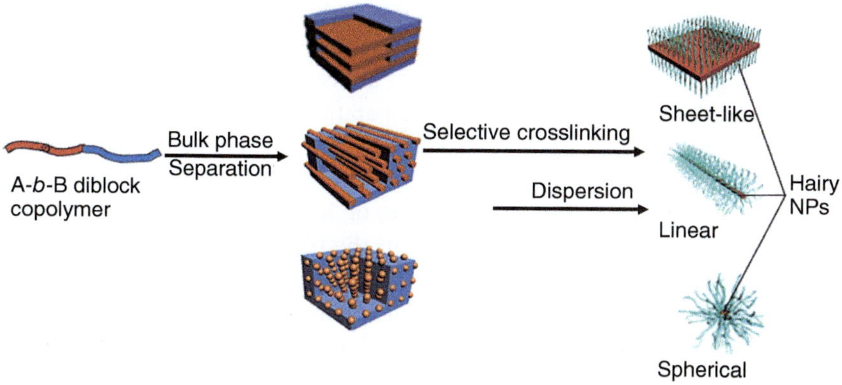

Figure 2.6 Schematic diagram of the preparation of hairy NPs with different shapes from reactive A-*b*-B diblock copolymers by bulk microphase separation, selective cross-linking, and dispersion process.

2.2.3 Preparation of Hairy NPs with Different Shapes

The basic idea of preparing hairy NPs based on bulk microphase separation of block copolymers is as follows: first, the expected phase separation structure is obtained by bulk microphase separation of block copolymers with pendant reactive functional groups, and then chemical reactions or the physical interaction between polymer chains could be selectively cross-linked to fix the discontinuous phase microdomains in the microphase separation structure, and then disperse the cross-linked samples in a good solvent for the continuous phase component to obtain hairy NPs with a certain degree of stability. The preparation process of NPs with different shapes and functions (taking the diblock copolymer as an example) is shown in Figure 2.6. If the cross-linking conditions are mild, the microphase separation morphology will not be destroyed, so the final shape of the NPs depends on the microphase separation structure of the block copolymer. Up to now, dispersed NPs with various structures, such as sheet-like, wire (tube)-like, spherical, porous sheet-like, and Janus asymmetric structures can be obtained by using this method.

This method of preparing hairy NPs was first reported by Ishizu et al. They fixed the NPs core by quaternizing the cross-linking reaction between dibromoalkane and P2VP. Firstly, the discontinuous phase component P2VP in the structure was separated by selective *in situ* cross-linking of PS-*b*-P2VP copolymer with 1,4-dibromobutane and then dispersed in tetrahydrofuran, which is a good solvent for PS to obtain spherical hairy polymer NPs [31]. Further, they used the same method to obtain Janus asymmetric NPs with a cross-linked core utilizing PS-*b*-P2VP-*b*-PBMA (poly(butyl methacrylate) [PBMA]) triblock copolymer [32]. Polystyrene-*b*-poly(2-cinnamoylethyl methacrylate) (PS-*b*-PCEMA) that could self-assemble into star and crew-cut micelles with PCEMA as cores in solutions, under UV irradiation, the PCEMA are cross-linked by [2 + 2] cyclization of cinnamyl units, and hairy spherical NPs are obtained. Liu also designed and synthesized two- or three-block copolymers with photo-cross-linked side groups of PCEMA as one block to fabricate NPs with different shapes, such as nanospheres, nanowires, and nanotubes [33–36]. In addition, Müller et al. used the block copolymer containing polybutadiene chain for microphase separation, and through the chemical cross-linking of the carbon–carbon double bond of the PB main chain to obtain a series of NPs with different compositions and shapes [37–42].

The above methods used for the synthesis of hairy NPs with various morphologies have gained great achievements, but the block copolymers chosen are all synthesized by anionic polymerization, which limits the design choices of polymers. On the other hand, the synthesized copolymers and prepared NPs are not easy to be post-modified, which limits the functionalization and application of hairy NPs to a certain extent. Due to the wide range of suitable monomers for living/controlled radical polymerization, and the active functional groups at the end of the polymer chains, the hairy NPs prepared from these synthesized polymers still have active functional groups, which can be further post-modified. Based on this, by living/controlled radical polymerization, such as reversible addition-fragmentation chain transfer polymerization and atom transfer radical polymerization, in recent years, Chen et al. have carried out a series of research works on the synthesis of reactive block copolymers and the regulation and functionalization of hairy polymer NPs based on linear diblock and triblock copolymers containing poly(3-(triethoxysilyl)propyl methacrylate) (PTEPM) that can undergo gelation reaction of polyglycidyl methacrylate (PGMA) with pendant epoxy groups, or PS, which has a high glass transition temperature.

2.2.3.1 Diblock Copolymers with PTEPM or PGMA Components

Organosiloxanes can be hydrolyzed and condensed quickly to form Si-O-Si cross-linked bonds under alkaline conditions. A series of research achievements have been made in the preparation of hybrid hairy NPs with different shapes by bulk microphase separation utilizing reactive two-block copolymers containing siloxane side groups [43–45]. By designing and synthesizing copolymer molecules, different nanoperiodic structures were obtained by using the bulk microphase separation, and then the morphology was fixed by *in situ* gelation reaction, and finally the stable structure of different shapes of organic/inorganic hybrid NPs was obtained. In the early stage, PTEPM-*b*-PS was selected as the model polymer to study the microphase separation behavior and could obtain the sheet-like, linear, and spherical core/shell hairy NPs. The core of the NPs is fixed by the cross-linked polysiloxane, and the shell is PS, which is densely grafted on the surface of the core, as shown in Figure 2.7 [44]. Also, the block copolymers containing PGMA, which bear epoxy in every repeating units have been developed, and an atmosphere of amines may induce the PGMA cores to be fixed [46, 47].

2.2.3.2 Diblock Copolymers Containing PS

In addition to the abovementioned chemical cross-linking method to fix the core of NPs, the immobilization of the core morphology can also be achieved through physical interaction. For example, PS has a high glass transition temperature

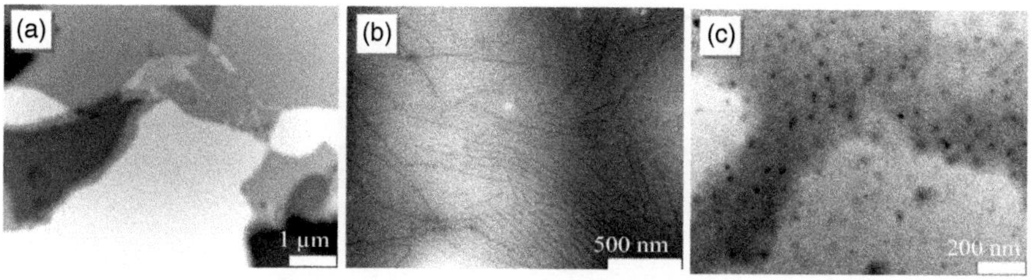

Figure 2.7 TEM results of NPs with different shapes prepared from PTEPM-*b*-PS block copolymers: (a) $PTEPM_{88}$-*b*-PS_{408}, sheet-like, (b) $PTEPM_{66}$-*b*-PS_{758}, linear; (c) $PTEPM_{46}$-*b*-PS_{1009}, spherical. Source: Zhang et al. [44], Reproduced with permission of American Chemical Society.

($T_g \sim 100\,°C$), which is in a glass state at room temperature, and the motions of the intertwined PS block chains and segments are frozen. If the phase-separated systems with PS as the discontinuous phase are stirred in a poor solvent of PS, they can maintain the original structure of the microphase separation. First, Chen et al. used PtBA-*b*-PS as a model diblock copolymer, and by changing the content of PS and dispersing the microphase separation samples in methanol (which is a poor solvent for PS and a good solvent for PtBA), nanosheets, nanowires, and nanospheres in which the glassy PS as the core and the PtBA as the shell could be obtained (Figure 2.8) [48]. If under external stimulus (heat, ultrasound, etc.), the kinetically restricted state of PS chains will be affected, which may lead to the destruction of the original morphology. In order to examine the stability of NPs with PS cores, the nanosheets were heated and refluxed in methanol at 65 °C and finally changed into nanospheres (Figure 2.9). This is because with the increasing temperature, the movement of PS chains is gradually enhanced and the interaction between block chains is weakened, then the original shape of NPs is destroyed, and nanospheres with PS as the core and PtBA as the shell were finally formed in the selective solvent. They then systematically studied the effect of external energy (such as stirring, heating, and ultrasound) on the morphology of NPs-containing glassy PS cores. By taking the nanowires formed using PS-*b*-PNIPAM copolymer in ethanol as an example, at room temperature with the prolonged stirring time, the nanowires would gradually be broken to form stable spherical structures.

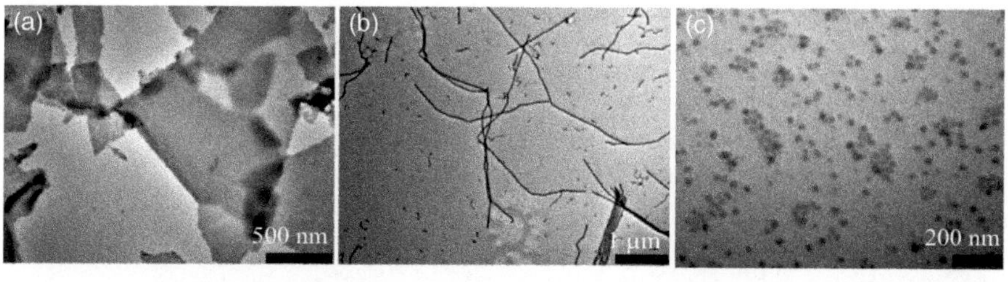

Figure 2.8 TEM results of hairy NPs prepared by bulk microphase separation of PtBA-*b*-PS copolymers with different block ratios: (a) $PtBA_{141}$-*b*-PS_{147}, sheet-like; (b) $PtBA_{310}$-*b*-PS_{189}, linear; (c) $PtBA_{310}$-*b*-PS_{90}, spherical. Source: Qin et al. [48], Reproduced with permission of American Chemical Society.

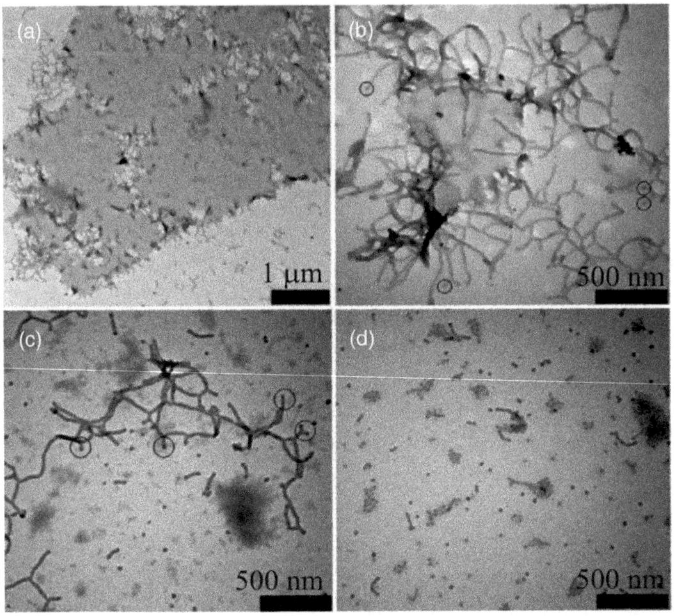

Figure 2.9 TEM results of samples obtained from methanol dispersions of nanosheets prepared from $PtBA_{141}$-*b*-PS_{147} copolymers when heated to reflux at 65 °C with different times: (a) 2 hours; (b) 5 hours; (c) 10 hours; (d) 24 hours. Source: Qin et al. [48], Reproduced with permission of American Chemical Society.

Its transformation into spherical shape can be accelerated with the assistance of heating or ultrasound [48].

Subsequently, Chen et al. prepared three shapes of hairy NPs with PS as the core and water-soluble PDMAEMA as the shell utilizing the PDMAEMA-*b*-PS block copolymer (Figure 2.10). The shell layer of these NPs contains a large number of functional amine groups, and the PS core can be removed by high-temperature calcination, which lays the foundation for the application of this material as a template to prepare hollow materials [49].

Recently, Chen et al. also obtained a hexagonally stacked porous-layered phase structure by bulk microphase separation of a PtBMA-*b*-PS diblock copolymer

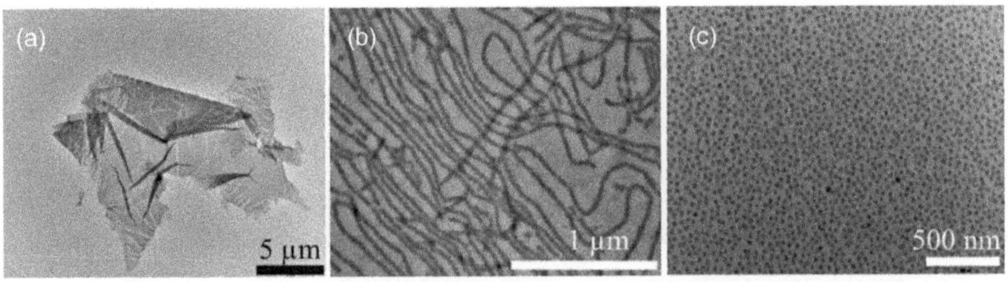

Figure 2.10 TEM characterization of NPs obtained by microphase separation of PDMAEMA-*b*-PS samples dispersed in an acidic aqueous solution with pH = 3 at room temperature: (a) $PDMAEMA_{271}$-*b*-PS_{339}, nanosheets; (b) $PDMAEMA_{234}$-*b*-PS_{164}, nanowires; (c) $PDMAEMA_{271}$-*b*-PS_{99}, nanospheres. Source: Yao et al. [49], Reproduced with permission of Royal Society of Chemistry.

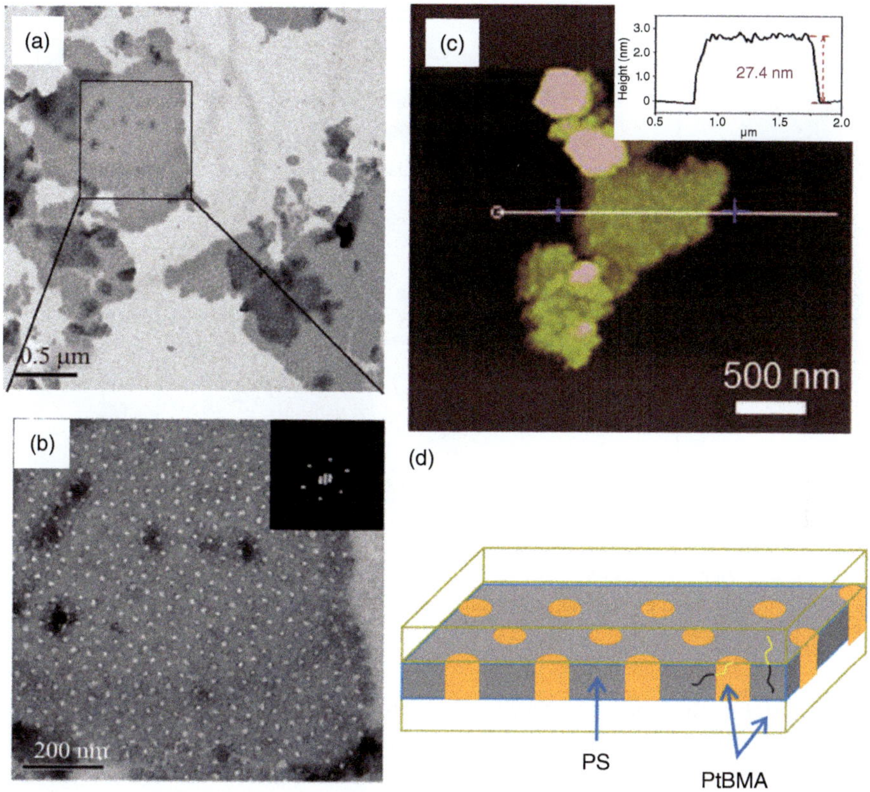

Figure 2.11 $PtBMA_{227}$-*b*-PS_{187} block copolymer bulk microphase separation samples obtained by dispersing the NPs in methanol at room temperature: (a)–(b) TEM images and partial magnification results; (c) AFM image, (d) schematic diagram of the porous nanosheets structure. Source: Yao et al. [50], Reproduced with permission of Elsevier Ltd.

(f_{PS} = 37.6 wt%) with a certain composition [50]. The phase-separated film was stirred in methanol to obtain well-dispersed porous nanosheets with a thickness of 27.4 nm and several regular PtBMA columnar domains with a diameter of about 12 nm, which are distributed in the PS continuous phase to form the core of the nanosheets, and another part of the PtBMA chains constitute the shell of the nanosheets (Figure 2.11). The PtBMA component in the dispersed porous nanosheets was further hydrolyzed with trifluoroacetic acid to obtain carboxyl-rich functional NPs, which disperse well in water. These materials with porous lamellar structures have potential applications in the fields of ion transport and lithography.

2.2.3.3 Triblock Copolymer System with PS Components

Compared with the preparation of hairy NPs by microphase separation of diblock copolymers, microphase separation of triblock copolymers can form more complex structures and more functional NPs after selective cross-linking and dispersion process.

Janus Asymmetric Hairy NPs Janus NPs are used to describe materials that have dual properties or chemical components at nano- or micro-scale. The essential feature of

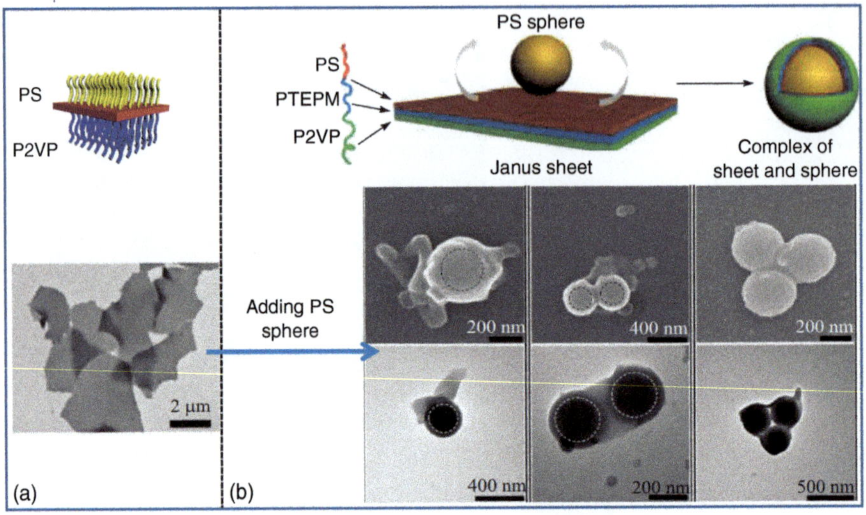

Figure 2.12 (a) TEM image and structural schematic diagram of Janus nanosheets prepared from $P2VP_{310}$-*b*-$PTEPM_{58}$-*b*-PS_{322}. (b) PS nanospheres wrapped by Janus nanosheets. Source: Gao et al. [51], Reproduced with permission of American Chemical Society.

these materials is their non-centrosymmetric properties, and they have been applied in many fields. Chen et al. obtained a three-component alternating multi-layered phase structure by bulk microphase separation of $P2VP_{310}$-*b*-$PTEPM_{58}$-*b*-PS_{322} [51]. After *in situ* cross-linking of PTEPM microdomains under acidic conditions and dispersion process, functional Janus nanosheets (about 35 nm in thickness and several microns in size) having a sheet-like PTEPM cross-linked core and PS and P2VP chains are grafted on both sides of the core with high density could be obtained, as shown in Figure 2.12a. Monodisperse PS spheres with a diameter of about 226 nm were added to the THF solution of Janus nanosheets, and then a large amount of acidic aqueous solution with pH = 2 was slowly dropped (acidic aqueous solution is a good solvent for P2VP and a poor solvent for PS). The PS block chain collapses, and the P2VP on the other side of the Janus nanosheet is still in an extended state, which causes the Janus nanosheets to curl toward the PS component side; when the PS nanosphere exists, the PS chains will wrap the PS nanosphere through hydrophobic interaction, and a dumpling-like nanocomposite is finally formed. The schematic diagram of the interaction process and the TEM results of the formed dumpling-like nanocomposite are shown in Figure 2.12b. The number of PS nanospheres in each Janus NP that can be wrapped by nanosheets is related to the size of the nanosheets: the larger the size of nanosheets is, the more PS nanospheres can be wrapped.

In addition, Müller et al. selected polystyrene-*b*-polybutadiene-*b*-polymethyl methacrylate and polystyrene-*b*-polybutadiene-*b*-poly(*tert*-butyl methacrylate) to prepare Janus NPs with spherical, linear, and sheet-like shapes [37, 38, 40]. In their system, the morphology of the assembly can be fixed by vulcanization or free radical cross-linking of PB domains.

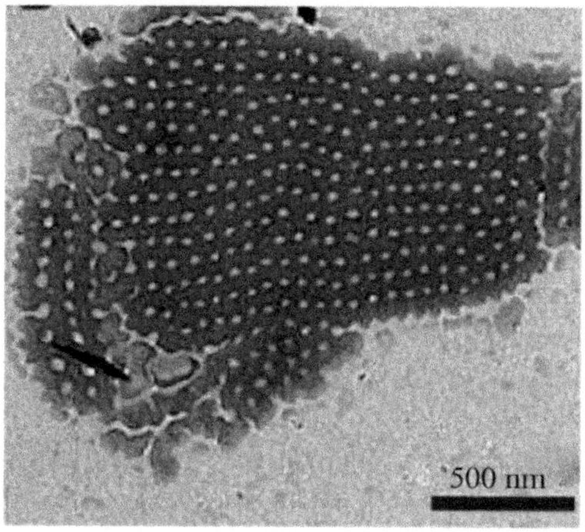

Figure 2.13 TEM images of core-cross-linked porous nanosheets prepared from PB_{22}-*b*-$P2VP_{29}$-*b*-$PtBMA_{49}$ triblock copolymerx. Source: Gao et al. [51], Reproduced with permission of American Chemical Society.

Porous Nanosheets Porous nanosheets can also be obtained by selective cross-linking and dispersion processes. As discussed in Section 2.1.2, porous-layered phase structures are not easy to obtain, and only account for a small proportion in the thermodynamic phase diagram. Müller et al. got a porous-layered phase structure by separating the bulk microphase of PB_{22}-*b*-$P2VP_{29}$-*b*-$PtBMA_{49}$ triblock copolymer with a certain composition and then cross-linked the PB microphase [41, 42]. The bulk sample was then immersed in THF to obtain dispersed porous nanosheets. Part of the PtBMA chains passed through the PB/P2VP layer, forming regular nanocolumnar microdomains in the PB/P2VP layer, and another part of the PtBMA chains was connected on both sides of the core and shell (Figure 2.13).

The microphase separation of block copolymers and small molecules or homopolymers with specific structures can also be used to precisely control the morphology of the assemblies. The principle is mainly that in the process of microphase separation, through physical interaction, small molecules or homopolymers with a specific structure selectively form a microphase with a block of the polymer, which changes the volume fraction of different microphases, and when the amount of change reaches a certain level, the corresponding phase structure transition occurs.

2.3 Hairy NPs via the Self-assembly of Block Copolymer in Solution

Molecular self-assembly is a process in which molecules spontaneously form stable and structured aggregates through non-covalent bonds [52]. Traditional amphiphilic molecules, such as surfactants, contain a hydrophilic head and a hydrophobic tail. It self-assembles into different aggregates in aqueous or organic solutions [22]. Block polymers consist of two or more chemically distinct polymer blocks connected by covalent bonds. Similar to the self-assembly behavior of amphiphilic small molecules, in a selective solvent, the interaction of the solvophilic block and solvophobic block with the solvent makes the block copolymer self-assemble into a variety of morphological structures, such as spherical micelles, rod-like (worm-like) micelles, vesicles, and large composite micelles [53–59]. The formation of these hairy NPs is primarily a result of the inherent molecular curvature and how this influences the packing manners of the copolymer chains [60]. The formation of a specific self-assembly structure can be targeted according to a dimensionless "packing parameter," p, which is defined as: $p = v/a_0 l_c$, where v is the volume of the hydrophobic chains, a_0 is the optimal area of the hydrophilic head group, and l_c is the length of the hydrophobic tail. The packing parameters of a given molecule usually indicate its most likely self-assembly morphology [61, 62]. General rules: when $p \leq 1/3$, it forms spherical micelles; when $1/3 \leq p \leq 1/2$, rod-like micelles are easily formed; and when $1/2 \leq p \leq 1$, polymer vesicles and double molecular layer are formed. Compared with small molecular aggregates, hairy polymer NPs have higher stability and durability due to their mechanical and physical properties, which makes them useful in a wide range of fields, including biomedicine, biomaterials, and catalysts.

The generation of block polymer NPs and the morphological structure of the NPs mainly depend on the contributions of three factors related to the free energy of the polymer aggregation: the stretched degree of the block copolymer, which forms the NPs core; the interaction degree between the NPs core and the solvent; and the interaction between the shell-forming blocks. What could affect the balance between the three factors above will have an effect on the final morphology of the polymer NPs. Now it has been found that many factors could influence the assembly morphology, such as the composition of the block copolymer, the concentration of the polymer, the composition, and properties of the solvent, additives, and temperature. This chapter will summarize the morphology of block copolymer self-assembly in solution and main related factors that have an effect on the morphology of the obtained hairy polymer NPs.

2.3.1 Morphology of Block Copolymers Assembly

The solution assembly behavior of block copolymers is similar to that of small molecular surfactants. The different physical properties of the copolymers eventually lead to the formation of various aggregates with different properties. A series of symmetric and asymmetric diblock copolymers can be obtained through the control of the polymerization process, and the ratio of the insoluble block to the soluble block can also be adjusted. So, for diblock copolymers in solution, asymmetric amphiphilic block copolymers with longer hydrophilic blocks eventually form star-like micelles with large shells and small cores in solution. In contrast, asymmetric diblock copolymers containing long hydrophobic blocks will form crew-cut micelles with large cores and small shells [63]. So far, most of the star-like micelles prepared are spherical [64], and the crew-cut micelles of different morphologies have also been observed [65–68].

2.3.1.1 Spherical Micelles

Simple spherical micelles consist of a spherical core surrounded by coronal chains. The formation of spherical micelles requires that the radius of the core cannot exceed the longest hydrophobic chains in their planar zigzag configuration and is shorter than the length of the hydrophobic chain. The spherical micelles are usually formed first when a selective solvent such as water is added into the cosolvent of diblock copolymers. The hydrophilic shell makes the micelles disperse stably in water, while the hydrophobic core provides an ideal encapsulation place for loading hydrophobic drugs [69], fluorescent probes [70], and other molecules. Therefore, spherical micelles have a wide range of applications in drug delivery and bioimaging. Up to now, the research about spherical micelles has gone far beyond the simple core–shell coronal structure, with some complex structures, such as polyionic complex micelles [71], shell cross-linked micelles [72], and Janus micelles with two significantly different hemispheres [73]. All these novel and functional micelles have gained more and more attention.

2.3.1.2 Rod-Like Micelles

Rod-like micelles are composed of a cylindrical core and a corona surrounding the core. Adding extra water to the equilibrium solution containing spherical micelles will increase the surface energy of the micelles. The size of the micelles will increase while the number of micelles will decrease, resulting in a decrease in the total area of the micelles in order to ensure a lower surface energy. The nucleation growth leads the nucleation segment in the nucleus to further extend, which in turn leads to the formation of low-free energy rod-like NPs (Figure 2.14). The study of rod-like (worm-like) NPs has been widely reported, including giant and short worm-like micelles, star-connected micelles, and worm-like micelles forming a network structure. Making use of the intra- and interaction among the block copolymers, more complex assembly structure could be designed and prepared. Zhu and Jiang reported that the ABC triblock copolymer aggregates to form spherical micelles through primary self-assembly [74]. The most important feature of these spherical micelles is that they can further self-assemble into more advanced structures, eventually leading to a large number of giant segmented worm-like micelles. Since rod-like micelles can provide excellent templates for metalloids, semiconductors, or magnetic NPs, controlling the length of rod-like micelles has become a research hot spot nowadays. Recently, based on polyferrocenyldimethysilane (PFS)-containing BCPs, Manners and Winnick reported a developed "living self-assembly" strategy to achieve excellent control of the rod length [75].

2.3.1.3 Bilayer Structure

Vesicles are the most common morphology with bilayer structures. Eisenberg et al. first observed the formation of polymer vesicles morphology during the polystyrene-*b*-polyacrylic acid (PS-*b*-PAA) self-assembly process [22]. Up to now, many different polymers could be chosen as excellent candidates to produce vesicles with various structures. The size of vesicles can be regulated by changing the water content of the system, and a higher concentration of polymer will produce some multilayer structures. Figure 2.14d shows a small amount of lamellar structures coexisting with vesicles, and Figure 2.14e shows a large lamellar structure. Compared with lamellar structure, vesicles are much easier to be observed due to their

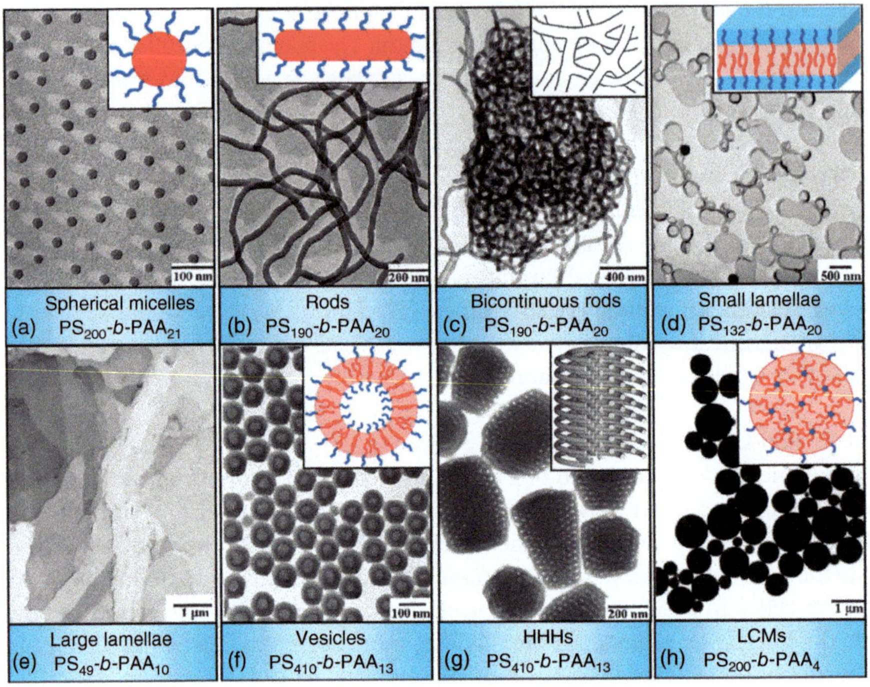

Figure 2.14 Transmission electron microscope and corresponding schematic diagrams of the various morphologies formed using amphiphilic block copolymer PS_m-*b*-PAA_n, where m and n are the polymerization degrees of PS and PAA, respectively. HHHs: hexagonally hollow hoops; LCMs: large composite micelles. Source: Mai and Eisenberg [22], Reproduced with permission of Royal Society of Chemistry.

thermodynamically stable ability [76]. The hexagonal hollow hoops structure is the inverse morphology of rod-like micelles (Figure 2.14g), and the large composite micelles with a big size and high size distribution are the aggregates of inverse micelles (Figure 2.14h). Recently, there have been many reports on the formation of vesicle structures from different types of block copolymers, and they have great potential applications due to their special bilayer structure, especially in the fields of drug delivery, diagnostic imaging, and nanoreactors [77–79]. In-depth research on vesicle structure in both academic and application aspects will promote the rapid development of these bilayer assemblies.

2.3.1.4 New Morphologies

By controlling the synthesis and self-assembly process of block copolymers, some new types of morphologies can be obtained, such as multi-compartment micelles, disc-shaped micelles, cyclic micelles, and bicontinuous micelles [80]. Liu et al. obtained multi-compartment micelles with a pH-response miktoarm star block terpolymer composed of one hydrophobic block polystyrene, one hydrophilic block polyethylene oxide (PEO), and poly[2-(dimethylamino)ethyl acrylate] (PDMAEMA)by changing the pH value [81]. By changing the content of ethylenediamine (EDA) and THF during the self-assembly process of triblock polymers (PAA-PMA-PS), Wooley and coworkers first discovered the ring-shaped micelle [82].

2.3.2 Preparation of Hairy Copolymer NPs

There are many methods for preparing hairy block copolymer NPs with various different morphologies using the solution self-assembly method. Generally, at room temperature when the hydrophobic chain of the block copolymer is in a glassy state, including PS-*b*-PAA and PS-*b*-PEO, the cosolvent method is a commonly used method for their assembly preparation. At first, dissolve the amphiphilic block copolymer in the cosolvent of all the blocks (the cosolvent is good for all blocks). Dioxane, DMF, and THF are three commonly chosen cosolvents. Then a selective solvent, such as water, is continuously added until its content is much higher than the needed amount of water (generally 25–50 wt%) when the aggregates are formed. The aggregates are then quenched using excess water to freeze their kinetic processes, also their morphology. Finally, the cosolvent is dialyzed out from the micellar solution in a selective solvent to obtain a structured, stable assembly solution. For diblock copolymer systems with relatively flexible hydrophobic chains, such as PB-*b*-PEO, the aggregates can be prepared directly through thin-film hydration method [83]. Generally, regardless of whether it is accompanied by mechanical mixing, sonic degradation, extrusion molding, electric field, and other conditions, this preparation method involves the rehydration of the block copolymer film in a selective solvent. This method is especially suitable for the fabrication of micron-scale giant vesicles. There are also other methods for the preparation of aggregates that have emerged during the past decade, among, which the latest development of microfluidics techniques provides an eye-catching way to prepare monodisperse vesicles with controllable size [84].

2.3.3 Major Factors Influencing the Morphology of Hairy NPs

2.3.3.1 Block Copolymer Composition

The formation of thermodynamically stable assembly is governed by three contributions to the free energy of the system: the stretching degree of the core-forming blocks, the interfacial tension between the micellar core and the solvent, and the repulsive interactions among corona-forming chains. The relative length of the core and shell-forming segments of the block copolymer can affect its assembly morphology. Usually, PS-*b*-PAA diblock copolymers with a relatively long PAA block will form spherical micelles. Taking the three contributions that can change the free energy of the system into consideration, three key parameters should be paid more attention: the average radius of core (R_{core}), the extension degree of the PS chain (S_c), and the area occupied by each shell chain on the core surface (A_c). In DMF-water mixed solution, with the increase of PAA block length, the aggregates formed with PS_{200}-*b*-PAA_n changed from spherical micelles to rod-like micelles, and then into vesicles. And the S_c and A_c gradually decrease since $S_c \sim N_{\text{PS}}^{-0.1}\ N_{\text{PAA}}^{-0.15}$. The N_{PS} and N_{PAA} represent the polymerization degrees of the two blocks. Therefore, the stretching extent of the chain decreases as the length of any block increases. When the aggregate changes its morphology, the stretch extent of the block polymers will also change. In the PS-*b*-PAA system, the stretching extent of the chain segment in the spherical micelles is higher than that of the rod-like micelles, and the extent of the chains in the core of the rod-like micelles is higher than that of the vesicle system. A_c is also affected by the relative length of the two blocks because $A_c \sim N_{PS}^{0.6} N_{PAA}^{0.15}$. Since the repulsive force between the shell block chains is inversely proportional to A_c, increasing the chain length of any block may reduce the shell repulsive force. Taking the nucleation segment as an example, keeping the length of the PAA segment unchanged, the longer nucleation segment PS forms a larger micellar core. Although the chain length of the shell-forming chains remains unchanged, the increase in the radius of the micellar core increases the surface area contained in each shell-forming chain, which reduces the repulsive force between the shell chains and drives more molecules into the aggregates resulting in the increasing of the core size. The aggregation number of molecular chains increases, and the degree of extension of molecular chains in the core increases.

2.3.3.2 Block Copolymer Concentration

The morphology of the assembly is also related to the concentration of the block copolymer. It has been reported that the aggregates formed by PS_{190}-*b*-PAA_{20} in the DMF-water mixed solution, when the concentration of the copolymer increases from 1 to 3.5 wt%, the morphology of the aggregate changes from spherical micelles to rod-like micelles, and then into vesicles [85]. The influence of the concentration of the copolymer on the morphology of the aggregate can be explained by the number of chain aggregation (N_{agg}). From the theory of small molecular surfactants: $N_{agg} = 2(C/CMC)^{1/2}$, in this equation, C represents the polymer concentration and CMC represents the critical micellar concentration, which is the copolymer concentration at the critical water content [86]. As the water content in the solution increases, the CMC decreases due to the increase of the Flory–Huggins parameter between the hydrophobic blocks and water, thus N_{agg} increases. The aggregation number is proportional to the concentration of the polymer, and the aggregation number increases with the increase of the polymer concentration, resulting in an increase in the repulsive force between the shells and the extension of the chain in the core. Therefore, increasing the concentration of the polymer can have a morphology effect on the aggregates similar to that of decreasing the chain length of corona chains.

2.3.3.3 The Nature of the Solvent

The nature of the solvent directly affects the size of the hydrophobic and hydrophilic chain within the aggregates, which has an important influence on the morphology of the aggregate [87]. When PS_{200}-*b*-PAA_{18} uses DMF as the solvent, spherical micelles are formed, large composite vesicles are formed in THF solvent, and rod-like micelles and vesicles are formed in the mixed solvent of DMF and THF. The closer the solubility parameter (δ) of the polymer and the solvent, the better the solubility of the polymer. When the δ value deviates, the size of the polymer chain decreases [88]. For PS-based copolymers, the solubility parameter of PS (δ = 16.6–20.2) is closer to THF (δ = 18.6) and dioxane (δ = 20.5) than DMF (δ = 24.8), therefore, in DMF or dioxane solvents, the swelling degree and the mobility of the PS chains in the core are better than that in the solvent of THF, which will promote the growth of micelles. The repulsive force between the chains of the copolymer shell depends on the size of polymer coil and the charge density of the polymer chain, while the coil size of the shell polymer chain and their interaction with space depend on the Flory–Huggins parameters of the solvent and the hydrophobic segment. If the solvent has a high dielectric constant, such as DMF (ε = 38.4), the ionizable PAA is partially charged, and the electrostatic repulsion between the shell molecular chains is relatively strong. In contrast, for solvents with low dielectric constants, such as dioxane (ε = 7.5) and THF (ε = 7.5), the charge on the PAA block can be ignored, and the electrostatic repulsion of the shell segment is weak and the repulsion effect is weak. The effective volume of each chain on the surface of the micelles will decrease as the repulsive force reduces, which is similar to the reduction in the PAA block length. Therefore, when the solvent changes from DMF to dioxane or THF, the morphology of the aggregates changes from spherical micelles to rod-like micelles and then to vesicles.

2.3.3.4 Additives

Adding additives such as salts, acids, bases, or surfactants to the solution will affect the morphology of the aggregates [89, 90]. In the presence of ions, the interaction between the molecular chains of the shell will change, causing the assembly morphology to change. In the PS-*b*-PAA copolymer system, the PAA block is partially ionized during the micellization process in a near-neutral environment, and the electrostatic effect becomes the main repulsive force between the shell molecular chains. The addition of salt can cause electrostatic shielding of the shell in the aggregate, and the addition of acid can protonate the shell of the aggregate (eliminate the charge), both can reduce the repulsive force between the molecular chains of the shell and the effective volume of the shell chains. With the increase of the salt or acid amount, the morphology of the aggregates changed from spherical micelles to rod-like micelles and then to vesicles. In contrast, the addition of alkali deprotonates the micellar shell, which leads to an increase in the degree of ionization and the repulsive force between the molecular chains of the shell, which has the opposite effect to the addition of acid. With the increase of alkali concentration, the morphology of aggregates changed from vesicles to spherical micelles. Small molecular surfactants, such as sodium dodecyl sulfonate (SDS) can also have an effect on the morphology of PS-*b*-PAA copolymers. SDS can affect the aggregation behavior of micelles from two aspects: the hydrophobic tail is inserted into the core of the micelles, which increases the distance between the shell (PAA chains); the counterion of the hydrophilic head can shield some ionized charges on the PAA chains, the repulsive force between the shell molecular chains is weakened, so that more polymer chains participate in the aggregation, therefore, under the condition of constant water content, the addition of SDS causes the changes in the extension of molecular chains in the core, which will change the final morphology of aggregates. As shown in Figure 2.15, in the

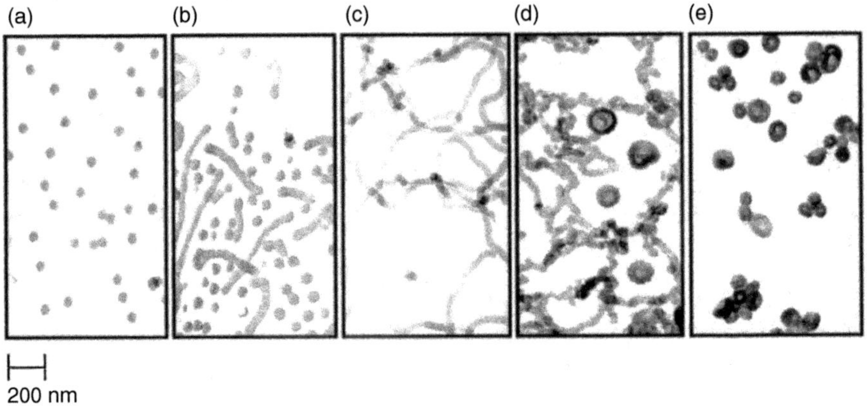

Figure 2.15 The effect of SDS content on the aggregation morphology of PS_{310}-*b*-PAA_{52} in dioxane/water mixed solution. Polymer concentration is 1.0 wt%; SDS concentrations are: (a) 2.0 mmol/L; (b) 5.1 mmol/L; (c) 9.2 mmol/L; (d) 11.0 mmol/L; (e) 11.7 mmol/L. Source: Burke and Eisenberg [90], Reproduced with permission of American Chemical Society.

dioxane/water solution with a water content of 11.5%, adding different amounts of SDS, the morphology of PS_{310}-*b*-PAA_{52} aggregates changes accordingly.

2.3.3.5 Other Factors

Temperature can also affect the morphology of the self-assembly aggregates. Just like adding water to the re-aggregation system [91, 92], the temperature change of the system has an effect on the polymer-solvent parameters [93]. Researchers chose small molecular alcohols (methanol, ethanol, and *n*-butanol) to study. Under pressure, the system is heated to above 140 °C. As the temperature rises, the solubility of the hydrophobic block increases, resulting in the formation of aggregates. When the temperature drops, the aggregates are frozen and the shape is fixed. For example, PS_{386}-*b*-PAA_{79} can form spherical micelles when heated to 160 °C in *n*-butanol, and form a mixture of microspheres and vesicles when heated to 115 °C [94]. What is more, Terreau et al. reported that increasing the polydispersity coefficient of PAA will transform the aggregates from spherical micelles into columnar micelles, and eventually vesicles [76]. In their experiments, they increased the polydispersity of PAA segments by mixing a series of copolymers with a fixed PS chain length but different PAA chain lengths that can make the PAA in the mixture have a bimodal or trimodal distribution.

Based on the above multiple influencing factors, people usually utilize two to three of them to control the final morphology of the resulting aggregates and draw the morphological phase diagram of the aggregates [95]. The phase diagram is very important in the study of polymer self-assembly behavior. From the phase diagram, not only the thermodynamic information of the aggregates can be obtained quantitatively, but also the relevant self-assembly conditions can be adjusted according to the phase diagram in order to quickly obtain the aggregate with desired morphology. With the continuous, in-depth research on the self-assembly of block copolymers, their potential industrial application values are constantly being discovered, especially in the fields of drug delivery, nanomaterials, and catalysis.

2.4 Summary

The preparation of hairy NPs by bulk microphase separation and self-assembly of block copolymers in solution is based on the incompatibility between different polymer chains in block copolymers to build ordered nanostructures. The periodic nanostructures formed by the bulk microphase separation of block copolymers can be realized by *in situ* cross-linking of the discontinuous phase region of the phase structure first and then dispersing the phase separation structure in the selective solvent of the continuous phase polymer. By changing the composition of the copolymer or adding foreign molecules (small molecules or polymers) with a specific structure, it is easy to obtain hairy NPs with spherical, linear, and sheet-like core/shell structures. This preparation method is simple and easy to operate and especially has unique advantages in the preparation of NPs with sheet-like and Janus asymmetric structures. This chapter also summarizes the self-assembly behavior of block copolymers in solution, and summarizes the various hairy NPs formed by block self-assembly in solution. The morphological structure of the hairy NPs mainly depends on the contributions of three factors related to the free energy of the polymer aggregation: the stretched degree of the block copolymer forming the NPs core; the interaction degree between the NPs core and the solvent; and the interaction between the shell-forming blocks. The morphology of the formed NPs is also affected by the composition of the block copolymer, the nature of the selective solvent, the concentration of the block copolymer, additives, and other factors. The complexity of the block copolymer self-assembly system currently studied is much lower than that presented in nature. In order to meet the needs of functional materials and biomimetic materials, it is necessary to increase the complexity of the self-assembly structure, which greatly promotes the development of block copolymer self-assembly. Although the preparation of hairy NPs with different shapes by the method presented in this chapter has many advantages, there are still some problems that need to be overcome. It is difficult for those who are not familiar with polymer synthesis and assembly to master; monodisperse spherical NPs can be obtained, but the length of linear NPs and the peripheral regularity of sheet-like NPs cannot be controlled; in the development of hairy NPs, it still faces the problem of how to effectively combine the functional components with the synthesis and self-assembly process at the same time. Although researchers have done a lot of work in this field, there is still a lot of work worthy of further exploration in enriching the morphology and structure of hairy NPs and their functionalization and applications.

References

1 Liu, L., Xu, K., Wang, H. et al. (2009). *Nat. Nanotechnol.* 4: 457–463.

2 Yuan, W., Wei, J., Lu, H. et al. (2012). *Chem. Commun.* 48: 6857–6859.

3 Yuan, J.-J., Mykhaylyk, O.O., Ryan, A.J., and Armes, S.P. (2007). *J. Am. Chem. Soc.* 129: 1717–1723.

4 Du, J. and Armes, S.P. (2008). *Langmuir* 24: 13710–13716.

5 Wang, G., Henselwood, F., and Liu, G. (1998). *Langmuir* 14: 1554–1559.

6 Miyamura, H., Akiyama, R., Ishida, T. et al. (2005). *Tetrahedron* 61: 12177–12185.
7 Li, G., Guo, L., Ma, S., and Liu, J. (2009). *J. Polym. Sci., Part A: Polym. Chem.* 47: 1804–1810.
8 Forster, S. and Plantenberg, T. (2002). *Angew. Chem. Int. Ed.* 41: 689–714.
9 Cheng, J.Y., Ross, C.A., Chan, V.Z.H. et al. (2001). *Adv. Mater.* 13: 1174–1178.
10 Wu, N.L.Y., Zhang, X., Murphy, J.N. et al. (2012). *Nano Lett.* 12: 264–268.
11 Urbas, A.M., Maldovan, M., DeRege, P., and Thomas, E.L. (2002). *Adv. Mater.* 14: 1850–1853.
12 Kang, C., Kim, E., Baek, H. et al. (2009). *J. Am. Chem. Soc.* 131: 7538–7539.
13 Orilall, M.C. and Wiesner, U. (2011). *Chem. Soc. Rev.* 40: 520–535.
14 Helfand, E. and Wasserman, Z.R. (1976). *Macromolecules* 9: 879–888.
15 Likhtman, A.E. and Semenov, A.N. (1994). *Macromolecules* 27: 3103–3106.
16 Bates, C.M., Maher, M.J., Janes, D.W. et al. (2014). *Macromolecules* 47: 2–12.
17 Leibler, L. (1980). *Macromolecules* 13: 1602–1617.
18 Huh, J. and Jo, W.H. (2004). *Macromolecules* 37: 3037–3048.
19 Matsen, M.W. and Bates, F.S. (1996). *Macromolecules* 29: 1091–1098.
20 Cochran, E.W., Garcia-Cervera, C.J., and Fredrickson, G.H. (2006). *Macromolecules* 39: 4264–4264.
21 Phillip, W.A., Hillmyer, M.A., and Cussler, E.L. (2010). *Macromolecules* 43: 7763–7770.
22 Mai, Y. and Eisenberg, A. (2012). *Chem. Soc. Rev.* 41: 5969–5985.
23 Bates, F.S. and Fredrickson, G.H. (1999). *Phys. Today* 52: 32–38.
24 Foerster, S., Khandpur, A.K., Zhao, J. et al. (1994). *Macromolecules* 27: 6922–6935.
25 Yi, X., Fan, X., Wan, L. et al. (2004). *Macromolecules* 37: 7610–7618.
26 Matsen, M.W. and Schick, M. (1994). *Macromolecules* 27: 187–192.
27 Wang, Z., Li, B., Jin, Q. et al. (2008). *Macromol. Theory Simul.* 17: 301–312.
28 Muñoz-Bonilla, A., Cerrada, M.L., and Fernández-García, M. (2007). *Macromol. Chem. Phys.* 208: 2654–2664.
29 Chen, J., Frisbie, C.D., and Bates, F.S. (2009). *J. Phys. Chem. C* 113: 3903–3908.
30 Zheng, W. and Wang, Z.-G. (1995). *Macromolecules* 28: 7215–7223.
31 Ishizu, K., Hosokawa, T., and Tsubaki, K. (2000). *Eur. Polym. J.* 36: 1333–1338.
32 Saito, R., Fujita, A., Ichimura, A., and Ishizu, K. (2000). *J. Polym. Sci., Part A: Polym. Chem.* 38: 2091–2097.
33 Underhill, R.S. and Liu, G.J. (2000). *Chem. Mater.* 12: 2082–2091.
34 Underhill, R.S. and Liu, G.J. (2000). *Chem. Mater.* 12: 3633–3641.
35 Yan, X.H., Liu, F.T., Li, Z., and Liu, G.J. (2001). *Macromolecules* 34: 9112–9116.
36 Yan, X.H., Liu, G.J., Liu, F.T. et al. (2001). *Angew. Chem. Int. Ed.* 40: 3593–3596.
37 Xu, H., Erhardt, R., Abetz, V. et al. (2001). *Langmuir* 17: 6787–6793.
38 Liu, V.A. and Müller, A.H.E. (2003). *Macromolecules* 36: 7894–7898.
39 Erhardt, R., Zhang, M.F., Boker, A. et al. (2003). *J. Am. Chem. Soc.* 125: 3260–3267.
40 Walther, A., André, X., Drechsler, M. et al. (2007). *J. Am. Chem. Soc.* 129: 6187–6198.

41 Schacher, F.H., Sugimori, H., Hong, S. et al. (2012). *Macromolecules* 45: 7956–7963.

42 Betthausen, E., Dulle, M., Hanske, C. et al. (2014). *Macromolecules* 47: 6289–6301.

43 Zhang, K., Gao, L., and Chen, Y. (2008). *Macromolecules* 41: 1800–1807.

44 Zhang, K., Gao, L., and Chen, Y. (2007). *Macromolecules* 40: 5916–5922.

45 Yao, D., Guo, Y., Chen, S. et al. (2013). *Polymer* 54: 3485–3491.

46 Qin, J., Jiang, X., Gao, L. et al. (2010). *Macromolecules* 43: 8094–8100.

47 Li, Y., Peng, B., and Chen, Y. (2013). *J. Mater. Chem. B* 1: 5694–5701.

48 Qin, J., Chen, Y., Yan, D., and Xi, F. (2010). *Macromolecules* 43: 10652–10658.

49 Yao, D., Chen, Y., and Jin, R. (2015). *J. Mater. Chem. B* 3: 5786–5794.

50 Yao, D., Zhang, K., and Chen, Y. (2016). *Polymer* 94: 1–7.

51 Gao, L., Zhang, K., and Chen, Y. (2012). *ACS Macro Lett.* 1: 1143–1145.

52 Whitesides George, M., Mathias John, P., and Seto Christopher, T. (1991). *Science* 254: 1312–1319.

53 Zhang, L. and Eisenberg, A. (1996). *J. Am. Chem. Soc.* 118: 3168–3181.

54 Riess, G. (2003). *Prog. Polym. Sci.* 28: 1107–1170.

55 Sun, J.-T., Hong, C.-Y., and Pan, C.-Y. (2012). *Soft Matter* 8: 7753–7767.

56 Tan, H., Wang, Z., Li, J. et al. (2013). *ACS Macro Lett.* 2: 146–151.

57 Lv, L., Zhang, Z., Li, H. et al. (2019). *ACS Macro Lett.* 8: 1222–1226.

58 Zhang, Z., Li, H., Huang, X., and Chen, D. (2017). *ACS Macro Lett.* 6: 580–585.

59 Zhang, Z., Zhou, C., Dong, H., and Chen, D. (2016). *Angew. Chem. Int. Ed.* 55: 6182–6186.

60 Blanazs, A., Armes, S.P., and Ryan, A.J. (2009). *Macromol. Rapid Commun.* 30: 267–277.

61 Förster, S., Zisenis, M., Wenz, E., and Antonietti, M. (1996). *J. Chem. Phys.* 104: 9956–9970.

62 Lim Soo, P. and Eisenberg, A. (2004). *J. Polym. Sci., Part B: Polym. Phys.* 42: 923–938.

63 Zhang, L. and Eisenberg, A. (1995). *Science* 268: 1728–1731.

64 Halperin, A., Tirrell, M., and Lodge, T.P. (1992). *Macromolecules: Synthesis, Order and Advanced Properties*. Berlin, Heidelberg: Springer Berlin Heidelberg.

65 Zhang, L., Yu, K., and Eisenberg, A. (1996). *Science* 272: 1777–1779.

66 Xu, J., Yin, Y., Wang, Z. et al. (2013). *J. Chem. Phys.* 138: 114905–114905.

67 Esselink, F.J., Dormidontova, E.E., and Hadziioannou, G. (1998). *Macromolecules* 31: 2925–2932.

68 Lam, Y.-M., Grigorieff, N., and Goldbeck-Wood, G. (1999). *Phys. Chem. Chem. Phys.* 1: 3331–3334.

69 Kabanov, A.V., Batrakova, E.V., and Alakhov, V.Y. (2002). *J. Controlled Release* 82: 189–212.

70 Maysinger, D., Lovrić, J., Eisenberg, A., and Savić, R. (2007). *Eur. J. Pharm. Biopharm.* 65: 270–281.

71 Harada, A. and Kataoka, K. (2006). *Prog. Polym. Sci.* 31: 949–982.

72 Read, E.S. and Armes, S.P. (2007). *Chem. Commun.* 43: 3021–3035.

73 Walther, A. and Müller, A.H.E. (2008). *Soft Matter* 4: 663–668.

74 Zhu, J. and Jiang, W. (2005). *Macromolecules* 38: 9315–9323.
75 Wang, X., Guerin, G., Wang, H. et al. (2007). *Science* 317: 644–647.
76 Terreau, O., Bartels, C., and Eisenberg, A. (2004). *Langmuir* 20: 637–645.
77 Zhu, Y., Yang, B., Chen, S., and Du, J. (2017). *Prog. Polym. Sci.* 64: 1–22.
78 Rideau, E., Dimova, R., Schwille, P. et al. (2018). *Chem. Soc. Rev.* 47: 8572–8610.
79 Idrissi, M.E., Meyer, C.E., Zartner, L., and Meier, W. (2018). *J. Nanotechnol.* 16: 63–63.
80 Holder, S.J. and Sommerdijk, N.A.J.M. (2011). *Polym. Chem.* 2: 1018–1028.
81 Liu, C., Hillmyer, M.A., and Lodge, T.P. (2009). *Langmuir* 25: 13718–13725.
82 Pochan Darrin, J., Chen, Z., Cui, H. et al. (2004). *Science* 306: 94–97.
83 LoPresti, C., Lomas, H., Massignani, M. et al. (2009). *J. Mater. Chem.* 19: 3576–3590.
84 Shum, H.C., Kim, J.-W., and Weitz, D.A. (2008). *J. Am. Chem. Soc.* 130: 9543–9549.
85 Zhang, L. and Eisenberg, A. (1999). *Macromolecules* 32: 2239–2249.
86 Israelachvili, J.N. (2011). *Intermolecular and Surface Forces*, 3e. San Diego: Academic Press.
87 Bhargava, P., Zheng, J.X., Li, P. et al. (2006). *Macromolecules* 39: 4880–4888.
88 Yu, Y., Zhang, L., and Eisenberg, A. (1998). *Macromolecules* 31: 1144–1154.
89 Shen, H. and Eisenberg, A. (2000). *Macromolecules* 33: 2561–2572.
90 Burke, S.E. and Eisenberg, A. (2001). *Langmuir* 17: 8341–8347.
91 Zhang, L., Shen, H., and Eisenberg, A. (1997). *Macromolecules* 30: 1001–1011.
92 Shen, H.W. and Eisenberg, A. (1999). *J. Phys. Chem. B* 103: 9473–9487.
93 Desbaumes, L. and Eisenberg, A. (1999). *Langmuir* 15: 36–38.
94 Zhang, L. and Eisenberg, A. (1998). *Polym. Adv. Technol.* 9: 677–699.
95 Discher Dennis, E. and Eisenberg, A. (2002). *Science* 297: 967–973.

3

Hairy Nanoparticles via Unimolecular Block Copolymer Nanoreactors

Wenjie Zhang[1] *and Xinchang Pang*[1,2]

[1] *Zhengzhou University, School of Materials Science and Engineering, No.100, Kexuedadao Road, Zhengzhou 450001, China*
[2] *School of Materials Science and Engineering, Henan University of Science and Technology, Luoyang 471023, PR China*

3.1 Background

Nanoparticles (NPs) have attracted widespread attention owing to their size and shape-dependent properties, which distinguished them from their bulk counterparts for the wide application in optics, optoelectronics, catalysis, energy conversion and storage, luminescence, and biology areas, which have shown up deep influence on our daily life [1–7]. The basic characteristics of inorganic NPs, such as morphology, architecture, components, size, surface properties, and assembly properties, largely influence the properties and corresponding applications of inorganic NPs. The past 30 years have witnessed great progress in the synthesis of inorganic NPs. However, these strategies require extremely strict experimental conditions, are difficult to generalize, or necessitate tedious multistep reactions and purification procedures [2, 3, 8–10]. Therefore, it is vital to develop a general and robust synthetic strategy to obtain high-quality NPs with excellent precisely control over their basic features.

In general, there are two strategies for the synthesis of NPs: top-down and bottom-up strategies [11–14]. For the top-down strategy, NPs are broken off from bulk materials through strong external forces at high price, like ball-milling, making this method unable to precisely tailor the morphology, architecture, size, and surface properties of the prepared NPs. The commonly used method is the bottom-up strategy, in which NPs are synthesized via nucleation and growth process of precursors with control of the abovementioned feature. The bottom-up strategies can be divided into several types, including hydrothermal and solvothermal method, thermal decomposition method, sol–gel processes, sonochemical synthesis, and template method, which are all wet chemical processes to synthesize inorganic NPs in a much easier, better controllable, more general, and robust way compared to the above mentioned top-down strategies.

Hairy Nanoparticles: From Synthesis to Applications, First Edition. Edited by Zhiqun Lin and Yijiang Liu.

The template method is a representative bottom-up strategy for the precisely controlled fabrication of inorganic NPs [5, 15–17]. The template method for synthesizing NPs involves the confinement of NP precursors within small volumes and the subsequent reduction or pyrolysis and aggregation of those precursors into NPs. These spatially isolated volumes can be termed as nanoreactors, and they impose barriers that not only restrict the movement of metal atoms and other reactants but also provide reaction conditions that are distinct from those of the surrounding environment [18–23]. Nanoreactors for NPs synthesis are mainly prepared by wet chemistry, which generally falls under the category of solution-based. Solution-based nanoreactors are broadly defined as stereo capsules that can be manipulated in solution. Nanoreactors change the basic chemical nature of molecules and moieties within them and alter how they behave in chemical reactions. In this way, nanoreactors can be exploited not only to make a new type of NPs, but also to gain new fundamental understanding of a chemical system or process or to develop an analytical tool based upon this insight [7, 11, 24–27]. With this template method, a pre-existing hard or soft template with desired nanostructures was used to direct the production of NPs with expected and novel features, which are inherited from the templates and difficult to obtain with other methods. Using predesigned nano-structural carbon or silica as hard templates, novel nanomaterials with unique nanostructures, various compositions, and highly thermal stabilities have been fabricated in a controlled way [28, 29]. However, these finely tailored hard templates are usually fabricated from soft templates. Soft templates, such as surfactants and polymers, are very powerful tools for nanofabrication.

Compared with supramolecular nanoreactors assembled from multiple molecules or linear block copolymers with dynamic stability affected by environmental variations, unimolecular nanoreactors with covalently stable and predesigned nanostructures are one of the most attractive types of soft templates to fabricate inorganic NPs in a programmable way [28, 29]. Due to the rapid development of polymer chemistry, various types of unimolecular micelles nanoreactors with diverse compositions and nanostructures have been prepared, and their application in the nanofabrication of NPs with expected features within these unimolecular nanoreactors has shown charming prospects. Among various unimolecular micelles to prepare inorganic NPs, the nonlinear unimolecular block copolymer nanoreactor technique has recently emerged as a general and robust strategy for crafting a rich diversity of NPs of interest with precisely controlled dimensions, compositions, architectures, and surface chemistry [30–33]. Reversible deactivation radical polymerization (RDRP) has been developed since the early 1990s, including atom-transfer radical polymerization (ATRP), reversible addition-fragmentation chain transfer (RAFT), and nitroxide-mediated free-radical polymerization (NMP) [34–38]. All these RDRP methods opened up an effective and powerful platform for fabricating advanced functional polymers with precisely controlled molecular architectures owing to the dynamic equilibrium established in the radical polymerization process [39, 40]. Block copolymers with controlled molecular weight, low polydispersity, controlled macromolecular architectures (e.g. star-like, bottlebrush-like, cyclic), and chemical compositions (block, alternating, and gradient copolymers) could be prepared via RDRP method and can be utilized as templates to precisely direct the growth of nanostructured materials with complex morphologies and diverse functionalities [35, 41–45]. Most of the block copolymer unimolecular micelles templates in this chapter were obtained by this technique.

It is notable that nonlinear block copolymers are unimolecular micelles, where each block copolymer arm of a nonlinear unimolecular block copolymer nanoreactor is covalently connected to a central core or polymer backbone [28, 31–33, 45–49]. Therefore, their structures are static and stable, representing a class of functional polymers with complex architecture for directing the synthesis of NPs [28, 30]. Rapid development in polymer chemistry has promoted remarkable progress in the controlled synthesis of unimolecular nanoreactors with diverse but uniform topologies, architectures, components, and sizes. The unimolecular nanoreactors with well-defined and stable nanofeatures can direct the fabrication of various NPs with prospective nanostructures. The inside cavity concentrated plenty of precursors and the outside tethered polymer chains provided dispersion stability of the gained NPs [30, 32]. In this way, NPs with expected topologies (sphere, rod, tube, and ring), architectures (compact, hollow, core@shell, and necklace-like), compositions (metal, metal oxide, semiconductor, doping, alloy, silica, and composite), sizes (generally 1–100 nm), surface properties (hydrophilic, hydrophobic, chemical reactivity, and stimuli responsivity), and assemblies (oligomer, chain, and aggregate) can be fabricated easily within reasonably designed unimolecular nanoreactors [18, 28, 30–32].

In this chapter, we provided a brief summary of the fabrication of inorganic NPs within unimolecular block copolymer micelles nanoreactors, which mainly focus on two sets of unimolecular nonlinear block copolymers as nanoreactors. The star-liked block copolymers for preparing 0D spherical NPs, including plain, hollow, core–shell, ring NPs, and the NPs assemblies; and the bottlebrush-like block copolymers for synthesizing 1D plain, nanotube, and core@shell nanorods (NRs) (even nanowires [NWs]). First, the rational design and synthesis of nonlinear unimolecular block copolymer micelles mainly via controlled/"living" radical polymerizations is introduced. Subsequently, their application as nanoreactors to synthesize monodisperse NPs or NRs with judiciously adjusted dimensions, compositions, and surface chemistry is presented. Furthermore, the applications of these polymer-ligated NPs for energy conversion, storage, and catalysis are then discussed. Finally, challenges and opportunities in this rapidly developing field are presented, and explored the many ways they can deepen our understanding of the basic microworld.

3.2 Synthesis and Properties of Block Copolymer Unimolecular Micelles

3.2.1 Properties of Unimolecular Block Copolymer Micelles

Unimolecular block copolymer micelles nanoreactors are composed of covalent bonds within one nanoscale macromolecule exhibiting low viscosity and

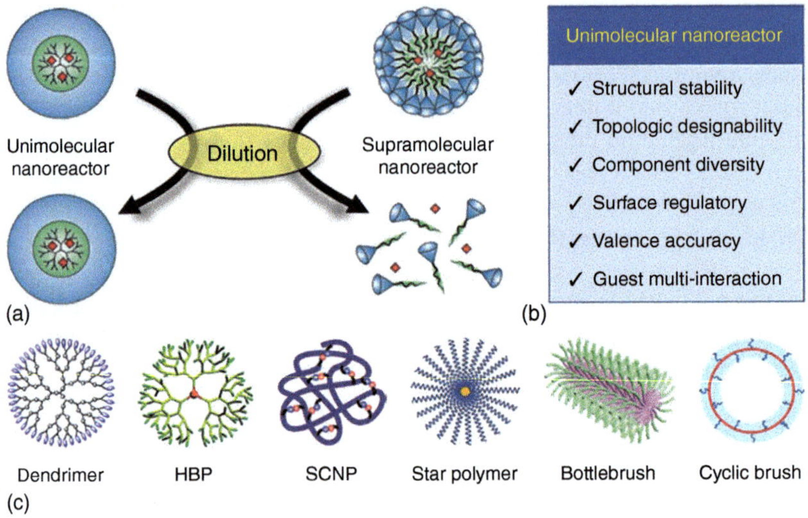

Figure 3.1 (a) Different behaviors of unimolecular nanoreactors and supramolecular nanoreactors under dilution. (b) The structural advantages of unimolecular nanoreactors. Source: Wang and Zhu [28]/Royal Society of Chemistry.

high solubility due to their rigid architectures, avoiding chain entanglements [28, 30, 32, 33, 47]. The advantages of unimolecular nanoreactors for the synthesis of inorganic NPs are listed as follows (Figure 3.1):

(1) The structural stability of unimolecular block copolymer nanoreactors is the most dominant feature compared to the linear block copolymer nanoreactors, which are the assemblies of multiple amphiphilic block copolymers via weak interactions under special conditions [28, 47]. The stability of these linear block copolymer nanoreactors could be destroyed seriously by slightly changing the environmental conditions, such as solvent, concentration, pH, ionic strength, and temperature. However, unimolecular block copolymers nanoreactors, such as star-like polymers, by contrast, can maintain their original architectures regardless of the variation of the outside solution conditions. The soft but steady structures of unimolecular nanoreactors can interact with metal precursors at the molecular level and direct the formation of NPs in a robust way.
(2) The structural characters of unimolecular block copolymer nanoreactors could be extensively regulated, including morphology, architecture, composition, size, and surface chemistry due to the rapid development of polymer chemistry and select special types of functional monomers [50]. All these delicately predesigned structural features of unimolecular block copolymer nanoreactors will be inherited to the fabricated NPs with diverse morphologies, various architectures, various compositions, aplenty but uniform sizes, abundant guest interactions, and ample surface chemistry.
(3) Furthermore, there are a large variety of metal precursors that can be effectively encapsulated into the cavities of unimolecular micelles nanoreactors with or without strong interaction to achieve the function of loading and conversion

of the metal precursors [6, 51, 52]. After metal precursors enrichment through strong interaction and subsequent reduction or pyrolysis, then NPs will form within unimolecular micelles nanoreactors. The unique properties and various types of unimolecular micelles nanoreactors with finely structural regulations make them powerful nanoreactors for templated nanofabrication of inorganic NPs.

3.2.2 Synthesis and Features of Star-Liked Block Copolymers

Star-like block copolymers represent a class of widely investigated nonlinear polymers consisting of many linear polymer arms with the chain-end functionalities fused at a central core. On the basis of the chemical composition and sequence distribution of polymer arms, star-liked polymers can be classified into a homoarm star-like polymer with identical arm types and a miktoarm star-like polymer having dissimilar arms. There have been tremendous advanced methods, such as controlled/"living" radical polymerization, in the synthesis of nonlinear star-liked polymers during the past several decades, and it was first synthesized by living anionic polymerization in the 1950s. As noted above, unimolecular star-like polymers have an architecture similar to that of self-assembled linear block copolymers micelles, yet with many arms tethered onto a central core and exhibiting markedly improved stability and robustness against the external stimuli, such as pH, heat, solvent, and salt.

Star polymers containing linear "arms" radiating from a "core" exhibit highly regular structures and unique properties owing to the spatially defined while compact architecture and controlled components and sizes. The synthetic approaches of star polymers could be roughly categorized as core-first, arm-first, and grafting-onto approaches [47]. To date, various star-like block copolymers with well-defined MWs, compositions, sequence structures, and functionalities have been synthesized by controlled/"living" radical polymerization techniques (i.e. ATRP, RAFT, and NMP) via core-first strategy. In addition, click chemistry is also a widely used method to produce star-like polymers via arm-first strategy that cannot be achieved otherwise by controlled/"living" radical polymerization mainly.

3.2.2.1 Synthesis of Star-Liked Block Copolymers via Core-First Method

Generally, the core-first approach has better structural controllability for the star polymer construction. The core-first strategy to prepare star-liked block copolymers carries several advantageous attributes: (i) the ability to precisely control the number of arms by judiciously choosing or selectively functionalizing the initiators core; (ii) the capability of facilely tailoring the chemical compositions of arms by identifying the monomer species; and (iii) the easy purification of the crude product by precipitation. However, this approach also faces some challenges, including the difficulties in precise characterization of the arm structure (i.e. MW, PDI) as well as synthesis of miktoarm star-like polymers [53–56]. Particularly, in the core-first strategy, multifunctional initiators are synthesized first, and polymer arms are then grafted sequentially via controlled/"living" radical polymerization to yield star-like polymers.

It is notable that the synthesis of a multifunctional initiators represents the key step in the core-first strategy. The initiators are often multifunctional molecules, such as cyclodextrins (β-CD, 21 arms), resorcinarene (growth of four or eight arms), calixarene (eight arms), stilbeneamine (three arms), 1,3,5-trivinylbenzene (three arms), polyhedral oligomeric silsequioxane (POSS, eight arms), hyperbranched polymers (i.e. hyperbranched polyester, hyperbranched polyglycerol, and hyperbranched conjugated polymer) [6, 39, 47, 57]. Thanks to the intrinsic branched structures of cyclodextrins (CDs), CDs-cored star polymers can be facilely obtained by using aforementioned polymerization techniques. CDs (including α-CD, β-CD, and γ-CD) are a family of cyclic oligosaccharides, which are composed of 6, 7, and 8 D-glucose units linked by α-1,4-glycosidic bonds, respectively (Figure 3.2). Depending on the number of D-glucose units, CDs have an upper rim in the range of 0.45–0.77 nm, a lower rim in the range of 0.57–0.95 nm, and a height of 0.78 nm [45, 56, 58]. CDs have abundant hydroxyls in the exterior surface, the hydroxyls

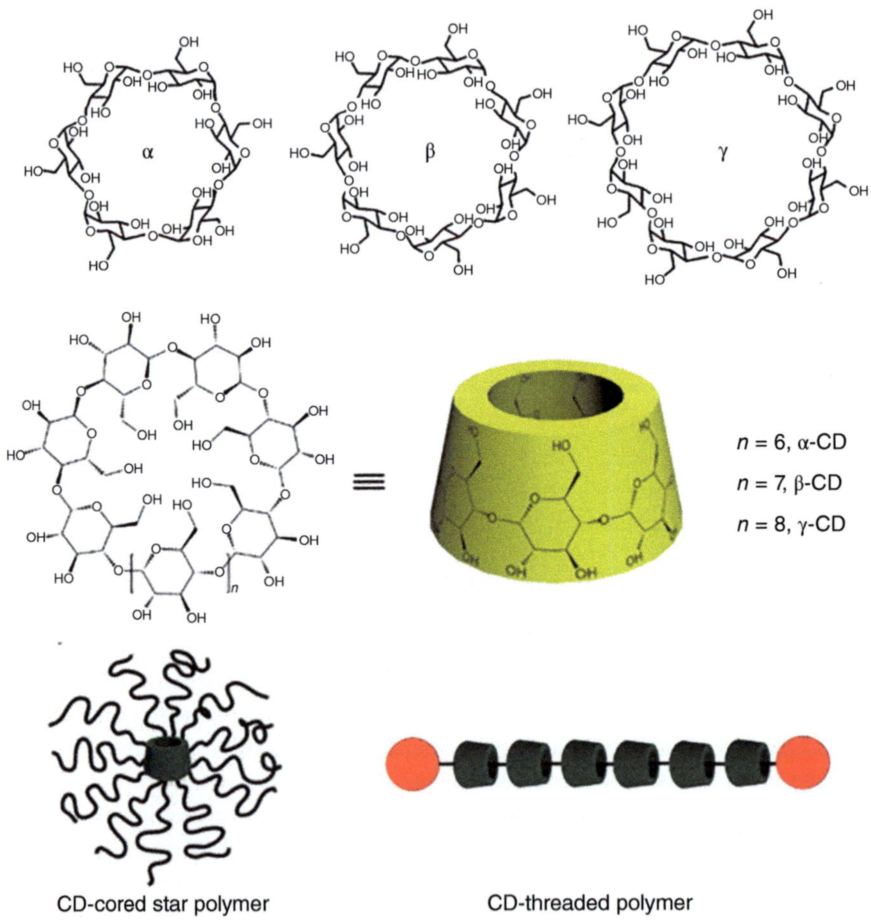

Figure 3.2 The chemical structures and parameters of common cyclodextrins. Source: Yao et al. [45]/Elsevier.

can be categorized as primary hydroxyls in the upper rim and secondary hydroxyls in the lower rim. These abundant hydroxyls at specific positions can be selectively modified with various functional moieties to impart the molecules with tailored functionalities. Taking advantage of CDs' properties, i.e. abundant hydroxyls in the exterior, CDs can be modified with initiators and then prepared as CD-based star polymers via controlled/"living" polymerization with controllable molecular weight and molecular weight distribution. What's more, based on CD-guest interactions, diverse supramolecular structures, e.g. polyrotaxanes (PRs), pseudo-block polymers, highly branched polymers, and comb-like brush polymers, can be built [59]. As a result, CD-based polymers have been widely investigated in the past decades. In this section, we mainly choose β-CD as an example to demonstrate its application as a multifunctional core in synthesizing a series of functional star-like block copolymer unimolecular micelles.

With regard to ATRP method, it has been frequently used to produce polymers with controlled compositions, structures, and functionalities. CD-cored star polymers with desired properties have been readily synthesized via ATRP method [39, 60]. β-CD possessing 21 hydroxyl groups (–OH) and can be modified by esterization reaction to endow the ATRP initiation sites (e.g. converting –OH into –Br functionality and thus forming a 21Br-β-CD macroinitiators). By adjusting the feeding ratio of 2-bromoisobutyryl bromide (BIBB) to CD, CD-based derivatives with desired initiating sites, e.g. 4, 7, and 21, can be obtained [58, 61]. Then, through the sequential ATRP process, star-liked diblock/triblock copolymer will be prepared. For example, Lin et al. synthesized 21-arm CD-cored amphiphilic block copolymers (PAA-*b*-PS) by sequential ATRP using 21-Br-β-CD as macroinitiators. The chain length of each block was tailored by tuning the ATRP reaction time. This approach can be extended to synthesize 21-arm β-CD-cored random copolymers by copolymerization of two monomers via one-pot ATRP. Similarly, a star-like β-CD-g-poly(acrylic acid)-*b*-polystyrene diblock copolymer (β-CD-g-PAA-*b*-PS) was prepared by sequential ATRP of *tert*-butyl acrylate (*t*BA) and St, and then followed by the hydrolysis of poly(*tert*-butyl acrylate) (P*t*BA) block to poly(acrylic acid) (PAA). Star-liked diblock copolymers composed of hydrophilic inner blocks (e.g. PAA; via coordination interaction between carboxyl groups of PAA and metal moieties of precursors) are often used as nanoreactors for crafting plain polymer-ligated NPs.

Traditional ATRP requires a copper catalyst in a low oxidation state (CuX) together with an alkyl halide as the ATRP initiator, while recently developed photoinduced ATRP (photoATRP) features the utilization of air-stable Cu(II) at a low ppm-level. The photoATRP was initially reported under UV irradiation in 2011, and the polymerization conditions were optimized using the ppm level of Cu(II) under visible light irradiation even in the near-infrared (NIR) range recently. He et al. reported the work in which photoATRP was applied to synthesize the pH-responsive and thermosensitive star-like copolymers PAA-*b*-P(MEA-*co*-$OEGA_{480}$) from β-cyclodextrin (β-CD) as the star core [62]. P(MEA-*co*-$OEGA_{480}$) was designed as the second outer block arm that possessed thermosensitive properties of LCST-type transition in water and UCST-type transition in ethanol. It especially has huge potentials in the field of life medicine and also could be used as a nanoreactor to prepare thermally responsive nanomaterials and extend the green photoATRP technology to the application of synthesizing star-liked polymers. Furthermore, star-shaped amphiphilic polymers can also be prepared via a simplified electrochemically mediated ATRP (seATRP) procedure under both potentiostatic and pseudo-galvanostatic conditions, utilizing only 50 ppm of Cu(II) complex, much less than previously reported concentrations of catalyst.

By directly taking advantage of abundant hydroxyls, CD-cored star-liked polymers can be obtained by ring-opening polymerization (ROP) method using CD-based derivatives as macroinitiators. For example, Wu et al. synthesized 21-arm β-CD-cored star polymer poly(L-lactide) (β-CD-PL-LA) by using native β-CD as macroinitiator. The ^{1}H NMR spectrum showed that all the 7 primary hydroxyls and 14 secondary hydroxyls were consumed to initiate the ROP of L-LA monomers. By adjusting the ratio of L-LA to β-CD, five β-CD-cored star polymers with different MWs were synthesized. Furthermore, GPC results showed monosymmetric peaks, indicating the absence of linear homopolymers as impurity in the product.

As mentioned above, CDs have two kinds of hydroxyls in the upper rim (7 primary hydroxyls) and lower rim (14 secondary hydroxyls), respectively. By selecting appropriate protection group (e.g. benzyl, acetyl, and trimethylsilyl groups), the arm number of CD-cored star-polymers can also be precisely controlled. Per-2,3-acetyl-β-CD molecule with exactly seven initiating sites can be precisely synthesized and used as macroinitiators to construct 7-arm CD-cored polymers through three protection/deprotection steps. In this case, all the primary hydroxyls of β-CD are first protected selectively by *tert*-butylchlorodimethylsilane (TBDMSCl). All the secondary hydroxyls of β-CD are then acetylated by acetic anhydride. In the end, per-2,3-acetyl-β-CD molecule is obtained by desilylation of per-6-TBDMS-β-CD using boron trifluoridedimethyl etherate ($BF_3{\bullet}Et_2O$).

All the primary hydroxyls of β-CD can be functionalized with RAFT agents for the synthesis of 7-arm β-CD-cored star polymers. One of the first such examples is 7-arm poly(*N*-vinylpyrrolidone) (β-CD-PVP_7) synthesized by RAFT polymerization using CD-based macromolecular chain transfer agent (macroCTA). This macroCTA was synthesized by copper-catalyzed azide–alkyne cycloaddition (CuAAC) reaction between alkyne-modified CTA and per-6-azide-β-CD. It is worth noting that seven hydroxyls in the upper rim of β-CD can be selectively converted to azide groups at the same time. After click reaction, the complete consumption of all the seven azide groups can be achieved. Afterward, β-CD-PVP_7 with desired MWs was synthesized by bulk RAFT polymerization at varying reaction times.

By choosing suitable functional monomers, the corresponding functional star-liked block copolymers can be realized. Such as external stimuli polymer, chemical reactive polymer, hydrophilic, and hydrophobic. Recently, thermo-responsive star-like β-CD-g-poly(acrylic acid)-*b*-poly(*N*-isopropylacrylamide) (denoted β-CD-g-PAA-*b*-PNIPAM) and photoresponsive star-like β-CD-g-poly(acrylic acid)-*b*-poly (7-methylacryloyloxy-4-methylcoumarin) (β-CD-g-PAA-*b*-PMAMC) diblock copolymers were designed and synthesized by employing *N*-isopropylacrylamide

(NIPAM) and 7-methylacryloyloxy-4-methylcoumarin (MAMC) as the second monomers, respectively. The MW of each block (i.e. the arm length) can be readily tuned by varying the polymerization time. Wang et al. prepared 21-arm β-CD-cored copolymers by copolymerization of (3-(4-formylphenoxy) propylacrylate) (FPPA) and oligo(ethylene glycol)methyl ether methacrylate (OEGMA). The anticancer drug DOX was then covalently conjugated with benzaldehyde of the copolymers. The release of DOX was triggered under mild acidic conditions resulting from the breakdown of pH-labile Schiff-base bonds. Moreover, the same group demonstrated another pH-sensitive CD-cored block copolymer by one-pot ATRP of pH-sensitive 2-(diisopropylamino) ethyl methacrylate (DPA) and OEGMA. These CD-cored copolymers were used to load both anticancer drugs (DOX) and photothermal agents (BBT-2FT) simultaneously for combined chemo and photothermal therapy. Under acidic conditions or NIR light irradiation, the payload was quickly released, exhibiting superior therapeutic effects.

Yang et al. synthesized CD-cored block/random star polycarbonates by sequential/one-pot ROP of benzyl chloride and mannose-functionalized cyclic carbonate monomers using per-2,3-acetyl-β-CD as macroinitiator and *N*-(3,5-trifluoromethyl) phenyl-*N*′-cyclohexylthiourea (TU)/1,8-diazabicyclo[5,4,0]undec-7-ene) (DBU) as cocatalysts. Subsequent deprotection of ipman groups and quaternization of benzyl chloride groups with various *N*,*N*-dimethylalkylamines produced mannose-targeting β-CD-cored star polycarbonates. These β-CD-cored star polymers exhibited efficient antibacterial activity and negligible hemolysis effect through mannose-mediated pathway.

Notably, β-CD-cored star polymers obtained by this ROP method have 21 hydroxyls on the exterior, which can be utilized for further functionalization. Shen et al. synthesized β-CD-cored 7-arm star poly(-caprolactone)s (CDSPCLs) by ROP method of monomers using per-2,3-acetyl-β-CD as macroinitiators and stannous octoate ($Sn(Oct)_2$) as catalyst. By further coupling reaction with carboxyl-capped poly(ethylene glycol) (PEG), amphiphilic star BCPs of poly(caprolactone-*b*-ethylene glycol)s (CDSPCL-*b*-PEGs) were synthesized. These star CDSPCL-*b*-PEGs BCs could form uniform spherical micelles in aqueous solution. The diameter of the micelles was controlled in the range of approximately 10–40 nm by tuning the length of hydrophobic PCL. Furthermore, all the primary hydroxyls of β-CD can be converted to amino, iodine, or thiol groups as initiating sites for the production of CD-cored 7-arm star polymers by ROP method. It was reported that CD-cored 7-arm POx was facilely synthesized via cationic ROP method using per-6-iodine-CD as macroinitiators. Moreover, per-6-iodine-β-CD can be further converted to per-6-thiol-β-CD by reacting with thiourea under an alkaline condition. For example, Yu et al. synthesized cell-penetrating star poly(disulfide)s by ROP of guanidine-containing disulfide monomers (CPDs) using per-6-thiol-β-CD as macroinitiators. The PDI of obtained star polymer (CD–CPD) was determined to be 1.32 by GPC. The CD–CPDs were demonstrated for loading camptothecin (CPT) and miRNA simultaneously through host–guest interaction between CPT and β-CD and electrostatic interaction between positive CPD and negative miRNA. Furthermore, by sequential thiol-ene click reaction and ROP method, per-6-thiol-β-CD can be conjugated with vinyl-terminated polymers and subsequently be used as macroinitiators for CL polymerization.

Feng et al. prepared unimolecular core@shell and hollow polymer NP with well-defined dimensions using spherical core@shell star-liked diblock copolymers as templates [63]. Monodisperse and structurally stable star-like diblock copolymers composed of inner degradable core blocks and outer photo-cross-linkable shell blocks were synthesized via a combination of two living polymerization techniques, namely, coordination-insertion ROP followed by RAFT. Subsequently, uniform unimolecular core@shell NPs were successfully produced by photo-cross-linking the shell blocks of star-like diblock copolymers (Figure 3.3). The core diameter and shell thickness of NPs are determined by molecular weights of inner core block and outer shell block, respectively, thereby rendering NPs with tunable structural characteristics. The cross-linking density of NPs can be readily controlled by varying the exposure time of star-like diblock copolymer templates to UV illumination. The selective degradation of inner core blocks yielded hollow polymer NPs, which retained structural integrity. The dye encapsulation and release studies revealed

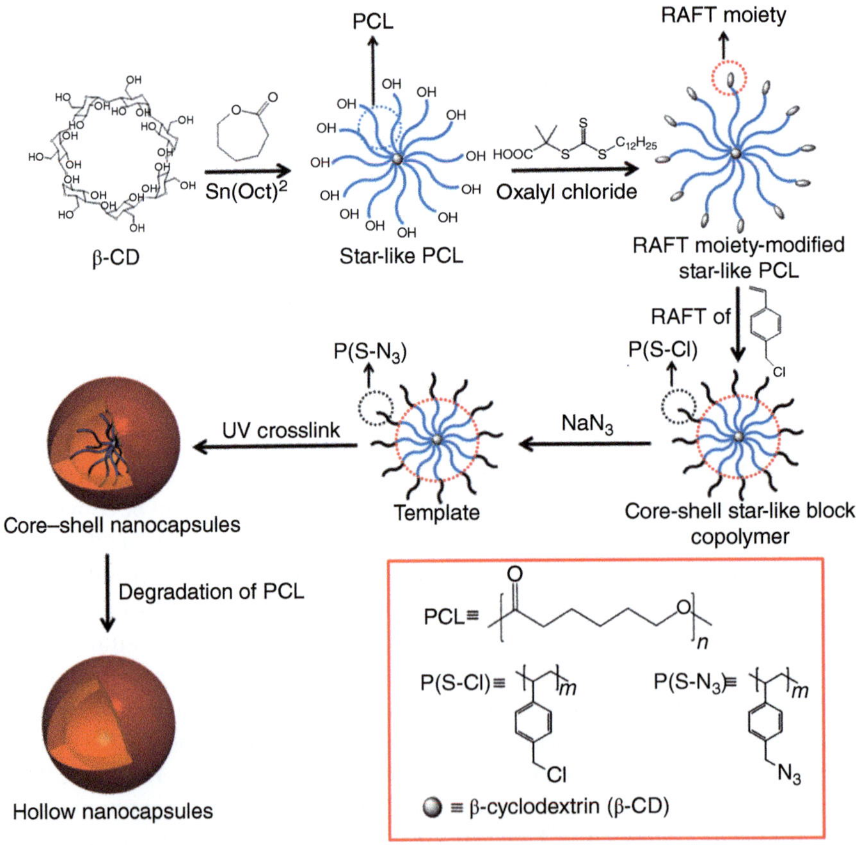

Figure 3.3 Synthetic strategy for unimolecular polymeric core@shell and hollow nanocapsules. Source: Feng et al. [63]/American Chemical Society.

that unimolecular core@shell NPs may be exploited as a new class of nanocarriers and promising drug nanovehicles.

3.2.2.2 Synthesis of Star-Liked Block Copolymers via Arm-First Method

It is widely known that several functional polymers cannot be achieved by controlled/"living" radical polymerization. An alternative strategy to overcome these disadvantages is to implement the arm-first strategy, where star-like polymers are yielded via the polymerization reaction between a macroinitiator (i.e. premade polymer with reactive end groups) and a polymerizable monomer or coupling reaction between reactive linear polymer and multifunctional core. Generally, click reaction is usually utilized for arm-first approach to prepare star-liked polymers. As to arm-first approach, linear polymers with reactive end groups (e.g. alkynyl group) are generally synthesized and then reacted with the cores, which possess several corresponding reactive sites (e.g. $-N_3$), such as CD-based derivatives to obtain well-defined star polymers. By coupling ATRP with click reaction, star-like triblock copolymers with functional outer block (e.g. conjugated block) are also attainable. For example, recently, star-like triblock copolymers of β-CD-g-polystyrene-*b*-poly(acrylic acid)-*b*-poly(3,4-ethylenedioxythiophene) (β-CD-g-[PS-*b*-PAA-*b*-PEDOT]) were synthesized. First, the star-like PS-*b*-P*t*BA diblock copolymer was obtained by sequential ATRP of St and *t*BA, followed by the conversion of the terminal bromine functional groups at the P*t*BA blocks into azide functionalities. Afterward, the click reaction between azide-functionalized star-like PS-*b*-P*t*BA and ethynyl-terminated poly(3,4-ethylenedioxythiophene) (PEDOT) synthesized by the Grignard metathesis reaction yielded star-like β-CD-g-[PS-*b*-P*t*BA-*b*-PEDOT] containing the outer conductive PEDOT block. On the other hand, for some special species of polymers that cannot be prepared by single controlled/"living" radical polymerization method, such as β-CD-g-poly(acrylic acid)-*b*-poly(3-hexylthiopene) (β-CD-g-[PAA-*b*-P3HT]), poly(acrylic acid)-*b*-poly-(ethylene oxide) (β-CD-g-[PAA-*b*-PEO]) [32], poly(acrylic acid)-*b*-poly(vinylidene fluoride) (β-CD-g-[PAA-*b*-PVDF]), and poly(acrylic acid)-*b*-poly(3,4-ethylenedioxy-thiophene) (β-CD-g-[PAA-*b*-PEDOT]), a combination of ATRP and click reaction was then implemented. The precise control over the MW of each arm (size of the unimolecular micelles, including the inner and outer size), composition of each block of polymers (Function).

Miktoarm star-like copolymers with various arm numbers and molar ratios were developed by ATRP of two or more different species of premade linear macroinitiators in the presence of divinyl cross-linker (e.g. divinylbenzene [DVB], ethylene glycol diacrylate [EGDA], etc.). On the other hand, by coupling linear polymer arms with reactive end groups to a multifunctional coupling agent, star-like polymers can be derived. Furthermore, amphiphilic CD-cored block copolymers can be synthesized by the combination of click reaction and ATRP method. Clearly, the integration of controlled/"living" radical polymerization with click reaction represents a robust strategy to develop star-like polymers with desired compositions and functions that cannot be accessed solely by controlled/"living" radical polymerization. Nonetheless, the synthetic strategies briefly summarized here have their own advantages and limitations. As noted above, star-like block copolymers produced by the core-first strategy stand out as favorable and reliable polymeric nanoreactors for directing the growth of NPs, which will be detailed later.

Liu et al. synthesized 7-arm and 21-arm β-CD-cored star polymers by click reaction between alkyne-terminated poly(*N*-isopropylacrylamide) (PNIPAM) and azide-functionalized CDs. Alkyne-terminated PNIPAM was synthesized by ATRP of NIPAM monomers, while azide-functionalized β-CDs with 7 and 21 azide groups were prepared by selective modification of 7 primary hydroxyls and all 21 hydroxyls of β-CDs with iodine and BIBB, respectively. After the click reaction, the peak corresponding to azide groups of β-CD disappeared completely. By varying PD of linear alkyne-PNIPAM from 10 to 40, the MWs of 7-arm and 21-arm β-CD-cored star polymers were finely adjusted. And the arm-first method can also be used to prepare block copolymers by combining with controlled/"living" polymerization. Pang et al. prepared β-CD-g-(PAA-*b*-P3HT) [50], β-CD-g-(PAA-*b*-PVDF), β-CD-g-(PAA-*b*-PEG), etc. by synthesis of β-CD-g-P*t*BA-Br via ATRP, then replace the –Br by azide group, then by reacting with P3HT, PVDF, and PEG all with alkyne-terminated group, the second block polymer was prepared by arm-first method. As P3HT, PVDF, and PEG cannot be prepared via ATRP, thus, the block copolymers were unable to be prepared via sequential ATRP process. Thus, the combination of ATRP and click chemistry is a useful chemistry tool for synthesis of block copolymers that cannot be prepared via single ATRP or click chemistry method.

It should be noted that arm-first method requires the fabrication of linear functional polymers with reactive group at the end of the polymer chain. Afterward CD-cored star polymers can be obtained by coupling reactions between these linear polymers and CD-based derivatives. In this method, linear polymers can be characterized easily to obtain some information about the arms. In contrast, the characterization of the final complex CD-cored star-liked polymers is much more difficult. However, excess linear polymers should be added to ensure all the reactive sites of CDs are consumed and free linear polymers present in the final product must be removed to obtain pure CD-cored star polymers. The purification procedure can be tedious because the MWs of the linear polymers are usually high. Moreover, the distribution of arm numbers of obtained star polymers may not be uniform, due to unreacted sites of CD-based derivatives [47].

3.2.3 Synthesis of Bottle Brush-Liked Block Copolymer

Bottlebrush-liked polymers, also known as "molecular bottlebrushes," can be regarded as a class of cylindrical unimolecular supermoleculars consisting of a long polymer backbone with densely grafted side block copolymer chains. Due to the steric hindrance of heavily tethered side chains, the polymer backbone is forced to adopt a stretched, worm-like conformation. Interestingly, the extent of backbone stretching and the aspect ratio of the backbone/arm are dictated by the backbone length and grafting density of side chains. Such an architecture endows bottlebrush-liked polymers with a set of peculiar properties, like decreased chain entanglement, relatively rigid backbone, tailorable side chains functionalities, etc. over their linear polymer counterparts [3]. For example, the greatly reduced chain entanglement of bottlebrush-like polymers can be used as building blocks to self-assemble into complex architecture with large domain size for use in photonic crystals, lithographic patterning, etc. [28]. According to the composition and distribution of side chains, they can be categorized into bottlebrush-liked homopolymer, bottlebrush-liked random polymer, bottlebrush-liked diblock copolymers, and bottlebrush-liked triblock copolymers. Bottlebrush-liked polymers with well-defined compositions and functionalities are generally synthesized by grafting-from, grafting-onto, and grafting-through strategies. In this section, our focus is set on the synthesis of bottlebrush-liked polymers via the grafting-from strategy as the resulting unimolecular bottlebrushes function as the appropriate nanoreactors for precision synthesis of 1D NRs, core@shell NRs, nanotubes, and NWs.

As a polymer backbone, the three substitutable hydroxyl groups on each anhydroglucose unit of cellulose render the grafting of dense polymer side chains from the cellulose backbone. As such, heavily grafted polymer brushes together with the inherent rigidity of the cellulose backbone force the formed bottlebrush-like polymers to adopt a well-defined, straight, cylindrical conformation. Such architecture provides this type of bottlebrush-like polymers with some properties not accessible for flexible backbone analogs. In the grafting-from method, the shorter side chains are directly grown from the polymer backbone with predesigned and arranged initiator sites. Likewise, advanced polymerization techniques, including living anionic polymerization, ring-opening metathesis polymerization (ROMP), ATRP, RAFT, and NMP have been employed to construct bottlebrush-like polymers with good control over chemical compositions, MWs, and PDI for both backbone and side chains. The synthesis of polymer backbone with multi-sites for initiation is crucial for the grafting-from strategy. The key feature of the polymer backbone lies in the presence of reactive groups along the backbone, which can be further converted into initiator sites.

For example, poly(hydroxyethyl methacrylate) (PHEMA), with numerous reactive hydroxyl groups tethered to its main chain, is a widely used backbone for constructing unimolecular bottlebrushes micelles. When hydroxyl groups of PHEMA are transformed into ATRP initiators after the esterification reaction with BIBB, a commonly used macroinitiator, poly(2-(2-bromoisobutyryloxy) ethyl methacrylate) (PBIEM) is generated. It is notable that various bottlebrush-liked polymers have been synthesized via ATRP of different monomers based on the PHEMA backbone, including a bottlebrush-liked homopolymer, diblock copolymers, and triblock copolymers. For example, a bottlebrush-like homopolymer of PHEMA-g-PGMA was obtained by ATRP of glycidyl methacrylate (GMA) from the PHEMA backbone, which was further evolved into single molecular hybrid nanocylinders after the reaction between a monothiol-functionalized polyhedral silsesquioxane (POSS-SH) and epoxy groups of PGMA. Likewise, a bottlebrush-liked homopolymer of PHEMA-g-PDMAEMA was synthesized by ATRP of 2-(dimethylamino)-ethyl methacrylate (DMAEMA), followed by sequential quaternization with iodomethane.

In contrast to bottlebrush-liked homopolymers, bottlebrush-liked diblock copolymers are attractive when exploited as nanoreactors for crafting 1D NRs and NWs. Recently, a variety of bottlebrush-liked diblock copolymers were synthesized via the sequential ATRP of two different monomers grafted from PHEMA backbone. For instance, PHEMA was first synthesized by anionic polymerization of 2-(trimethysilyloxy)ethyl methacrylate (TMS-HEMA) and then converted into PBIEM by cleavage of trimethylsilyl groups and esterification with BIBB. Polymer side chains of P*t*BA-*b*-PnBA were anchored along the PHEMA backbone through sequential ATRP of *t*BA and nBA. Similarly, consecutively grafting the monomer pairs of *tert*-butyl methacrylate (*t*BMA) and oligo(ethylene glycol) methacrylate (OEGMA), *t*BA and DMAEMA, and HEMA and OEGMA by ATRP from the PHEMA backbone yielded bottlebrush-liked diblock copolymers of PHEMA-g-[P*t*BMA-*b*-POEGMA], PHEMA-g-[P*t*BA-*b*-PDMAEMA], and PHEMA-g-[PHEMA-*b*-POEGMA], respectively.

Furthermore, a stimuli-responsive bottlebrush-liked diblock copolymer of poly(hydroxyethyl methacrylate) grafted poly(3-acryloylpropyltrimethoxysilane)-*b*-poly(2-(dimethylamino)ethyl methacrylate) (denoted PHEMA-g-[PAPTS-*b*-PDMAEMA]) was synthesized by the similar method with a salt-sensitive outer block of PDMAEMA. By combining ROP of ε-caprolactone (CL) and ATRP of DMAEMA, bottlebrush-like PHEMA-g-[PCL-*b*-PDMAEMA] diblock copolymers with degradable inner PCL block were obtained. Specifically, the polyinitiator of PBIEM with 3200 initiating sites was first synthesized by ATRP and then employed to sequentially grow 3-acryloylpropyltrimethoxysilane (APTS) and OEGMA catalyzed by CuBr. The resulting bottlebrush-like diblock copolymer of PHEMA-g-[PAPTS-*b*-POEGMA] was used to produce a water-soluble silica NW by the hydrolysis and condensation of the trimethoxysilyl groups in the inner PAPTS block. It is interesting to note that, by changing the ATRP initiators from one-site to four-site during the synthesis of PHEMA backbone, a four-arm PHEMA backbone was derived, rendering the creation of a four-arm molecular bottlebrush with PAA-*b*-PS side chains.

Recently, brushes of PDMAEMA and PAA were grown from the cellulose backbone via ATRP of DMAEMA and *t*BA, respectively (i.e. bottlebrush-liked homopolymers). It is notable that a rich variety of bottlebrush-liked diblock and triblock copolymers based on the cellulose backbone were recently rationally designed and synthesized. The compositions and functionalities of the resulting cylindrical unimolecular bottlebrushes can be readily tuned by simply selecting the monomers of interest and varying the sequence of each block grown. Specifically, cellulose-based macroinitiators (i.e. cellulose-Br) of different lengths were first synthesized by a two-step esterification and purified by fractional precipitation [30]. Subsequently, different polymer side chains (i.e. brushes), including diblocks of P*t*BA-*b*-PS and P*t*BA-*b*-PEG, triblocks of P4VP-*b*-P*t*BA-*b*-PS, P4VP*b*-P*t*BA-*b*-PEG, PS-*b*-P*t*BA-*b*-PS, and PS-*b*-P*t*BA-b-PEG were then grafted from the cellulose backbone via sequential ATRP of respective monomers. Finally, after hydrolysis of the P*t*BA into PAA blocks, amphiphilic cellulose-g-[PAA-*b*-PS], cellulose-g-[P4VP-*b*-PAA-*b*-PS], cellulose-g-[PS-*b*-PAA-*b*-PS], and cellulose-g-[PS-*b*-PAA-*b*-PEG] were derived, and

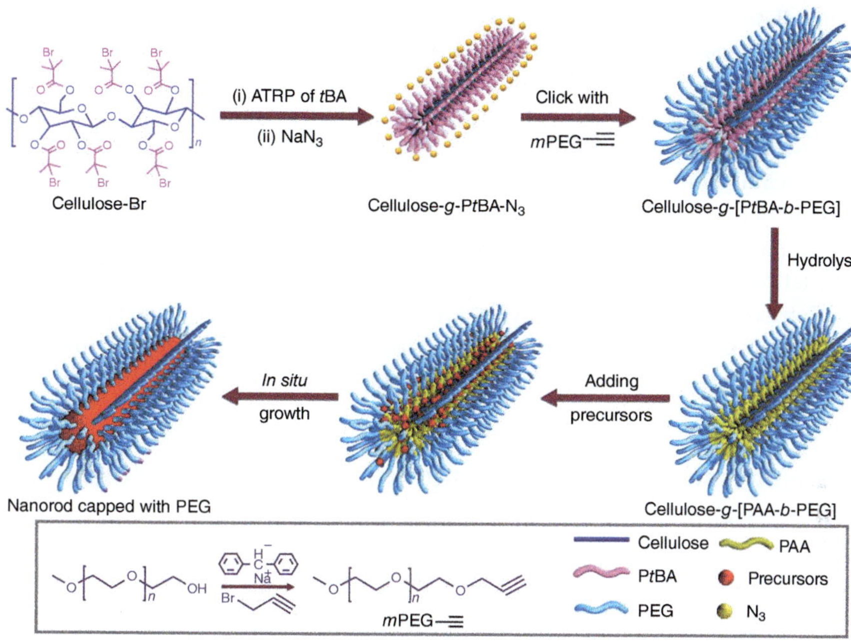

Figure 3.4 Schematic representation of synthesis of cellulose-*g*-[poly(acrylic acid)-*b*-poly(ethylene glycol)] (cellulose-*g*-[PAA-*b*-PEG]) with linear all hydrophilic PAA-*b*-PEG diblock copolymers as side chains and its use as a cylindrical nanoreactor for the preparation of PEG-capped plain 1D nanocrystals (nanorods).

double hydrophilic cellulose-g-[PAA-*b*-PEG] and triple-hydrophilic cellulose-g-[P4VP-*b*-PAA-*b*-PEG] can also be obtained. These unimolecular bottlebrush-liked micelles can be employed as nanoreactors for crafting 1D NRs of different architectures (plain, core/shell, and hollow), as depicted in Figure 3.4.

We note that similar to star-like polymers, the MWs and compositions of polymer side chains in bottlebrush-like block copolymers can be readily controlled by varying the polymerization time and using disparate monomers of interest. As discussed above, controlled/"living" radical polymerizations and/or ROP are frequently used in the grafting-from strategy, enabling the control over chemical compositions, MWs, PDI, and functionalities of grown side chains. Nevertheless, owing to the steric hindrance of massively grafted side chains along the backbone, the density of the initiator sites to be grafted is low. In contrast, in the grafting-onto approach, various efficient coupling reactions (e.g. esterification reactions, thiol-epoxy coupling, and CuAAC reaction) are utilized to anchor presynthesized side chains onto the backbone with higher grafting density. Notable, it is also convenient to graft functional side chains (e.g. conjugated polymers; PEDOT and P3HT) onto the backbone via CuAAC. Finally, for the grafting-through method, bottlebrush-like block copolymers with different blocks in the backbone and compartmentalized brushes attached to the same backbone can be attained by ROMP, controlled/"living" radical polymerization (i.e. ATRP), and ROP of the corresponding macromonomers.

It is also feasible to tune the length of side chains and the grafting density by regulating the respective MWs of side chains and backbone, respectively.

Similar to a star polymer, a bottlebrush-like polymer has radiating linear "arms" from a linear core instead of a compact core of a star polymer. Bottlebrush polymers show high density of side chains and robust 3D architectures because of one or more side chains radiating from every repeating unit in the linear core. The bottlebrush polymer is also a powerful nanoreactor for nanofabrication, which is similar to the star polymer. Due to the novel topology, the bottlebrush polymer can template 1D NRs fabrication, such as NRs and nanotubes, which have shown myriad applications, such as catalysis, electronics, photonics, sensing, and fundamental research in crystallization kinetics of NRs. The synthesis of NRs templated from unimolecular block copolymer nanoreactors will be presented in the next section.

3.3 Synthesis of Monodispersed Nanoparticles via Block Copolymer Unimolecular Micelles Nanoreactors

Colloidal nanocrystals exhibit a wide range of size- and shape-dependent properties and have found application in myriad fields, including optics, electronics, mechanics, drug delivery, and catalysis, to name but a few. Synthetic protocols that enable the simple and convenient production of colloidal nanocrystals with controlled size, shape, and composition are therefore of key general importance. In this section, unimolecular block copolymer micelles nanoreactors methods to precisely controlled synthesis of inorganic NPs are summarized.

3.3.1 Star-Like Block Copolymers as Unimolecular Nanoreactors

3.3.1.1 Plain Nanoparticles

Due to their intriguing optical, electronic, and catalytic properties, inorganic metal and semiconductor NPs hold great promise in many applications ranging from optoelectronics and sensors to catalysis and medicine. Since these unique properties are closely related to the dimension of inorganic NPs, it is of critical importance to precisely control the size, shape, and polydispersity of these nanostructure materials. To date, many synthetic approaches have been developed for the preparation of various types of metal and semiconductor NPs with desired shapes, compositions, and morphologies. Among these methods, unimolecular micelles-directed template-synthesis of inorganic NPs is a relatively new approach but has become an active area of research due to the easy control in the size and composition of the materials. This section will summarize the recent progress in the synthesis of inorganic NPs by using unimolecular micelles as templates.

Compared with conventional methods, the nanomaterials obtained within the amphiphilic star-liked copolymer nanoreactors are tightly and permanently covered by a layer of polymer chains, which results in good solubility of NPs in nonpolar solvents with prolonged stability. The size and composition of nanomaterials can be easily controlled by manipulating the molecular weight of the polymer template and selecting appropriate precursors, respectively. As mentioned above, the metal precursors occupied the specific domain of the unimolecular nanoreactors via strong interactions, such as coordination and electrostatic force, and subsequent pyrolysis or reduction to form functional inorganic NPs inheriting the nanostructures from the unimolecular nanoreactors in a precisely controlled way. If there is no strong interaction between metal precursors and unimolecular nanoreactors, irregular inorganic NPs without expected nanostructures will form. Notably, the outer domains of the unimolecular nanoreactors are intimately and permanently tethered on the obtained NPs surface to provide stability, solubility, biocompatibility, responsivity, assembly, reactivity, and valences of the resulting nanohybrids. The integrated organic–inorganic hybrid NPs exhibit varied properties and broad applications.

Recently, Minko and coworkers reported a unimolecular micelle system constituted by the star-shaped PS_7-$P2VP_7$ polystyrene/poly(2-vinylpyridine) block copolymer and applied it for the deposition of nanosized palladium clusters. Unimolecular micelles with core@shell structures were obtained with noticeable segregation into the collapsed core and the extended shell formed by stretched polymer arms of P2VP and polystyrene (PS) in water and toluene, respectively. Metallization of P2VP arms of the copolymer could result in the localization of 1–3 nm palladium clusters in the outer shell of unimolecular micelles, forming star-like structures with metallized arms. With the presence of palladium clusters, the obtained organic–inorganic nanocomposite exhibited significantly improved contrast for AFM imaging. In a similar strategy, the core PEGDMA-(P2VP-*b*-PS)$_n$ amphiphilic star copolymer was also applied as a unimolecular template for the synthesis of Au NPs in toluene. In this regard, the Au3+ precursors would selectively coordinate with the inner blocks of P2VP in the star copolymers to confine the formation of Au NPs within the polymer interior. When the reducing agent N_2H_4 was added into the solution, Au NPs were thus deposited within the micelle core. The outer PS shell provided protection of the synthesized Au NP and enabled the easy dispersion of the particles in toluene. When core PDVB-(PAA-*b*-PMMA)$_n$ (DVB = divinylbenzene) polymers were used as unimolecular templates, the selective binding between carboxylic acid groups in the inner PAA blocks and Ag cations was also successfully demonstrated for the synthesis of Ag NPs.

Taking advantage of recent progress in polymer chemistry, Pang et al. prepared star-liked macromolecules by growing block copolymer arms from a β-cyclodextrin core (Figure 3.5). The β-CD-based star polymers used for nanofabrication were prepared through the core-first approach via successive ATRP steps. β-CD equipped with 21 bromides, 21Br-β-CD, was synthesized and employed as an ATRP macroinitiator, resulting in star-like polymers with well-defined and compact architectures, controlled molecular weights and distributions, and regulatable ratios of two or more radiating blocks. Finally, the obtained star polymers were utilized as nanoreactors to fabricate NPs resulting from the strong coordination interaction between the specific polymer blocks and metal precursors. A carefully chosen solvent mixture ensures that the inorganic precursor selectively accumulates in the hydrophilic core of these unimolecular micelles. Subsequent condensation at elevated temperatures produces single-crystalline NPs that are the same size as the

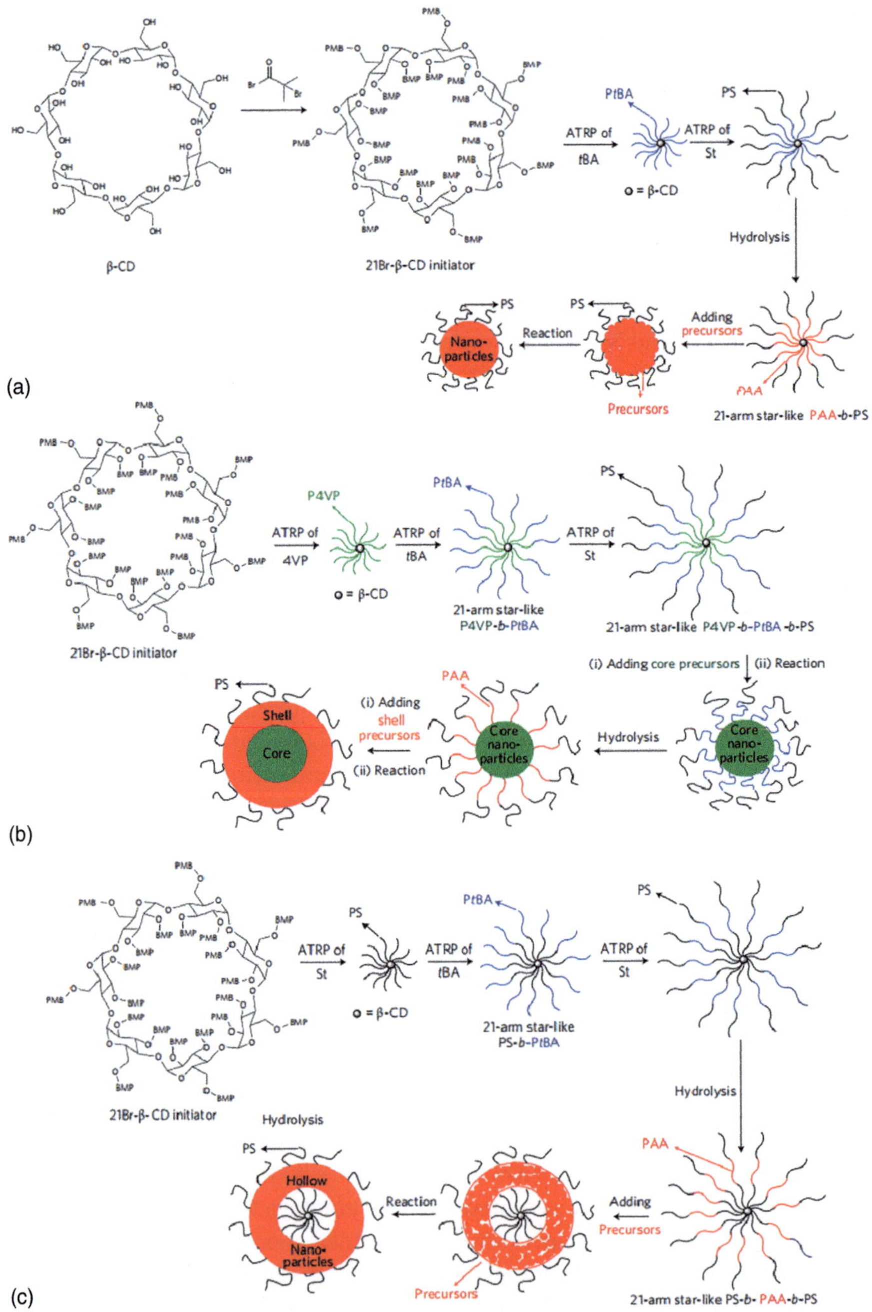

Figure 3.5 Schematic representation of synthetic strategies for nanoparticles with different architectures (plain, core@shell, and hollow) using amphiphilic star-like block copolymers as nanoreactors. (a–c), Formation of plain nanoparticles (a), core@shell nanoparticles (b), and hollow nanoparticles (c). CD, cyclodextrin; BMP, 2-bromo-2-methylpropionate; St, styrene. Source: Pang et al. [32]/Springer Nature.

micelle core. The most remarkable feature of this approach is its generality. Nanostructures of just about any chemical composition can be prepared; examples include metallic, semiconductor, magnetic, ferroelectric, and luminescent single-crystalline NPs. This level of versatility is unprecedented in the literature and illustrates the robustness of unimolecular micelles during the growth of an inorganic crystal inside their core. Although the exact nanocrystal growth mechanism is still not well understood, high-resolution transmission electron microscopy images convincingly demonstrate that the final material is not polycrystalline.

With this method, not only compact NPs, but core@shell and hollow NPs could also be fabricated. Taking core@shell NPs as an example, β-CD-based star polymers armed with triblock copolymers, P4VP-*b*-P*t*BA-*b*-PS, have been utilized as unimolecular nanoreactors. The core metal precursor can be occupied in the P4VP block via strong interaction (coordination or electrostatic interaction) and then be reduced to form the core. After hydrolysis of the P*t*BA block to the PAA block, the shell metal precursor can be further occupied in the PAA domain via strong interaction and then be reduced to create the shell. The sizes of the core and shell in resulting NPs could be tailored by adjusting polymerization time during the ATRP process. The surface-tethered PS block, which can be designed as other polymers, could provide intimately and permanently surface features of the obtained NPs. Furthermore, the capped polymer tethered to the surface of the NPs eliminates the aggregation of NPs and ensures the solution' stability and dispersity in polar or nonpolar solvents. Thus, polymer layer-capped NPs are fabricated successfully in a controlled way.

The star polymer as the unimolecular nanoreactor for nanofabrication is a general and robust approach for the controlled and effective preparation of nearly uniform NPs with regulated architectures, compositions, sizes, and surfaces. With this method, NPs with various compositions, such as noble metals, metal oxides, semiconductors, silica, and composites can be obtained by choosing appropriate precursors and subsequent growth conditions. A β-CD-based star polymer armed with triblock copolymer, PS-*b*-P*t*BA-*b*-PS, was developed to craft colloidal Au or Ag hollow NPs. The internal cavity diameter and intermediate shell thickness in the obtained hollow NPs could be precisely tailored by controlling the degrees of polymerization of internal PS and intermediate P*t*BA blocks, respectively, during the star polymer preparation. It is worth mentioning that the plasmonic properties of these hollow NPs could be easily tailored by altering the outer shell diameter or shell thickness of the hollow NPs, which was verified well with simulations. Compared to those of the compact counterparts with the same outer diameter, the plasmonic peaks of the hollow NPs exhibited increasing redshift with decreased shell thicknesses.

Zhang et al. recently reported multifunctional polymer unimolecular micelles, which are used as templates to fabricate stable gold NPs in a one-step reaction without the addition of external reductants. In this study, the unimolecular micelles nanoreactors were made from the 21-arm star-like block copolymer β-cyclodextrin-g-(poly(lactide))-*b*-poly(2-(dimethylamino)ethyl methacrylate)-*b*-poly[oligo(2-ethyl-2-oxazoline)methacrylate]$_{21}$ (β-CD-(PLA-*b*-PDMAEMA-*b*-PEtOxMA)$_{21}$), in which both β-CD and PLA formed the hydrophobic core of the unimolecular micelles and

PEtOxMA with short side chains formed the shell. The tertiary amine groups of the PDMAEMA block could act as a reducing agent to reduce the $AuCl_4^-$ precursor to zero-valent gold in aqueous solution, and these gold atoms combined mutually to form the final Au NPs. It showed that the sizes and morphologies of the gold NPs were well controlled by adjusting the PDMAEMA length and the concentrations of the star-like polymer and $HAuCl_4$. Significantly different from previous methods, the *in situ* generation of Au NPs in this study could avoid the use of organic solvents and other reducing reagents. Together with the stabilization behaviors and low cytotoxicity, the gold NPs developed in this study could adapt to further biological applications.

Also, the confinement of the nanoreactors can be used to synthesize NPs. Yan et al. reported the synthesis of a well-defined hollow polymer NP derived from star-shaped unimolecular micelles. β-Cyclodextrin was first applied as an efficient macroinitiator to prepare a star-shaped PCL via ROP (Figure 3.6). Then, the star-shaped PCL was modified to be a macro-RAFT agent for photoinduced electron/energy transfer-reversible addition-fragmentation chain transfer (PET-RAFT) polymerization of S-Cl monomers. The prepared unimolecular micelles can be photo-cross-linked under UV irradiation after a simple nucleophilic substitution reaction, which made –Cl groups to be $-N_3$ groups. After the selective removal of the PCL core, hollow polymer NPs were achieved and exhibited to be a general nanoreactor strategy for the fabrication of nanocrystals with well-controlled architectures. Compared with unimolecular micelle templates, the nanocrystals prepared by hollow templates are absolutely pure, as no polymer chains are embedded in the inorganic nanocrystals. In addition, by changing the concentration of the precursor,

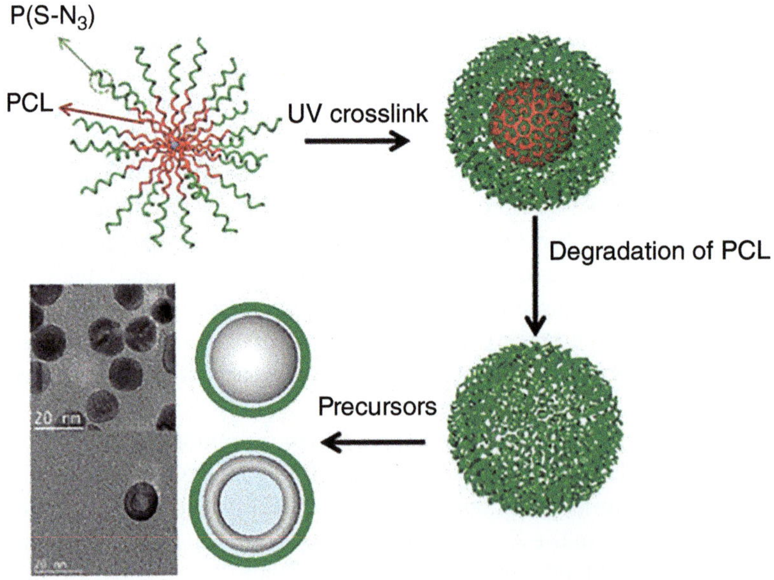

Figure 3.6 Scheme illustration of the synthesis of nanoparticles via the star-liked polymers nanoreactors.

the structure of the nanocrystal can be changed from a normal spherical structure to a hollow structure.

To date, only a few multifunctional initiators have been investigated for crafting unimolecular micelles. Silver and platinum NPs were made via dendrimer-initiated unimolecular micelles. PbS NPs were grown from unimolecular micelles grafting from cross-linked microgel cores. Gold NPs were prepared from PS-*b*-P2VP star-like block copolymer micelles using anionic polymerization coupled with ethylene glycol dimethacrylate. Palladium NPs were generated from amphiphilic hyperbranched polyglycerols and star-shaped polystyrene-*b*-poly(2-vinylpyridine) block copolymer micelles. However, none of these studies can provide monodisperse inorganic NPs. One reason is that the synthesized NPs may aggregate if they are not well encapsulated by the outer block polymers. Moreover, in a selective solvent, the amphiphilic block copolymer chains of star-like unimolecular micelles may bunch together and further assemble into bigger micelles, similar to the self-assembled micelles of linear amphiphilic block copolymers, resulting in polydisperse micelles. The terminal bromines on β-CD-based star polymers after ATRP can be converted to azide groups, which can be modified with acetylene-terminated polymers (PEDOT, PVDF, and so on) via the effective click reaction. Notably, ferroelectric PVDF-capped ferroelectric $BaTiO_3$ fabricated within star polymers with uniform and stable nanocomposites were promising when used in electronics, catalysis, energy conversion and storage, and biotechnology.

The tethered polymer on the obtained NPs can also exhibit reactivity or responsivity and further affect the assembly behavior and properties. Au NPs can be fabricated within the β-CD-based star polymer armed with PAA-*b*-poly(7-methylacryloyloxy-4-methylcoumarin), PAA-*b*-PMAMC (Figure 3.7). The coumarin unit in PMAMC tethered on the surface of Au NPs showed photo-cross-linking and photocleavage behaviors and then induced the reversible assembly and disassembly of the Au NPs triggered through light. Under irradiation with 365 nm light, the coumarin units underwent a photodimerization process, resulting in gradually formed Au NP aggregates with irregular shape and size. Remarkably, using 254 nm light as a trigger, the aggregates gradually disassembled and finally went back to the isolated NP state with their original shape and size. The plasmonic absorptions of the assembled and isolated Au NPs were also tuned reversibly under different light irradiations. It is also worth noting that functional polymer can be grown outside as the shell of the nanocrystal when diblock and triblock star-like copolymers are employed to yield inorganic nanocrystals with the potential for further chemical reaction too. Thermo-responsive PNIPAM-capped Au NPs can induce redshift and increase the intensity of plasmonic peak through increasing temperature because of the collapse of PNIPAM on the surface of Au NPs restricting the assembly of Au NPs. Interestingly, when adding linear PNIPAM, the Au NPs assembled to form aggregates through increasing temperature, and the plasmonic peak showed greater redshift and decreased intensity. All the above thermally triggered processes were reversible. Markedly, the absence and presence of free PNIPAM induced the switchable catalytic and optical performance of PNIPAM-capped Au NPs.

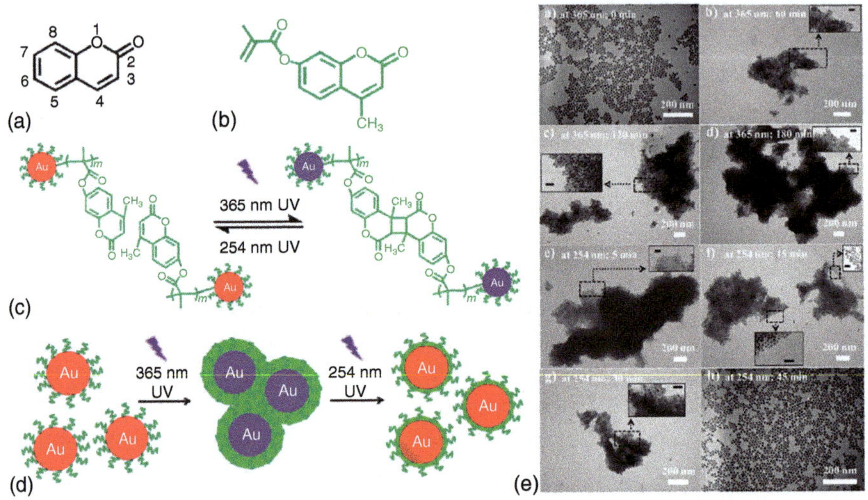

Figure 3.7 Light-enabled reversible self-assembly of Au NPs intimately and permanently capped with photoresponsive polymer chains. Chemical structures of (A) coumarin and (B) MAMC. (C) Schematic illustration of photodimerization (photo-cross-linking) and photocleavage behaviors of PMAMC chains situated on the surface of Au NPs when subjected to 365 and 254 nm UV light irradiation, respectively. (D) Stepwise representation of light-enabled self-assembly and disassembly of PMAMC-capped Au NPs upon 365 and 254 nm UV light irradiation, respectively. (E) TEM images of PMAMC-capped Au NPs after the PMAMC-capped Au NPs CH_2Cl_2 solution was exposed to (a–d) 365-nm UV light and (e, f) 254-nm UV light for a given time.

The unimolecular micelles formed from a new class of copolymers, such as 21-arm β-CD-P4VP-*b*-PAA-*b*-PS, β-CD-PAA-*b*-PS, and β-CD-PAA-*b*-PEO were structurally stable and can overcome the intrinsic instability of linear block copolymer micelles. The permanent connection between the NPs and the respective hydrophobic or hydrophilic polymer chains rendered these polymers soluble in either organic or aqueous environments, respectively. Interestingly, a mixed solvent was used in this work to facilitate the encapsulation of inorganic precursors, resulting in spherically-shaped inorganic nanocrystals. This could further facilitate the easy synthesis of various sizes and architectures of metallic, ferroelectric, magnetic, semiconductor, and luminescent colloidal NPs. The improved size control afforded by star-liked block copolymer nanoreactors has enabled a series of investigations into their specific size-dependent properties. The size-dependent properties reviewed include the surface plasmon resonance (SPR) absorbance of Au NPs and magnetic properties of Fe_3O_4. Nanocomposites with advanced dielectric and ferroelectric properties for applications, such as catalysis, solar cells, and lithium-ion batteries (LIBs), were also detailed, with each property benefiting greatly from the well-defined nanocomposite structures.

3.3.1.2 Core@Shell Nanoparticles

As an important class of nanomaterials, core@shell NPs that integrate two dissimilar materials with distinct functionalities have garnered considerable attention as they enable the synergistic coupling of the two constituents and thus offer opportunities for exploring the intriguing properties of their interface [63–66]. To effectively produce high-quality core@shell NPs, a moderate lattice mismatch between the two different materials is often required. Notably, such conditioned material combinations significantly limit the core and shell material choices. Clearly, it is of great interest to develop alternative means of forming core@shell NPs with a large lattice mismatch. Moreover, it is highly desirable to be capable of precisely tailoring the diameter of the core and the thickness of the shell, as well as the surface properties of the core@shell NPs.

Star-like amphiphilic triblock copolymers were rationally designed and synthesized by combining two sequential ATRP reactions with a click reaction (Figure 3.8) [32]. Subsequently, a family of uniform magnetic@plasmonic core@shell NPs was crafted by capitalizing on these triblock copolymers as nanoreactors. The diameter of the magnetic core and the thickness of the plasmonic shell could be independently and accurately controlled by varying the molecular weights of the

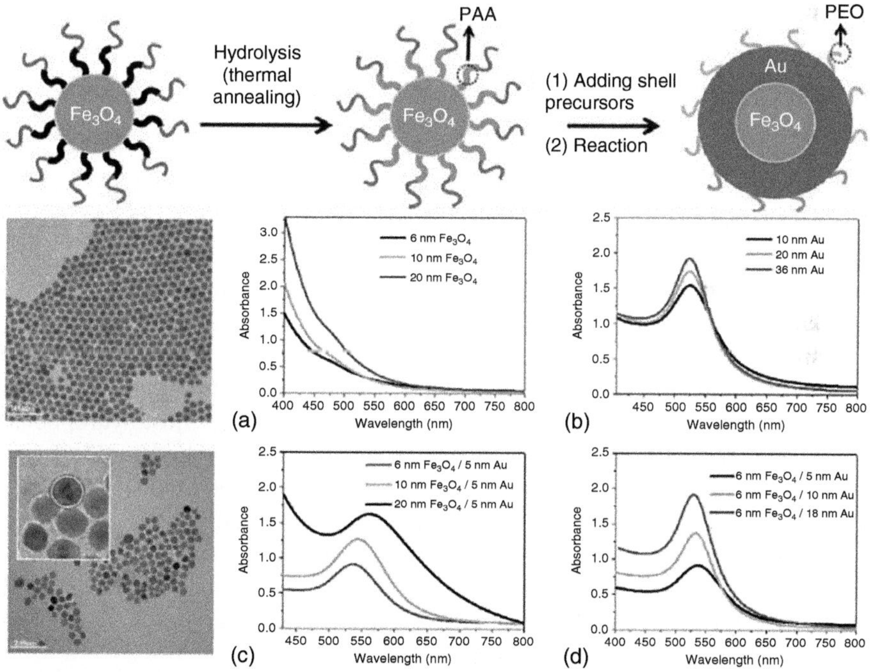

Figure 3.8 Synthetic strategy for magnetic/plasmonic Fe_3O_4@Au core@shell nanoparticles based on star-like P4VP-*b*-P*t*BA-*b*-PEO triblock copolymers as nanoreactors. Formation of magnetic Fe_3O_4 cores capped with P*t*BA-*b*-PEO. (d) Formation of water-soluble PEO-capped Fe_3O_4@Au core@shell nanoparticles. TEM images of PEO-capped Fe_3O_4@Au core/shell nanoparticles with various dimensions. Visible-light absorption spectra of various nanoparticles. (a) Pure Fe_3O_4 nanoparticles with diameters of 6, 10, and 20 nm. (b) Pure Au nanoparticles with diameters of 10, 20, and 36 nm. (c) Fe_3O_4@Au core@shell nanoparticles with Fe_3O_4 core diameters of 6, 10, and 20 nm and a fixed Au shell thickness of 5 nm. (d) Fe_3O_4@Au core@shell nanoparticles with Au shell thicknesses of 5, 10, and 18 nm and a fixed Fe_3O_4 core diameter of 6 nm.

inner and intermediate blocks of the star-like triblock copolymers, respectively. The surface plasmonic absorption of core@shell NPs with different core diameters and shell thicknesses was systematically studied and theoretically modeled. This robust strategy provides easy access to a large variety of multifunctional NPs with large lattice mismatches for use in optics, optoelectronics, catalysis, or bioimaging.

ATRP of 4-vinylpyridine, *t*-butyl acrylate, and styrene in sequential order from a β-cyclodextrin core yielded an amphiphilic star-like triblock copolymer, poly(4-vinylpyridine)-*b*-poly(*t*-butyl acrylate)-*b*-polystyrene (P4VP-*b*-P*t*BA-*b*-PS). Subsequently, star-like triblock copolymer was composed of inner hydrophilic P4VP blocks, central hydrophobic P*t*BA blocks, and outer hydrophobic PS blocks with well-defined molecular architecture, and molecular weight of each block was judiciously exploited as nanoreactor for synthesis of precisely shaped, hairy plasmonic@semiconductor Au@TiO_2 core@shell NPs. The resulting Au@TiO_2 NPs were intimately and permanently tethered with outer PS chains that enabled the superior solubility of NPs in nonpolar solvents. The PS chains on the surface of these bifunctional NPs were carbonized by annealing in an inert atmosphere. In comparison to a widely used TiO_2 network film (i.e. P25)-based device, dye-sensitized solar cells (DSSCs) assembled by incorporating a thin layer of carbonized Au/TiO_2 NPs on the top of P25 film as photoanode exhibited largely improved short-circuit current density, JSC (18.4% increase), and power conversion efficiency, PCE (13.6% increase), respectively. Such improvements were attributed to the surface plasmon-enabled light harvesting enhancement of Au core and fast electron transport promoted by the carbon layer coating on Au/TiO_2 NPs.

Another noteworthy feature of this technique is that both core@shell nanocrystals can be synthesized. Lin and coworkers prepared star-liked polymers with triblock copolymer arms containing two different hydrophobic blocks in the shell and a hydrophilic block as the core [32]. After the growth of the first inorganic nanocrystal in the core, the structural integrity of these unimolecular micelles is retained, allowing for a subsequent hydrolysis step of the innermost hydrophobic shell and selective accumulation of a second inorganic precursor in the midsection. As a result, Lin et al. were able to prepare unusual core@shell heterostructures, such as Fe_3O_4@$PbTiO_3$ core@shell particles, which would be difficult to synthesize by conventional epitaxial growth because of the large lattice mismatch (nearly 20%) between the two phases. Furthermore, the chemical composition of the shell and the core can be inverted, as demonstrated by the preparation of well-defined Au@CdSe and CdSe@Au core@shell NPs.

High-quality perovskite NPs have witnessed rapid advances over the past few years, however, the stability of perovskite NPs is a critical factor that promotes further progress in optoelectronic fields. Using β-CD-based star polymers armed with triblock copolymers, P4VP-*b*-P*t*BA-*b*-PEO or P4VP-*b*-P*t*BA-*b*-PS, as unimolecular nanoreactors, a methylammonium lead bromide ($MAPbBr_3$) core was prepared within the P4VP block, and the SiO_2 shell was formed via hydrolyzing the tetramethyl orthosilicate occupying the PAA block pyrolyzed from the P*t*BA block. The delicately designed star polymers as nanoreactors can template the controlled fabrication of core@shell NPs with tailored perovskite diameter, SiO_2 thickness, and tethered polymers. The PS- or PEO-tethered perovskite@SiO_2 core@shell nanocomposite exhibited improved stabilities (colloidal stability, water stability, and photostability) and appealing solution processability. Uniform organo–silica hybrid NPs with compact and hollow architectures were also crafted within β-CD-based star polymers armed with a Si-containing homopolymer and a diblock copolymer with a Si-containing block as the outer layer, respectively. β-CD-based star polymers can also be used as nanoreactors for organic NP fabrication, including conjugated or nonconjugated polymer NPs with compact or hollow architectures via the cross-linking process. This class of soft NPs might be developed as a new branch of nanocarriers and drug nanovehicles. Not only can the hydrophobic and hydrophilic features of templated NPs be adjusted as PS- and poly(ethylene oxide) (PEO)-capped perovskite@SiO_2 core@shell nanocomposites mentioned above, because of the intimate and permanent connection between the outer block and the obtained NPs, the stability, dispersity, reactivity, and responsivity of the formed NPs can also be controlled by choosing an appropriate polymerization approach and component.

3.3.1.3 Hollow Nanoparticles

The ability to tailor the size and shape of NPs enables the investigation into the correlation between these parameters and optical, optoelectronic, electrical, magnetic, and catalytic properties. Despite several effective approaches available to synthesize NPs with a hollow interior, it remains challenging to have a general strategy for creating a wide diversity of high-quality hollow NPs with different dimensions and compositions on demand. To further demonstrate the universal nature of this approach, the researchers synthesized a number of hollow nanocrystals by using star-liked molecules containing a hydrophilic mid-shell between the hydrophobic core and the outer shell. The overall size and the thickness of the resulting inorganic shell are controlled by changing the degree of polymerization in the first two blocks of the arms, as demonstrated by electron microscopy images of hollow Au NPs and Cu_2O nanocrystals. All these combinations demonstrate that the covalent nature of star-like amphiphilic polymers offers an opportunity for robust nanoreactors for the templated synthesis of inorganic nanocrystals.

Chen et al. reported a general and robust strategy for *in situ* crafting of monodisperse hairy hollow noble metal NPs by capitalizing on rationally designed amphiphilic star-like triblock copolymers as nanoreactors (Figure 3.9). The intermediate blocks of star-like triblock copolymers can associate with metal precursors via strong interaction (i.e. direct coordination or electrostatic interaction), followed by reduction to yield hollow noble metal NPs. Notably, the outer blocks of star-like triblock copolymers function as ligands that intimately and permanently passivate the surface of hollow noble metal NPs (i.e. forming hairy permanently ligated hollow NPs with superior solubility in nonpolar solvents). More importantly, the diameter of the hollow interior and the thickness of the shell of NPs can be readily controlled. As such, the dimension-dependent optical properties of hollow NPs are scrutinized by combining experimental studies and theoretical modeling. The dye encapsulation/release studies indicated that hollow NPs may be utilized as

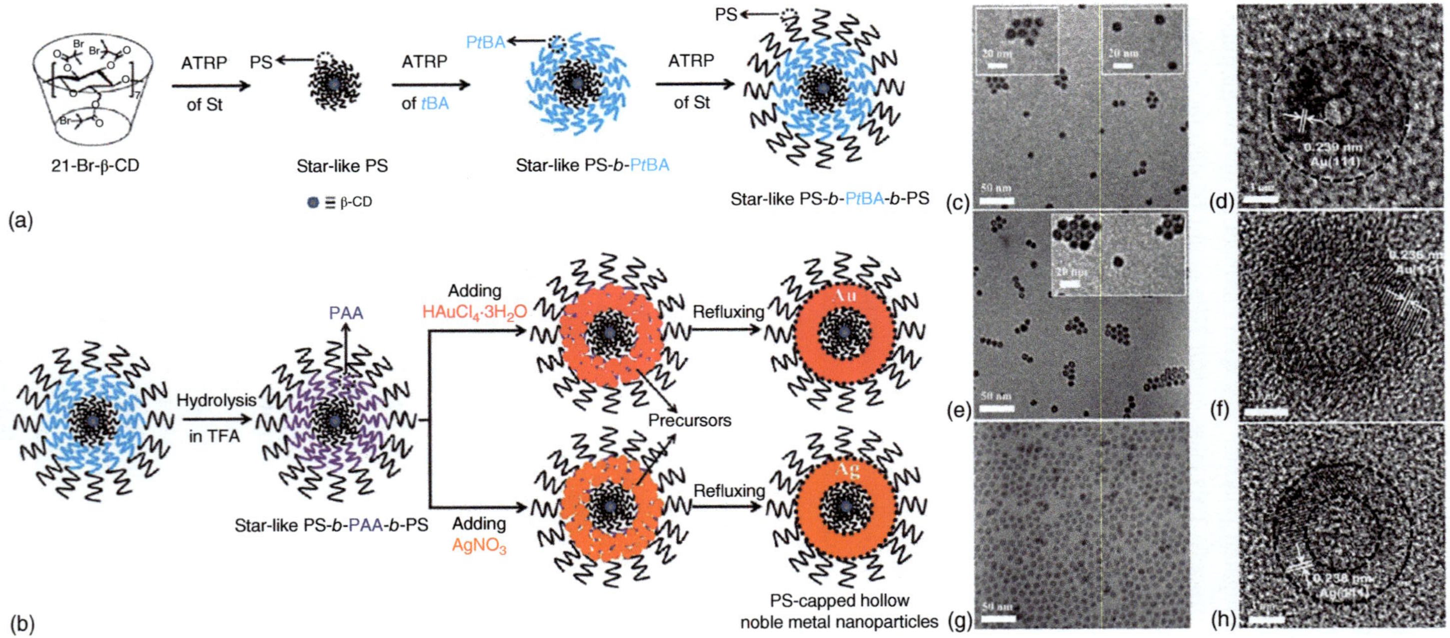

Figure 3.9 Synthetic route to hairy hollow plasmonic nanoparticles (i.e. PS-capped hollow Au and Ag nanoparticles [NPs]) crafted by capitalizing on amphiphilic star-like PS-*b*-PAA-*b*-PS triblock copolymers as nanoreactors with synthesis of (a) Star-like PS-*b*-P*t*BA-*b*-PS and (b) PS-capped hollow plasmonic NPs; (c, e, g) TEM and (d, f, h) HRTEM images of PS-capped hollow noble metal NPs crafted by employing amphiphilic star-like PS-*b*-PAA-*b*-PS triblock copolymers as nanoreactors.

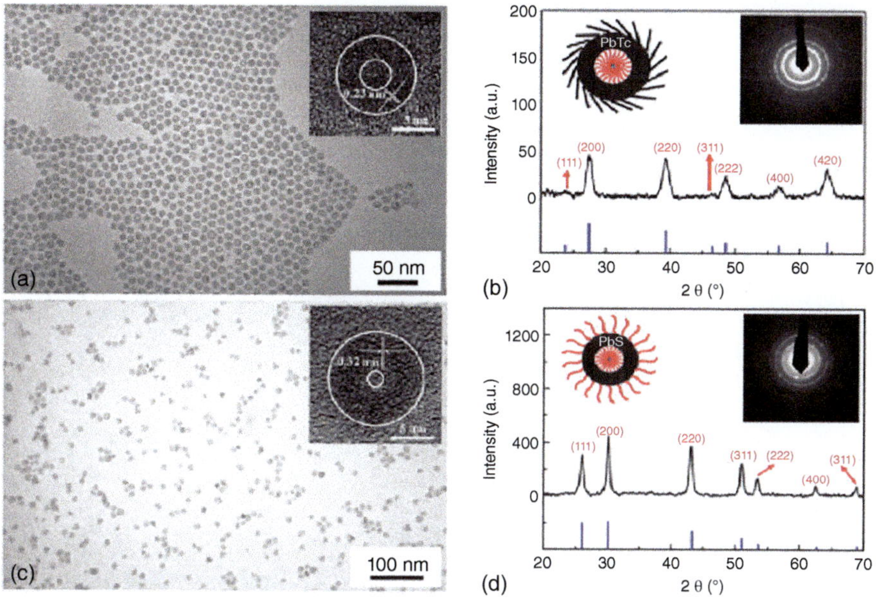

Figure 3.10 (a, c) TEM image and (b, d) XRD profile of PbTe HNPs capped with conjugated PEDOT crafted by employing star-like PS-*b*-PAA-*b*-PEDOT as nanoreactors.

attractive guest molecule nanocarriers. As the diversity of precursors is amenable to this star-like triblock copolymer nanoreactor strategy, it can conceptually be extended to produce a rich variety of hairy hollow NPs with different dimensions and functionalities for applications in catalysis, water purification, optical devices, lightweight fillers, and energy conversion and storage.

Conjugated-polymer-capped Pb chalcogenide NPs with compact and hollow architectures were fabricated within β-CD-based star polymers armed with PAA-*b*-PEDOT and PS-*b*-PAA-*b*-PEDOT copolymers (Figure 3.10). PEDOT-capped PbTe NPs with long-term stability and dispersion were promising in thermoelectric materials after doping with polystyrene sulfonate, which could function as heat-to-electricity converters and provide an alternative route for power generation and refrigeration. PEDOT-capped PbS and PbTe hollow NPs showed an absorption maximum blueshift fitted with theoretical predictions.

3.3.1.4 Nanoring

Moreover, this process can be further generalized to construct nanoring nanocrystal morphologies by varying the type, number, and length of copolymer arms. Yin et al. report the design and use of a novel class of dendritic amphiphilic core-double-shell macromolecules as templates for the fabrication of well-organized, nanoporous thin TiO_2 films (Figure 3.11) [67]. The macromolecules consist of a central rigid polyphenylene dendrimer core and a flexible amphiphilic double-polymer shell with twelve arms, in which a PS block serves as the hydrophobic segment and PAA forms the hydrophilic moiety. The shape-persistent second-generation polyphenylene

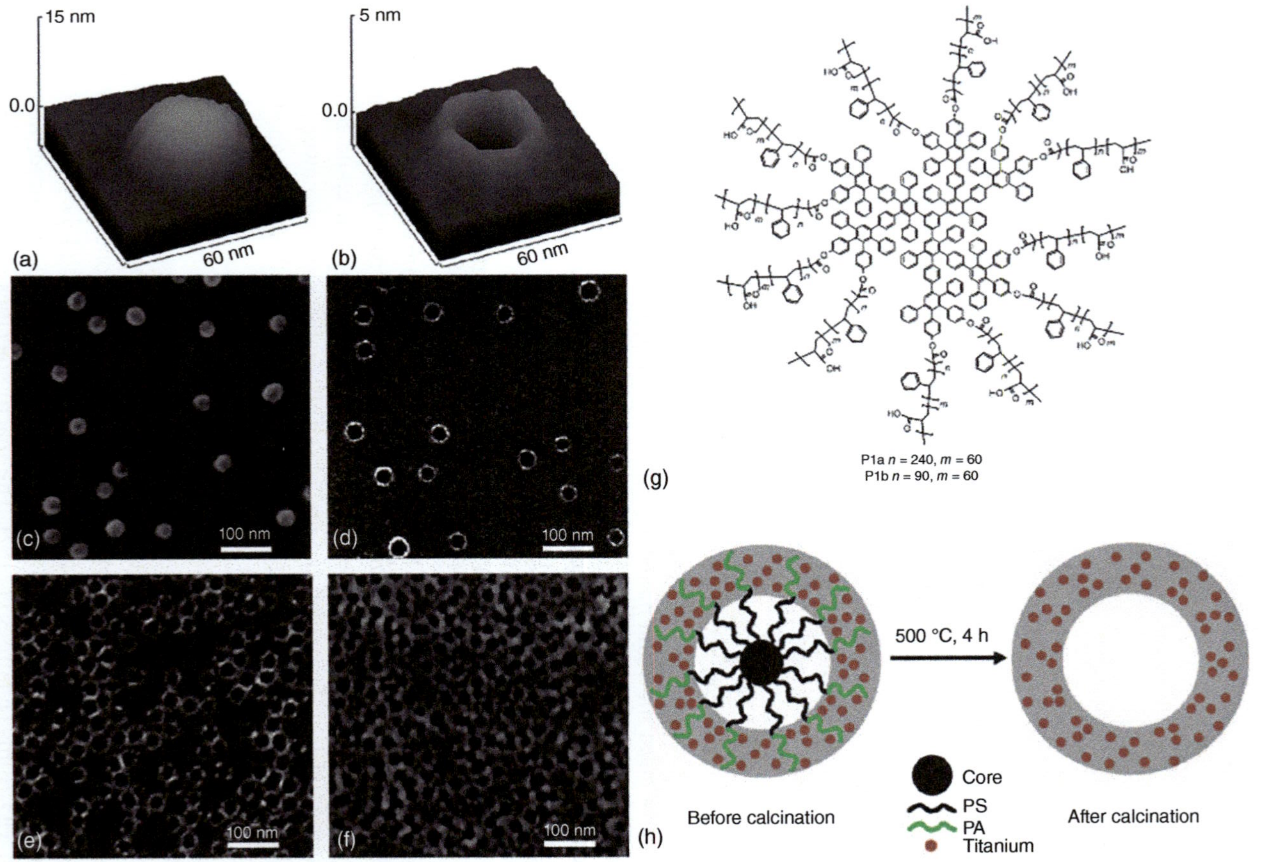

Figure 3.11 (a) Plot surface of AFM image of TiO_2 templated by P1. (a) before and (b) after calcination. SEM image of TiO_2 templated by (c) P1 in DMF before calcination; (d) P1 in DMF after calcination; (e) P1 in DMF after calcination; (f) P1 in DMF after calcination; (g) P1 in DMF after calcination; (h) P2 in DMF after calcination. (g) Chemical structural formula of dendritic amphiphilic diblock copolymers; Schematic illustration of the TiO_2 cyclization process templated by amphiphilic core-double-shell macromolecules; (h) Dendritic amphiphilic core-double-shell macromolecules serve as templates to direct the morphogenesis of crystal TiO_2

dendrimer core was selected as a scaffold as it provides a globular shape, a perfectly branched structure, and the availability of a defined number of functional groups at the surface that can undergo a further "grafting-from" procedure. The target of our research was the reproducible formation of TiO_2 thin films with specific morphologies by the systematic adjustment of the length of hydrophobic PS blocks in these dendritic amphiphilic core-double-shell macromolecules.

Liang et al. reported a facile and efficient synthesis method for preparing anatase titanium dioxide (TiO_2) nanorings based on multiarm, star-liked amphiphilic polystyrene-*b*-poly(acrylic acid) (PS-*b*-PAA) diblock copolymers as nanoreactors, which were prepared via a sequential ATRP technique followed by the conversion of polystyrene-*b*-poly(*tert*-butyl acrylate) (PS-*b*-P*t*BA) to PS-*b*-PAA (Figure 3.12) [68].

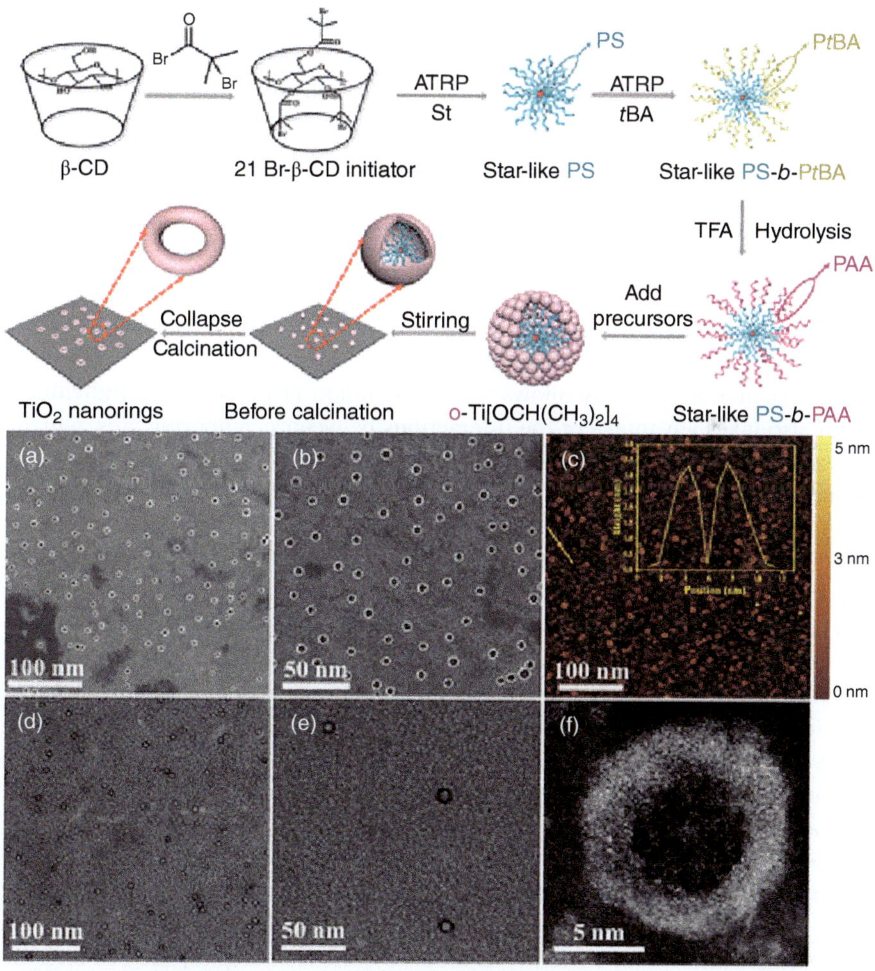

Figure 3.12 Schematic representation of synthesis strategies for TiO_2 nanorings using multiarm, starlike amphiphilic diblock PS-*b*-PAA copolymers as nanoreactors; (a, b) SEM images of TiO_2 nanorings with different magnification; (c) AFM images of the nanorings. (d–f) TEM images of nanorings with different magnifications.

The outer PAA block of nanoreactors possessed carboxylic acid groups, which could coordinate with a titanium precursor followed by high-temperature calcination to form crystalline TiO_2 nanorings. The living nature of ATRP enabled the precise preparation of star-like diblock copolymer nanoreactors with a controlled length of each block (i.e. P*t*BA and PS), thereby tailoring the inner diameter and wall thickness of the resulting TiO_2 nanorings, which were inaccessible to conventional routes. The obtained multi-arm, star-like amphiphilic diblock PS-*b*-PAA copolymers were then used as nanoreactors to prepare homogeneous TiO_2 nanorings. After stirring for the desired time, metal precursor titanium isopropoxide (TTIP) was added to the multiarm, starlike amphiphilic diblock PS-*b*-PAA copolymers dissolved in DMF solution. The precursors interacted strongly with the carboxyl group of PAA, so they were preferred to occupy in the region of the PAA chains rather than the inner PS chains. The nucleation and growth of TiO_2 nanostructures resulted from the hydrolysis reaction. The obtained TiO_2 nanostructures were calcinated at 500 °C, and the nanoreactors were removed at high temperature. Because of the lack of the supporting effect of PS, the TiO_2 nanostructures collapsed, and the collapsed apical shell could cover the side walls of the nanorings, thus forming the crystalline TiO_2 nanorings.

3.3.1.5 Colloidal Nanoparticles Assemblies

Recently, the focus of synthetic efforts has been directed toward the creation of secondary structures of colloidal NPs, which hold great promise for the development of advanced materials with novel integrated functions. Clustering NPs into secondary structures not only allows the combination of properties of individual NPs but also takes advantage of the interactions between neighboring NPs, which can result in new properties not present in the original constituents [69]. A well-known example is the assembly of noble metal NPs into secondary structures, which induces near-field coupling of surface plasmon between adjacent particles. As a result, new optical properties can be obtained, inducing shifts of plasmonic peaks and the generation of "hot spots" that are very useful for enhancing Raman scattering.

The past two decades witnessed the rapid advancement of synthetic strategies for preparing various colloidal nanocrystals with controlled chemical compositions, dimensions, morphologies, and architectures. Currently, one of the research hot spots in the field of colloidal nanocrystals has been directed toward the creation of secondary nanostructured organization or superstructures consisting of several colloidal nanocrystals of approximately several to tens of nanometers in one ensemble or cluster, which is termed as colloidal nanocrystal assemblies (CNAs). Traditional preparation methods of CNAs can be classified into two categories (i.e. bottom-up method and top-down method). However, there were still inherent limitations and drawbacks associated with currently available synthetic strategies, such as lack of general applicability and precise controllability.

An effective process for the fabrication of one-dimensional (1D) colloidal superparamagnetic magnetite (Fe_3O_4) nanocrystal clusters was developed by combining magnetic assembly and photoinduced cross-linking under an external

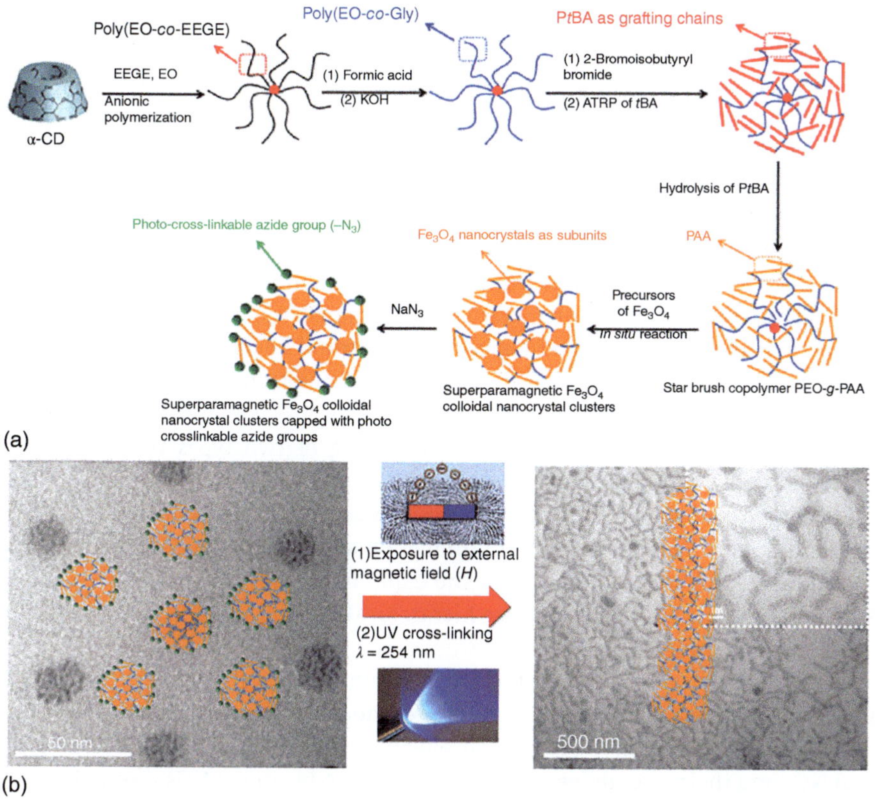

Figure 3.13 Scheme (a) Preparation of spherical superparamagnetic Fe_3O_4 colloidal nanocrystal assemblies coated with an azide end group, multiarm star brush copolymer PEO-*g*-PAA as templates. (b) Fabrication of 1D superparamagnetic Fe_3O_4 colloidal nanocrystal assemblies by combining magnetic assembly with photoinduced cross-linking.

magnetic field (Figure 3.13) [18]. First, a series of water-soluble multiarm star brush copolymers with different molecular weights using α-cyclodextrin as a multifunctional initiator, star brush copolymer PEO-g-PAA, which consisted of PEO as the main chain and PAA as functional side chains, were prepared by combining anionic ring-opening copolymerization with ATRP. Sequentially, star brush copolymer PEO-g-PAA as a spherical unimolecular template was utilized to structure-direct the *in situ* formation of spherical superparamagnetic Fe_3O_4 colloidal nanocrystal clusters by loading precursors of Fe_3O_4 ($FeCl_3$ and $FeCl_2$) into the PAA compartment. Surface-tethered PAA grafting chains of spherical Fe_3O_4 colloidal nanocrystal clusters, produced by ATRP, retain terminal bromine atoms that were subsequently converted into photo-cross-linkable azide groups ($-N_3$) through nucleophilic substitution. The spherical Fe_3O_4 colloidal nanocrystal clusters coated with photo-crosslinkable azide groups were assembled into temporary 1D secondary nanostructures by being exposed to external magnetic fields. The resulting temporary 1D nanoclusters (NCs) were efficiently cross-linked by exposure to UV light to stabilize the 1D secondary nanostructures of Fe_3O_4

nanocrystals. The dimension of the resulting stable and cross-linked 1D clusters can be tuned by adjusting the strength of the external magnetic field.

In an effort to overcome the limitations and synthetic dilemmas encountered in conventional CNAs preparation methods, in this study, we reported a novel and general synthetic strategy for crafting a large variety of CNAs by employing rationally designed star-like poly(4-vinylpyridine)-*b*-polystyrene (P4VP-*b*-PS) diblock copolymers as nanoreactors prepared by sequential ATRP. The molecular weight of P4VP and PS blocks (i.e. chain length of polymers) was precisely regulated by ATRP reaction time, leading to the formation of CNAs with tunable overall diameter. Hierarchical nanostructure of the CNAs, including nanocrystal number density and component nanocrystal size, could be programmed by simply varying metal precursor loaded into copolymer nanoreactors and reaction time, which rendered it an effective route to the various CNAs compared to traditional approaches. Furthermore, hybrid CNAs consisting of two types of component nanocrystals with different chemical compositions and functionalities in one CNAs were accomplished via nanoreactors strategy, which was inaccessible by traditional fabrication routes. This study concerns fundamental polymer chemistry (i.e. ATRP) and inorganic nanomaterial design in the exploration of a universal strategy for preparing inorganic CNAs with precise hierarchical control, which was inaccessible by conventional routes.

Ultrasmall noble metal nanoclusters (NCs, <2 nm) with high surface atomic ratio and tunable electronic structure have attracted extensive interest in the past decades [16, 70–72]. Although there are several effective approaches for the synthesis of metal NCs and their assemblies, the preparation of stable metal NCs assemblies for the application in catalysis is still a challenge, owing to their high surface energy [73–75]. Zhang and coworkers reported that the star-liked diblock copolymer β-CD-g-poly(4-vinyl pyridine)-*b*-polystyrene (β-CD-g-P4VP-*b*-PS) was used as a general and robust unimolecular micelles scaffolds for the *in situ* synthesis of metal NCs assemblies by combining with thiol ligands, which is considered to be the key to limit the size of individual NCs to ultrasmall sub-2 nm regime (Figure 3.14) [62]. In addition, the length of the P4VP block can be precisely controlled through the polymerization process of ATRP, so that the size of NCs assemblies can be precisely controlled via ATRP process. By changing the amount of precursors and the P4VP block segments of the polymers, the number (10–160) of encapsulated metal NCs contained in the micelles could also be highly controlled. Supported by the successful preparation of various metal NCs assemblies (e.g. Ag, Pt, Cu, and AuAg alloy) and the good structural stability during catalytic application and recycling, we deduced that this facile strategy will provide an efficient platform for the preparation of metal NCs assemblies, which have broad application prospects in the fields of catalysis, optics, and biomedicine.

3.3.2 Cylindrical Polymer Brushes as Unimolecular Nanoreactors

1D rod-like nanostructures are of fundamental interest for examining size- and shape-dependent phenomena and can have applications that include

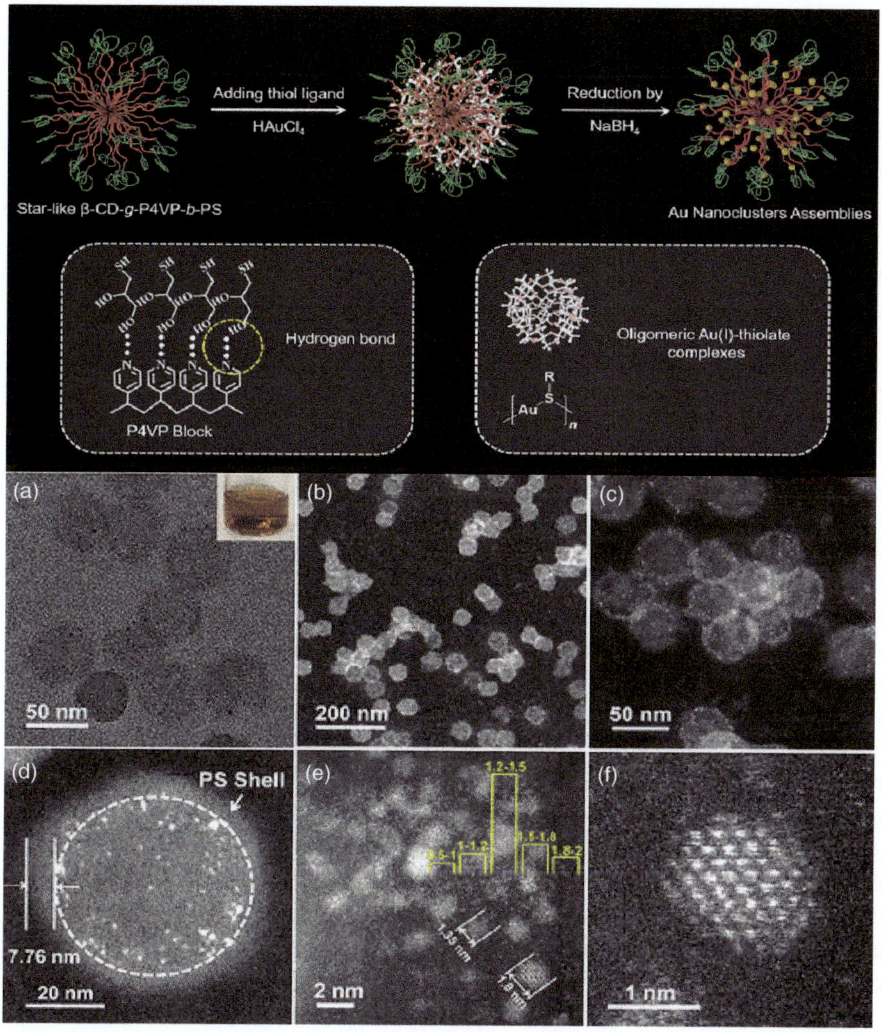

Figure 3.14 Illustration of the designed process *in situ* synthesis of the Au NCs assemblies in star-like β-CD-*g*-P4VP-*b*-PS polymer scaffolds.

next-generation electronics and sensing elements. The growth of such nanostructures poses considerable challenges for synthetic chemists and materials scientists. However, because few materials naturally grow in such an anisotropic manner, linear pores (molds) or surface templates (such as DNA) are used to guide their formation. Assembling NPs into 1D nanostructures with precisely controlled size and shape renders the exploration of new properties and construction of 1D miniaturized devices possible. The physical properties of such nanostructures depend heavily on the size, chemical composition, and surface chemistry of NPs constituents, as well as the close proximity of adjacent NPs within the 1D nanostructure. Chemical synthesis provides an intriguing alternative means of creating 1D nanostructures composed of self-assembled NPs in terms of material diversity,

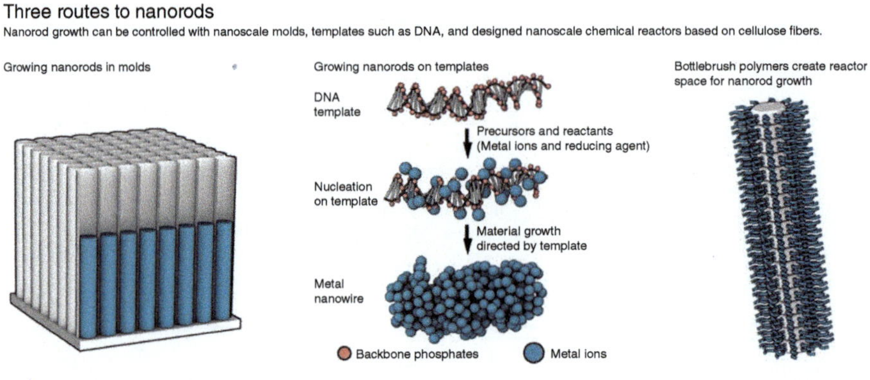

Figure 3.15 Three routes to nanorods.

size controllability, shape regularity, and low-cost production. However, this is an area where progress has been slower.

Methodologies for 1D material synthesis must direct growth by defining the reaction space in which product formation occurs (Figure 3.15). A common approach is to use porous matrices, such as alumina and polycarbonate membranes as molds. The linear pore channels in these matrices are filled with soluble precursors, such as metal ions that are chemically transformed into insoluble material that conforms to the pore shape. However, the matrix needs to be removed to isolate individual rods. Diameters have been limited by the range of available pore sizes (tens of nanometers), but the recent development of molecular-based molds may overcome some of these problems. The smallest-diameter NRs have been prepared by using naturally occurring linear biopolymers, such as DNA. The rod-like double helix of individual DNA molecules is just 2 nm in diameter and available in precise lengths with sub-nanometer resolution. Although DNA intrinsically lacks many useful optoelectronic properties, its ability to bind a wide range of chemical precursors enables templated synthesis of 1D materials. This mechanism relies on precursor nucleation on the DNA, followed by growth guided by the template shape. The approach is adaptable to a wide range of material types, including metals, inorganic compounds, and even soft molecular-based conductors. However, the limited solubility of DNA in nonaqueous solvents restricts the types of chemical reactions, and hence materials, compatible with this method. Furthermore, the growing material entombs the template, so the surface chemistry of the final nanostructure is dictated by its growing material and not the template, a situation that is not always desirable [3].

When considering 1D unimolecular block polymer templates, cylindrical polymer brushes (CPBs) are ideal candidates. These structures are composed of linear or dendritic side polymer arms grafted from a linear polymer backbone [76]. Recently, the preparation of 1D NRs was reported by applying the same nanoreactor concept and substituting β-CD with cellulose as a macroinitiator [30]. Interestingly, inorganic nanocrystals can grow from either the backbone or the side chain, depending on the specific CPB structure. If the polymer backbone is selected for binding metal precursors, the binding sites along the backbone are typically not frequent enough to grow continuous metal NWs. As a result, 1D inorganic NP assemblies were generated. It is also observed that the structure of CPBs changes dramatically after the loading of metal precursors, and a pearl necklace-like morphology formed with metal ions locally aggregates into "beads." By contrast, the inorganic materials grown on the side chains will be reduced to a uniform and continuous inorganic NWs.

Recently, Pang et al. proposed a general strategy for synthesizing a wide range of NRs using an amphiphilic cellulose polymer brush as a template. Such method offers a high degree of control over the size and architecture of 1D inorganic nanocrystals enabled by narrow molecular weight of grafted polymer brushes. They used cellulose as a rigid backbone with intensive hydroxyl groups that allow the grafting of dense block copolymer side chains through bromine modification and sequential ATRPs (Figure 3.16). Similar to the β-CD strategy, robust 1D NRs with compact, core@shell, and hollow architectures have been fabricated using a cylindrical cellulose-based bottlebrush polymer as the unimolecular nanoreactor. Taking a compact NR as an example, the cellulose-g-(PAA-*b*-PS) bottlebrush can be synthesized and used as the nanoreactor, in which the inner PAA block directs NR growth through metal concentration and reduction and the outer PS block provides dispersity and stability. Various NRs with precisely controlled architectures, compositions, sizes, and surfaces could also be obtained in a similar method. The length of the obtained NR can be tuned from the nanoscale to the micron scale by choosing cellulose with different molecular weights as linear backbones of bottlebrush polymers.

The approach is readily adaptable to synthesizing nanostructures with desirable properties (such as metallic, semiconducting, magnetic, or luminescent) and of increased complexity (for example, multimaterial core–shell structures). By increasing the block count and introducing protection-deprotection strategies, different chemical functionalities can be sequentially revealed to enable multistep syntheses. The final and perhaps most farsighted aspect of this approach is that the NRs have a surface, or exterior sheath that is defined from the outset by the outermost polymer block. This control of surface chemistry can provide stable suspensions of NRs in different solvents, which both facilitates their use in experiments as well as in the development of processing routes.

The bottlebrush-liked polymer can also be used to fabricate 1D NPs assemblies. Using cellulose-g-PAA as the nanoreactor, after the concentration of SnO_2 precursors and template growth of polydopamine (PDA), the PDA-coated SnO_2 nanocomposites were crafted and used as electrode materials for LIBs (Figure 3.17) [77]. The PDA performed as a passivating solid electrolyte interphase layer. Compared with uncoated SnO_2 electrode-based batteries, excellent electrochemical performance and superior long-term cycling stability (over 300 cycles) have been achieved after coating the protective PDA to enhance the Sn/Li_2O interface fraction of SnO_2-based electrodes.

The bottlebrush-liked polymers can also be prepared using a linear polymer with a bromine-containing repeating unit as the backbone and subsequent ATRP processes. These bottlebrush-liked polymers have lower brush density compared to cellulose-based bottlebrush polymers. Generally, 1D soft nanotubes are usually

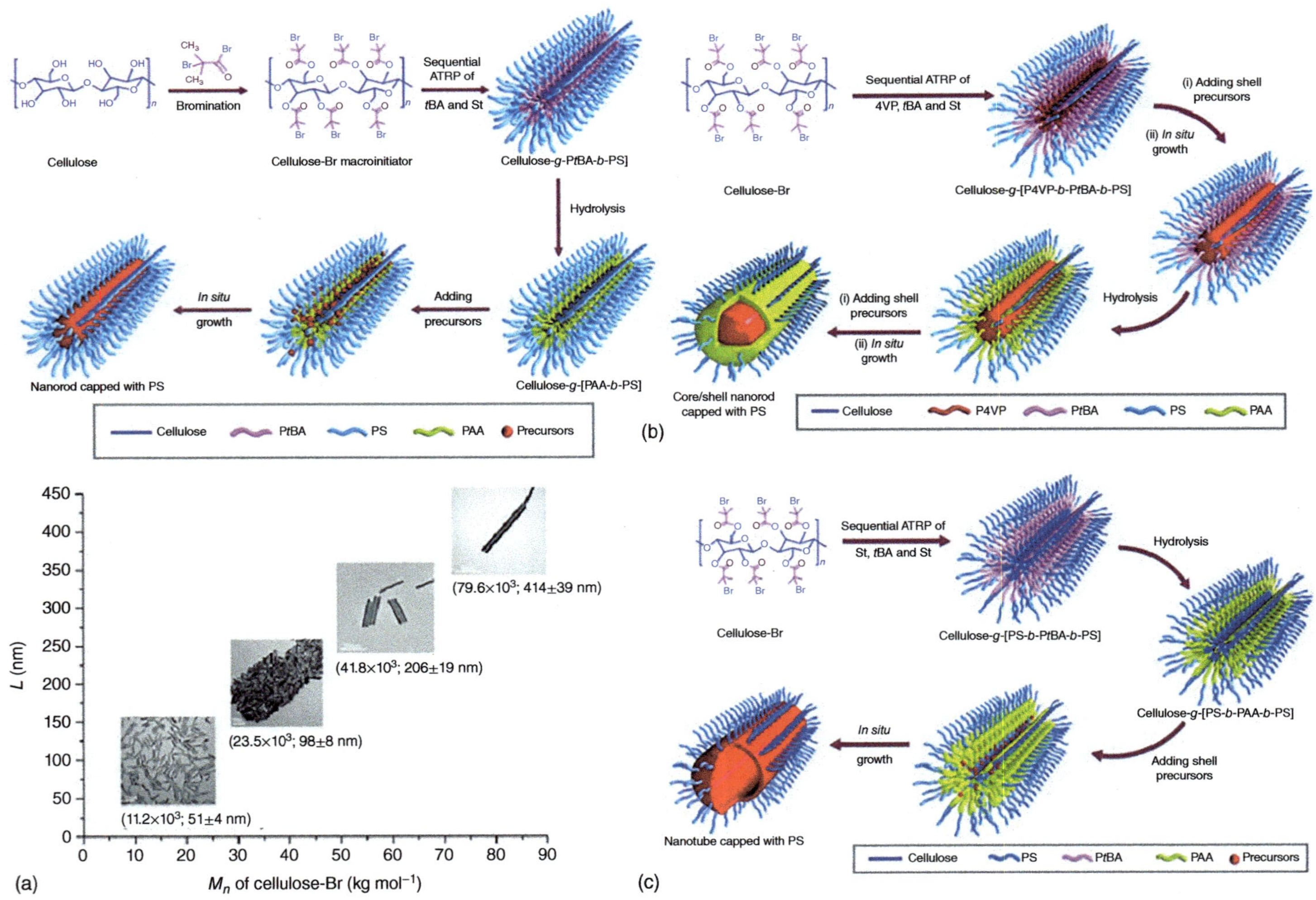

Figure 3.16 Synthetic strategies for 1D nanocrystals, using amphiphilic cylindrical BBCPs as nanoreactors.

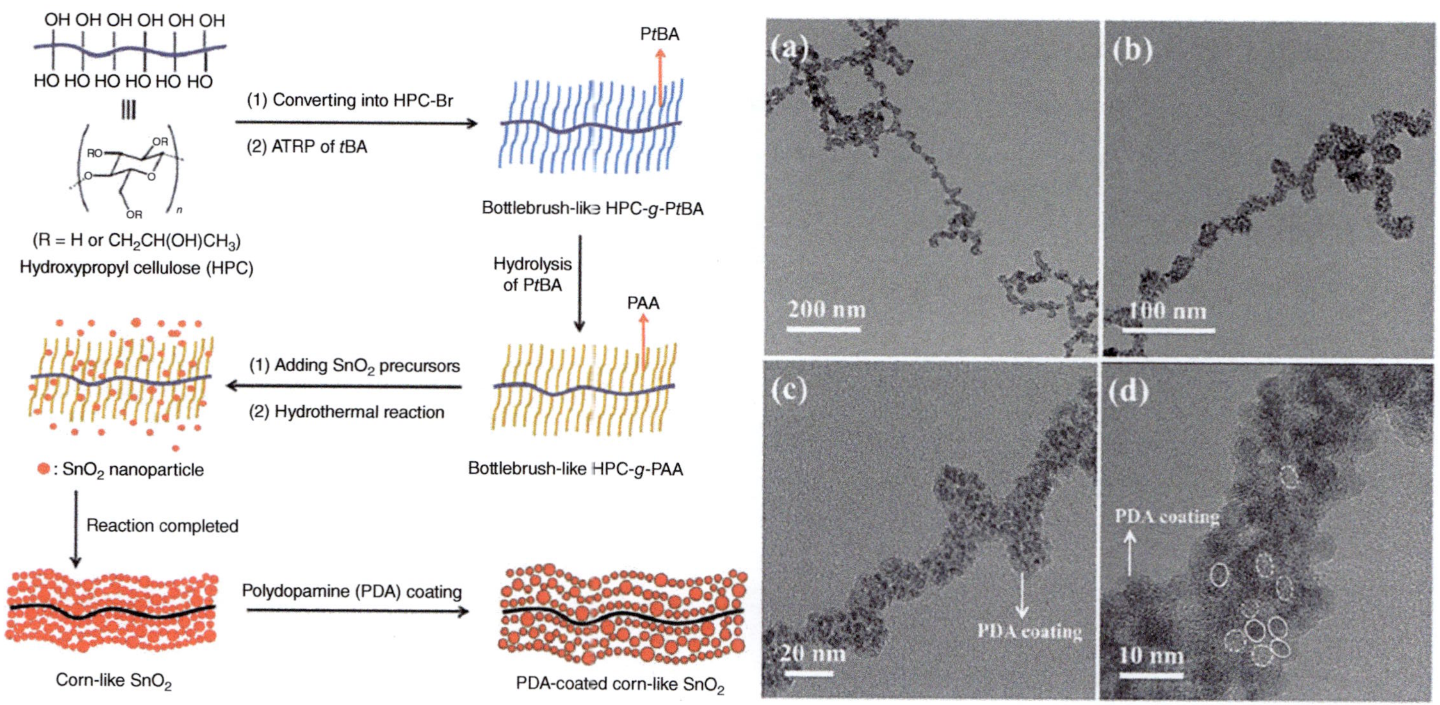

Figure 3.17 Synthesis of the hydrophilic bottlebrush-like HPC-*g*-PAA polymer template and templated growth of PDA-coated corn-like SnO_2 nanocrystals; (a, b) Lower magnification and (c, d) higher magnification TEM images of PDA-coated SnO_2 nanocrystals.

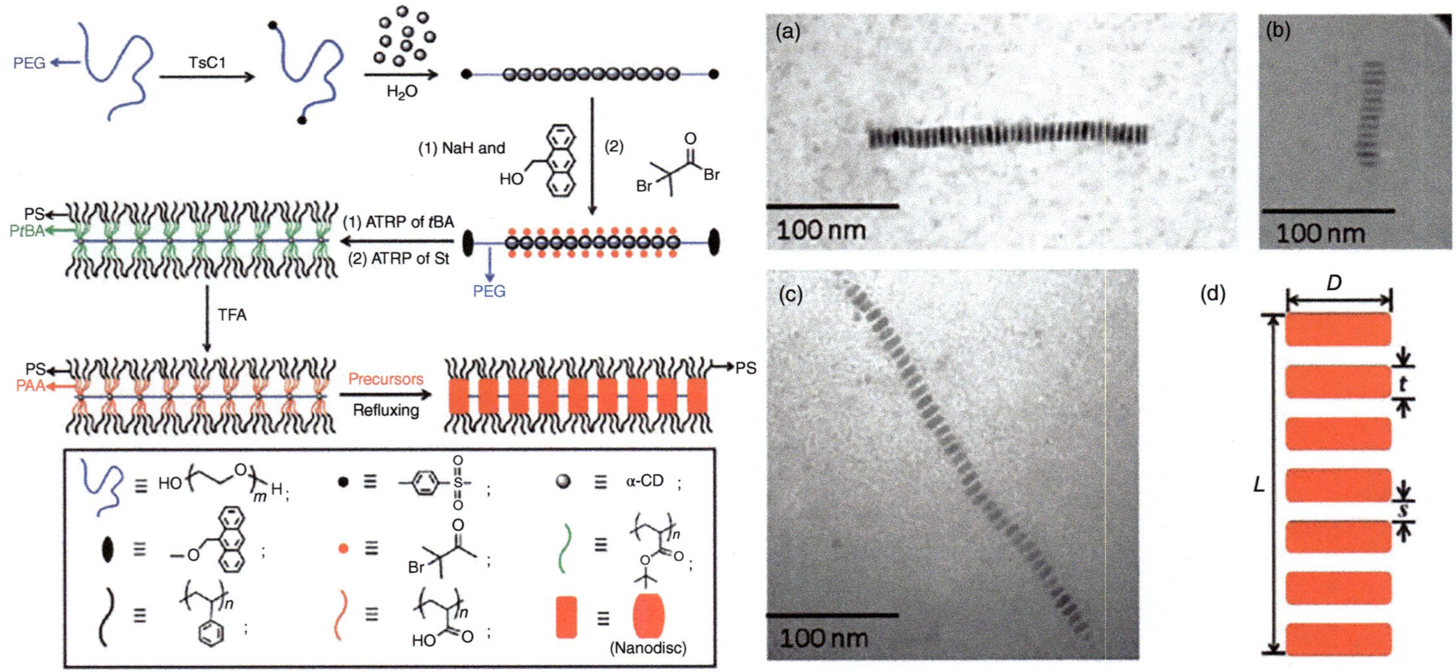

Figure 3.18 (a) Schematic diagram for the synthesis of hairy PR and its use as template for 1D necklace-like structure, and TEM images of CdSe (b), Fe_3O_4 (c), and $BaTiO_3$ (d) necklaces.

obtained using these bottlebrush polymers as unimolecular nanoreactors. In this way, water-soluble organo–silica hybrid nanotubes and organo-soluble organic nanotubes were fabricated with good controllability in dimensions and compositions. Organic nanotubes with well-defined one or two open ends were also fabricated via cross-linking of specific bottlebrush blocks within a multiblock bottlebrush along the backbone. Star-like bottlebrush polymers, bottlebrush polymers with multiarms, are also synthesized and used for nanofabrication. The three-armed bottlebrush polymer, amphiphilic PHEMA-g-(PAA-*b*-cPS) was synthesized via ATRP followed by partial cross-linking of PS. After nanofabrication within this armed bottlebrush polymer, cPS-capped $CsPbBr_3$ perovskite NPs were obtained with greatly improved stabilities over those obtained through conventional approaches due to the cross-linked hydrophobic PS on the surface. The internal PAA block directed the critical accumulation of metal ions and then formation of $CsPbBr_3$ NPs. Concurrently, the outer cPS shell enhanced the passivation of NPs against external stimuli, consequently endowing the final perovskite NPs with excellent structural stability (UV light, heat, water, moisture, and oxygen). The obtained $CsPbBr_3$ NPs were nearly spherical instead of the expected three-armed topology (branch), which may be due to the low brush density allowing the collapse of the PAA block during nanofabrication and the low length ratio of the PHEMA backbone to the PAA brush. In a similar way, a three-armed organic nanotube was fabricated by cross-linking the outer layer of the three-armed bottlebrush polymer with a larger length ratio of the backbone to the brush.

Lin and coworkers designed an intriguing 1D necklace-like nanostructure comprising α-CD nanodiscs modified with Br substitutes to initiate sequential ATRPs and then metal ion gathering and reduction (Figure 3.18) [59]. Different from the normal morphology of star polymer nanoreactors, the assembled uniform functional nanodiscs along the PEG flexible polymer chain assumed the rectangle shape due to the crush between these unimolecular nanoreactor units. Such 1D organic–inorganic nano-necklaces are composed of uniform inorganic NPs, capped with hydrophobic polymer chains and connected with PEG chain, offering good solubility of the nano-necklaces in common solvents (Figure 3.18). Nano-necklaces with various compositions, such as semiconductor CdSe, magnetic Fe_3O_4, and ferroelectric $BaTiO_3$ have been prepared. These novel structures of nano-necklaces may be developed for some new applications in optoelectronics, catalysis, ferroelectrics, plasmonics, magnetic devices, and other areas.

3.4 Application of Polymer-Capped Nanoparticles

Diverse properties of the obtained NPs templated from unimolecular nanoreactors have led to a large number of potential applications. This kind of robust and unconventional strategy is promising for ingenious nanofabrication and has a positive influence on the fields of electronics, photonics, sensing, catalysis, biomedicine, etc. [6, 11, 29, 78]. The most attractive characteristic of NPs prepared with block copolymer templates is their high stability. In addition, the outer polymer chains can enhance their performance in various applications. To date, NPs generated with block copolymer templates have been applied in several fields, including catalysis, solar cells, and lithium-ion batteries. In this section, the applications of functional NPs templated by block copolymers are discussed.

3.4.1 Solar Energy Conversion

The polymer chains on the outside of the NPs can play an important role in solar cell applications. $NaYF_4$:Yb/Er upconversion nanoparticles (UCNPs) prepared using star-liked PAA-*b*-PEO templates have been used in solar cells [8, 79]. Specifically, PEO-capped UCNPs were employed as mesoporous electrodes with tunable pore size in organolead halide perovskite solar cells (Figure 3.19). The performance of devices with an UCNPs layer is better than that of devices with a conventional TiO_2 layer under NIR irradiation at 980 nm. It is likely that, in conjunction with the growth and crystallization of perovskite facilitated by the $NaYF_4$:Yb/Er UCNPs, the upconversion effect of $NaYF_4$:Yb/Er UCNPs enhanced the light harvesting and

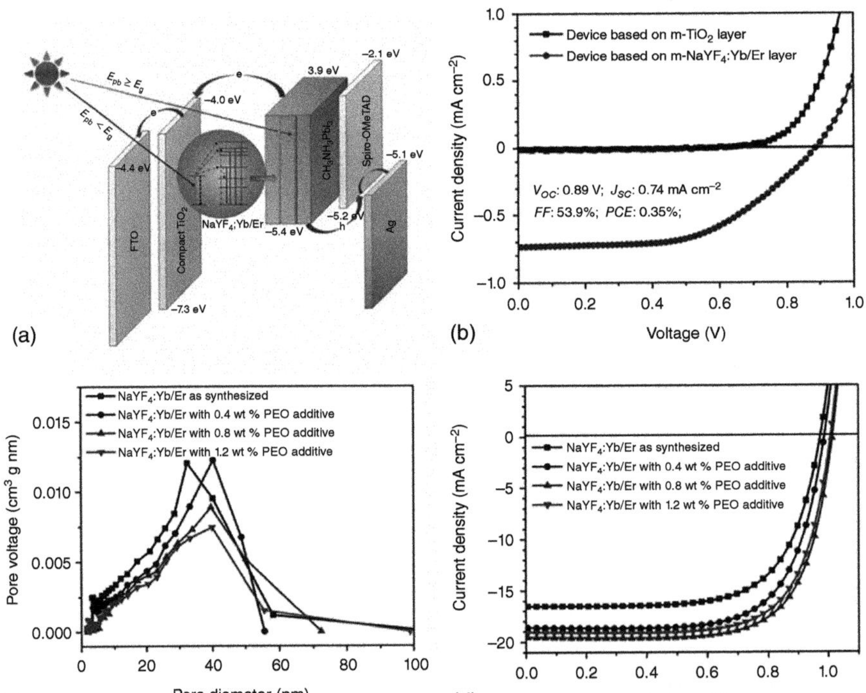

Figure 3.19 (a) Energy levels and charge-transfer processes in perovskite solar cells with$NaYF_4$:Yb/Er UCNPs as the mesoporous electrode. (b) The current density–voltage (J–V) characteristics of perovskite solar cells using a $NaYF_4$:Yb/Er UCNP layer and a TiO_2 layer under irradiation with NIR light at 980 nm. (c) Change in the pore diameter of the mesoporous structure with different amounts of added linear PEO. (d) Current density of perovskite solar cells with varied PEO additives. Source: He et al. [8]/John Wiley & Sons.

improved the device efficiency. It is worth noting that the outer PEO chains impart not only good solubility and long-term stability to the NPs but also facilitate the formation of the mesoporous structure. The pore sizes of the mesoporous structures can be tuned by adding PEO homopolymer (Figure 3.19). Mesoporous structures with larger pores were advantageous for the penetration of perovskite into these structures. This penetration promotes absorption of the incident photons and transport of charge carriers throughout the electrodes.

PS-capped Au@TiO_2 core–shell NPs combined with widely used TiO_2 network films were utilized as photoanodes in DSSCs. The PS layer on the Au@TiO_2 NPs was removed by calcination in air or carbonization in an inert atmosphere to afford uncoated Au@TiO_2 NPs and C-Au@TiO_2 NPs. Both of them exhibited improved performance in DSSCs compared to devices with commercial P25 films. The addition of uncoated Au@TiO_2 core@shell NPs improves the short-circuit current density (Jsc) by 10% and the PCE by 7.4%. This is explained by two factors: (i) the light-harvesting efficiency is improved by the SPR effect of Au cores and (ii) sintering of the TiO_2 shell at high temperature resulted in it converting from an amorphous into an anatase phase. C-Au@TiO_2 NPs enabled an even greater enhancement in the performance (increase in Jsc by 18.4% and a PCE increase of 13.6% compared to commercial P25 TiO_2 films). The performance enhancement is attributed to the conductive carbon layer increasing the electron collection efficiency by reducing the electron-transfer time and extending the lifetime of the electrons.

3.4.2 Light-Emitting Diodes

Recent impressive development in the synthesis of perovskite QDs (ABX_3; both organic–inorganic and all-inorganic) as building blocks for miniaturized materials and devices have emerged as an important research activity. They have found promising applications in photovoltaics, light-emitting diodes (LEDs), scintillators, lasers, etc. The practical applications of ABX_3 QDs have been primarily limited by their long-term stability due to their ionic crystal nature and tendency to decompose under external stimuli (e.g. UV, water, oxygen, polar solvents, heat, etc.). In this context, recently, uniform all-inorganic perovskite $CsPbBr_3$ QDs were prepared using a judiciously designed amphiphilic star-like diblock copolymer of β-CD-g-PAA-*b*-PS as a molecularly engineered nanoreactor. The resulting $CsPbBr_3$ QDs were permanently ligated with hydrophobic PS chains, offering outstanding colloidal stability against water. Owing to their good optical properties, a white-light-emitting diode (WLED) device consisting of blue-emitting GaN blue chip, green-emitting PS-ligated $CsPbBr_3$ QDs, and red-emitting CdSe/Cd1−xZnxSe1−ySy/ZnS QDs was fabricated. The PS-capped $CsPbBr_3$ QD-based WLED had a color coordinate of (0.31, 0.32), very close to the white color coordinate of (0.33, 0.33), and a color gamut broader than that of NTSC and sRGB standards. Most importantly, the PS-ligated $CsPbBr_3$ QD-based WLED demonstrated excellent long-term stability under ambient conditions and high thermal cycling stability between 25 and 100 °C owing to the permanently ligated PS hairs.

3.4.3 Lithium-Ion Batteries

On their own, compact NPs do not offer especially attractive performances in LIBs. The addition of an outer carbon layer can improve their electrochemical performance. It is for this reason that block copolymer-templated NPs are employed in LIB applications. $ZnFe_2O_4$ NPs crafted using PS@PAA nanospheres were implemented in LIB applications. PS@PAA nanospheres not only serve as reactors for the growth of $ZnFe_2O_4$ NPs, but also provide a latent carbon source in terms of the outer PS layer after thermal decomposition in an inert atmosphere (Figure 3.20) [80]. The $ZnFe_2O_4$/carbon nanocomposites exhibit the best performance (79.3 wt% carbon). After 70 cycles at a current density of 100 mAg@1, an ultrahigh capacity of 1500 mAhg@1 was observed. This is approximately three times the capacity of uncoated $ZnFe_2O_4$ NPs. The carbon coating stabilizes the structure and improves the conductivity of the inner NPs. In addition, the continuous carbon network maintained the integrity of the $ZnFe_2O_4$ nanostructures during cycling.

The well-defined morphologies and sizes of hollow nanostructures play a critical role in enhancing the cycling and rate performance of LIB electrodes. There are two advantages afforded by hollow structures from block copolymer templates: (i) Hollow architectures can provide a larger contact area with the electrolyte and effectively accommodate the volume changes that occur during lithiation–delithiation processes. The volume change during charging and discharging is the single biggest hurdle limiting the improvement of battery lifetime and

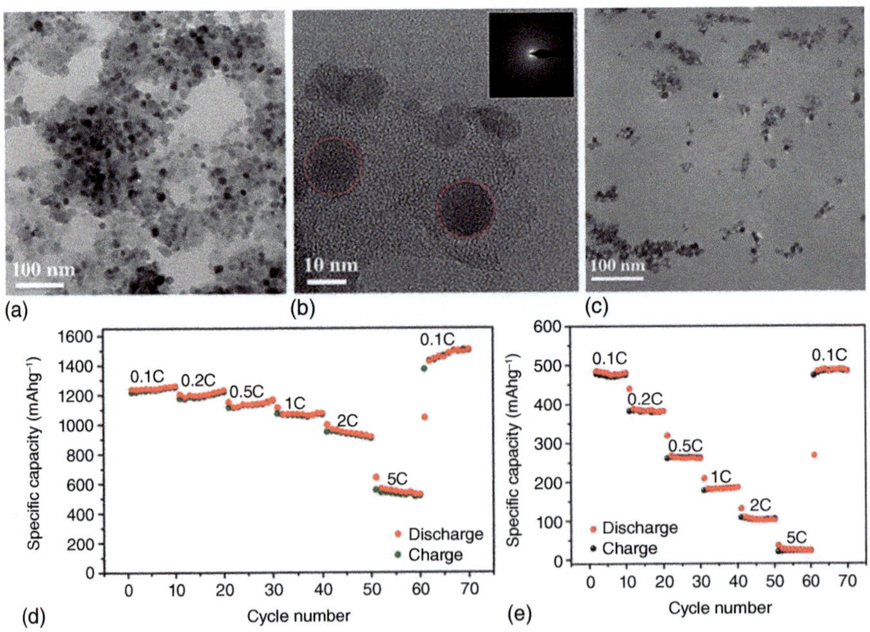

Figure 3.20 (a) TEM and (b) HRTEM images of $ZnFe_2O_4$/carbon nanocomposites. (c) TEM image of uncoated $ZnFe_2O_4$ nanoparticles. (d) Rate performance of $ZnFe_2O_4$/carbon nanocomposites and (e) $ZnFe_2O_4$ nanoparticles.

storage capacity. (ii) The small grain size of the hollow spheres shortens the diffusion path during cycling and creates more active sites for charge transfer reactions. This effectively shortens the lithium-ion transport distance and improves the structural stability of the active materials [81]. Hollow silica nanospheres template from triblock copolymer micelles (both PS-*b*-PMAPTAC-*b*-PEO and PS-*b*-P2VP-*b*-PEO) enable high cycling performance through an alloying/dealloying process [82]. After 100 cycles at a charging rate of 1 C, the capacity retention of the hollow silica nanospheres (93.4%) was much higher than that of bulk Si powder (70.7%).

3.4.4 Catalysis

Functional NPs have been applied in catalysis because of their enhanced surface areas compared to bulk materials. A common problem is that NPs tend to aggregate into larger clusters with small molecular ligands [72, 83–88]. This leads to a reduction in the active surface area, which lowers the catalytic performance. It is for this reason that noble metal NPs stabilized with block copolymers have been applied as catalysts. Pd colloids templated within PS-*b*-P4VP micelles were first used as a catalyst in Heck coupling reactions. In these studies, several traditional catalysts in addition to PS-*b*-P4VP stabilized Pd NPs were applied for the coupling of 4-bromoacetophenone and styrene. The yields from the reaction using each catalyst were compared. Several Pd catalysts stabilized with small-molecule ligands tend to form Pd black, an aggregation of Pd colloids, which leads to moderate product yields (30–35%), whereas templated Pd NPs display much higher catalytic performance (73–92% for different concentrations), and excellent stability (remaining dispersible for several months). It was also found that the catalytic efficiencies of Pd NPs were significantly affected by the size of the NPs and the length of the outer polymer chains. The smallest NPs exhibited the best catalytic activity because of their enhanced surface areas. In contrast, Pd NPs with longer outer polymer chains showed lower activity, since longer polymer chains lead to steric hindrance, which reduces the contact between the Pd NPs and reactants. Interestingly, the catalytic performance of Pd NPs prepared from water-soluble block copolymer templates can be tuned by adjusting the pH value, thereby enabling a simple and typical method of catalytic control. Pd NPs prepared using PHI-*b*-PCEMA-*b*-PAA core@shell@corona micelles (the outer PI block was hydroxylated to form water-soluble PHI) were employed to catalyze hydrogenation reactions. In the case of vinylacetic acid (VAA), the rate of hydrogenation was relatively low at pH 10 and gradually increased up to pH 3. This is attributed to the repulsion between the deprotonated PAA blocks and deprotonated VAA substrate at basic pH values.

Water-soluble Au (also Pd and Pt) NPs with an average diameter of 10 nm synthesized from linear PEO-*b*-PAADHBC templates were used as a reducing agent for the decomposition of nitroarenes. The catalytic performance was assessed by the turnover frequency (TOF), which represents the degree of reaction per unit time per amount of catalyst used in the reaction. It was noted that the DHBC-templated Au NPs (800 h@1) displayed a higher TOF than that of Au NPs prepared by reduction with citrate (570 h@1). The surrounding polymer chains likely play an important role in enhancing the catalytic performance by keeping the particles from aggregating over time at both elevated temperatures and extreme pH values. Similar observations were also reported for dendrimer-templated NP catalysts [89]. The enhanced performance was attributed to the steric confinement effect of the dendrimer templates on the NPs. Similarly, polymer-stabilized Pd and Au/Pd bimetallic NPs were used as hydrogenation catalysts. In each case, higher stability and selectivity were observed with polymer-capped catalysts. The role of the outer polymer chains in the enhancement of the catalytic reactivity still requires further investigation.

Coating a shell layer onto NPs is another way to avoid activity-reducing aggregation. In addition, favorable properties can be derived from the interaction between compatible core and shell materials. For example, early research suggests that metal NPs deposited on TiO_2 nanostructures undergo equilibration of the Fermi level following excitation with UV light and possess enhanced efficiency in charge-transfer processes [66]. Lin and coworkers synthesized Au@TiO_2 core@shell NPs with well-controlled core diameters and shell thicknesses by using a star-liked P4VP-*b*-PAA-*b*-PEO template and they were employed in photocatalysis (Figure 3.21) [90]. The photocatalytic rate was evaluated by degradation of rhodamine B (RhB). The performance of three NP catalysts block copolymer-templated Au@TiO_2 core@shell NPs, block copolymer-templated TiO_2 NPs, and commercial TiO_2 (P25) were compared. The block copolymer-templated TiO_2 NP catalysts had a similar activity to P25. The Au@TiO_2 core@shell catalysts exhibited much higher activity. This is attributed to two factors. First, the enhanced light harvesting by the SPR-active Au core enabled greater energy uptake. Second, the Schottky junction

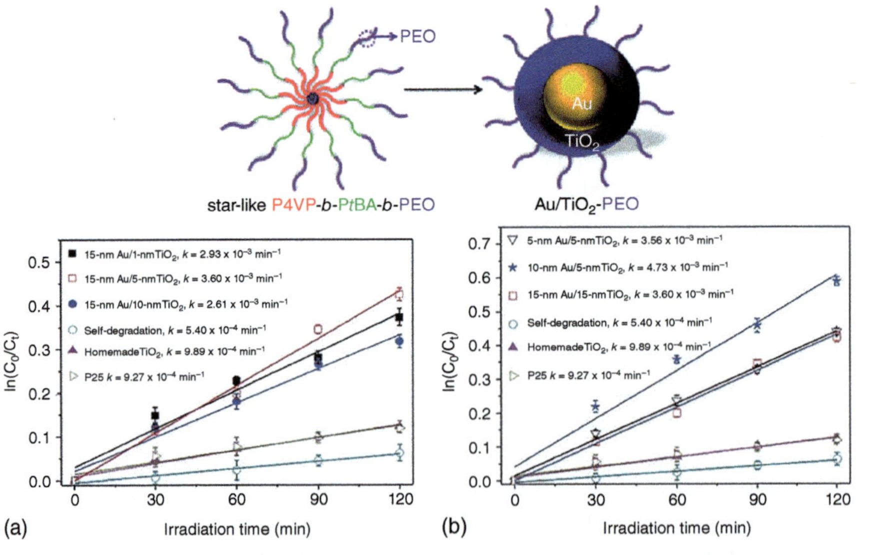

Figure 3.21 Catalytic performance of Au@TiO_2 core@shell nanoparticles with varied core diameters and shell thicknesses. (a) Fixed Au core diameter (15 nm) and varied TiO_2 shell thicknesses. (b) Fixed TiO_2 thickness (5 nm) and varied Au core diameters. Source: Wang et al. [90]/Royal Society of Chemistry.

formed at the interface between the conductive Au core and semiconducting TiO_2 shell led to enhanced transfer of photogenerated electrons and holes from the Au core to the TiO_2 shell. Of the core@shell structures with fixed TiO_2 shells (5 nm), Au@TiO_2 with a 10 nm Au core showed the highest reaction rate. This is likely because the smaller Au cores (5 nm) show a relatively weak SPR and the larger Au cores (15 nm) have less surface contact with the TiO_2 shell.

3.5 Conclusions and Perspectives

3.5.1 Conclusion

In this chapter, we provided a brief summary of the fabrication of inorganic NPs within unimolecular nanoreactors. Rapid development in polymer chemistry has promoted remarkable progress in the controlled synthesis of unimolecular block copolymers micelles nanoreactors with diverse but uniform topologies, architectures, components, and sizes. The unimolecular micelles nanoreactors with well-defined and stable features can precisely control the fabrication of various NPs with prospective nanostructures. The inside cavity concentrated plenty of precursors, and the outside tethered polymer chains provided dispersion stability of the gained NPs. In this way, NPs with expected topologies (sphere, rod, tube, branch, and ring), architectures (compact, hollow, ring, core–shell, and necklace-like), compositions (metal, metal oxide, semiconductor, doping, alloy, silica, and composite), sizes (generally 1–100 nm), surface properties (hydrophilic, hydrophobic, reactivity, valence, and responsivity) and assemblies (oligomer, chain, and aggregate) can be fabricated easily within reasonably designed unimolecular micelles nanoreactors. Diverse properties of the obtained NPs templated from unimolecular micelles nanoreactors have led to a large number of potential applications. This kind of robust and unconventional strategy is promising for ingenious nanofabrication and has a positive influence on the fields of electronics, photonics, sensing, catalysis, biomedicine, etc.

3.5.2 Perspectives

As an ingenious inorganic nanomaterials synthesis method, the further development of NPs templated from unimolecular block copolymers nanoreactors may open new horizons in the fabrication of inorganic NPs with novel but inaccessible nanostructures. Firstly, the topologies of the NPs can be further expanded. Branched NPs can be fabricated within easily synthesized multiarmed bottle-brush polymers. Nanoplates can be fabricated within 2D brush-liked polymers, which can be potentially synthesized from the fast-developing 2D polymers or surface-initiated controlled/“living” polymerization. Porous nanomaterials can be fabricated from compacted oligomers of unimolecular micelles nanoreactors as the pores template if their outer layers have strong interaction with metal precursors. Janus nanomaterials can also be prepared via the Janus nanoreactors, such as Janus star-liked block copolymers used as the nanoreactors to prepare 0D Janus NPs. And 2D triblock copolymer brush-liked polymer can also be used to prepare 2D Janus nanoplate. The morphology of NPs produced using block copolymer templates is not only limited to spherical shapes. The block copolymers' unimolecular micelles can also be extended to anisotropic rod-like, Janus, and 2D structures. As previously discussed, Janus star-liked polymers have been realized through selective modification of β-CD. However, to date, strictly biphasic Janus NPs have not been realized. And this work has been realized by our group and will be published soon.

Secondly, the compositions of the inorganic NPs can be enriched. All the unimolecular nanoreactors could template the fabrication of high-entropy nanoalloys with or without phase separation. The nanocomposites containing inorganic nonmetals can also be fabricated easily. The unimolecular nanoreactors can also be used to control the synthesis of nanohydrogels via the cross-linking reactions confined inside the nanoreactors.

Thirdly, the valences of the templated NPs can be controlled precisely. The unimolecular nanoreactors with precise functional groups (number, type, and location) can be maintained after nanofabrication and function as the valences of the obtained NPs for further nanoreactions between NPs-like small molecules to form a complex and designable NP assembly. The new horizons of this robust and unconventional nanofabrication strategy will show striking fascination in nanostructure crafting and application expansion.

Fourthly, the surface of the templated NPs can be designed intelligently. The intimately and permanently connected polymers on the surface of NPs can direct their assembly to superstructures or responsivity to external stimuli. The directed self-assembly of functional inorganic nanomaterials to fabricate hierarchical structure is a new hot field. By precisely controlling the reactive or electrostatic interaction polymers of the outside block polymer, the nanomaterials prepared by using the unimolecular micelles nanoreactors can be an assembly to control the number of the particles assembled. Thus, the double precisely controlled synthesis of nanoassemblies can be realized.

Fifthly, the size of NPs is mainly determined by the molecular weight of the functional blocks, and the MW is precisely controlled by the controlled/ "living" polymerization method. By investigating the scaling theory of star-liked polymers in solution and the polymerization kinetics. Thus, we could build the connections between the reaction time and the size of NPs to truly realize the precisely size-controlled synthesis of NPs.

Ultimately, the goal is to develop a simple route to making monodisperse nanocrystals of any desired shape and size on a large scale and at a low cost. The directed binding, confinement, and growth of nanocrystals by organic templates is a broad approach to achieving these goals. Several specific templating strategies, each with its own fundamentally different physics, chemistry, advantages, and disadvantages, were investigated in turn. In so doing, general trends regarding the overall applicability of each approach could be made.

References

1 Grzelczak, M., Vermant, J., Furst, E.M., and Liz-Marzán, L.M. (2010). *ACS Nano* 4: 3591–3605.
2 Hamley, I.W. (2003). *Nanotechnology* 14: R39.
3 Houlton, A. (2016). *Science* 353: 1204–1205.
4 Kang, X. and Zhu, M. (2019). *Chem. Soc. Rev.* 48: 2422–2457.
5 Kim, B.H., Hackett, M.J., Park, J., and Hyeon, T. (2013). *Chem. Mater.* 26: 59–71.
6 Liu, Y., Wang, J., Zhang, M. et al. (2020). *ACS Nano* 14: 12491–12521.
7 Monson, C.F. and Woolley, A.T. (2003). *Nano Lett.* 3: 359–363.
8 He, M., Pang, X., Liu, X. et al. (2016). *Angew. Chem. Int. Ed.* 55: 4280–4284.
9 Heise, A., Hedrick, J.L., Frank, C.W., and Miller, R.D. (1999). *J. Am. Chem. Soc.* 121: 8647–8648.
10 Karaky, K., Reynaud, S., Billon, L. et al. (2005). *J. Polym. Sci., Part A: Polym. Chem.* 43: 5186–5194.
11 Liu, Y., Goebl, J., and Yin, Y. (2013). *Chem. Soc. Rev.* 42: 2610–2653.
12 Peng, X., Manna, L., Yang, W. et al. (2000). *Nature* 404: 59–61.
13 Prieto, G., Tüysüz, H., Duyckaerts, N. et al. (2016). *Chem. Rev.* 116: 14056–14119.
14 Radloff, C., Vaia, R.A., Brunton, J. et al. (2005). *Nano Lett.* 5: 1187–1191.
15 Xie, J.P., Zheng, Y.G., and Ying, J.Y. (2009). *J. Am. Chem. Soc.* 131: 888–889.
16 Zhang, L. and Wang, E. (2014). *Nano Today* 9: 132–157.
17 He, Y.J., Pang, X.C., Jiang, B.B. et al. (2017). *Angew. Chem. Int. Ed.* 56: 12946–12951.
18 Bai, J., Wang, X., Fu, P. et al. (2016). *J. Phys. Chem. C* 120: 705–714.
19 Bosman, A.W., Janssen, H.M., and Meijer, E.W. (1999). *Chem. Rev.* 99: 1665–1688.
20 Braun, E., Eichen, Y., Sivan, U., and Yoseph, G.B. (1998). *Nature* 391: 775–778.
21 Crooks, R.M., Zhao, M., Sun, L. et al. (2000). *Acc. Chem. Res.* 34: 181–190.
22 Cung, K., Han, B.J., Nguyen, T.D. et al. (2013). *Nano Lett.* 13: 6197–6202.
23 Rica, R.D. and Matsui, H. (2008). *Angew. Chem. Int. Ed.* 47: 5415–5417.
24 Mertig, M., Colombi Ciacchi, L., Seidel, R. et al. (2002). *Nano Lett.* 2: 841–844.
25 Müllner, M., Lunkenbein, T., Breu, J. et al. (2012). *Chem. Mater.* 24: 1802–1810.
26 Müllner, M., Lunkenbein, T., Schieder, M. et al. (2012). *Macromolecules* 45: 6981–6988.
27 Padalkar, S., Hulleman, J.D., Kim, S.M. et al. (2008). *Nanotechnology* 19: 275602.
28 Wang, Y. and Zhu, X. (2020). *Nanoscale* 12: 12698–12711.
29 Zhou, T., Zhu, J., Gong, L. et al. (2019). *J. Am. Chem. Soc.* 141: 2852–2856.
30 Pang, X.C., He, Y.J., Jung, J., and Lin, Z. (2016). *Science* 353: 1268–1272.
31 Pang, X.C., Zhao, L., Akinc, M. et al. (2011). *Macromolecules* 44: 3746–3752.
32 Pang, X., Zhao, L., Han, W. et al. (2013). *Nat. Nanotechnol.* 8: 426–431.
33 Qiu, H., Hudson Zachary, M., Winnik Mitchell, A., and Manners, I. (2015). *Science* 347: 1329–1332.
34 Kato, M., Kamigaito, M., Sawamoto, M., and Higashimura, T. (1995). *Macromolecules* 28: 1721–1723.

35 Wang, J.S. and Matyjaszewski, K. (1995). *J. Am. Chem. Soc.* 117: 5614–5615.
36 Braunecker, W.A. and Matyjaszewski, K. (2007). *Prog. Polym. Sci.* 32: 93–146.
37 Cao, Y., Xu, Y., Zhang, J. et al. (2015). *Polymer* 61: 198–203.
38 Chen, M., Zhong, M., and Johnson, J.A. (2016). *Chem. Rev.* 116: 10167–10211.
39 Matyjaszewski, K. (2012). *Macromolecules* 45: 4015–4039.
40 Matyjaszewski, K. and Xia, J. (2001). *Chem. Rev.* 101: 2921–2990.
41 Dadashi-Silab, S. and Matyjaszewski, K. (2018). *Macromolecules* 51: 4250–4258.
42 Heise, A., Hedrick, J.L., Trollsås, M. et al. (1998). *Macromolecules* 32: 231–234.
43 Pan, X., Fantin, M., Yuan, F., and Matyjaszewski, K. (2018). *Chem. Soc. Rev.* 47: 5457–5490.
44 Pan, X., Tasdelen, M.A., Laun, J. et al. (2016). *Prog. Polym. Sci.* 62: 73–125.
45 Yao, X., Huang, P., and Nie, Z. (2019). *Prog. Polym. Sci.* 93: 1–35.
46 Nag, A. and Pradeep, T. (2022). *ACS Nanosci. Au* 2 (3): 160–178.
47 Ren, J.M., McKenzie, T.G., Fu, Q. et al. (2016). *Chem. Rev.* 116: 6743–6836.
48 Wu, Z., Yao, Q., Zang, S., and Xie, J. (2019). *ACS Mater. Lett.* 1: 237–248.
49 Yi, C., Yang, Y., Liu, B. et al. (2020). *Chem. Soc. Rev.* 49: 465–508.
50 Pang, X., Feng, C., Xu, H. et al. (2014). *Polym. Chem.* 5: 2747–2755.
51 Loo, L., Guenther, R.H., Lommel, S.A., and Franzen, S. (2007). *J. Am. Chem. Soc.* 129: 11111–11117.
52 Lu, Z. and Yin, Y. (2012). *Chem. Soc. Rev.* 41: 6874–6887.
53 Zhang, M. and Müller, A.H.E. (2005). *J. Polym. Sci., Part A: Polym. Chem.* 43: 3461–3481.
54 Zhao, Y., Shuai, X., Chen, C., and Xi, F. (2004). *Macromolecules* 37: 8854–8862.
55 Zhao, Y., Chen, Y., Chen, C., and Xi, F. (2005). *Polymer* 46: 5808–5819.
56 Zubarev, E.R. (2013). *Nat. Nanotechnol.* 8: 396–397.
57 Mountrichas, G., Mpiri, M., and Pispas, S. (2005). *Macromolecules* 38: 940–947.
58 Zhang, W., Kong, H., Wu, Z. et al. (2021). *Chem. Mater.* 33: 5067–5075.
59 Xu, H., Xu, Y., Pang, X. et al. (2015). *Sci. Adv.* 1: e1500025.
60 Matyjaszewski, K., Tsarevsky, N.V., and Nicolay, V. (2009). *Nat. Chem.* 1: 276–288.
61 Kong, H., Zhang, W., Shi, G. et al. (2021). *Langmuir* 37: 10461–10468.
62 He, J., Zhang, W., Lv, C. et al. (2021). *Polymer* 215: 123345.
63 Feng, C., Pang, X., He, Y. et al. (2014). *Chem. Mater.* 26: 6058–6067.
64 Djalali, R., Li, S.-Y., and Schmidt, M. (2002). *Macromolecules* 35: 4282–4288.
65 Song, Y., Li, Y., Zhou, M. et al. (2021). *Sci. Adv.* 7: eabd2091.
66 Subramanian, V., Wolf, E.E., and Kamat, P.V. (2004). *J. Am. Chem. Soc.* 126: 4943–4950.
67 Yin, M., Cheng, Y., Liu, M. et al. (2008). *Angew. Chem. Int. Ed.* 47: 8400–8403.
68 Liang, T.C., Wang, Y., Shi, G. et al. (2021). *J. Phys. Chem. Lett.* 12: 3456–3463.
69 Boles, M.A., Engel, M., and Talapin, D.V. (2016). *Chem. Rev.* 116: 11220–11289.
70 Chen, P.C., Liu, Y., Du, J.S. et al. (2020). *J. Am. Chem. Soc.* 142: 7350–7355.
71 Deng, Y., Zhang, Z., Du, P. et al. (2020). *Angew. Chem. Int. Ed.* 59: 6082–6089.
72 Germano, L.D., Marangoni, V.S., Mogili, N.V.V. et al. (2019). *ACS Appl. Mater. Interfaces* 11: 5661–5667.
73 Chakraborty, I. and Pradeep, T. (2017). *Chem. Rev.* 117: 8208–8271.

74 Gao, C., Lyu, F., and Yin, Y. (2021). *Chem. Rev.* 121: 834–881.
75 Goswami, N., Yao, Q., Luo, Z. et al. (2016). *J. Phys. Chem. Lett.* 7: 962–975.
76 Sheiko, S.S., Sumerlin, B.S., and Matyjaszewski, K. (2008). *Prog. Polym. Sci.* 33: 759–785.
77 Jiang, B., He, Y., Li, B. et al. (2017). *Angew. Chem. Int. Ed.* 56: 1869–1872.
78 Yi, C., Liu, H., Zhang, S. et al. (2020). *Science* 369: 1369–1374.
79 Li, X., Iocozzia, J., Chen, Y. et al. (2018). *Angew. Chem. Int. Ed.* 57: 2046–2070.
80 Jiang, B., Han, C., Li, B. et al. (2016). *ACS Nano* 10: 2728–2735.
81 Zhao, S., Wang, Z., He, Y. et al. (2017). *ACS Energy Lett.* 2: 111–116.
82 Sasidharan, M., Nakashima, K., Gunawardhana, N. et al. (2011). *Nanoscale* 3: 4768–4773.
83 Min, B.K., Wallace, W.T., and Goodman, D.W. (2004). *J. Phys. Chem. B* 108: 14609–14615.
84 Bihani, M., Ansari, T.N., Finck, L. et al. (2020). *ACS Catal.* 10: 6816–6821.
85 Corrigan, N., Shanmugam, S., Xu, J., and Boyer, C. (2016). *Chem. Soc. Rev.* 45: 6165–6212.
86 Hu, J., Yuan, H., Peng, L. et al. (2016). Sci. *China Chem.* 59: 277–281.
87 Ishida, T., Murayama, T., Taketoshi, A., and Haruta, M. (2020). *Chem. Rev.* 120: 464–525.
88 Li, G. and Jin, R. (2013). *Acc. Chem. Res.* 46: 1749–1758.
89 Zhao, M. and Crooks, R.M. (1999). *Angew. Chem. Int. Ed.* 38: 364–366.
90 Wang, M., Pang, X., Zheng, D. et al. (2016). *J. Mater. Chem. A* 4: 7190–7199.

4

Environmentally Responsive Hairy Inorganic Particles

Caleb A. Bohannon[1], Ning Wang[1,2], and Bin Zhao[1]

[1] *University of Tennessee, Department of Chemistry, Knoxville, Tennessee 37996, United States*
[2] *China Medical University, School of Forensic Medicine, Department of Chemistry, Shenyang, Liaoning 110122, China*

4.1 Introduction

Polymer brush-grafted particles, also called hairy particles, are a class of nanostructured hybrid materials composed of a layer of polymer chains densely grafted on the surface of a core particle, commonly inorganic or metallic, by a covalent bond [1–5]. The grafting density is sufficiently high that the distance between neighboring polymer chains is smaller than the unperturbed size of the grafted polymer, forcing the polymer chains to stretch away from the core particle's surface and exhibit deformed conformations under equilibrium conditions [6]. These hybrid particles can exhibit the properties of both core particles (e.g. optical, magnetic, and mechanical properties) and grafted polymers (e.g. environmental compatibility and responsive property), often in a synergistic fashion. Hairy particles have received tremendous interest in the past two decades; significant effort has been invested to develop synthetic methods for preparing well-defined structures, understand their behavior and properties at various conditions, and explore their uses in technology [1–37]. The (potential) applications of hairy particles have been demonstrated in a wide variety of areas, including fabrication of advanced polymer nanocomposites, sensing, aqueous and oil lubrication, supported catalysts, delivery of substances, such as drugs, pickering emulsion stabilizers, and self-cleaning materials [8–15]. In addition, well-defined multicomponent hairy particles provide an excellent platform for studying the self-assembly and responsive behavior of multicomponent polymer systems in a confined geometry, with the curvature (radius) of core particles as a new parameter for controlling the behavior of surface-tethered polymer brushes and hairy particles [2, 9, 16–19].

Polymer brush-grafted particles are commonly prepared by three methods: (i) grafting-to, (ii) grafting-from, and (iii) polymer micelle-templated synthesis [1–5, 7]. "Grafting-to" is the most straightforward among the three general methods, where a functional group at the end or in the middle of polymer chains (e.g. –COOH)

Hairy Nanoparticles: From Synthesis to Applications, First Edition. Edited by Zhiqun Lin and Yijiang Liu.

reacts with a complementary functional group on the surface of particles (e.g. –OH group), producing covalently tethered polymer chains on the surface of the particles [18]. This method has been used to prepare homopolymer- and binary mixed polymer-grafted particles [16–19]. However, because of the steric hindrance, the grafting density of the tethered polymer chains on the surface can be low, which may affect, e.g. the dispersibility of hairy particles in polymer matrix or the stability in a solvent medium. Note that besides the conventional grafting of functionalized polymer chains onto the surface of pre-synthesized particles, one can also use end-functionalized polymers as stabilizing ligands in the synthesis of core nanoparticles (NPs) to produce polymer-grafted particles directly [16], where the size and geometry of core NPs can be controlled by the stabilizing polymer chains.

"Grafting-from" is also called surface-initiated polymerization [1–5, 7]. In this approach, an initiator is first immobilized on the surface of the particles through, e.g. silane chemistry, followed by the in situ surface-initiated polymerization of monomer molecules to grow end-tethered polymer chains. This method has been shown to be able to overcome the steric issue encountered in the "grafting-to" process and allow for the synthesis of polymer brushes with high grafting densities because monomer molecules are significantly smaller in size than macromolecules and can diffuse freely to the growing chain ends. Since Prucker and Rühe reported in 1998 the growth of polystyrene (PS) chains from silica gels that were surface-modified with an azo radical initiator via conventional free radical polymerization [20, 21], "grafting-from" has become the predominant method for the synthesis of hairy particles [20–34]. A wide variety of "living"/controlled polymerization techniques, particularly reversible deactivation radical polymerization, such as atom transfer radical polymerization (ATRP), nitroxide-mediated radical polymerization (NMRP), and reversible addition-fragmentation chain transfer (RAFT) polymerization, have been employed in the surface-initiated polymerization. These polymerization methods allow for the synthesis of hairy particles with controlled molecular weights, narrow molecular weight distributions, high grafting densities, and well-defined architectures. Under the optimized conditions, hairy particles with grafting densities of 0.3–1.2 chains nm^{-1} [2] can be routinely achieved by surface-initiated "living"/controlled polymerization [5]. Block copolymer brush-grafted particles can be readily synthesized from well-defined homopolymer brush-grafted particles via chain extension with different monomers [1, 32]. Scheme 4.1 shows the synthesis of hairy particles by surface-initiated ATRP (SI-ATRP), NMRP, and RAFT polymerization with a surface-immobilizable initiator or chain transfer agent (CTA) for each polymerization method as an example. Despite all the benefits, one disadvantage of "grafting-from" is that the grafted polymer chains have to be cleaved off from the particles for determining the molecular weights and dispersities (Đ). A more convenient method is the addition of a free initiator (or a free CTA) to produce a free polymer analog. It has been shown by many research groups that the molecular weight and Đ of the free polymer formed in the solution from the free initiator/CTA are essentially the same as those of the grafted polymer on the particles [5, 23, 27, 32].

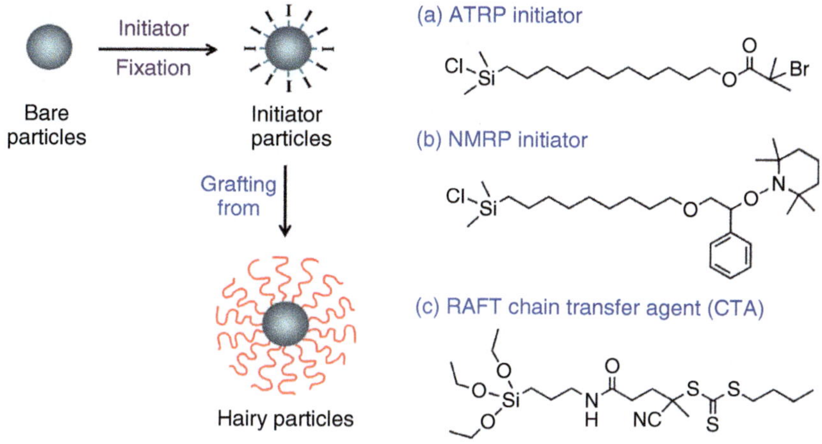

Scheme 4.1 Schematic illustration for the synthesis of hairy particles by "grafting-from" using (a) ATRP, (b) NMRP, and (c) ARFT polymerization, with a surface-immobilizable initiator or chain transfer agent for each polymerization method as an example.

The third general method is the polymer micelle-templated synthesis of hairy particles, where the core of polymer micelles, either multimolecular block copolymer or unimolecular star block micelles, is loaded with a metal salt or an organometallic compound, which is subsequently reduced or reacted with a later-added reactant to produce particles [35–37]. By using unimolecular star block copolymer micelles, this method has been shown to be successful in the synthesis of hairy particles with uniform core particle sizes, various particle shapes, and uncommon particle morphologies (e.g. hollow hairy NPs) [37].

Environmentally responsive hairy particles can exhibit different structures, properties, and behaviors under different conditions. This chapter presents two types of such hairy inorganic particles, binary mixed homopolymer brush- (Section 4.2) and thermoresponsive polymer brush-grafted silica particles (Section 4.3) [2, 5, 32, 38–57]. Mixed polymer brushes are composed of two or more chemically distinct polymers, usually homopolymers, randomly or alternately immobilized by one end on a solid substrate via a covalent bond with a sufficiently high grafting density for both polymers [2, 58]. These multicomponent brushes can undergo self-reorganization in a confined geometry to minimize the system's free energy upon, e.g. solvent changes, and exhibit distinct and reversible morphologies. Note that the grafted polymers in mixed brushes are not necessarily stimuli-responsive polymers in a conventional sense [2], which are usually defined as the polymers that exhibit relatively large and abrupt conformational changes [59], often with a dramatic solubility change, in liquid media upon application of external stimuli (e.g. temperature or pH changes, light, etc.). In contrast, in thermoresponsive hairy particles, the grafted polymers are thermosensitive polymers that exhibit either a lower critical solution temperature (LCST) or an upper critical solution temperature (UCST) transition commonly in water or aqueous solution, resulting in distinctive thermoresponsive behaviors for hairy particles [47]. In Section 4.2, we first present the unique synthetic strategy developed in our laboratory for the preparation of well-defined binary mixed homopolymer brushes on silica particles and then their responsive behaviors as revealed by ^{1}H NMR spectroscopy analysis and transmission electron microscopy. Section 4.3 focuses on thermoresponsive hairy particles, from the synthesis to the LCST transition of the grafted polymer chains on the surface of particles, and the responsive behavior and application of these particles. A brief summary and outlook are provided in Section 4.4.

4.2 Environmentally Responsive Well-defined Binary Mixed Homopolymer Brush-grafted Silica Particles

4.2.1 Introduction to Mixed Polymer Brushes

Mixed polymer brushes have received tremendous interest in the past decades due to their fascinating responsive behavior and a rich variety of morphologies under different environments. These multicomponent brushes have been widely investigated for potential applications, such as surface-responsive materials and self-cleaning materials [2, 10, 60]. For example, if one polymer is hydrophilic and the other is hydrophobic, the surface property of the binary mixed brushes could be easily switched from hydrophobic to hydrophilic and vice versa simply by washing the brushes with different solvents [60]. The intriguing behavior of binary mixed homopolymer brushes on a flat solid substrate is governed by a number of factors, including (i) degree of polymerization (DP or N), dispersity (Đ), and grafting density (σ) of each type of polymer chains (N_A, $Đ_A$, and σ_A; N_B, $Đ_B$, and σ_B), (ii) distribution of the grafting sites of two polymers on the substrate, and (iii) Flory–Huggins interaction parameters (χ) between two grafted polymers and between each grafted polymer and environment (e.g. solvent or polymer matrix) [2]. Unlike block copolymers in the bulk state, where the junction point that covalently links different blocks together is mobile at the interface of microdomains, the grafting sites of polymer chains in the mixed brushes are immobilized on the surface of the solid substrate. In addition, because the thicknesses of polymer brushes are typically less than 100 nm, the environmental factors play a very important role in determining the structure of mixed brushes.

Under the equilibrium melt conditions, there are two possible ways that symmetric binary mixed homopolymer brushes on a flat substrate can microphase segregate: lateral phase separation, producing a "rippled" nanostructure, and vertical microphase separation, forming a "layered" morphology (Scheme 4.2) [61–74]. The theoretical study by Marko and Witten has predicted that the lateral microphase separation should be the one that takes place, and the critical molecular weight for microphase separation is expected to be 2.27 times that for the same homopolymers in a symmetric blend at its demixing threshold (i.e. $\chi_{AB}N = 4.54$, where $N = N_A = N_B$) [61]. The periodicity of the nanopattern formed from the lateral phase separation is predicted to be 1.97 times the chain root-mean-square end-to-end distance ($<R_{rms}>$). The lateral phase separation of symmetric mixed brushes under melt conditions predicted by Marko and Witten has been repeatedly

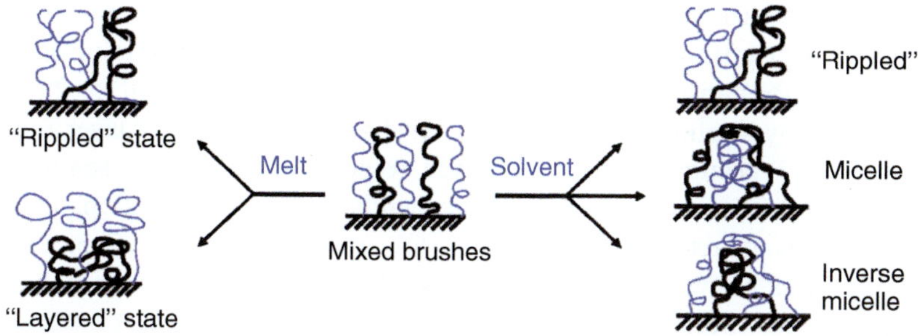

Scheme 4.2 Simplified schematic illustration of self-assembly of binary mixed homopolymer brushes on a flat substrate under equilibrium melt conditions and in various solvents. Source: Reproduced with permission from Ref. [2]; © 2009, American Chemical Society.

corroborated by other researchers in the theoretical and simulation studies. Dong showed that the spatial period of the pattern is independent of the interaction strength between the two grafted polymers [63].

In solvents, mixed homopolymer brushes display even richer behavior than in melt because the solvent quality for both polymers can be readily and continuously tuned [58, 64–71]. Zhulina and Balazs studied a special type of mixed brushes theoretically, which is constructed by grafting a symmetric diblock copolymer onto a solid substrate via a reactive group at the junction point of the two blocks [58]. By varying the quality of nonselective solvents, grafting density, and chain length, a variety of self-assembled nanostructures within the brush layer were observed. Müller later also investigated the behavior of binary mixed brushes on a flat surface in nonselective solvents [66]. At large incompatibilities or asymmetric compositions, two interesting "dimple" phases, where different types of polymer chains form clusters arranged on a quadratic (the checkerboard structure) or hexagonal lattice, were discovered theoretically. Wenning et al. found that the density fluctuation of the grafting points of two polymers in the mixed brushes led to irregular nanostructures and prevented the formation of a long-range order [68]. In selective solvents, surface tethered micelles with the solvophobic polymer chains forming a dense core and the solvophilic polymer chains constituting the corona have been predicted (Scheme 4.2), and the brushes exhibit the surface properties of the solvophilic polymer [69–71]. The surface properties of mixed brushes can be reversibly switched by treating the brushes with different selective solvents. These theoretical and simulation studies showed that a large variety of surface nanostructures could be achieved in the mixed brushes by tuning various parameters, such as N_A, N_B, σ_A, σ_B, chemical compositions, and solvent quality. In addition, different morphologies formed from the same mixed brushes are reversible due to the covalent grafting of the polymer chains on the solid substrate, making mixed polymer brushes robust surface-responsive materials.

The first experimental study of mixed brushes was reported by Sidorenko et al. who prepared mixed PS/poly(2-vinylpyridine) (PVP) brushes on flat substrates by a two-step surface-initiated conventional free radical polymerization process from a surface-immobilized azo initiator [60]. The switching properties of the obtained mixed brushes were beautifully demonstrated by exposing the brushes to toluene, a selective solvent for PS, and an acidic aqueous solution, a selective solvent for PVP, making the brushes hydrophobic with PS chains residing at the outermost layer and hydrophilic with PVP occupying the upper layer, respectively. Since this seminal work, considerable efforts were made on the synthesis, characterization, and potential applications of mixed polymer brushes [67, 71–86]. Besides the aforementioned "grafting-from," new methods such as "grafting-to" and surface-initiated "living"/controlled polymerization were also developed for the synthesis of mixed brushes. For example, Julthongpiput et al. synthesized Y-shaped brushes by grafting a diblock copolymer to silicon wafers at the junction point of the two blocks [78]. However, the brush thickness is rather low, typically <5 nm. We developed a unique "grafting-from" method for the synthesis of well-defined binary mixed homopolymer brushes by sequential SI-ATRP and SI-NMRP from asymmetric difunctional initiator (Y-initiator)-functionalized silica wafers (Scheme 4.3) [83–85]. ATRP and NMRP are two different types of reversible deactivation radical polymerization performed under different conditions [87, 88]. Y-initiators were designed to ensure that the two initiators were well-mixed in the initiator layer and, thus, the two grafted polymers in the brush layer. We confirmed that the NMRP initiator was thermally stable under the chosen conditions for SI-ATRP [83]. Therefore, the use of SI-ATRP and SI-NMRP allowed us to independently control the chain lengths of the two grafted polymers while achieving narrow dispersities of the grafted polymer chains.

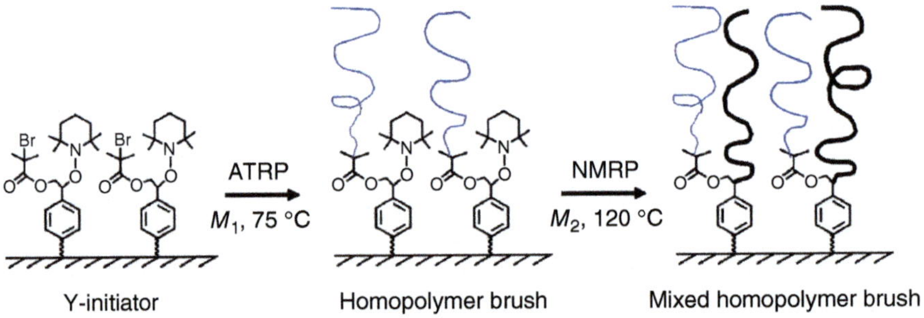

Scheme 4.3 Synthesis of well-defined mixed homopolymer brushes by sequential surface-initiated atom transfer radical polymerization (ATRP) and nitroxide-mediated radical polymerization (NMRP) from a Y-initiator-functionalized silicon wafer in a two-step process. Source: Reproduced with permission from Ref. [2]; © 2009, American Chemical Society.

Although the environmentally responsive properties of mixed brushes on flat solid substrates have been well documented [67, 71–86], less efforts were made to elucidate how different polymers were organized in the brush layers. Using step-by-step oxygen plasma etching and atomic force microscopy (AFM) ("AFM tomography"), Minko and coworkers characterized the structure of mixed PS/poly(methyl methacrylate) (PMMA) brushes on silicon wafers [89], which were synthesized by two-step surface-initiated conventional free radical polymerizations.

The brushes exhibited a rippled structure upon exposure to toluene, a nonselective solvent, and a dimpled structure after the treatment with acetone, a selective solvent for PMMA. It was concluded that the rippled morphology comprised of depressed PS-rich and elevated PMMA-rich domains; for the dimpled morphology, the core of the clusters was made up of PS chains, and the very top layer and valleys between clusters were formed by PMMA chains. Despite these studies, the nanostructures of mixed brushes under different conditions were not directly visualized, largely because of the challenges in the synthesis of well-defined mixed brushes and the application of common tools used in the study of block copolymers, such as transmission electron microscopy (TEM), differential scanning calorimetry (DSC), and scattering methods.

4.2.2 Mixed Polymer Brushes Grafted on Particles

When two distinct polymers are densely grafted on inorganic or metallic particles, an intriguing class of hybrid particles is obtained [2]. These hairy particles are expected to exhibit both the environmentally responsive properties of mixed brushes and the properties of core particles (e.g. mechanical, optical, catalytic, and magnetic properties) that are not commonly associated with the organic polymers. They also offer a unique opportunity to study by TEM the self-assembled morphologies of mixed brushes under different conditions because TEM can be directly used to visualize the nanostructures of the mixed brushes grafted on particles with a size of <200 nm and larger silica particles can be microtomed to thin sections for TEM observation [39]. We note here, however, that the mixed brushes grafted on spherical particles are not just simply a curved version of the brushes on planar substrates. A new fundamental parameter, the curvature of the particle surface, is introduced into the mixed brush system, which is expected to play a significant role in the self-assembly of mixed brushes on solid substrates and lead to new morphologies. It is especially interesting when the size of the particles is comparable to the sizes of polymer chains [90–94]; the segments of the densely grafted polymer chains become more relaxed with increasing distance from the grafting surface. In addition, the higher surface area of the particles compared with flat substrates allows for the cleavage of the surface-tethered polymer chains for thorough characterization of the molecular weights, dispersities, and grafting densities of the grafted polymers if the brushes are made by surface-initiated "living"/controlled polymerizations. The molecular characteristics of the grafted polymers are critical for understanding the behavior of mixed brushes under different conditions.

Roan studied by computer simulations the microphase separation of binary mixed homopolymer brushes grafted on spherical NPs with a size comparable to those of the grafted polymer chains [90]. By tuning the chain lengths, grafting densities, and grafting patterns of the two polymers, he observed the formation of well-ordered islanded nanostructures with a well-defined number of islands (3, 6, 8, 9, 10, and 12 islands) from the lateral microphase separation of mixed brushes in nonselective good solvents. The equilibrium morphology can be changed from rippled to 12-islanded and then layered upon increasing the chain length disparity when the grafting densities for the two polymers are maintained identical. Based on these simulation results, Roan suggested the use of microphase-separated mixed brush-grafted NPs for the fabrication of multivalent NPs, the NPs with controlled numbers and locations of functional sites. Wang et al. subsequently conducted a systematic study of the microphase separation of binary mixed brushes on nanospheres with a radius (R) comparable to the polymer sizes by numerical implementation of the self-consistent field theory [91]. The effects of overall grafting density, composition, chain length asymmetry between two grafted homopolymers, and sphere radius on the formation of islanded structures on the surface of nanospheres were thoroughly investigated, and a series of islanded nanostructures with a well-defined number of islands ranging from 2 to 12 except for 11 were observed. This is in contrast to the hexagonal and quadratic arrangement of microphase-separated nanodomains of mixed brushes on planar substrates [66]. It was pointed out that the radius of the NPs plays a critical role in determining the morphology of mixed brushes on nanospheres. Besides the phase separation behavior of mixed brushes on NPs, self-assembled hierarchical structures of symmetric binary mixed brush-grafted NPs were also studied using a self-consistent field theory [94]. When the ratio of the sphere radius (R) to the polymer size ($N^{1/2}b$, where N and b are Kuhn monomer number and length, respectively) was smaller than 1, an unusual non-close-packed simple cubic crystal lattice was observed, and some unconventional microphases, such as tetragonal cylinder phases, were formed in the polymer matrix. When the size ratio increased to above 1, the most stable NP arrangement changed to a body-centered cubic structure with the formation of a bicontinuous nanostructure in the polymer matrix. Additionally, calculations showed that the mixed brush NPs preferred to arrange into the lattice where they could maintain the free energy-minimizing morphology as an isolated particle.

These computer simulation studies revealed intriguing morphologies of individual mixed brush-grafted NPs and self-assembled hierarchical superstructures of mixed brush NPs, governed by a number of factors, including chain lengths, chain length disparity, overall and individual grafting densities, grafting pattern, substrate geometry, and sphere radius [90–94]. To study the responsive properties and the morphologies of binary mixed homopolymer brushes grafted on particles, it is necessary to develop robust synthetic methods that allow for the control of those molecular parameters of binary mixed homopolymer brushes.

4.2.3 Synthesis of Well-defined Binary Mixed Homopolymer Brushes on Silica Particles

We applied the method developed for the synthesis of mixed brushes on flat substrates (Scheme 4.3) to grow well-defined binary mixed homopolymer brushes on silica particles (Scheme 4.4) [38]. Silica particles with an average diameter of 180 nm used in this work were synthesized by the Stöber process, which involves the hydrolysis and condensation of tetraethoxysilane in ethanol in the presence of ammonia and is well known to produce spherical silica particles with tunable sizes and a relatively narrow size distribution [95, 96]. A monochlorosilane-terminated

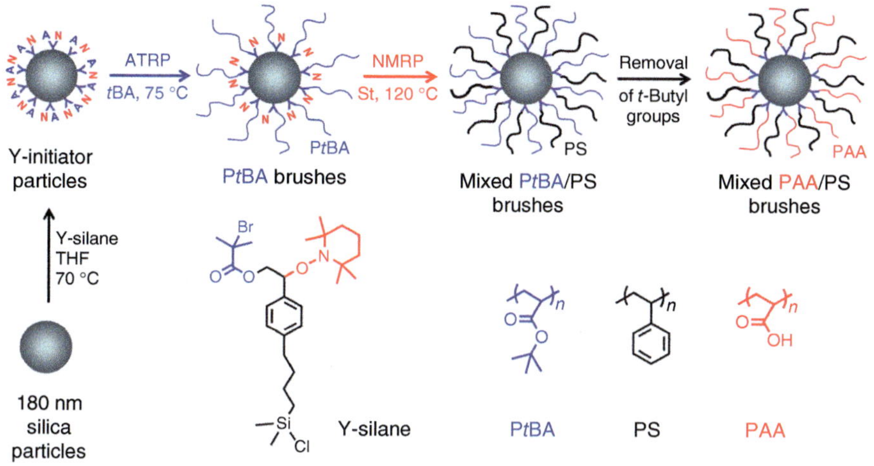

Scheme 4.4 Synthesis of well-defined mixed poly(*tert*-butyl acrylate)/polystyrene brushes and amphiphilic mixed poly(acrylic acid)/polystyrene brushes on 180 nm silica particles.

Y-initiator (Y-silane) was immobilized on the surface of silica particles via the reaction between the silanol groups on the surface of silica particles and the Si–Cl bond of the Y-initiator [38]. Y-silane was synthesized through a four-step procedure as shown in Scheme 4.5. The synthesis started with a reaction of 4-vinylbenzyl chloride with the free radicals formed from the decomposition of benzoyl peroxide (BPO) at 80 °C in the presence of 2,2,6,6-tetramethylpiperidinoxy (TEMPO). The product from the first step was then reacted with allylmagnesium chloride to install a double bond at the end, followed by the esterification with 2-bromo-2-methylpropionyl bromide to introduce the ATRP initiating group, producing the Y-initiator. Y-silane was synthesized by subsequent hydrosilylation of the Y-initiator with dimethylchlorosilane in the presence of a platinum catalyst and used immediately to functionalize the surface of 180 nm silica particles dispersed in tetrahydrofuran (THF). After the surface immobilization was conducted in a nitrogen atmosphere at 70 °C for 40 hours, the Y-initiator-functionalized silica particles were purified by multiple cycles of dispersion in THF and centrifugation. On the basis of the particle size determined from scanning electron microscopy and the thermogravimetric analysis (TGA) data, the Y-initiator density on the surface of the initiator particles was estimated to be 0.33 nm^2/initiator [38].

Mixed poly(*tert*-butyl acrylate) (P*t*BA)/polystyrene (PS) brushes were then grown from the Y-initiator-functionalized 180 nm silica particles in two steps by sequential SI-ATRP of *tert*-butyl acrylate (*t*BA), conducted at 75 °C, and SI-NMRP of styrene at 120 °C. We chose to grow P*t*BA chains first because the radical formation from the initiator and catalyst in ATRP is a bimolecular process, while the thermal decomposition of the NMRP initiating moiety is a unimolecular process [87, 88]. A unimolecular initiator activation mechanism is advantageous for the synthesis of the second type of polymer chains from homopolymer brush-grafted silica particles because of the steric hindrance presented by the existing chains in the second step.

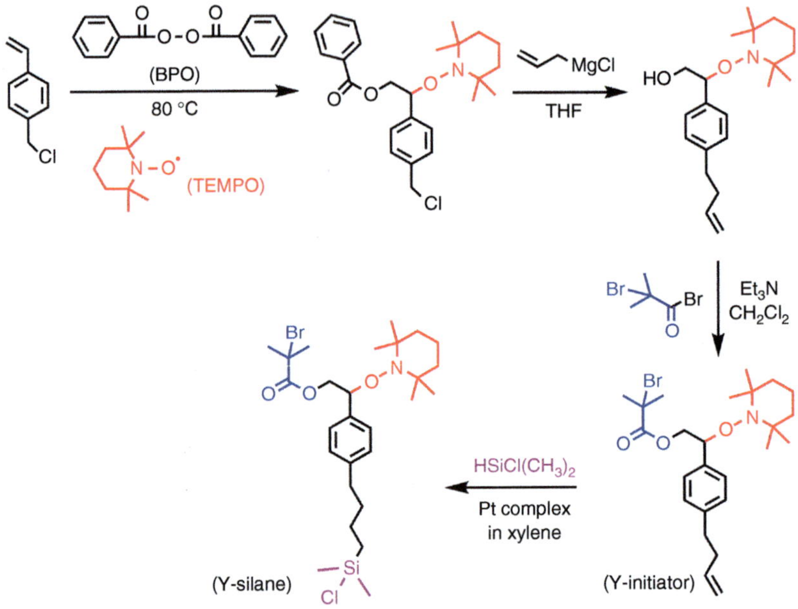

Scheme 4.5 Synthesis of a surface-immobilizable monochlorosilane-functionalized asymmetric difunctional initiator (Y-silane).

As mentioned earlier, the NMRP initiating group in the Y-initiator was confirmed to be stable under the chosen ATRP conditions. A sacrificial ATRP initiator, ethyl 2-bromoisobutyrate (EBiB), was added into the polymerization mixture to better control the surface-initiated polymerization. The monomer conversion was monitored by ^{1}H NMR spectroscopy analysis; kinetic analysis showed a linear relationship between $\ln([M]_0/[M])$ versus reaction time (Figure 4.1), indicating that the polymerization was of first order with respect to the monomer concentration and the number of the growing chains remained constant during the reaction. This meant that the polymerization was "living." The P*t*BA brush-grafted silica particles were isolated by centrifuge and repeatedly washed to remove any physically attached polymer chains. The successful growth of P*t*BA chains from the Y-initiator immobilized on the surface of silica particles was confirmed by ^{1}H NMR analysis of P*t*BA hairy particles in $CDCl_3$, which showed the characteristic peaks of P*t*BA, and TGA, which revealed a large decrease in weight retention at 800 °C compared with the Y-initiator-functionalized silica particles.

To determine the molecular weight and the dispersity of the grafted polymer, the silica core of P*t*BA brush-grafted particles was etched with hydrofluoric acid (HF). The cleaved polymer was collected and characterized by ^{1}H NMR spectroscopy and size exclusion chromatography (SEC) analysis. The molecular weights of the grafted polymer were 23.2 kDa ($M_{n,SEC}$) and 26.6 kDa ($M_{w,SEC}$) from SEC with respect to PS standards, essentially the same as those of the free polymer ($M_{n,SEC}$ = 24.2 kDa and $M_{w,SEC}$ = 26.4 kDa). However, the dispersity of the grafted P*t*BA was slightly broader

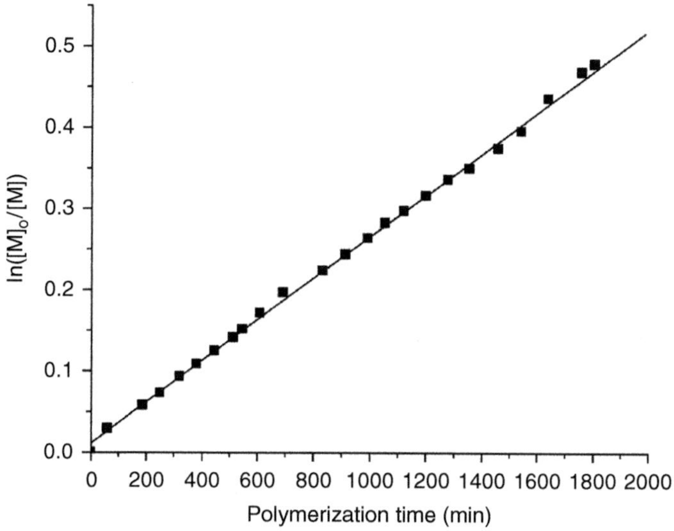

Figure 4.1 Plot of $\ln[M]_0/[M]$ versus time for the kinetic analysis of ATRP of *t*-BA at 75 °C in the synthesis of P*t*BA brushes from Y-initiator-functionalized silica particles. The ratio of $[M]_0$: [free ATRP initiator]$_0$: $[CuBr]_0$: $[CuBr_2]_0$: [*N*,*N*,*N*′,*N*″,*N*″-pentamethyldiethylenetriamine]$_0$ was 553 : 1 : 1.07 : 0.033 : 1.05. Source: Reproduced with permission from Ref. [38]; © 2005, American Chemical Society.

($Đ$ = 1.15) than that of the free polymer ($Đ$ = 1.09). From the DP of the polymer, the average diameter of silica particles, and the TGA data, the P*t*BA grafting density (σ_{PtBA}) was calculated to be 0.40 chain nm^{-2}, assuming that the silica particles are spherical and the density of silica particles is 2.07 g cm^{-3}. The high grafting density indicated that the polymer chains were highly stretched away from the grafting sites on the surface of silica particles.

In the second step for the synthesis of mixed brush-grafted silica particles, SI-NMRP of styrene from P*t*BA brush-grafted silica particles was carried out at 120 °C with the addition of a free NMRP initiator, 1-phenyl-1-(2′,2′,6′,6′-tetramethyl-1′-piperidinyloxy)-2-benzoyloxyethane (PTEMPO). Kinetic analysis showed that the polymerization was "living," as reflected by a nearly linear relationship between $\ln([M]_0/[M]$ versus time. After the polymerization proceeded to a point at which the molecular weight of PS was comparable to that of P*t*BA, the reaction was stopped, and the particles were isolated by centrifugation and purified by repeated dispersion in $CHCl_3$ and centrifugation. ^{1}H NMR spectroscopy analysis of the obtained hairy particles in $CDCl_3$ showed the presence of the characteristic peaks of P*t*BA at 1.42 and 2.20 ppm and the distinctive aromatic peaks of PS in the range of 6.2–7.2 ppm, indicating the success in the growth of PS chains from the P*t*BA brush-grafted silica particles by SI-NMRP. From the TGA data, the grafting density of PS was calculated to be 0.37 chains/nm^2/chain, which was similar to that of P*t*BA (0.40 chains nm^{-2}). To determine the M_n and the $Đ$ of the grafted PS chains, HF was used to etch the silica core of the mixed brush-grafted silica particles. SEC analysis showed that the $M_{n,SEC}$ of the polymer mixture cleaved from the mixed P*t*BA/PS brush particles was 37.5 kDa. This value is significantly higher than that of either free P*t*BA or PS but lower than the sum of two free polymers (M_n = 47.2 kDa), suggesting that a portion of the mixture was the diblock copolymer of P*t*BA and PS formed from the Y-initiator. This implied that the P*t*BA and PS chains in the brush layer were well mixed and the grafting points were nearly uniformly distributed, making the mixed brushes well suited for a morphology study. Since the Y-initiator contained a hydrolyzable ester bond, we hydrolyzed the polymer mixture cleaved from the mixed brush particles in a solution of $NaOCH_3$ in a mixture of THF and CH_3OH with a volume ratio of 13 : 1. After the removal of the solvents, cyclohexane was used to extract PS from the residual solid. The extracted polymer was confirmed to be PS by ^{1}H NMR spectroscopy, and SEC analysis showed that the $M_{n,SEC}$ of the grafted PS was 23.8 kDa, essentially the same as that of the free polymer ($M_{n,SEC}$ = 23.0 kDa). This confirmed that the molecular weights and dispersities of the grafted and free polymers were essentially the same, and the molecular characteristics of a free polymer could be used for the corresponding grafted polymer on silica particles synthesized by surface-initiated reversible deactivation radical polymerization.

4.2.4 Responsive Properties of Binary Mixed Homopolymer Brush-grafted Silica Particles

Binary mixed polymer brushes on planar substrates have been shown to exhibit switching properties in response to environmental changes [60, 84, 85]. To demonstrate the solvent-responsive properties of mixed brush-grafted silica particles, we synthesized amphiphilic mixed poly(acrylic acid) (PAA)/PS brush particles from mixed P*t*BA/PS brush-grafted silica particles with a P*t*BA $M_{n,SEC}$ of 24.2 kDa and a PS $M_{n,SEC}$ of 23.0 kDa by removing the *tert*-butyl groups of P*t*BA [38]. The complete cleavage of the *tert*-butyl groups was confirmed by ^{1}H NMR spectroscopy analysis. $CHCl_3$ is a selective solvent for PS, whereas methanol is a selective solvent for PAA. To study the dispersibility of mixed PAA/PS brush particles in these two solvents, mixed PAA/PS brush particles (~1.0 mg) were ultrasonicated in $CHCl_3$ (2 ml) and CH_3OH (2 ml), respectively. For comparison, PS brush-grafted silica particles with a $M_{n,SEC}$ of 18.7 kDa, synthesized from the same Y-initiator particles by SI-NMRP, and PAA brush-grafted silica particles, prepared by the removal of the *tert*-butyl groups of P*t*BA brush-grafted silica particles with a $M_{n,SEC}$ of 17.6 kDa, were also ultrasonicated in these two solvents. As can be seen in Figure 4.2, PS brush particles can be readily dispersed in $CHCl_3$ but not in methanol, whereas PAA brush particles can be dispersed in CH_3OH but not in $CHCl_3$. In contrast, mixed PAA/PS brush particles can be dispersed in both solvents, forming stable dispersions. Strong light scattering was observed in the dispersions of PS brush particles in $CHCl_3$, PAA brush particles in CH_3OH, and mixed PAA/PS brush particles in both $CHCl_3$ and CH_3OH when a laser light passed through the dispersions of hairy particles (Figure 4.2).

To further investigate the responsive property of mixed PAA/PS brush particles in selective solvents, ^{1}H NMR spectroscopy analysis of mixed PAA/PS brush particles in $CDCl_3$, *N,N*-dimethylformamide-d7 (DMF-d7), and CD_3OD was performed (Figure 4.3). DMF-d7 is a good solvent for both PAA and PS. To facilitate the

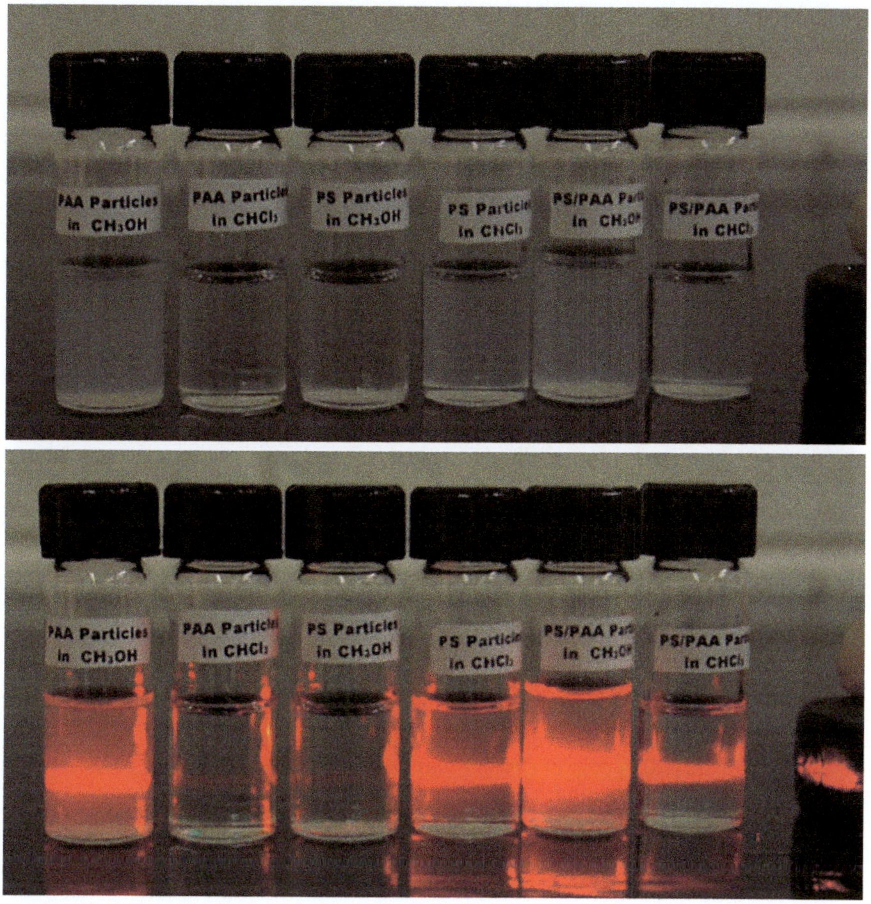

Figure 4.2 A Tyndall light scattering experiment shows the stable dispersions of PAA brush particles in CH_3OH but not in $CHCl_3$, PS brush particles in $CHCl_3$ but not in CH_3OH, and mixed PAA/PS brush particles in both $CHCl_3$ and CH_3OH. Source: Reproduced with permission from Ref. [38]; © 2005, American Chemical Society.

dispersion of mixed PAA/PS brush particles and the chain reorganization in selective solvents, a drop of DMF-d7 was added prior to $CDCl_3$ and CD_3OD. In DMF-d7, the characteristic peaks of PS at 6.0–7.4 ppm (from the aromatic ring) and PAA at 2.45 ppm (from backbone –CH–) were visible, indicating that both PS and PAA chains were highly solvated and mobile. In $CDCl_3$, the aromatic peaks were visible, but the peak of PAA at 2.45 ppm disappeared, suggesting that the PAA chains collapsed and the PS chains were mobile. On the other hand, in CD_3OD, the aromatic peaks of PS were absent while the peak of PAA at 2.45 ppm was observed, indicating that the PS chains collapsed and the PAA chains were solvated. These observations demonstrated the responsive behavior of mixed brush-grafted particles through chain reorganization according to the environmental conditions.

Although the above Tyndall experiment and ^1H NMR spectroscopy analysis unequivocally showed the responsive property of mixed brush-grafted

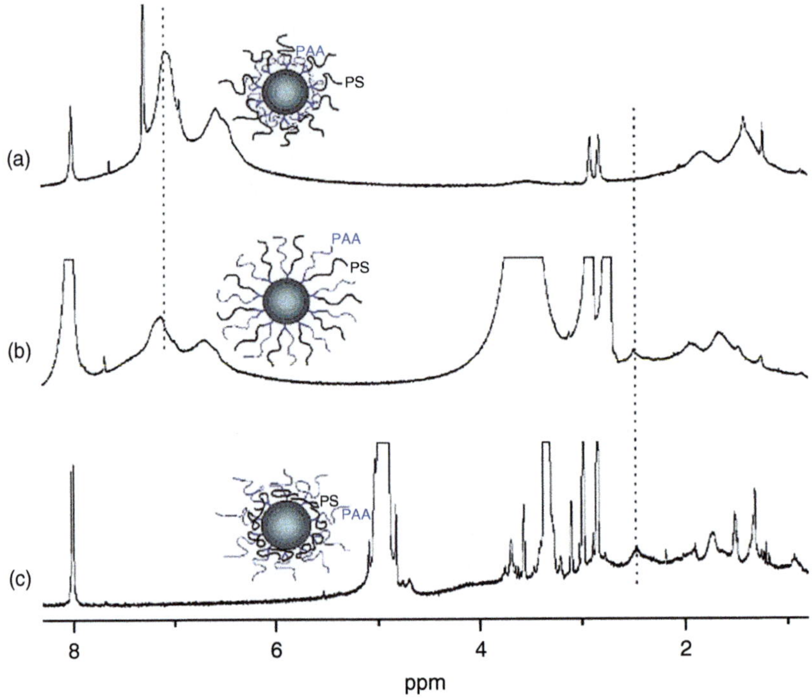

Figure 4.3 ^1H NMR spectra of mixed PAA/PS brush particles dispersed in (a) $CDCl_3$, (b) DMF-d7, and (c) CD_3OD. A drop of DMF-d7 was added into the mixed brush particles prior to $CDCl_3$ and CD_3OD to increase the concentration of the dispersed hairy particles. Source: Reproduced with permission from Ref. [38]; © 2005, American Chemical Society.

particles, it remained unclear how the two polymers self-reorganized when the environmental conditions changed. As mentioned earlier, TEM is well suited for the characterization of the morphology of mixed brushes on particles. For the first time, TEM revealed that the mixed P*t*BA/PS brushes grafted on 180 nm silica particles with a P*t*BA $M_{n,SEC}$ of 24.2 kDa and a PS $M_{n,SEC}$ of 23.0 kDa phase separated laterally under equilibrium melt conditions [39], resulting in a "rippled" nanostructure with a feature size of ~10 nm. This length scale was on the order of the $<R_{rms}>$ values of the two grafted polymers in the unperturbed state, as predicted by Marko and Witten for symmetric binary mixed homopolymer brushes in melt [61]. In solvents, mixed brushes on flat substrates have been shown by theoretical and simulation studies to exhibit a rich variety of self-assembled morphologies, including rippled, layered, and dimpled phases depending on the molecular characteristics of the grafted polymers and the solvent quality [58, 64–71]. To examine the morphologies of mixed brushes grafted on silica particles cast from different solvents by TEM, a mixed P*t*BA/PS brush-grafted 180 nm silica particle sample with P*t*BA $M_{n,SEC} = 24.2$ kDa, $\sigma_{PtBA} = 0.40$ chains nm^{-2}, PS $M_{n,SEC} = 23.0$ kDa, and $\sigma_{PS} = 0.37$ chains nm^{-2} was selected. The TEM sample was prepared by solution casting a 1.0 mg ml^{-1} dispersion of the hairy particles in $CHCl_3$, a good solvent for both P*t*BA and PS, on carbon-coated TEM grids and staining with RuO_4 vapor. As

(a) (b)

Figure 4.4 Bright-field TEM micrographs of (a) a top view and (b) a cross-sectional view of mixed P*t*BA/PS brush-grafted 180 nm silica particles with P*t*BA $M_{n,SEC} = 24.2$ kDa, $\sigma_{PtBA} = 0.40$ chains nm^{-2}, PS $M_{n,SEC} = 23.0$ kDa, and $\sigma_{PS} = 0.37$ chains nm^{-2} that were solution cast at a concentration of 1.0 mg ml^{-} from $CHCl_3$, a good solvent for both P*t*BA and PS. The thin section (~50 nm) in (b) was obtained by cross-sectional microtoming with a diamond knife at ambient temperature. The TEM samples were stained with RuO_4 vapor for 30 minutes. Source: Reproduced with permission from Ref. [40];

can be seen from the bright field TEM micrograph shown in Figure 4.4a, the hairy particles self-assembled into a hexagonal pattern, and a random worm-like, nearly bicontinuous brush nanostructure was observed [40], indicating that the mixed P*t*BA/PS brushes phase separated laterally as under equilibrium melt conditions [39]. Note that RuO_4 selectively stained PS, making it dark under TEM, while P*t*BA nanodomains were bright. Both grafted P*t*BA and PS chains appeared to be highly stretched, and the interstitial spaces between silica particles were filled by microphase-separated polymer chains. In addition, the wormlike dark PS and bright P*t*BA nanodomains from neighboring particles were bridged. Figure 4.4b shows a TEM micrograph of a microtomed thin section (~50 nm) of the $CHCl_3$-cast mixed brush particles embedded in a standard epoxy, which revealed a "rippled" morphology formed from lateral microphase separation. The thickness of the laterally phase-separated brush layer was ~30 nm, in contrast to ~20 nm observed in the melt state, indicating that the polymer chains were highly stretched in chloroform. Image analysis showed that the average widths of dark PS and bright P*t*BA stripes were 13.5 ± 4.7 and 8.2 ± 3.6 nm, respectively [40]. These values were similar to the $<R_{rms}>$ values of the two grafted polymers (10.2 nm for PS and 9.2 nm for P*t*BA in the unperturbed state), as predicted by the theoretical studies [58, 61]. We also synthesized a lower molecular weight mixed P*t*BA/PS brush-grafted 180 nm silica particles with P*t*BA $M_{n,SEC}$ of 10.4 kDa, $\sigma_{PtBA} = 0.32$ chains nm^{-2}, PS $M_{n,SEC}$ of 11.9 kDa, and $\sigma_{PtBA} = 0.34$ chains nm^{-2}. No strong microphase separation was observed after drop casting from chloroform (Figure 4.5), as in the melt state [39]. However, this mixed brush particle sample also self-assembled into a hexagonal pattern during the solution casting process.

Figure 4.5 Bright-field TEM micrograph of mixed P*t*BA/PS brush-grafted 180 nm silica particles with PtBA $M_{n,SEC}$ of 10.4 kDa, $\sigma_{PtBA} = 0.32$ chains nm^{-2}, PS $M_{n,SEC}$ of 11.9 kDa, and $\sigma_{PtBA} = 0.34$ chains nm^{-2} cast from $CHCl_3$, a good solvent for both P*t*BA and PS. The sample was stained with RuO_4 vapor for 30 minutes. Source: Reproduced with permission from Ref. [40]; © 2008, American Chemical Society.

The morphology of mixed P*t*BA/PS brush-grafted silica particles cast from chloroform depends on a number of factors, including the molecular weights of the two polymers as shown in Figures 4.4 and 4.5, grafting density, and chain length disparity [64, 69, 90]. NMRP is a reversible deactivation radical polymerization technique, allowing for the synthesis of mixed brush particles with different PS molecular weights from the same batch of P*t*BA brush-grafted silica particles in one-pot polymerization while maintaining similar grafting densities for the two polymers [41]. Figure 4.6 shows TEM images of four mixed P*t*BA/PS brush samples grafted on 160 nm silica particles with the same P*t*BA $M_{n,SEC}$ of 24.5 kDa but different PS chain lengths after solution casting from chloroform (the grafting densities are included in the Figure caption) [42]. For the sample with a PS M_n of 14.8 kDa, much smaller than that of P*t*BA (24.5 kDa), the surface-tethered PS chains segregated into isolated, nearly spherical nanodomains with a typical size of 13 nm (Figure 4.6a). With increasing PS M_n to 18.7 kDa, the PS nanodomains began to merge into short cylinders (Figure 4.6b). For the brush sample with P*t*BA $M_{n,SEC}$ of 24.5 kDa, and PS M_n of 24.9 kDa, a nearly bicontinuous nanostructure was observed (Figure 4.6c), similar to the morphology shown in Figure 4.4a [40] and also similar to the computer simulation results from Wang and Müller [[69]]. Upon further increasing the PS M_n to 30.4 kDa, a morphological transition was observed, where dark PS nanodomains were more connected while bright P*t*BA nanodomains became increasingly isolated (Figure 4.6d). Thus, at a fixed P*t*BA $M_{n,SEC}$, the morphology of mixed P*t*BA/PS brushes evolved from the isolated PS domains in the P*t*BA matrix, to a bicontinuous wormlike morphology, and then a two-layered nanostructure where the laterally microphase-separated mixed brush bottom layer was covered by longer PS chains with increasing PS M_n from below to above that of P*t*BA.

To study the surface micellar structures in selective solvents predicted by theoretical and simulation studies, *n*-octane, a good solvent for P*t*BA but a poor solvent

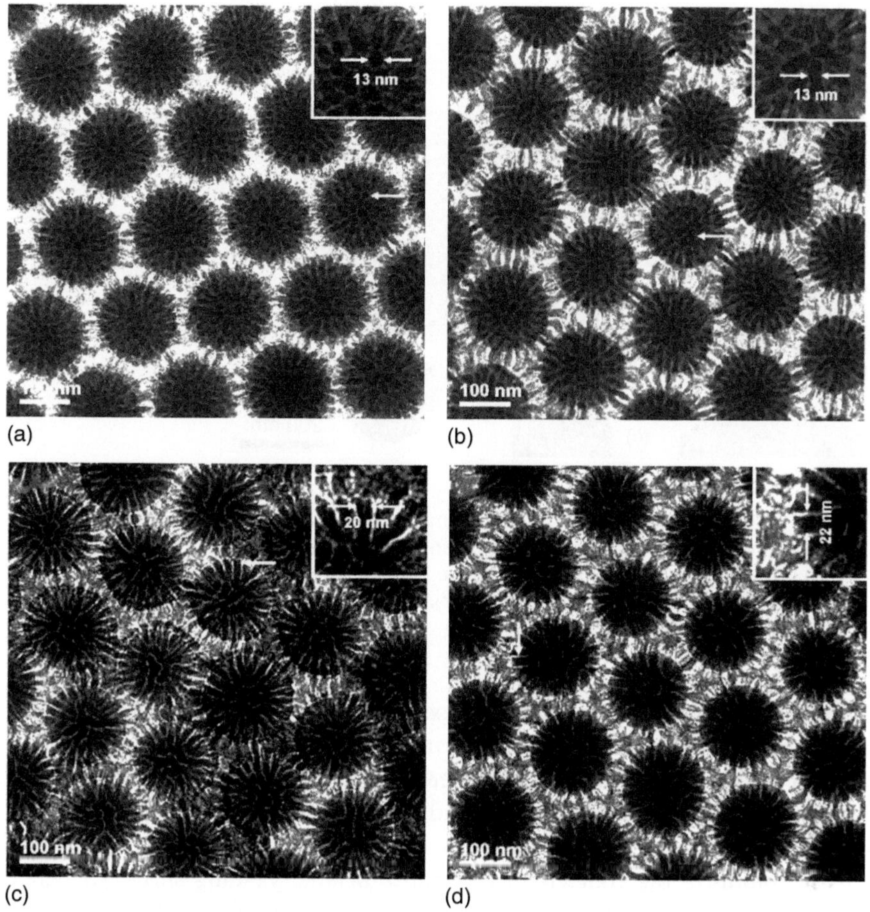

Figure 4.6 Bright-field TEM images of mixed P*t*BA/PS brush-grafted 160 nm silica particles with the same P*t*BA M_n = 24.5 kDa and σ_{PtBA} = 0.36 chains nm^{-2} but different PS molecular weights after casting from $CHCl_3$, annealed with $CHCl_3$ vapor, and stained with RuO_4 at room temperature for 20 minutes. (a) PS M_n = 14.8 kDa and σ_{PS} = 0.21 chains nm^{-2}; (b) PS M_n = 18.7 kDa and σ_{PS} = 0.26 chains nm^{-2}, (c) PS M_n = 24.9 kDa and σ_{PS} = 0.27 chains nm^{-2}, and (d) PS M_n = 30.4 kDa, σ_{PS} = 0.32 chains nm^{-2}. The inset in each TEM image shows the enlarged area marked by the white arrow. Source: Reproduced with permission from Ref. [42]; © 2010 American Chemical Society.

for PS, was chosen [40]. Once again, the mixed P*t*BA/PS brush-grafted, 180 nm silica particle sample with P*t*BA $M_{n,SEC}$ of 24.2 kDa, σ_{PtBA} = 0.40 chains nm^{-2}, PS $M_{n,SEC}$ of 23.0 kDa, and σ_{PS} = 0.37 chains nm^{-2} was used. *n*-Octane was slowly added into a dispersion of the hairy particles in $CHCl_3$ under strong stirring conditions. $CHCl_3$ was then gradually removed under a reduced pressure at room temperature by exploiting the large difference of the two solvents in boiling point (b.p.) ($CHCl_3$, b.p. = 61.2 °C; *n*-octane, b.p. = 125.6 °C). Figure 4.7 shows a TEM image of the mixed brush particles cast from *n*-octane. The grafted polymer chains were no longer highly stretched; instead, the hairy particles became tighter and

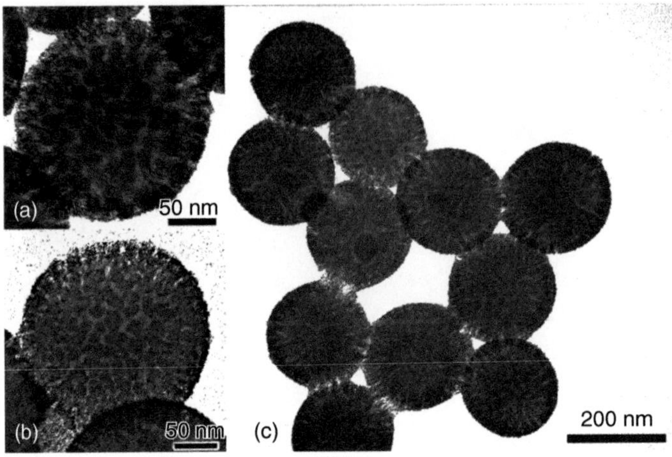

Figure 4.7 TEM images of mixed P*t*BA/PS brush-grafted silica particles with a P*t*BA $M_{n,SEC}$ of 24.2 kDa and a PS $M_{n,SEC}$ of 23.0 kDa cast from *n*-octane, a selective solvent for P*t*BA. (a) and (b) are magnified images showing the isolated dark PS domains in the bright P*t*BA matrix. (c) is a lower magnification image of the hairy particles. The samples were stained with RuO_4 vapor for 30 minutes. Source: Reproduced with permission from Ref. [40]; © 2008 American Chemical Society.

did not self-assemble into an ordered hexagonal pattern. Clearly, the wormlike morphology formed under the good solvent conditions (Figure 4.4) was changed to a surface micellar structure with isolated dark PS domains buried in the bright P*t*BA matrix (Figure 4.7). This is attributed to the collapse and aggregation of the grafted PS chains in *n*-octane, which was shielded by the solvated grafted P*t*BA chains to minimize the contacts of PS from *n*-octane, yielding surface-tethered micelles. However, the dark PS domains were not perfectly spherical, likely due to the difficulty in achieving equilibrium surface micellar structures in the solvent exchange process. Image analysis showed an average diameter of 18.6 ± 3.0 nm for PS domains, which was larger than the average width of the PS domains (13.5 nm) from $CHCl_3$ (Figure 4.4a).

In short, we developed a unique method for the synthesis of well-defined binary mixed homopolymer brushes on silica particles by using two different reversible deactivation radical polymerization techniques, ATRP and NMRP, to grow two chemically distinct polymers from Y-initiator-functionalized silica particles. The responsive properties of mixed P*t*BA/PS brushes grafted on silica particles were demonstrated through a Tyndall scattering experiment and ^1H NMR spectroscopy analysis in different solvents, and the morphologies of mixed brushes in good and selective solvents were revealed by TEM. These studies clearly showed that mixed polymer brushes could undergo spontaneous reorganization according to the environmental conditions and exhibit distinct nanostructures. These unique properties make mixed brush particles suitable for potential applications, such as encapsulation and release of substances and self-cleaning surfaces [10, 97, 98].

4.3 Thermoresponsive Polymer Brush-grafted Silica Particles

Thermoresponsive water-soluble polymers exhibit a large and abrupt conformational change from expanded coil-like to collapsed globular in water when the temperature passes through a certain point [59, 99]. Macroscopically, when the polymer concentration is sufficiently high, the clear aqueous solution turns cloudy, and the critical temperature is called the cloud point. There are two types of thermoresponsive water-soluble polymers, those exhibiting a lower LCST in the polymer-water phase diagram [59, 99–105] and those displaying a UCST [104–106]. Examples of the former include poly(*N*-isopropylacrylamide) [59, 99], poly(oligo[ethylene glycol] methacrylate)s [100], poly(oligo[ethylene glycol] acrylate)s [102, 103], poly(vinyl caprolactam), and more recently, poly(sulfobetaine methacrylate)s with *N*-alkyl substituents of an appropriate length [104, 105]. It has been commonly accepted that the LCST transition stems from the hydrophobic effect [99]; above the cloud point, the dissolution of an LCST-type thermoresponsive polymer in water becomes unfavorable. On the other hand, UCST-type water-soluble polymers usually contain either neutral pendant groups that can form multiple hydrogen bonds or charged zwitterionic side chains; heating overcomes the hydrogen bonding or zwitterionic interactions, causing the polymers to be dissolved in water and display a UCST [104–106]. This section presents the synthesis and properties of LCST-type thermoresponsive water-soluble polymer brush-grafted silica particles from our research laboratory [32, 47–57]. We first discuss the LCST transition of thermoresponsive polymer brushes grafted on silica particles in water, then the temperature-induced phase transfer of thermoresponsive hairy particles between water and an immiscible organic or ionic liquid phase, then the thermally induced gelation of thermosensitive hairy NPs in water in a moderate concentration regime, and finally the use of thermoresponsive hairy NPs to enhance the mechanical properties of thermoresponsive water-soluble ABC linear triblock copolymer hydrogels.

4.3.1 Synthesis and Thermally Induced LCST Transition of Thermoresponsive Polymer Brushes Grafted on Silica Particles

Using SI-ATRP, we synthesized thermoresponsive poly(methoxydi[ethylene glycol] methacrylate) (PDEGMMA) brush- and poly(methoxytri[ethylene glycol] methacrylate) (PTEGMMA) brush-grafted silica particles with a core diameter of 133 or 135 nm and studied their LCST transitions in water [47]. The molecular structures of DEGMMA and TEGMMA are shown in Scheme 4.6. The cloud points of PDEGMMA and PTEGMMA in water are 26 and 52 °C, respectively, according to the report in the literature [100]. The bare silica particles in this study were made by the Stöber process [95, 96] and functionalized with a monochlorosilane-terminated ATRP initiator, 11-(2-bromo-2-methyl)propionyloxy)undecyldimethylchlorosilane (Scheme 4.1a). The SI-ATRP of DEGMMA or TEGMMA was performed at 40 °C in THF with the addition of a free initiator, benzyl 2-bromoisobutyrate, using

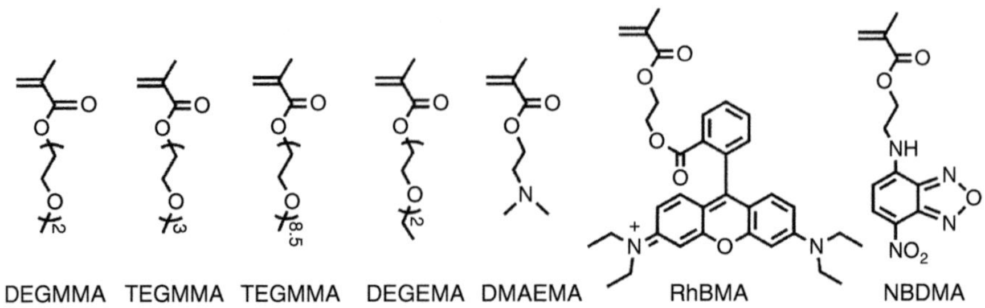

Scheme 4.6 Molecular structures of methoxydi(ethylene glycol) methacrylate (DEGMMA), methoxytri(ethylene glycol) methacrylate (TEGMMA), methoxypoly(ethylene glycol) methacrylate with M_n of 475 g mol^{-1} (PEGMMA), ethoxydi(ethylene glycol) methacrylate (DEGEMA), 2-(*N,N*-dimethylamino)ethyl methacrylate (DMAEMA), RhBMA, and NBDMA used in the synthesis and study of thermoresponsive hairy silica (nano)particles.

CuBr/*N,N,N′,N′,N″*-pentamethyldiethylenetriamine (PMDETA) as catalyst. When the molecular weight reached a desired value (for PDEGMMA, $M_{n,\mathrm{SEC}}$ = 19.9 kDa relative to PS standards and DP = 152; for PTEGMMA, $M_{n,\mathrm{SEC}}$ = 27.5 kDa with respect to PS calibration and DP = 193), the polymerizations were stopped and the hairy particles were separated and repeatedly washed with THF by centrifugation. From TGA analysis of hairy particles, the grafting densities of PDEGMMA and PTEGMMA brushes were calculated to be 0.28 and 0.18 chains nm^{-2}, respectively. The cloud points of the purified PDEGMMA and PTEGMMA free polymers in water at a concentration of 0.5 wt% were 25 and 48 °C, respectively, determined from the plots of optical transmittance versus temperature obtained using a UV–vis spectrometer. These values are slightly lower than those reported in the literature for the two polymers, likely due to the factors, such as molecular weight, dispersity, end groups, and concentration that are known to influence the cloud points. Note that the clouding transition from the plot of transmittance versus temperature was sharp (<2 °C) for both polymers.

The LCST transitions of PDEGMMA and PTEGMMA brushes grafted on silica particles were investigated by variable temperature ^{1}H NMR spectroscopy analysis and dynamic light scattering (DLS) [47]. For PDEGMMA hairy particles, upon heating from 15 to 30 °C, the peak in the range of 0–1 ppm in the ^{1}H NMR spectra, which was from the methyl group on the backbone, disappeared at ~25 °C, indicating an LCST transition. For PTEGMMA hairy particles, all peaks were visible in the temperature range of 35 to 60 °C; only the peaks at ~2 ppm (–CH_2-of the backbone) and 4.2 ppm (–$COOCH_2$–) decreased slightly and gradually when the temperature was raised from 45 to 55 °C, suggesting that the LCST transition of PTEGMMA brushes was not a sharp but a gradual process [47]. DLS measurements showed that the apparent hydrodynamic diameter (D_h) of PDEGMMA hairy particles in water at a concentration of 0.05 mg ml^{-1} decreased from 203 to 192 nm when the temperature was increased from 21 to 25 °C and leveled off at temperatures above 25 °C. The LCST transition began at 21 °C, about 4 °C lower than the cloud point of the corresponding free polymer in water, and continued in the temperature range from 21 to 25 °C. For PTEGMMA brush particles, the onset temperature of the LCST transition

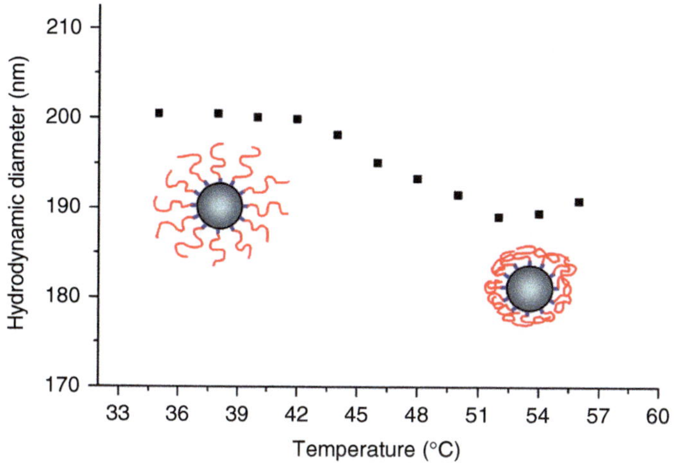

Figure 4.8 Average hydrodynamic diameter of PTEGMMA brush-grafted silica particles as a function of temperature in the temperature range of 35–55 °C measured by dynamic light scattering. The concentration of the PTEGMMA hairy particles in the dispersion was 0.05 mg ml⁻. Source: Reproduced with permission from Ref. [47]; © 2006, American Chemical Society.

(42 °C) was even lower compared with the clouding transition of the free PTEGMMA in water (48 °C), and the D_h decreased continuously over a much wider temperature range, from 200 nm at 42 °C to 189 nm at 52 °C (Figure 4.8). Thus, for both thermoresponsive hairy particle samples, the thermally induced LCST transition of polymer brushes began at a lower temperature compared to the free polymer in water and occurred in a broader temperature range (4–10 °C). These observations could be caused by the interactions between neighboring polymer chains in the brush layer. The lower onset temperature for the LCST transition of hairy particles likely resulted from the higher local concentration of the grafted polymer chains in the brushes. In addition, the interchain contacts reduced the water-segment interactions, making the collapse of the brushes weaker than that of free polymer chains in a dilute solution.

It should be noted here that for thermosensitive polymer brushes grafted on even smaller particles, e.g. with a size of tens of nanometers, the substrate curvature may play a significant role, and the LCST transition could manifest itself in a two-stage process [107, 108] For example, Wu et al. reported that the hydrodynamic radius of thermosensitive poly(*N*-isopropylacrylamide)-grafted, 70 nm silica NPs in water decreased continuously and gradually in the temperature range of 20–29 °C, followed by a sharper reduction from 31 to 37 °C [107]. The two-stage LCST transition was also observed from the plot of radius of gyration (R_g) versus temperature. The first stage was believed to result from the *n*-cluster-induced collapse of the inner zone of PNIPAM brushes, and the second transition was attributed to the collapse of the outer region of the brushes, which had a lower segment density than the inner zone. Using micro-differential scanning calorimetry, Shan et al. also found that the PNIPAM brushes grafted on small gold NPs exhibited double transitions [108].

4.3.2 Thermally Induced Phase Transfer of Thermoresponsive Hairy Particles Between Two Immiscible Liquid Phases

4.3.2.1 Thermally Induced Phase Transfer of Thermoresponsive Hairy Particles Between Water and Immiscible Organic Solvents

Transport of substances, such as molecules, supramolecular assemblies, and (nano)particles, across the interfaces between immiscible liquid phases, is critical for a wide variety of applications, including heterogeneous reactions, phase transfer catalysis, separation, and delivery of drugs and biomolecules [109–111]. Although one-way transport is widely practiced [111], reversible and complete transfer of substances between two immiscible phases is highly desired, for example, for developing next-generation recyclable catalysts. We discovered that thermoresponsive hairy particles could be induced to transfer reversibly and quantitatively between water and an immiscible liquid phase [48, 51–53]. Above the LCST, thermoresponsive polymer brushes grafted on silica particles collapse, and water becomes a poor solvent for the brushes. We reasoned that under this condition, if a water-immiscible organic solvent, a good solvent for the brushes, was present, thermoresponsive hairy particles could cross the interface and enter into the organic phase. To test this hypothesis, two batches of PTEGMMA brush-grafted 141 nm silica particles were synthesized, one with $M_{n,SEC}$ of 29.3 kDa and σ of 0.32 chains nm^{-2}, and the other with $M_{n,SEC}$ of 24.5 kDa and σ of 0.31 chains nm^{-2}. We found that the PTEGMMA brush particles dispersed in pure water at a concentration of 1.0 mg in an ice/water bath spontaneously and completely moved from the aqueous phase to the ethyl acetate (EA) phase upon heating at 60 °C after 24 hours (Figure 4.9a) [48]. Similarly, the hairy NPs dispersed in pure EA at 63 °C were completely transferred to the aqueous phase upon cooling in an ice/water bath under the stirring conditions (Figure 4.9b). The thermally induced phase transfer was reversible; Figure 4.9c shows the pictures of the biphasic system from the 7th cooling to the 9th heating in a reversible transfer experiment. In general, the transfer of hairy particles from water to EA upon heating at 60 °C was faster than cooling in the ice/water bath.

Interestingly, we found that when pre-mutually saturated water and EA were used, the transfer rates of PTEGMMA brush particles between two liquid phases increased significantly, by 40 times or even larger, under the stirring conditions [48]. For example, when the two solvents were mutually saturated in an ice/water bath and the EA-saturated water was used to disperse the hairy NPs, 95% of the particles were transferred from the aqueous phase to the EA phase in 10 minutes upon heating in a 60 °C oil bath, which was significantly shorter than the time (435 minutes) required for the transfer of the same percentage of hairy particles with the use of pure water and EA. For the cooling-induced transfer in an ice/water bath, the two solvents were mutually saturated at 60 °C. We observed that 73% of PTEGMMA hairy particles crossed the water–EA interface in 15 minutes under stirring in an ice/water bath, in contrast to 575 minutes for a similar percentage of hairy particles in the system using the originally pure solvents. The much faster transfer of PTEGMMA hairy particles in pre-mutually saturated solvent systems

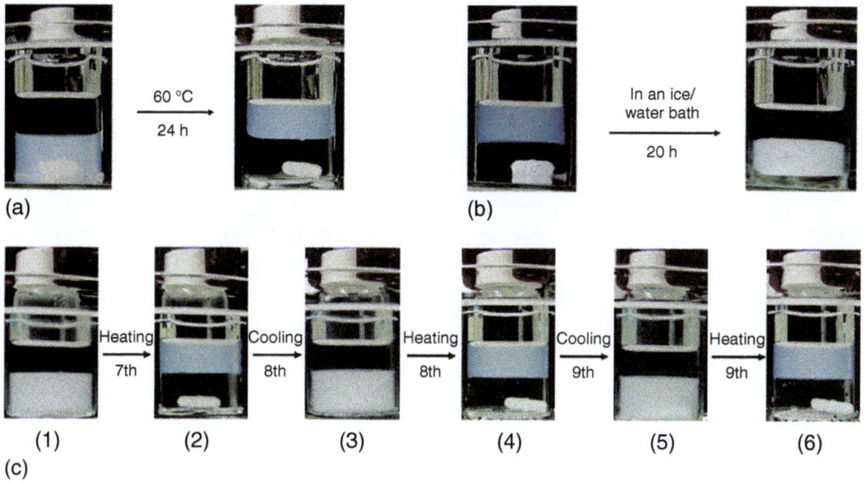

Figure 4.9 (a) Optical photos of the initial state of the aqueous (bottom)/ethyl acetate (EA) (upper) biphasic system with 1.0 mg ml⁻ PTEGMMA brush particles dispersed in pure water at 0 °C and (b) the final state after stirring at 60 °C for 24 hours. (b) Optical photos of the initial state of the aqueous/EA biphasic system at 60 °C with 1.0 mg ml⁻ PTEGMMA brush particles dispersed in pure EA at 63 °C and the final state after stirring in an ice/water bath for 20 hours. (c) Optical images of the biphasic system after 7th cooling in an ice/water bath (1), 7th heating in a 60 °C oil bath (2), 8th cooling (3), 8th heating (4), 9th cooling (5), and 9th heating (7). The PTEGMMA brush particles were initially dispersed in water-saturated EA at 63 °C at a concentration of 1.0 mg ml⁻. Source: Reproduced with permission from Ref. [48]; © 2007, American Chemical Society.

can be attributed to the combined effects of solvent phase separation and the LCST transition of PTEGMMA brushes [48]. The solubility of EA in water decreases from 9.71% to 6.06% upon heating from 0 to 60 °C, while the solubility of water in EA decreases from 4.23% to 2.13% upon cooling from 60 to 0 °C [112]. Therefore, the aqueous phase saturated with EA at 0 °C undergoes phase separation upon heating in a 60 °C oil bath. On the other hand, the EA phase saturated at 60 °C separates upon cooling in an ice/water bath. The pre-mutually saturated solvents phase separated in response to temperature changes, resulting in the formation of liquid drops (Scheme 4.7). The particles have a strong tendency to adsorb to the interface [113–115], and the increased interfacial area facilitated the transfer of PTEGMMA hairy particles from one phase to another. At 0 °C, the PTEGMMA hairy particles exhibited a significantly higher affinity to water. However, at 60 °C, above the LCST transition of PTEGMMA brushes, the affinities towards the two solvents were reversed, causing the particles to move from water and EA.

We also studied other organic solvents, including toluene and 1-butanol. The PTEGMMA hairy NPs were found to undergo quantitative transfer from 1-butanol and toluene to water upon cooling in an ice/water bath. However, only a portion of hairy NPs (60%) spontaneously moved across the interface from water to 1-butanol and no NPs transferred into the toluene phase upon heating at 60 °C [48]. Presumably, the temperature-induced affinity changes for these two biphasic

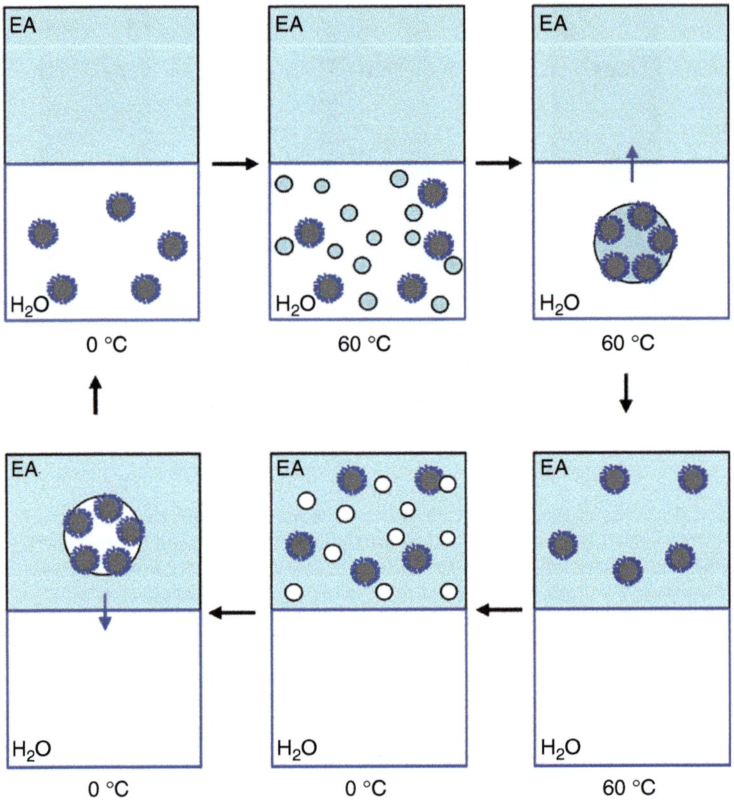

Scheme 4.7 Schematic illustration of phase transfer of thermosensitive PTEGMMA brush-grafted nanoparticles between water and ethyl acetate phases facilitated by the solvent phase separation upon heating and cooling. Source: Reproduced with permission from Ref. [48]; © 2007, American Chemical Society.

systems were not large enough to drive quantitative and reversible phase transfer of thermoresponsive PTEGMMA hairy NPs.

4.3.2.2 Thermally induced Phase Transfer of Thermoresponsive Hairy Particles Between Water and a Hydrophobic Ionic Liquid

Room-temperature ionic liquids (ILs), with melting points below or around room temperature, exhibit many attractive characteristics, such as extremely low vapor pressure, tunable hydrophilicity/hydrophobicity, adjustable solvation ability, and excellent thermal and chemical stability [116]. ILs have been intensively explored for use as media for reactions and separation. Lodge and coworkers reported that PEO-based block copolymer micelles and vesicles can undergo thermally induced phase transfer between water and hydrophobic ILs, which were named as micelle and vesicle shuttles [117–119]. These block copolymer micelle and vesicle shuttles with PEO coronas preferentially stay in water at room temperature, are transferred to the IL phase upon heating, and then moved back to the aqueous phase upon cooling. Schubert and coworkers found that poly(2-nonyl-2-oxazoline)-*b*-poly(2-ethyl-2-oxazoline) block copolymer micelles could also be induced to transfer reversibly between water and a hydrophobic IL by changing temperature [120]. Similar to the thermally induced phase transfer of PTEGMMA brush particles between water and EA, the phase transfer of block copolymer micelles and vesicles was driven by the changes in the relative affinity of the corona polymers to the two liquid phases.

To study the effect of LCST transitions of thermoresponsive polymers on the phase transfer of the corresponding polymer brush particles between water and a common hydrophobic IL, 1-ethyl-3-methylimidazolium bis(trifluoromethylsulfonyl)imde ([EMIM][TFSI]), we synthesized six hairy particle samples with different thermoresponsive properties by SI-ATRP of methoxypoly(ethylene glycol) methacrylate with M_n of 475 g mol^{-1} (PEGMMA, Scheme 4.6), DEGMMA, TEGMMA, or their mixtures from initiator-functionalized, 205 nm silica particles [51]. These samples were silica particles grafted with PPEGMMA with two different molecular weights ($M_{n,SEC}$ of 23 and 40 kDa), PTEGMMA with a $M_{n,SEC}$ of 31 kDa, PDEGMMA with a $M_{n,SEC}$ of 18 kDa, a copolymer of PEGMMA and TEGMMA with a PEGMMA molar content of 59% (P[PEGMMA-*co*-TEGMMA]-82, $M_{n,SEC}$ = 29 kDa), and a copolymer of PEGMMA and TEGMMA with a PEGMMA molar content of 40% (P[PEGMMA-*co*-TEGMMA]-74, $M_{n,SEC}$ = 23 kDa). The calculated grafting densities of the brushes in these hairy particles were in the range of 0.23–0.29 chains nm^{-2}. The cloud points of free PPEGMMA with $M_{n,SEC}$ of 23 and 40 kDa, P(PEGMMA-*co*-TEGMMA)-82, P(PEGMMA-*co*-TEGMMA)-74, PTEGMMA, and PDEGMMA in pure water were 98, 98, 82, 74, 52, and 25 °C, respectively. The cloud points of thermoresponsive water-soluble polymers in water are known to be affected by organic or inorganic additives [48]. Since the phase transfer of thermoresponsive hairy NPs involved two phases, we also measured the cloud points of the free polymers in the [EMIM][TFSI] saturated water at a concentration of 0.5 wt% and they were 95, 94, 80, 72, 43 °C, and < 0 °C, respectively. (PDEGMMA was found to be insoluble in ionic liquid-saturated water.) While the presence of [EMIM][TFSI] decreased the cloud points of all free polymers in water, it had a larger effect for PTEGMMA and PDEGMMA, with a decrease of cloud point by 9 °C and > 25 °C, respectively.

When the LCST of the thermoresponsive polymer was sufficiently high, the hairy particles were found to spontaneously transfer between water and [EMIM][TFSI] upon temperature changes [51]. The PDEGMMA brush particles moved spontaneously from the water phase to the IL layer upon heating but did not transfer back to the aqueous phase even after stirring for a long time at 0 °C, likely because the cloud point of the corresponding thermosensitive PDEGMMA polymer in the [EMIM][TFSI]-saturated water was below 0 °C, as mentioned earlier. Using UV–vis absorbance measurements, we determined the transfer temperature (T_{tr}) of each hairy particle sample from water to IL upon heating. T_{tr} was found to be a linear function of the cloud point of the corresponding free polymer in the IL-saturated water, as shown in Figure 4.10. Note that the molecular weight of PPEGMMA brushes had a negligible effect on the T_{tr} of PPEGMMA hairy particles. The discovered linear dependence allows for the design of phase transfer particle systems with

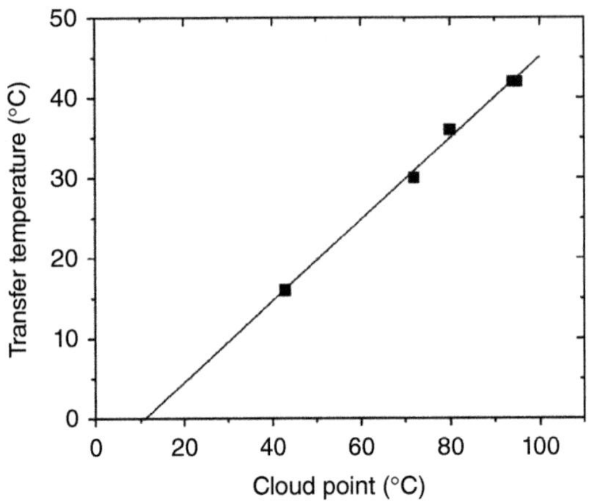

Figure 4.10 Transfer temperature of thermoresponsive polymer brush-grafted silica particles from water to [EMIM][TFSI] versus cloud point of the corresponding free polymer in the [EMIM][TFSI]-saturated water. Source: Reproduced with permission from Ref. [51]; © 2011 American Chemical Society.

desired transfer temperatures as the cloud point of thermoresponsive water-soluble polymers can be readily changed via copolymerization with a hydrophilic or hydrophobic comonomer. From the equilibrium distributions of hairy particles in the two phases at different temperatures, we calculated the standard enthalpy and entropy changes ($\Delta H°$ and $\Delta S°$) for the particles to move from water to the IL phase upon gradual heating. Both $\Delta H°$ and $\Delta S°$ were found to be positive for the particles, indicating that the thermally induced phase transfer of thermosensitive hairy particles between the two phases was entropically driven and governed by the LCST of the polymer brushes grafted on the surface of silica particles in water.

As shown above, to change the T_{tr} of thermoresponsive hairy particles from water to the IL phase, one has to perform a polymerization to make a new brush particle sample each time with a particular cloud point, which can be a tedious process. More conveniently, if a small amount of a weak acid or base is incorporated into a thermosensitive polymer, the cloud point of the resultant copolymer in water depends on the pH of the aqueous solution and can be changed continuously and reversibly over a broad temperature range [121]. Using SI-ATRP of a mixture of TEGMMA and *tert*-butyl methacrylate (*t*BMA) with a molar ratio of 100 : 11 and subsequently cleaving the *tert*-butyl groups with CF_3COOH, we prepared a dually responsive hairy particle sample, 180 nm silica particles grafted with thermo- and pH-responsive poly(TEGMMA-*co*-methacrylic acid) [P(TEGMMA-*co*-MAA)] brushes [52]. As expected, the cloud point of P(TEGMMA-*co*-MAA), made from the free polymer P(TEGMMA-*co*-*t*BMA) by treatment with CF_3COOH, in the [EMIM][TFSI]-saturated water increased with increasing pH and can be varied from 48 °C to about 100 °C in the pH range of 3.0–6.5 (Figure 4.11A). Similar to the hairy particles with nonionic thermoresponsive (co)polymer brushes, the thermo- and pH-responsive hairy particles spontaneously moved from the aqueous phase to the [EMIM][TFSI] layer when heated at 80 °C and returned back to the water layer when cooled at 10 °C regardless of whether the pH of the aqueous phase was 2.99, 5.00, or 7.02. UV–vis absorbance measurements were conducted at different temperatures in a gradual heating process to determine

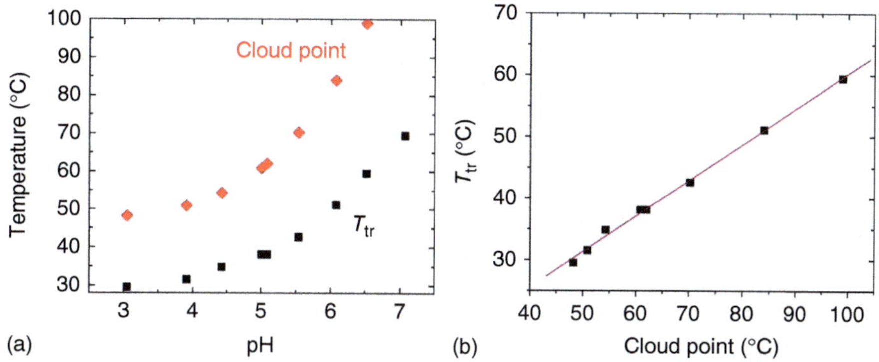

Figure 4.11 Plots of (a) the transfer temperature (T_{tr}) of P(TEGMMA-*co*-MAA) hairy particles and the cloud point of free polymer P(TEGMMA-*co*-MAA) in [EMIM][TFSA]-saturated water versus pH and (b) the T_{tr} of hairy particles versus the cloud point of free polymer P(TEGMMA-*co*-MAA) in [EMIM][TFSA]-saturated water. The cloud points of P(TEGMMA-*co*-MAA) presented here were calculated from a fit equation obtained from experimental data. The straight line in (b) was a linear fit with $R = 0.998$ ($y = 2.332 + 0.579x$, where y is T_{tr} and x is the cloud point). Source: Reproduced with permission from Ref. [52]; © 2011, American Chemical Society.

the T_{tr} of the hairy particles at pH values ranging from 3.05 to 7.07; the T_{tr} of hairy particles from water to [EMIM][TFSI] was found to exhibit the same trend as the cloud point with regard to pH (Figure 4.11a). The transfer temperature showed a linear dependence on the cloud point of the corresponding free polymer (Figure 4.11b), similar to the observation in Figure 4.10 for hairy particles grafted with different thermoresponsive polymers. The tunability of T_{tr} of this hairy particle sample allowed us to achieve pH-induced reversible phase transfer at a constant temperature by changing the pH of the aqueous phase (Figure 4.12) and multiple transfer processes between the two liquid phases by changing both the temperature and pH of the aqueous phase. This demonstrated the advantage of this system compared with thermoresponsive hairy particles with a fixed transfer temperature.

The studies described above showed that thermoresponsive poly(oligo(ethylene glycol) methacrylate) brush-grafted silica particles exhibited spontaneous, reversible, and quantitative phase transfer between water and EA or a hydrophobic ionic liquid, [EMIM][TFSI], upon temperature changes. The thermally induced phase transfer of hairy particles between two immiscible liquid phases was driven by the changes in the relative affinity of hairy particles for the two solvents caused by the LCST behavior of the polymer brushes in water. The T_{tr} of hairy particles from water to [EMIM][TFSA] upon heating was lower than the cloud point of the corresponding thermoresponsive polymer in [EMIM][TFSA]-saturated water, and there was a linear correlation between the T_{tr} and the cloud point. By incorporation of a small amount of carboxylic acid groups into thermoresponsive polymer brushes, the cloud point and thus the transfer temperature from water to the ionic liquid can be continuously and reversibly tuned.

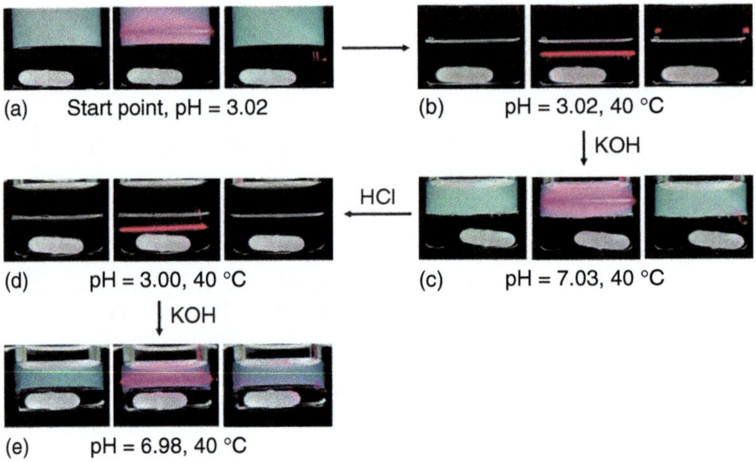

Figure 4.12 Phase transfer of thermo- and pH-sensitive P(TEGMMA-*co*-MAA) hairy particles between water and [EMIM][TFSI] by changing the pH of the aqueous layer between 3.0 and 7.0. (a) The pH at the starting point was 3.02. (b) After the system was stirred at 40 °C for 4 hours. (c) After the pH of the aqueous layer was changed to 7.03 and the mixture was stirred at 40 °C for 4 hours. (d) After the pH was adjusted to 3.00 and the system was stirred at 40 °C for 4 hours. (e) After the pH was changed to 6.98 and the system was stirred overnight at 40 °C. In each group of pictures, the left one shows the vial without a laser beam; the central and right pictures show the vial with a laser beam passing through the top and bottom layers, respectively. Source: Reproduced with permission from Ref. [52]. © 2011 American Chemical Society.

4.3.3 Thermoreversible Gelation of Thermoresponsive Diblock Copolymer Brush-grafted Silica Nanoparticles in Water

Injectable thermosensitive block copolymer hydrogels have received considerable interest in the past decades due to their ability to exhibit thermally induced in situ reversible sol–gel transitions and their potential in a variety of applications, e.g. drug delivery and wound dressing [122, 123]. In general, there are two types of stimuli-responsive block copolymer micellar hydrogels: (i) 3-dimensional network hydrogels formed by the bridging of the water-soluble middle block of an ABA or ABC triblock copolymer or a star diblock copolymer among micellar cores and (ii) packing- or jamming-based hydrogels consisting of discrete, commonly spherical, block copolymer micelles [122–132]. 3-D network hydrogels are more robust against dissolution in water, and the critical gelation concentration (CGC) can be as low as 5 wt% [124, 128]. In contrast, the CGC for packing-based block copolymer hydrogels is typically higher, ~20 wt%, and the hydrogels can be readily dissolved in water due to the lack of bridging chains between micellar cores [129–132].

Inorganic or metallic NPs densely grafted with thermoresponsive diblock copolymer chains, with either the inner or the outer block or both being thermoresponsive, are structurally similar to the spherical core–shell–corona triblock and star diblock copolymer micelles. Thus, the aqueous dispersions of such hairy NPs with a sufficiently high concentration are expected to be able to undergo thermally induced gelation through the aforementioned two mechanisms with pre-determined locations for the core NPs in the hydrogels. These hybrid hydrogels could exhibit additional functions from inorganic or metallic core NPs, such as magnetic or optical properties, which are not associated with conventional thermoresponsive block copolymer micellar hydrogels. The core NPs could be porous or hollow, and the interior surface could be modified to be hydrophobic or hydrophilic for the loading of different substances. The swelling–collapse transitions of thermoresponsive polymer brushes grafted on the NPs can be utilized to regulate the accessibility of functional groups or substances in the brushes or in the core NPs. Thus, the hydrogels achieved by thermoresponsive diblock copolymer-grafted NPs can effectively overcome the limitations of injectable block copolymer micellar hydrogels.

To demonstrate the formation of injectable hybrid hydrogels of stimuli-responsive hairy NPs, we first synthesized thermoresponsive diblock copolymer brush-grafted, 17 nm silica NPs [32]. The polymer brushes grafted on the NPs were PDEGMMA-*b*-poly(DEGMMA-*co*-2-[methacryloyloxy]ethyltrimethylammonium iodide) (PDEGMMA-*b*-P[DEGMMA-*co*-TMAEMA-I]). These hairy NPs were synthesized by a one-pot, two-step SI-ATRP polymerization using DEGMMA and 2-(*N*,*N*-dimethylamino)ethyl methacrylate (DMAEMA, Scheme 4.6) and subsequent quaternization of DMAEMA monomer units in the second block with iodomethane (Scheme 4.8). The SI-ATRP homopolymerization of DEGMMA was conducted first until a desired monomer conversion was reached, and then DMAEMA was added to the polymerization mixture. The quaternized DMAEMA monomer units were permanently charged, facilitating the dispersion of hairy NPs in water and granting stability against aggregation at temperatures above the LCST of PDEGMMA.

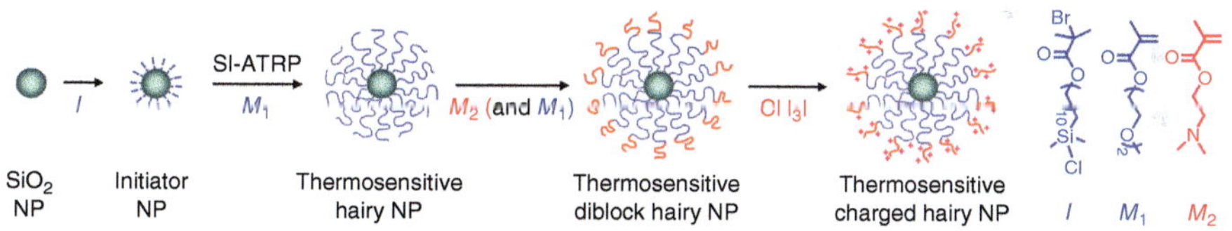

Scheme 4.8 Synthesis of PDEGMMA-*b*-P(DEGMMA-*co*-TMAEMA-I) brush-grafted silica NPs with PDEGMMA as the inner block by a one-pot, two-step surface-initiated atom transfer radical polymerization and subsequent quaternization of tertiary amine units with iodomethane. Source: Reproduced with permission from Ref. [32]. © 2015, Royal Chemical Society.

The 17 nm silica NPs were functionalized with 11-(chlorodimethylsilyl)undecyl 2-bromoisobutyrate, a chlorodimethylsilane-functionalized ATRP initiator (Scheme 4.1a). The one-pot, two-step SI-ATRP was conducted with the addition of a free initiator, ethyl 2-bromoisobutyrate, at 60 °C using $CuBr/CuBr_2$/PMDETA as the catalyst. After the DEGMMA monomer conversion reached 29.8%, a portion of the polymerization mixture was removed using a degassed syringe, and the second monomer, DMAEMA, which had already been degassed, was injected into the polymerization mixture for the synthesis of the second block, P(DEGMMA-*co*-DMAEMA). SEC analysis of the cleaved grafted PDEGMMA from silica NPs with HF showed that the SEC trace overlapped with that of the free PDEGMAA, indicating that the molecular weights and dispersities of the free and grafted polymers were essentially the same, similar to our previous observation [38]. The DP of PDEGMMA was 101, calculated from the monomer conversation. After the addition of the second monomer, three different block copolymer brush-grafted silica NP samples, designated as BHP-1, -2, and -3, were obtained at different polymerization times. Analysis showed that the DPs of the second block for the three samples were 42, 75, and 149, containing 17, 32, and 57 DMAEMA units, respectively. The calculated grafting densities of PDEGMMA brushes and diblock copolymer brushes on silica NPs were similar to each other, in the range of 0.60–0.61 chains nm^{-2}. The hairy NPs were then quaternized with iodomethane, and the obtained NPs were denoted as BHP-1-Q, -2-Q, and -3-Q. These charged thermoresponsive hairy NPs were then dispersed in Milli-Q water via ultrasonication.

Aqueous dispersions of BHP-1-Q showed reversible, cooling-induced gelation at low-to-moderate concentrations. For instance, at a concentration of 6.5 wt%, the dispersion was a free-flowing, slightly viscous liquid at and above room temperature, but it turned into a transparent stationary gel, and the sol–gel transition temperature ($T_{sol-gel}$) by visual inspection was 17 °C. The transition was reversible; heating converted the gel into a sol, with no change in transparency, suggesting the stabilization of hairy NPs by the outer block and the prevention of the irreversible aggregation of hairy silica NPs. Oscillatory shear rheological experiments were performed to quantitatively characterize the sol–gel transition, and the $T_{sol-gel}$ was found to be 18.1 °C, essentially the same as that from the visual inspection. The CGC for this sample was found to be between 5 wt% and 6 wt%.

The thermally induced reversible sol–gel transitions of the aqueous dispersions of BHP-1-Q originated from the LCST transition of the inner PDEGMMA block. At lower temperatures, the inner thermoresponsive PDEGMMA block was hydrated and was more extended, increasing the volume fraction of hairy NPs in the dispersion. When the volume fraction reached the critical value for packing, the hairy NPs were jammed, resulting in the formation of a hydrogel. At temperatures above the LCST transition, the inner block collapsed, driving the transition from a gel to a sol. This mechanism is similar to those for thermally induced sol–gel transitions of aqueous solutions of thermoresponsive diblock copolymers and has been confirmed by the DLS study of the hairy NPs in water at a concentration of 0.2 mg g^{-1} that showed a clear but broad LCST transition (Figure 4.13).

BHP-2-Q, which had a longer charged outer block than that of BHP-1-Q, also displayed thermally induced, reversible sol–gel transitions in water when the concentration was sufficiently high. At the same concentration, the $T_{sol-gel}$ of BHP-2-Q was higher than that of BHP-1-Q. DLS showed an even broader LCST transition of BHP-2-Q (Figure 4.13). BHP-3-Q had the longest charged outer block with a DP of 149 in comparison to 75 for BHP-2-Q and 42 for BHP-1-Q. This sample showed no sol–gel transition; at 4.0 wt% or higher, the dispersions remained a gel in the temperature range of 0 to 100 °C, while the 3.0 wt% dispersion was a liquid and no sol–gel transition was observed from 0 °C to 65 °C. DLS showed that the hydrodynamic size

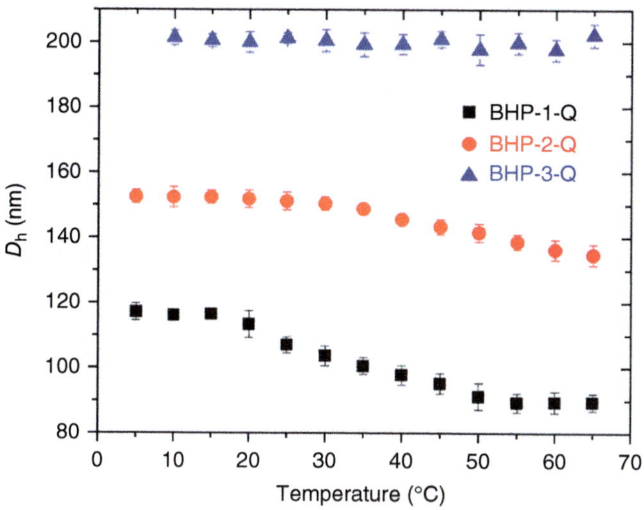

Figure 4.13 Average apparent hydrodynamic size D_h, obtained from DLS study of 0.2 mg g^{-1} aqueous dispersion of PDEGMMA-*b*-P(DEGMMA-*co*-TMAEMA-I) brush-grafted silica NPs with PDEGMMA as the inner block in Milli-Q water, along with the standard deviation as a function of temperature for BHP-1-Q (black square), BHP-2-Q (red circle), and BHP-3-Q (blue triangle). Source: Reproduced with permission from Ref. [32]; © 2015 Royal Chemical Society.

of BHP-3-Q stayed the same in the temperature range of 0–70 °C and no LCST transition was observed (Figure 4.13), likely because the long charged outer block was highly hydrated, opposing the contraction force arising from the LCST transition and making the LCST transition disappear.

We further synthesized thermosensitive diblock copolymer brush-grafted 23.5 nm silica NPs with a water-soluble polyelectrolyte PTMAEMA-I inner block and a thermosensitive PDEGMMA outer block in three steps [57]. The first step was the SI-ATRP of DMAEMA in the presence of a free initiator, ethyl 2-bromoisobutyrate, and the DP of PDMAEMA was calculated to be 271. In the second step, SI-ATRP of DEGMMA was conducted from the purified PDMAEMA brush-grafted 23.5 nm silica NPs with the addition of the free PDMAEMA polymer obtained from the first step as the free macroinitiator. Two samples were obtained, one with a PDEGMMA DP of 100 and another one with a PDEGMMA DP of 195. SEC analysis showed that the peak of the free copolymer shifted to the higher molecular weight side and remained relatively narrow (Đ ≤ 1.27). The formation of diblock copolymer brushes on silica NPs was further confirmed by the cleavage of the brushes from silica NPs using HF to etch the core [57]. The third step was the quaternization of hairy NPs with iodomethane, and the two samples were denoted as Q-100 (for the sample with PDEGMMA DP of 100) and Q-195 (for the sample with $DP_{PDEGMMA}$ of 195). From TGA analysis, the grafting densities of PDMAEMA brush-grafted NPs and two diblock copolymer brush NPs were calculated and were in the range of 0.43–0.47 chains nm^{-2}.

Both hairy NP samples were relatively easily dispersed in Milli-Q water under ultrasonication, and the obtained NP dispersions were transparent. Aqueous dispersions of Q-100 and Q-195 with moderate concentrations exhibited reversible sol–gel transitions upon heating (Figure 4.14 shows the photos of a 6.0 wt% dispersion of Q-195 in water at 20 and 40 °C). A 6.0 wt% aqueous dispersion of Q-100 was a flowing liquid at 20 °C but a self-supporting transparent hydrogel at 50 °C; this sol-gel transition was fully reversible. The $T_{sol-gel}$ was 43.5 °C, determined by the vial-inversion method. For Q-195, at 6.0 wt%, the $T_{sol-gel}$ from visual inspection was 32 °C, which was 11.5 °C lower than Q-100 at the same concentration. Rheological measurements performed in heating and cooling ramps showed $T_{sol-gel}$ of 29.3 °C upon heating and $T_{gel-sol}$ of 27.0 °C (Figure 4.15). For both samples, $T_{sol-gel}$ increased with decreasing concentration of hairy NPs in aqueous dispersions.

The thermally induced sol–gel transitions of Q-100 and Q-195 in water at sufficiently high concentrations were triggered by the LCST transition of the PDEGMMA outer block, similar to the thermally induced formation of 3-D network

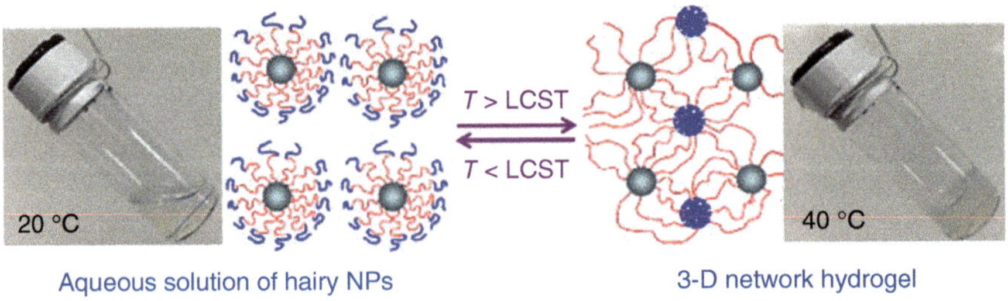

Figure 4.14 Optical photos of a 6.0 wt % aqueous dispersion of P(TMAEMA-I)-*b*-PDEGMMA brush-grafted silica NPs Q-195 with PDEGMMA as the outer block at 20 and 40 °C. Source: Reproduced with permission from Ref. [57]; © 2016, American Chemical Society.

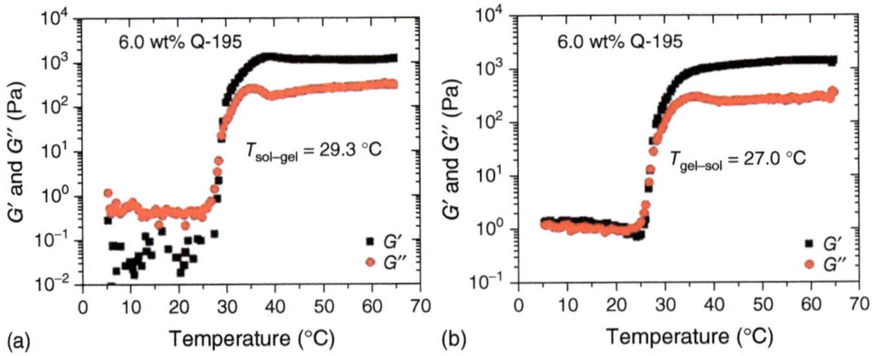

Figure 4.15 Plots of dynamic storage modulus G' and loss modulus G'' of a 6.0 wt% aqueous dispersion of P(TMAEMA-I)-*b*-PDEGMMA brush-grafted silica NPs (Q-195) obtained from an oscillatory shear experiment performed (a) in a heating ramp at a frequency of 1 Hz, a strain amplitude of 0.2%, and a heating rate of 3 °C min^{-1} and (b) in a cooling ramp using a frequency of 1 Hz, a strain amplitude of 0.2%, and a cooling rate of −3 °C min^{-1}. Source: Reproduced with permission from Ref. [57]; © 2016, American Chemical Society.

hydrogels of thermosensitive amphiphilic ABC triblock copolymers [56, 128], e.g. poly(ethylene-*alt*-propylene)-*b*-PEO-*b*-poly(*N*-isopropylacrylamide) ABC triblock terpolymer [128]. When the concentration was above the CGC and the temperature was higher than the LCST of PDEGMMA, the PDEGMMA outer block collapsed and self-assembled into micellar cores, which physically cross-linked the polymer brush-grafted NPs into a 3-D network. In contrast to traditional thermoresponsive ABA triblock copolymers, our diblock copolymer brush-grafted silica NPs resemble the core–shell–corona micelles of amphiphilic ABC with a preorganized hydrophobic core [56, 128]. As such, there were essentially no unimers present at temperatures far below the LCST of the outer block, making the thermally induced gelation sharp and efficient (Figure 4.15). Since these hairy NP hydrogels are formed by the association of the collapsed thermoresponsive outer blocks to form a 3D network of NPs and hydrophobic micellar domains bridged by the polyelectrolyte chains, they should be able to reform if the gel network is disrupted, e.g. by large amplitude shearing. To study the recoverability of these hairy NP hydrogels, the gels of 6.0 wt% aqueous dispersions of Q-100 and Q-195 at 55 °C were subjected to oscillating shear that alternated between 0.2% and 100% strain amplitude. As shown in Figure 4.16, at the 0.2% strain amplitude, the gels were maintained, whereas at the 100% strain amplitude, the network was disrupted, as indicated by the steep drop in both storage and loss moduli and a larger G'' value than G'. Upon returning to the 0.2% strain amplitude, the hydrogels of both hairy NPs reformed quickly, demonstrating their self-healing property.

These studies showed that well-defined thermoresponsive diblock copolymer brush-grafted silica NPs, synthesized by SI-ATRP, exhibited thermally induced, reversible clear solution-to-clear hydrogel transition in water at a sufficiently high concentration, with the gelation mechanism depending on the location of the thermoresponsive block in the diblock copolymer brushes. These hybrid hydrogels can overcome the limitations of conventional ABA, ABC, or star diblock polymer

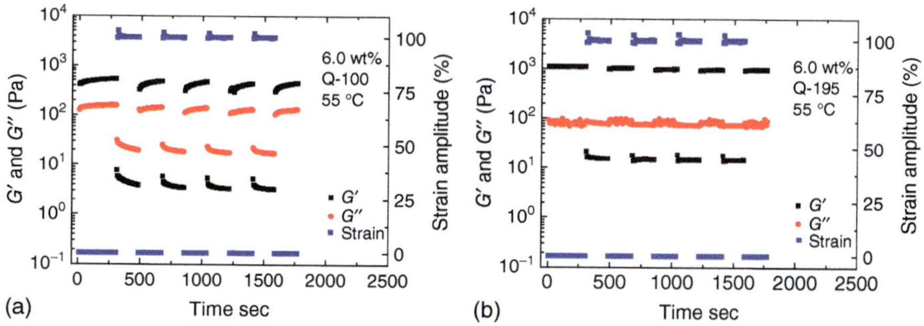

Figure 4.16 Plots of dynamic storage modulus G' and loss modulus G'' (left axis) of a 6.0 wt% aqueous dispersion of Q-100 (a) and Q-195 (b) obtained from oscillatory shear experiments performed at 55 °C using a frequency of 1 Hz. The strain amplitude oscillated between 0.2% and 100% amplitude (right axis) at given intervals. Source: Reproduced with permission from Ref. [57]; © 2016, American Chemical Society.

hydrogels by introducing more material functions through the choice of NPs and thus have potential in a wide array of applications, e.g. in biomedical areas.

4.3.4 Thermoresponsive Polymer Brush-grafted Nanoparticles for Enhancing Gelation of Thermoresponsive Linear ABC Triblock Copolymers in Water

As discussed in the preceding section, thermoresponsive block copolymers in water can exhibit temperature-induced reversible sol–gel transitions when the polymer concentration is sufficiently high. These hydrogels are termed injectable hydrogels because their liquid state enables the use of a syringe and needle for delivery and in situ gelation. Thermosensitive ABA and ABC triblock copolymer hydrogels are particularly attractive because the physically cross-linked 3D polymer networks display better stability against dissolution in water than diblock copolymer hydrogels, and their critical gelation concentrations are also typically lower than those of thermoresponsive diblock copolymers. However, the networks of ABA and ABC triblock copolymers in the hydrogels contain many defects, including loops and dangling blocks, which are not elastically active and cannot sustain the external stress applied to the hydrogels, thus lowering the mechanical strength of the hydrogels [56, 124–128, 133–136].

We reasoned that the addition of thermosensitive hairy NPs with an LCST similar to those of the thermosensitive outer blocks of a linear triblock copolymer with a hydrophilic middle block and thermoresponsive outer blocks, might be able to enhance the mechanical properties of the hydrogels because such hairy NPs can act as "seeds" to reduce defects and promote the formation of 3D networks (Scheme 4.9) [56]. To test this hypothesis, we synthesized a doubly thermoresponsive hydrophilic ABC linear triblock copolymer, PDEGEMA-*b*-P(TMAEMA-I)-*b*-(PDEGMMA-*co*-RhBMA) (ABC-Q), by RAFT polymerization [137, 138] and subsequent post-polymerization quaternization of the PDMAEMA middle block with iodomethane (Scheme 4.10). The cloud points of PDEGEMA and PDEGMMA in water are 4 and 26 °C, respectively [139]. The $M_{n,SEC}$ and $Đ$ of the precursor of ABC-Q (i.e. PDEGEMA-*b*-PDMAEMA-*b*-P(DEGMMA-*co*-RhBMA) were 84.5 kDa and 1.31, respectively, indicating that the three steps of RAFT polymerization were all controlled. The DPs of PDEGEMA, PDMAEMA, and P(DGEMMA-*co*-RhBMA) were calculated to be 75, 345, and 142, respectively. Note that the molar ratio of DEGMMA and RhBMA in the third-step RAFT polymerization was $1:6.3\times10^{-4}$, and thus the molar content of RhBMA per chain was very small. ABC-Q was then obtained by quaternization of the middle PDMAEMA block of the precursor ABC triblock copolymer with CH_3I, and the complete conversion was confirmed by ^{1}H NMR spectroscopy analysis. DLS study showed that ABC-Q in water at 2.0 mg g^{-1} underwent two transitions upon heating from 2 to 38 °C, although the first transition temperature could not be pinpointed because the LCST of PDEGEMA was only 4 °C. The onset temperature of the second transition from the P(DGEMMA-*co*-RhBMA) block was 31 °C from the scattering intensity plot and 27 °C from the apparent hydrodynamic size plot as a function

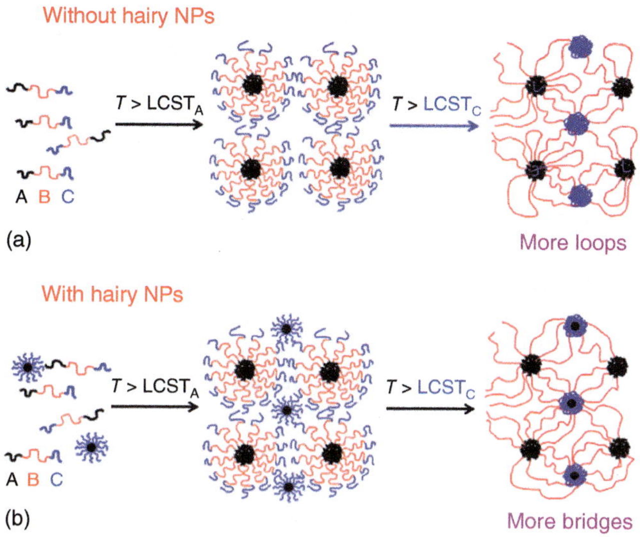

Scheme 4.9 Stepwise formation of a 3-D network hydrogel by a doubly thermosensitive ABC triblock copolymer in the absence (a) and presence (b) of thermoresponsive hairy nanoparticles with an LCST similar to that of the higher LCST block of ABC triblock copolymer. Source: Reproduced from Ref. [56]; © 2016 American Chemical Society.

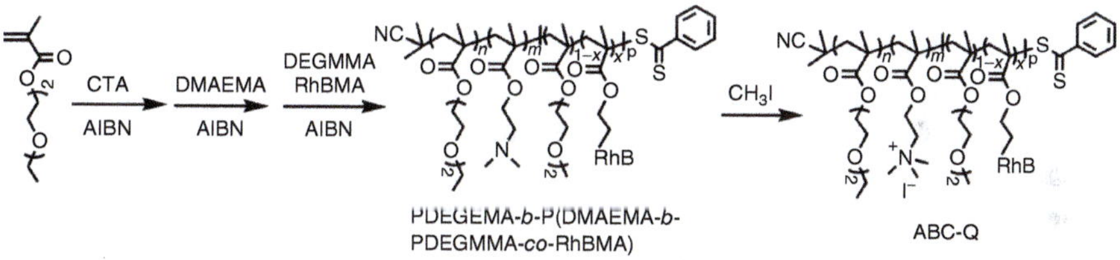

Scheme 4.10 Synthesis of doubly thermosensitive ABC linear triblock copolymer PDEGEMA-*b*-PDMAEMA-*b*-P(DEGMMA-*co*-RhBMA) by RAFT and subsequent quaternization with iodomethane to produce ABC-Q.

of temperature. Two thermosensitive hairy NPs were synthesized by SI-ATRP from initiator-functionalized 20 nm silica NPs, NP-29.5 with an LCST of 29.5 °C, similar to that of the P(DEGMMA-*co*-RhBMA) block of ABC-Q, and NP-51 with a significantly higher LCST of 51 °C. Fluorescence resonance energy transfer (FRET) was used to investigate the interactions between thermosensitive ABC copolymer micelles and hairy NPs [140]. While a small amount of RhBMA (Scheme 4.6), a FRET acceptor, was incorporated into the PDEGMMA block of ABC-Q, a similarly small amount of NBDMA (Scheme 4.6), a FRET donor, was introduced into the thermoresponsive brushes of hairy NPs.

As expected, aqueous solutions of ABC-Q with concentrations of 6.0 and 12.0 wt% underwent transitions from free-flowing clear liquids to free-standing clear gels upon heating from ~0 to 40 °C. The thermal gelation was reversion; cooling in an ice/water bath changed the gels to sols. The gelation behaviors of the two polymer solutions were investigated by oscillatory shear rheological measurements conducted in heating ramps using a constant frequency of 1 Hz, a strain amplitude of 1%, and a heating rate of 3 °C min^{-1}. In general, at lower temperatures, the values of both dynamic storage modulus G' and dynamic loss modulus G'' were small, with $G' < G''$, indicating a liquid state. Upon further heating, G' and G'' increased sharply, and G' eventually became larger than G''. The crossover point of G' and G'' curves is commonly taken as the sol–gel transition temperature ($T_{sol\text{-}gel}$). At higher temperatures, G' was significantly larger than G'', suggesting an elastic solid-like behavior (i.e. the gel state). To study the effect of incorporating NP-29.5 hairy NPs into the hydrogels of ABC-Q, a dispersion of NP-29.5 in water was added into the 6 wt% solution of ABC-Q, and the NP-to-polymer mass ratio was gradually increased from 20 : 100, 30 : 100, 40 : 100, 50 : 100 to 60 : 100. The concentration of ABC-Q was maintained at 6 wt% by removing a calculated amount of water from the sample. Rheological measurements were conducted using the aforementioned conditions, and for each sample, the measurement was repeated three times. The maximum values of G' (G'_{max}) and the $T_{sol\text{-}gel}$ from each rheological curve were determined. With increasing the mass ratio of NP-29.5 to polymer, the $T_{sol\text{-}gel}$ stayed at around 29.8 °C with changes less than 0.3 °C, suggesting that the addition of NP-29.5 did not adversely affect the formation of 3-D network hydrogels. In contrast, the G'_{max} value increased appreciably upon increasing the NP-to-polymer mass ratio from 0 : 100 to 40 : 100, and changed only slightly beyond that, as can be seen from the plot of G' versus temperature from representative heating ramps for all six samples on a linear scale in Figure 4.17a.

Figure 4.18a shows the plot of the average value of the G'_{max} from three repeated measurements versus NP-to-ABC-Q mass ratio. For the 6 wt% solution of ABC-Q in the absence of NP-29.5, the average G'_{max} value was 756 Pa. For the samples with the NP-to-ABC-Q mass ratios of 20 : 100, 30 : 100, 40 : 100, 50 : 100, and 60 : 100, the average G'_{max} values were 866 Pa, 940 Pa, and 1075 Pa, 1069 Pa, and 1099 Pa, respectively. Thus, when the NP-to-polymer ratio was 60 : 100, the G'_{max} increased by 45% compared with the sample without NP-29.5, indicating that the addition

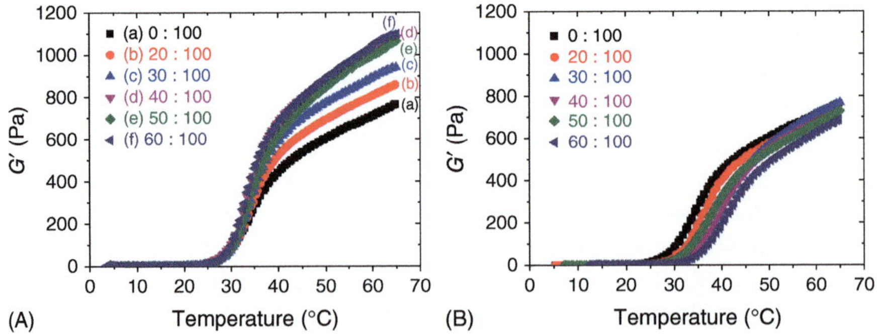

Figure 4.17 Plots of dynamic storage modulus G' versus temperature for (A) aqueous mixtures of ABC-Q and NP-29.5 and (B) aqueous mixtures of ABC-Q and NP-51 at various NP-to-polymer mass ratios. Source: Reproduced with permission from Ref. [56]; © 2016 American Chemical Society.

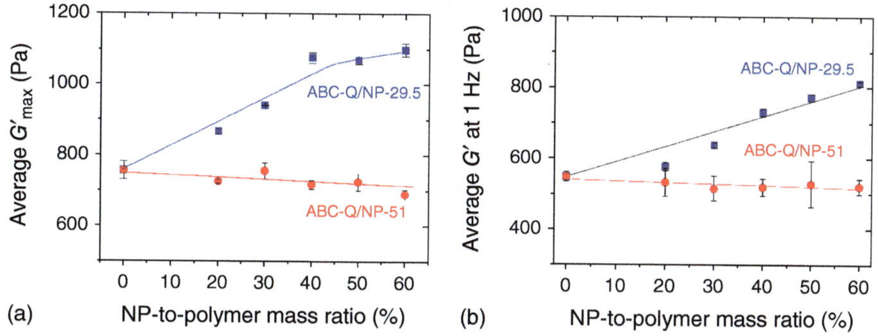

Figure 4.18 Plots of (a) average G'_{max} from three heating ramps and (b) average G'_{1Hz} from three frequency sweeps at 40 °C along with error bars versus NP-to-polymer mass ratio for hydrogels of 6 wt% ABC-Q containing NP-29.5 and NP-51. Lines are for guiding eyes. Source: Reproduced with permission from Ref. [56]; © 2016 American Chemical Society.

of NP-29.5 indeed improved the gel's strength considerably. We also conducted frequency sweeps at 40 °C for all samples and plotted the average value of G' at 1 Hz (G'_{1Hz}) from three runs as a function of the NP-to-ABC-Q mass ratio. Similar to the observations for G'_{max}, the G'_{1Hz} increased with increasing mass ratio of NP-29.5 to ABC-Q (Figure 4.18b). At the ratio of 60 : 100, the G'_{1Hz} was 48% larger compared with the solution in the absence of NP-29.5. These data confirmed that the addition of NP-29.5 improved the mechanical strength of the micellar hydrogel of ABC-Q.

The effect of adding NP-51 hairy NPs into the 6 wt% aqueous solution of ABC-Q was investigated in the same manner. Note that NP-51 exhibited an LCST of 51 °C in water, much higher than that of the P(DEGMMA-*co*-RhBMA) block of ABC-Q. Similar to the addition of NP-29.5, the $T_{sol-gel}$ exhibited little change in the presence of NP-51. Figure 4.17B shows the plots of G' versus temperature on a linear scale from representative heating ramps for all the samples; compared to the mixtures with NP-29.5, the G' value at 65 °C decreased slightly upon increasing the mass ratio of NP-51 to ABC-Q from 0 : 100 to 60 : 100. The average G'_{max} from heating ramps and the average G'_{1Hz} from frequency sweeps were plotted against the NP-to-polymer mass ratio (Figure 4.18). In contrast to the hybrid hydrogels of ABC-Q and NP-29.5, both G'_{max} and G'_{1Hz} seemed to decrease slightly with increasing mass ratio of NP-51 to ABC-Q. While the average G'_{max} decreased from 756 Pa in the absence of hairy NPs to 689 Pa at the mass ratio of 60 : 100 for NP-51 to ABC-Q, G'_{1Hz} decreased from 548 to 527 Pa. Thus, no positive effect was found by the addition of NP-51 with an LCST much higher than that of the P(DEGMMA-*co*-RhBMA) block of ABC-Q into the ABC-Q hydrogels.

The observed different effects of adding NP-29.5 and NP-51 into the 6 wt% aqueous solution of ABC-Q stemmed from their LCSTs. NP-29.5 had an LCST almost the same as that of the P(DEGMMA-co-RhBMA) outer block of ABC-Q and likely acted as "seeds" to absorb the collapsed P(DEGMMA-*co*-RhBMA) block of ABC-Q, resulting in more bridging chains between micellar cores in the network and thus an increase in G' (Scheme 4.9). Note that G' is proportional to the volume density of elastically active bridging chains between the micellar cores [126]. In contrast, when the P(DEGMMA-*co*-RhBMA) outer block of ABC-Q collapsed and associated into micellar cores upon heating, NP-51 were still highly hydrated and hence excluded from the hydrophobic micellar cores. To experimentally confirm the spatial locations of NP-29.5 and NP-51 in the network hydrogels of ABC-Q, FRET experiments were conducted for dilute aqueous mixtures of ABC-Q and hairy NPs at 15 and 40 °C. For the mixture of the ABC triblock copolymer and NP-29.5, FRET was observed between NBDMA in the brushes and RhBMA in ABC-Q at 40 °C, indicating the strong interactions between NP-29.5 and the collapsed P(DEGMMA-*co*-RhBMA) block of ABC-Q. In contrast, no FRET was observed between NP-51 and ABC-Q at 40 °C, suggesting that there was little or no interaction between hairy NPs and the hydrophobic micellar cores of P(DEGMMA-*co*-RHBMA) [56]. Thus, the rheological studies and the FRET experiment confirmed our hypothesis that hairy NPs with an appropriate LCST could promote the formation of 3D networks by reducing the number of loops and dangling chains and improving the mechanical properties of ABC linear triblock copolymer hydrogels. The method described here could be used to design injectable hybrid hydrogels with improved strengths and new functions via the use of polymer brush-grafted NPs (e.g. thermosensitive hairy mesoporous silica NPs).

4.4 Summary and Outlook

Environmentally responsive polymer brush-grafted particles undergo structural changes in response to external stimuli, exhibiting different properties and behaviors under different conditions. Two types of responsive hairy particles, binary mixed homopolymer brush- and thermoresponsive polymer brush-grafted silica (nano)particles, were presented in this Chapter. A unique and robust synthetic strategy was devised for the synthesis of well-defined binary mixed homopolymer brushes on silica particles by sequential SI-ATRP and SI-NMRP from Y initiator-functionalized silica particles. This method allowed for the preparation of mixed brushes with predetermined molecular weights, narrow dispersities, and controlled chain length disparity, making them well suited for physical chemistry study. TEM showed that the mixed P*t*BA/PS brushes grafted on silica particles underwent spontaneous reorganization upon exposure to different solvents and exhibited distinct morphologies. For mixed P*t*BA/PS brushes grafted on silica particles with similar molecular weights and grafting densities for the two polymers, wormlike, nearly bicontinuous nanostructures formed from lateral microphase separation were observed upon solution casting of hairy particles from a good solvent. On the other hand, surface-tethered, nearly spherical micelles were found when a selective solvent was employed. Despite the progress achieved in the understanding of the responsive behavior of mixed brushes, many fundamental issues remain unanswered (for example, the morphologies of binary mixed homopolymer brushes grafted on small NPs with a radius similar to the polymer sizes and on cylindrical NPs). Elucidation of the self-assembly behavior of binary mixed homopolymer brushes under various conditions not only improves our understanding of bicomponent brushes in a confined geometry but also facilitates the exploration of their potential applications and the study of more complex systems, such as tricomponent systems.

The second part of this chapter centers on thermoresponsive hairy silica (nano)particles prepared by SI-ATRP. DLS studies revealed that the LCST transition of thermoresponsive polymer brushes grafted on silica particles began at a lower temperature and occurred over a broader temperature range compared with the corresponding free polymer in water. The thermosensitive hairy particles were found to undergo spontaneous, quantitative, and reversible phase transfer between water and ethyl acetate or a hydrophobic ionic liquid, [EMIM][TFSI]. There existed a linear relationship between the transfer temperature of hairy particles from water to the ionic liquid upon heating and the cloud point of the corresponding free polymer in IL-saturated water, allowing for facile design of phase transfer systems with tunable transfer temperatures. When thermoresponsive diblock copolymer brushes were grafted on silica NPs, reversible sol–gel transitions of aqueous dispersions of hairy NPs were observed upon temperature changes when the concentration of hairy NPs was sufficiently high. The gelation mechanism of thermosensitive diblock copolymer brush-grafted NPs depended on the location of the thermoresponsive block in the diblock copolymer brushes (i.e. the inner or outer block). In addition, we found that the incorporation of thermosensitive hairy NPs into doubly thermoresponsive hydrophilic ABC linear triblock copolymer hydrogels improved the gel strength when the LCST of the hairy NPs and the higher LCST outer block of the ABC copolymer were the same or similar. Thermoresponsive polymer brush-grafted (nano)particles have potential uses in controlled encapsulation and delivery of substances, design of phase transfer catalysts and recyclable catalysts, aqueous lubrication, and biotechnology.

Acknowledgements

We thank the University of Tennessee Knoxville and NSF for the support of the research presented in this chapter and we are indebted to coworkers and collaborators, especially Professor Lei Zhu on TEM characterization of morphology of binary mixed homopolymer brush-grafted silica particles and Professor Timothy P. Lodge on stimuli-induced phase transfer of hairy particles between water and a hydrophobic ionic liquid.

References

1 Pyun, J. and Matyjaszewski, K. (2001). *Chem. Mater.* 13: 3436–3448.
2 Zhao, B. and Zhu, L. (2009). *Macromolecules* 42: 9369–9383.
3 Kumar, S.K., Jouault, N., Benicewicz, B., and Neely, T. (2013). *Macromolecules* 46: 3199–3214.
4 Hui, C.M., Pietrasik, J., Schmitt, M. et al. (2014). *Chem. Mater.* 26: 745–762.

5 Chancellor, A.J., Seymour, B.T., and Zhao, B. (2019). *Anal. Chem.* 91: 6391–6402.

6 Zhao, B. and Brittain, W.J. (2000). *Prog. Polym. Sci.* 25: 677–710.

7 Wu, L., Glebe, U., and Böker, A. (2015). *Polym. Chem.* 6: 5143–5184.

8 Fernandes, N.J., Koerner, H., Giannelis, E.P., and Vaia, R.A. (2013). *MRS Commun.* 3: 13–29.

9 He, J., Huang, X.L., Li, Y.C. et al. (2013). *J. Am. Chem. Soc.* 135: 7974–7984.

10 Motornov, M., Sheparovych, R., Lupitskyy, R. et al. (2007). *Adv. Funct. Mater.* 17: 2307–2314.

11 Zhao, B., Jiang, X.M., Li, D.J. et al. (2008). *J. Polym. Sci., Part A: Polym. Chem.* 46: 3438–3446.

12 Yavuz, M.S., Cheng, Y.Y., Chen, J.Y. et al. (2009). *Nat. Mater.* 8: 935–939.

13 Liu, G.Q., Liu, Z.L., Li, N. et al. (2014). *ACS Appl. Mater. Interfaces* 6: 20452–20463.

14 Wright, R.A.E., Wang, K.W., Qu, J., and Zhao, B. (2016). *Angew. Chem., Int. Ed.* 55: 8656–8660.

15 Zhang, L., Bei, H.P., Piao, Y. et al. (2018). *ChemPhysChem* 19: 1956–1964.

16 Shan, J., Nuopponen, M., Jiang, H. et al. (2005). *Macromolecules* 38: 2918–2926.

17 Zubarev, E.R., Xu, J., Sayyad, A., and Gibson, J.D. (2006). *J. Am. Chem. Soc.* 128: 4958–4959.

18 Zubarev, E.R., Xu, J., Sayyad, A., and Gibson, J.D. (2006). *J. Am. Chem. Soc.* 128: 15098–15099.

19 Wang, B.B., Li, B., Zhao, B., and Li, C.Y. (2008). *J. Am. Chem. Soc.* 130: 11594–11595.

20 Prucker, O. and Rühe, J. (1998). *Macromolecules* 31: 592–601.

21 Prucker, O. and Rühe, J. (1998). *Macromolecules* 31: 602–613.

22 von Werne, T. and Patten, T.E. (1999). *J. Am. Chem. Soc.* 121: 7409–7410.

23 Husseman, M., Malmstrom, E.E., McNamara, M. et al. (1999). *Macromolecules* 32: 1424–1431.

24 Pyun, J., Jia, S.J., Kowalewski, T. et al. (2003). *Macromolecules* 36: 5094–5104.

25 Bartholome, C., Beyou, E., Bourgeat-Lami, E. et al. (2003). *Macromolecules* 36: 7946–7952.

26 Ohno, K., Koh, K., Tsujii, Y., and Fukuda, T. (2003). *Angew. Chem., Int. Ed.* 42: 2751–2754.

27 Ohno, K., Morinaga, T., Koh, K. et al. (2005). *Macromolecules* 38: 2137–2142.

28 Mulvihill, M.J., Rupert, B.L., He, R.R. et al. (2005). *J. Am. Chem. Soc.* 127: 16040–16041.

29 Li, C., Han, J., Ryu, C.Y., and Benicewicz, B.C. (2006). *Macromolecules* 39: 3175–3183.

30 Hotchkiss, J.W., Lowe, A.B., and Boyes, S.G. (2007). *Chem. Mater.* 19: 6–13.

31 Tian, C.C., Bao, C.H., Binder, A. et al. (2013). *Chem. Commun.* 49: 8668–8670.

32 Wright, R.A.E., Hu, B., Henn, D.M., and Zhao, B. (2015). *Soft Matter* 11: 6808–6820.

33 Chmielarz, P., Yan, J.J., Krys, P. et al. (2017). *Macromolecules* 50: 4151–4159.

34 Seymour, B.T., Wright, R.A.E., Parrott, A.C. et al. (2017). *ACS Appl. Mater. Interfaces* 9: 25038–25048.

35 Guo, Y. and Moffitt, M.G. (2007). *Macromolecules* 40: 5868–5878.

36 Jiang, B.B., Pang, X.C., Li, B., and Lin, Z.Q. (2015). *J. Am. Chem. Soc.* 137: 11760–11767.

37 Chen, Y.H., Yoon, Y.J., Pang, X.C. et al. (2016). *Small* 12: 6714–6723.

38 Li, D.J., Sheng, X., and Zhao, B. (2005). *J. Am. Chem. Soc.* 127: 6248–6256.

39 Zhao, B. and Zhu, L. (2006). *J. Am. Chem. Soc.* 128: 4574–4575.

40 Zhu, L. and Zhao, B. (2008). *J. Phys. Chem. B.* 112: 11529–11536.

41 Jiang, X.M., Zhong, G.J., Horton, J.M. et al. (2010). *Macromolecules* 43: 5387–5395.

42 Jiang, X.M., Zhao, B., Zhong, G.J. et al. (2010). *Macromolecules* 43: 8209–8217.

43 Horton, J.M., Tang, S.D., Bao, C.H. et al. (2012). *ACS Macro Lett.* 1: 1061–1065.

44 Bao, C.H., Tang, S.D., Horton, J.M. et al. (2012). *Macromolecules* 45: 8027–8036.

45 Li, W.K., Bao, C.H., Wright, R.A.E., and Zhao, B. (2014). *RSC Adv.* 4: 18772–18781.

46 Bao, C.H., Tang, S.D., Wright, R.A.E. et al. (2014). *Macromolecules* 47: 6824–6835.

47 Li, D.J., Jones, G.J., Dunlap, J.R. et al. (2006). *Langmuir* 22: 3344–3351.

48 Li, D.J. and Zhao, B. (2007). *Langmuir* 23: 2208–2217.

49 Li, D.J., Dunlap, J.R., and Zhao, B. (2008). *Langmuir* 24: 5911–5918.

50 Jiang, X.M., Wang, B.B., Li, C.Y., and Zhao, B. (2009). *J. Polym. Sci. Part A: Polym. Chem.* 47: 2853–2870.

51 Horton, J.M., Bai, Z.F., Jiang, X.M. et al. (2011). *Langmuir* 27: 2019–2027.

52 Horton, J.M., Bao, C.H., Bai, Z.F. et al. (2011). *Langmuir* 27: 13324–13334.

53 Bao, C.H., Horton, J.M., Bai, Z.F. et al. (2014). *J. Polym. Sci., Part B: Polym. Phys.* 52: 1600–1619.

54 Hu, B., Henn, D.M., Wright, R.A.E., and Zhao, B. (2014). *Langmuir* 30: 11212–11224.

55 Hu, B., Wright, R.A.E., Jiang, S.S. et al. (2016). *Polymer* 82: 206–216.

56 Hu, B., Fu, W.X., and Zhao, B. (2016). *Macromolecules* 49: 5502–5513.

57 Wright, R.A.E., Henn, D.M., and Zhao, B. (2016). *J. Phys. Chem. B* 120: 8036–8045.

58 Zhulina, E. and Balazs, A.C. (1996). *Macromolecules* 29: 2667–2673.

59 Gil, E.S. and Hudson, S.M. (2004). *Prog. Polym. Sci.* 29: 1173–1222.

60 Sidorenko, A., Minko, S., Schenk-Meuser, K. et al. (1999). *Langmuir* 15: 8349–8355.

61 Marko, J.F. and Witten, T.A. (1991). *Phys. Rev. Lett.* 66: 1541–1544.

62 Marko, J.F. and Witten, T.A. (1992). *Macromolecules* 25: 296–307.

63 Dong, H. (1993). *J. Phys. II France* 3: 999–1020.

64 Lai, P.Y. (1994). *J. Chem. Phys.* 100: 3351–3357.

65 Singh, C., Pickett, G.T., and Balazs, A.C. (1996). *Macromolecules* 29: 7559–7570.

66 Müller, M. (2002). *Phys. Rev. E* 65: 030802.

67 Minko, S., Müller, M., Usov, D. et al. (2002). *Phys. Rev. Lett.* 88: 035502.

68 Wenning, L., Müller, M., and Binder, K. (2005). *Europhys. Lett.* 71: 639–645.

69 Wang, J.F. and Müller, M. (2009). *J. Phys. Chem. B* 113: 11384–11402.
70 Merlitz, H., He, G.L., Sommer, J.U., and Wu, C.X. (2009). *Macromolecules* 42: 445–451.
71 Uhlmann, P., Merlitz, H., Sommer, J.U., and Stamm, M. (2009). *Macromol. Rapid Commun.* 30: 732–740.
72 Minko, S., Usov, D., Goreshnik, E., and Stamm, M. (2001). *Macromol. Rapid Commun.* 22: 206–211.
73 Lemieux, M., Usov, D., Minko, S. et al. (2003). *Macromolecules* 36: 7244–7255.
74 Santer, S., Kopyshev, A., Yang, H.K., and Rühe, J. (2006). *Macromolecules* 39: 3056–3064.
75 Wang, J., Kara, S., Long, T.E., and Ward, T.C. (2000). *J. Polym. Sci. Part A: Polym. Chem.* 38: 3742–3750.
76 Minko, S., Luzinov, I., Luchnikov, V. et al. (2003). *Macromolecules* 36: 7268–7279.
77 Minko, S., Müller, M., Motornov, M. et al. (2003). *J. Am. Chem. Soc.* 125: 3896–3900.
78 Julthongpiput, D., Lin, Y.H., Teng, J. et al. (2003). *J. Am. Chem. Soc.* 125: 15912–15921.
79 Ionov, L., Sidorenko, A., Stamm, M. et al. (2004). *Macromolecules* 37: 7421–7423.
80 Wang, Y., Zheng, J.X., Brittain, W.J., and Cheng, S.Z.D. (2006). *J. Polym. Sci., Part A: Polym. Chem.* 44: 5608–5617.
81 Zhao, B. (2003). *Polymer* 44: 4079–4083.
82 Zhao, B. (2004). *Langmuir* 20: 11748–11755.
83 Zhao, B. and He, T. (2003). *Macromolecules* 36: 8599–8602.
84 Zhao, B., Haasch, R.T., and MacLaren, S. (2004). *J. Am. Chem. Soc.* 126: 6124–6134.
85 Zhao, B., Haasch, R.T., and MacLaren, S. (2004). *Polymer* 45: 7979–7988.
86 Motornov, M., Sheparovych, R., Katz, E., and Minko, S. (2008). *ACS Nano* 2: 41–52.
87 Matyjaszewski, K. and Xia, J.H. (2001). *Chem. Rev.* 101: 2921–2990.
88 Hawker, C.J., Bosman, A.W., and Harth, E. (2001). *Chem. Rev.* 101: 3661–3688.
89 Usov, D., Gruzdev, V., Nitschke, M. et al. (2007). *Macromolecules* 40: 8774–8783.
90 Roan, J.R. (2006). *Phys. Rev. Lett.* 96: 248301.
91 Wang, Y.Q., Yang, G.A., Tang, P. et al. (2011). *J. Chem. Phys.* 134: 134903.
92 Ma, X., Yang, Y.Z., Zhu, L. et al. (2013). *J. Chem. Phys.* 139: 214902.
93 Ma, X., Chen, C.Y., Yang, Y.Z., and Qiu, F. (2014). *Soft Matter* 10: 6005–6013.
94 Chen, C.Y., Zhang, T.C., Zhu, L. et al. (2016). *ACS Macro Lett.* 5: 718–723.
95 Philipse, A.P. and Vrij, A. (1989). *J. Colloid Interface Sci.* 128: 121–136.
96 Bogush, G.H., Tracy, M.A., and Zukoski, C.F. IV, (1988). *J. Noncryst. Solids* 104: 95–106.
97 Motornov, M., Sheparovych, R., Lupitskyy, R. et al. (2007). *J. Colloid Interface Sci.* 310: 481–488.
98 Fox, T.L., Tang, S.D., Zhang, G.Q. et al. (2016). *Macromol. Chem. Phys.* 217: 1767–1776.

99 Schild, H.G. (1992). *Prog. Polym. Sci.* 17: 163–249.

100 Han, S., Hagiwara, M., and Ishizone, T. (2003). *Macromolecules* 36: 8312–8319.

101 Zhao, B., Li, D.J., Hua, F.J., and Green, D.R. (2005). *Macromolecules* 38: 9509–9517.

102 Hua, F.J., Jiang, X.G., Li, D.J., and Zhao, B. (2006). *J. Polym. Sci. Part A: Polym. Chem.* 44: 2454–2467.

103 Zhao, B. (2021). *J. Phys. Chem. B* 125: 6373–6389.

104 Wang, N., Seymour, B.T., Lewoczko, E.M. et al. (2018). *Polym. Chem.* 9: 5257–5261.

105 Lewoczko, E.M., Wang, N., Lundberg, C.E. et al. (2021). *ACS Appl. Polym. Mater.* 3: 867–878.

106 Seuring, J. and Agarwal, S. (2012). *Macromol. Rapid Commun.* 33: 1898–1920.

107 Wu, T., Zhang, Y.F., Wang, X.F., and Liu, S.Y. (2008). *Chem. Mater.* 20: 101–109.

108 Shan, J., Chen, J., Nuopponen, M., and Tenhu, H. (2004). *Langmuir* 20: 4671–4676.

109 Cole-Hamilton, D.J. (2003). *Science* 299: 1702–1706.

110 Desset, S.L. and Cole-Hamilton, D.J. (2009). *Angew. Chem., Int. Ed.* 48: 1472–1474.

111 Chechik, V., Zhao, M.Q., and Crooks, R.M. (1999). *J. Am. Chem. Soc.* 121: 4910–4911.

112 Stephenson, R. and Stuart, J. (1986). *J. Chem. Eng. Data* 31: 56–70.

113 Aveyard, R., Binks, B.P., and Clint, J.H. (2003). *Adv. Colloid Interface Sci.* 100: 503–546.

114 Duan, H.W., Wang, D.Y., Kurth, D.G., and Möhwald, H. (2004). *Angew. Chem. Int. Ed.* 43: 5639–5642.

115 Wang, D.Y., Duan, H.W., and Möhwald, H. (2005). *Soft Matter* 1: 412–416.

116 Welton, T. (1999). *Chem. Rev.* 99: 2071–2083.

117 He, Y.Y. and Lodge, T.P. (2006). *J. Am. Chem. Soc.* 128: 12666–12667.

118 Bai, Z.F., He, Y.Y., and Lodge, T.P. (2008). *Langmuir* 24: 5284–5290.

119 Bai, Z.F., He, Y.Y., Young, N.P., and Lodge, T.P. (2008). *Macromolecules* 41: 6615–6617.

120 Guerrero-Sanchez, C., Wouters, D., Hoeppener, S. et al. (2011). *Soft Matter* 7: 3827–3831.

121 Yin, X., Hoffman, A.S., and Stayton, P.S. (2006). *Biomacromolecules* 7: 1381–1385.

122 Joo, M.K., Park, M.H., Choi, B.G., and Jeong, B. (2009). *J. Mater. Chem.* 19: 5891–5905.

123 He, C.L., Kim, S.W., and Lee, D.S. (2008). *J. Controlled Release* 127: 189–207.

124 O'Lenick, T.G., Jiang, X.G., and Zhao, B. (2010). *Langmuir* 26: 8787–8796.

125 Woodcock, J.W., Wright, R.A.E., Jiang, X.G. et al. (2010). *Soft Matter* 6: 3325–3336.

126 O'Lenick, T.G., Jin, N.X., Woodcock, J.W., and Zhao, B. (2011). *J. Phys. Chem. B.* 115: 2870–2881.

127 Woodcock, J.W., Jiang, X.G., Wright, R.A.E., and Zhao, B. (2011). *Macromolecules* 44: 5764–5775.

128 Zhou, C., Hillmyer, M.A., and Lodge, T.P. (2012). *J. Am. Chem. Soc.* 134: 10365–10368.
129 Jiang, X.G., Jin, S., Zhong, Q.X. et al. (2009). *Macromolecules* 42: 8468–8476.
130 Jin, N.X., Woodcock, J.W., Xue, C.M. et al. (2011). *Macromolecules* 44: 3556–3566.
131 Jin, N.X., Zhang, H., Jin, S. et al. (2012). *J. Phys. Chem. B.* 116: 3125–3137.
132 Jin, N.X., Zhang, H., Jin, S. et al. (2012). *Macromolecules* 45: 4790–4800.
133 Li, C., Buurma, N.J., Haq, I. et al. (2005). *Langmuir* 21: 11026–11033.
134 Sugihara, S., Kanaoka, S., and Aoshima, S. (2004). *J. Polym. Sci. Part A: Polym. Chem.* 42: 2601–2611.
135 Aoshima, S. and Kanaoka, S. (2008). *Adv. Polym. Sci.* 210: 169–208.
136 Zhou, C., Toombes, G.E.S., Wasbrough, M.J. et al. (2015). *Macromolecules* 48: 5934–5943.
137 Chiefari, J., Chong, Y.K., Ercole, F. et al. (1998). *Macromolecules* 31: 5559–5562.
138 Fu, W.X., Bai, W., Jiang, S.S. et al. (2018). *Macromolecules* 51: 1674–1680.
139 Ishizone, T., Seki, A., Hagiwara, M. et al. (2008). *Macromolecules* 41: 2963–2967.
140 Henn, D.M., Fu, W.X., Mei, S. et al. (2017). *Macromolecules* 50: 1645–1656.

5

Self-Assembly of Hairy Nanoparticles with Polymeric Grafts

Xiaoxue Shen, Huibin He, and Zhihong Nie

Fudan University, State Key Laboratory of Molecular Engineering of Polymers, Department of Macromolecular Science, Shanghai, 200438, P. R. China

5.1 Introduction

The term "self-assembly" refers to the spontaneous organization of building blocks (e.g. molecules, colloidal units, or even larger objects) into ordered or larger structures through noncovalent interactions, e.g. electrostatic interaction [1], hydrophobic/hydrophilic interaction [2], and hydrogen bonding [3]. Among others, inorganic nanoparticles (NPs) have attracted great attention due to their tunable size, shape, and composition, as well as intrinsic optical, electronic, and magnetic properties [4]. Self-assembly of these NPs enables fine-tuning of their emergent collective properties arising from the coupling interactions between plasmons, excitons, and magnetic moments of constituent NPs. The assemblies of these NPs are attractive for use in sensing [5], catalysis [6], nanomedicine [7], energy conversion and storage [8], and optical/electronic devices [9]. Typically, the assembly of inorganic NPs can be performed in solutions, at the interface (i.e. liquid–liquid, liquid–solid, liquid–gas), under external fields (e.g. light, electric, magnetic, and flow field), and on the soft or hard templates (e.g. carbon nanotubes (CNTs), polymers, proteins, and viruses) [7, 10]. The formation of well-defined assembly structures by these strategies often requires delicate control over the interactions between NP/NP, NP/solvent, NP/surface, and NP/template. Surface functionalization of NPs with different ligands (i.e. small molecules, DNAs, and polymers) offers an important route to regulate the aforementioned interactions and hence controllable assembly of NPs.

Hairy nanoparticles (HNPs) often refer to polymer-grafted NPs compromising a central organic or inorganic core and a shell of polymeric corona including synthetic polymers, biopolymers, DNAs, etc. [11] HNPs feature various highly tunable structural parameters, including (i) the chemistry of the core (e.g. metal, semiconductor, and magnetic NPs); (ii) the shape (e.g. spheres, rods, dumbbells, plates, and cubes) and size of the core; (iii) the composition of the polymer

Xiaoxue Shen and **Huibin He** are contributed equally to this chapter.

Hairy Nanoparticles: From Synthesis to Applications, First Edition. Edited by Zhiqun Lin and Yijiang Liu.

corona (e.g. homopolymers, block copolymers [BCPs], random copolymers, and the mixture of polymers); (iv) the length of polymer chain; and (v) the grafting density (σ) of polymer ligands on NPs [11]. Thus, HNPs have emerged as a unique class of assembly building blocks for constructing functional polymer/NP hybrid nanostructures [10b, 12].

This chapter summarizes recent advances in the self-assembly of HNPs, with an emphasis on tailoring the structures of NP assemblies. We mainly focus on discussing hybrid assembly structures including colloidal molecules (CMs), one-dimensional (1-D) chains (linear, branched et al.), two-dimensional (2-D) lattices, and three-dimensional (3-D) superstructures (e.g. clusters, vesicles, and crystals). We choose to limit our discussions to the building blocks of polymer-grafted inorganic NPs (PGNPs) consisting of inorganic NP cores (with dimensions in the range of ~1–100 nm) and polymer coronas. The self-assembly of HNPs with *polymeric cores* or *DNA coronas* is beyond the scope of this chapter. Furthermore, representative applications of PGNP assemblies in nanomedicine and optical/electronic devices are introduced to show off their great potential. The chapter ends with an outlook of challenges and future directions in the field.

5.2 Self-Assembly of PGNPs into Colloidal Molecules

CMs can be defined as discrete nonspherical clusters of a small number of particles with well-defined geometry, recapitulating the structure and symmetry of conventional molecules [13]. Various attractive forces (e.g. electrostatic attraction, hydrogen bonding, and solvophobic interactions) can be used to trigger the assembly of CMs [14]. The directionality in interparticle interactions is key to the formation of CMs with programmable architectures. This is generally achieved by functionalizing particles with a discrete number of attractive surface regions (patches) or packing particles in droplets with subsequent solvent evaporation [3c, 15]. For nanoscale particles, DNA-based strategies have enabled the formation of CMs with high precision and complexity [15c, 16]. These approaches either necessitate site-specific decorating of the surface of NPs with complementary single-strand DNAs or precise caging of NPs in DNA frames with defined binding sites. Moreover, the assembly of PGNPs has also offered a simple yet scalable route to the fabrication of CMs with different architectures.

5.2.1 Precisely Defined Assembly of Patchy NPs

Patchy NPs are a class of nanoscopic colloids composed of a core and a discrete number of attractive surface regions [17]. Existing strategies for the generation of polymeric patches on PGNPs can be divided into three categories: (i) selective attachment of polymers to certain crystal facets or regions of NPs (e.g. tips of nanorods and edges of nanoplates) owing to the differential binding affinity; (ii) localized coating of polymers by masking some parts of NP surfaces; and (iii) microphase separation or conformational changes of homopolymer or copolymers brushes on the surface of PGNPs. The well-defined number and spatial distribution of patches dictate the directional interactions between PGNPs, thus the controllable self-assembly of NPs to form defined CMs [15a, d, 18].

5.2.1.1 Isotropic NPs

Inorganic nanospheres are frequently used as the cores of PGNPs, but their symmetrical shape and isotropic surface composition make it challenging to site-specifically functionalize their surface with polymers. To tackle this challenge, Chen and coworkers developed a "mix-and-heat" method to synthesize patchy NPs [19]. Briefly, a mixture of citrate-stabilized gold nanoparticles (AuNPs), polystyrene-*b*-poly(acrylic acid) (PS-*b*-PAA) polymers, and hydrophobic (such as 2-dipalmitoyl-*sn*-glycero-3-phosphothioethanol) and hydrophilic (e.g. diethylamine, 2-mercaptoacetic acid, or 4-mercaptobenzoic acid) small molecular thiols in DMF/H_2O was heated at 110 °C for 2 hours and then slowly cooled down to induce polymer self-assembly. The process led to the formation of Janus-like patchy NPs with defined polymeric patches. The formation of the patches was attributed to the binding competition between a hydrophobic and a hydrophilic ligand on the surface of AuNPs, which guided the selective attachment of PS-*b*-PAA on one side of the AuNPs. When a basic NaCl solution was added to an aqueous dispersion of patchy NPs, the reduced electrostatic interactions triggered their self-assembly of NPs to form dimers at 40% yield.

Phase segregation of end-grafted polymers on NPs is a simple yet promising strategy for the generation of patches. Such induced anisotropy of NPs does not rely on regiospecific surface modifications, which can be tedious to realize in large quantities. For instance, Kumacheva et al. demonstrated the formation of polystyrene (PS) patches on AuNPs with a relatively low σ of polymers. When water is added to a dispersion of PS-grafted NPs in DMF, a uniform layer of end-grafted polymer brushes on NPs thermodynamically segregates into a discrete number of pinned micelles due to the reduced quality of solvents for polymers (Figure 5.1a) [20]. The number of patches formed on NPs is related to the size and shape of the NP core, as well as the length and σ of the PS brushes (Figure 5.1b and c). The patches on nanoscale surface result in anisotropic interactions (e.g. electrostatic repulsion forces and hydrophobicity) between NPs and hence their subsequent self-assembly into defined dimers, trimers, or chains (Figure 5.1d). In this process, the surface patterning and NP assembly are induced by the same hydrophobic interaction. As a result, the formation of patches and assembly of NPs occur simultaneously, which leads to poor control over the formation of CMs. In another work, Kumacheva and coworkers programmed the surface patterning and self-assembly processes of inorganic NPs as two separate stages by using two orthogonal stimuli (Figure 5.1e) [21]. Polymeric patches were generated on AuNPs capped with a BCP of polystyrene-*b*-poly(4-vinylbenzoic acid) [PS-*b*-P(4VBA)], denoted as NPs@PS-*b*-P(4VBA), when a non-solvent (water, stimulus 1) was added into the dispersion of NPs in *N,N*-dimethylformamide (DMF), a good solvent for the copolymers. In this process, NP-adjacent PS blocks served as segments for patches, whereas surface-remote P(4VBA) blocks effectively stabilized the NPs

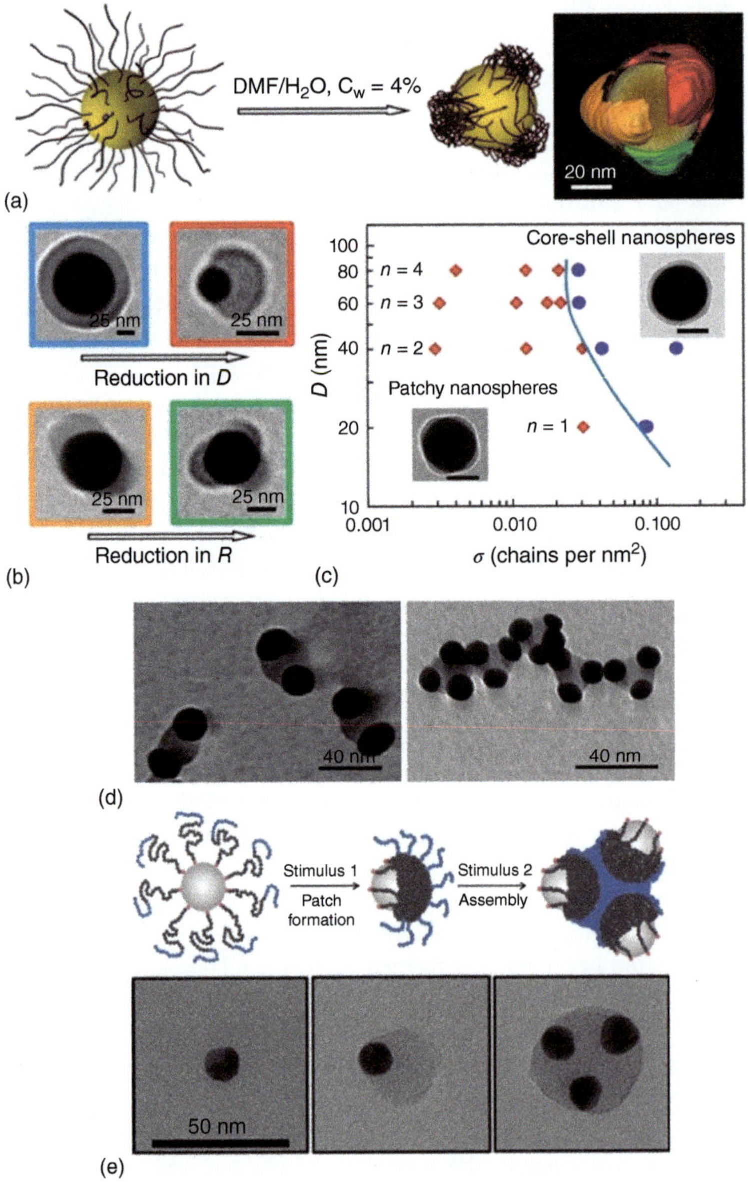

Figure 5.1 The generation of patchy nanospheres through the phase segregation of end-grafted polymers on NPs and the self-assembly of the patchy NPs into CMs. (a) Schematics of solvent-mediated formation of pinned polymer micelles (patches) on AuNPs. Right image is electron tomography reconstruction of AuNPs (60 nm diameter) with three patches of PS (molecular weight (MW) = 50 kDa; $\sigma = 0.02$ chains nm^{-2}). (b) Effect of NP size (top) and polymer dimensions (bottom) on the formation of PS patches (MW = 30 kDa in green box (bottom right) and 50 kDa for others (top low and bottom left); $\sigma = 0.03$ chains nm^{-2}). (c) Experimental diagram of patchy formation on AuNPs. Scale bars in insets are 50 nm. (d) TEM images of dimers of single patch AuNPs (left) and trimers of single patch AuNPs in chains (right). (e) Schematics of staged surface patterning and self-assembly of BCPs-grafted NPs. Bottom images are the representative TEM images of AuNPs at different stages. Source: (a–d) Choueiri et al. [20], Reproduced with permission from Nature Publishing Group. (e) Rossner et al. [21], Reproduced with permission from Wiley-VCH.

as individuals. Subsequent crosslinking of polyacid blocks by copper-(II) acetate (stimulus 2) drove the self-assembly of patchy NPs to form small NP clusters containing ~3–4 PGNPs.

5.2.1.2 Anisotropic NPs

As mentioned earlier, directional interactions are crucial for the formation of CMs. Compared to spherical NPs, the inherent chemical heterogeneity of anisotropic NPs allows site-specific functionalization of polymer brushes on NPs, thus facilitating directional clustering via specific molecular interactions. For instance, polymers (e.g. PS) can be selectively grafted onto the tips of Au nanorods (AuNRs) due to the low density of CTAB on these facets. Owing to their differential surface chemistry at the side and ends, these AuNRs end-grafted with PS brushes self-assembled into a variety of defined structures (e.g. chains, rings, bundles, vesicles), when the quality of the solvent was tuned [2b, 22]. As an example, the NR dimers with side-to-side orientation were obtained by adding water into a dispersion of the AuNRs in a mixture of tetrahydrofuran (THF)/DMF.

In addition to the directional molecular interactions arising from chemical patches, the steric constraints imposed by the shape of anisotropic NPs may also guide the NPs to assemble along certain direction. Thus, the combination of site-specific functionalization and shape anisotropy allows assembling of anisotropic NPs into CMs with geometries that are otherwise not readily accessible. For instance, NRs end-grafted with polymers usually refers not to form cross-like bundles due to the strong side-to-side attractive interactions [2b, 22]. To overcome this problem, Liz-Marzán and coworkers introduced a steric hindrance into the system by using PS_{509}-tip-functionalized gold nanodumbbells (AuNDs@PS_{509}) with spheroidal tips as building blocks [23]. The steric hindrance between the large tips overcame the side-to-side attraction between a pair of AuNDs, leading to the formation of bundles with cross-like arrangement of dumbbells (Figure 5.2). A two-step approach was used to generate such cross-like assemblies (Figure 5.2a). In the first step, AuND doublets with parallel orientation were induced by hydrophobic interactions of PS chains due to the increased solvent polarity. In the second step, AuND doublets were encapsulated within a shell of amphiphilic BCP of PS_{403}-*b*-PAA_{62} to inhibit further aggregation and promote the formation of cross-like assemblies. During thermal treatment of the solution, the compression force arising from the collapse of the BCP shell triggered the full conversion of dimeric clusters with side-to-side rod arrangement to clusters with cross-like arrangement (Figure 5.2b and Figure 5.2c).

When triangular nanoprisms were used as the core of PGNPs, small molecular ligands such as 2-naphthalenethiol (2-NAT) can be selectively functionalized at the prism tips to produce three patches of desired size (Figure 5.2d) [24]. Owing to the hydrophobicity of 2-NAT, the PS block of PS-*b*-PAA BCPs can be attached to the 2-NAT patches through hydrophobic interactions, while the negatively charged hydrophilic PAA block stuck outward. By increasing the ionic strength of aqueous solution, the steric hindrance and weaker electrostatic repulsion of PS-*b*-PAA patches led to twisted dimeric CMs including stars (Figure 5.2e) and slanting diamonds (Figure 5.2f).

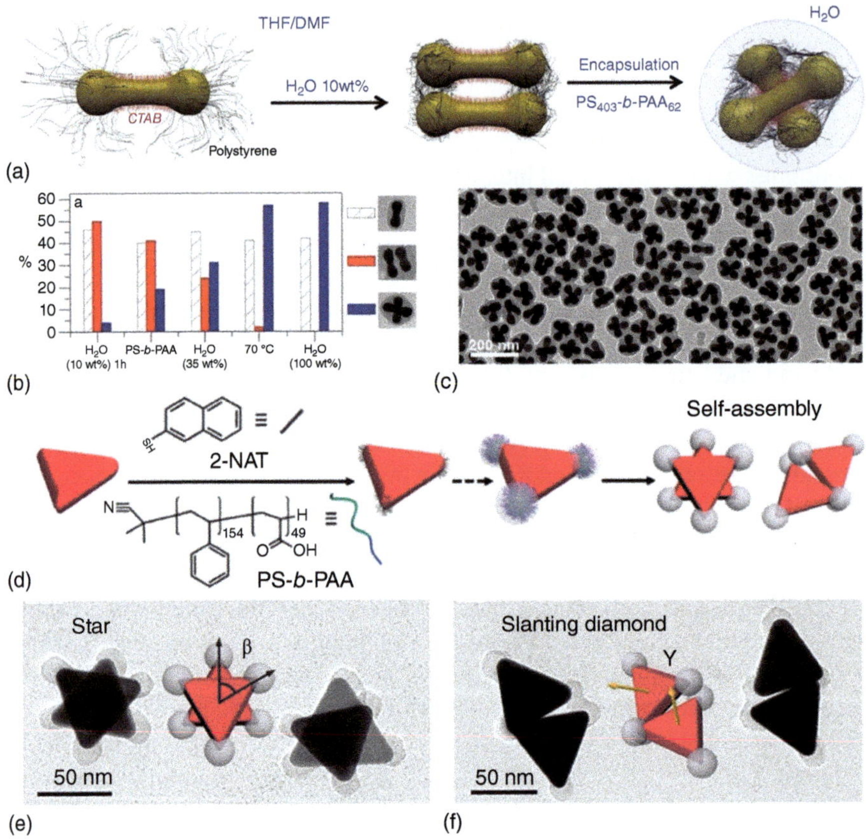

Figure 5.2 Self-assembly of PGNPs with anisotropic core into CMs. (a–c) Self-assembly of nanodumbbells end-grafted with PS. (a) Schematics of the formation of CMs with cross-like morphology by self-assembly of PS-functionalized AuNDs. (b) Distribution of morphologies at different experimental steps. Addition of water (10 wt%) leads to formation of parallel dimers (50%) which can be partly converted to cross-like dimers by encapsulation with PS-*b*-PAA. Extra amount of water and thermal treatment finally lead to conversion of all dimers into cross-like morphology. (c) TEM image of cross-like dimers. (d–f) Self-assembly of patchy nanoprisms. (d) Schematic representation of the formation of patchy nanoprisms and their self-assembly into stars and slanting diamonds. (e) TEM image of star-like assemblies. (f) TEM image of slanting diamond assemblies. Source: (a–c) Grzelczak et al. [23], Reproduced with permission from American Chemical Society. (d–f) Kim et al. [24], Reproduced with permission from American Chemical Society.

5.2.2 Polymer-Guided Assembly of NPs

Although assembly of patchy NPs is a powerful strategy for CMs fabrication, there are two main challenges: (i) the preparation of high-quality patchy NPs at high yield, especially from as-synthesized inorganic NPs; and (ii) the controllable self-assembly of patchy NPs to form target nanostructures at high yield. In addition to the formation of patches on NPs, the conformational changes of polymer brushes on NP surface may also induce anisotropic interparticle interactions to guide the directional self-assembly of NPs to form CMs.

Unlike the construction of CMs from patchy NPs, PGNPs without site-specific surface modification can assemble to form CMs. When inorganic NPs are functionalized with polymers, the resulted PGNPs inherit the solubility and assembly capability of the coating layer of polymer brushes. Binary polymers of poly(ethylene glycol) (PEG) and poly(methyl methacrylate) (PMMA) grafted NPs ($AuNPs_{42\text{-nm}}@PEG_{5\text{-kDa}}/PMMA_{23.6\text{-kDa}}$, $\sigma = 0.5\,\text{chains}\,nm^{-2}$) show good solubility in DMF, a common solvent for PEG and PMMA (Figure 5.3a) [25]. When water, a poor solvent for the polymers, was added to the DMF solution of NPs, these PGNPs were self-assembled into dimers at a yield of 60%. The driving force of the assembly is the hydrophobic effect of PMMA brushes. Both the size of the AuNPs and the relative molar ratio of hydrophobic brushes to hydrophilic brushes are critical for the preferential formation of the dimers. In addition to the mixed hydrophobic and hydrophilic polymers, the NP dimer can also be generated by using BCPs-coated NPs. Amphiphilic copolymer of PS_{154}-*b*-PAA_{49} encapsulated $AuNPs_{17\text{-nm}}@PS_{2\text{-kDa}}$ could form dimers at 65% yield by treating DMF/H_2O (V/V = 6 : 1) solution of PGNPs with HCl (final concentration of 5 mM) and heating at 60 °C for 2 h. Under such conditions, the morphological transformation of the polymer shells and charge repulsion between NPs led to the dimerization of PGNPs but prevented further aggregation of PGNPs from forming trimers or tetramers (Figure 5.3b) [26]. In another work, Halas and coworkers incubated $AuNPs_{15\text{-nm}}@PS_{12\text{-kDa}}$ with PS_{17}-*b*-PAA_{83} in DMF at 60 °C for 2 h [27]. The subsequent addition of a small amount of water into this mixture yielded a variety of uniform clusters containing predominantly 3 ~ 25 NPs.

Due to the nonrecognition interactions such as hydrophobic interaction and van der Waals force, the aforementioned CMs are mainly generated from one type of colloidal "atoms." Introducing recognition interactions (e.g. electrostatic interaction, hydrogen bond, host–guest interaction, and chemical reactions) into the assembly system enables the formation of CMs containing multiple distinctive NP "atoms." For instance, Nie and coworkers used the neutralization reaction between weak acid and base groups of copolymer ligands to achieve the directional bonding of binary NPs into CMs at high yield [28]. As illustrated in Figure 5.4a, poly(ethelyene oxide)-*b*-(acrylic acid-random-styrene) [PEO-*b*-P(AA-*r*-St)] and poly(ethelyene oxide)-*b*-poly(N,N-dimethylaminoethyl methacrylate-random-styrene) [PEO-*b*-P(DMAEMA-*r*-St)] were used to functionalize NP-A and NP-B of binary NPs, respectively. After mixing the two types of PGNPs in solution, the neutralization reaction between –COOH and $-NMe_2$ groups of copolymers drove the directional colloidal bonding of NPs. The precise assembly process generated CMs with symmetric AB_x nanostructures (x is the coordination number of B onto A) at high yield via (Figure 5.4b and Figure 5.4c). The value of x is governed by the ratio of the total number of acid groups to the total number of base groups on single NP-A and NP-B (Figure 5.4d). Steric and Coulombic repulsions between NPs are crucial to determining the symmetry of the CMs. This method is applicable

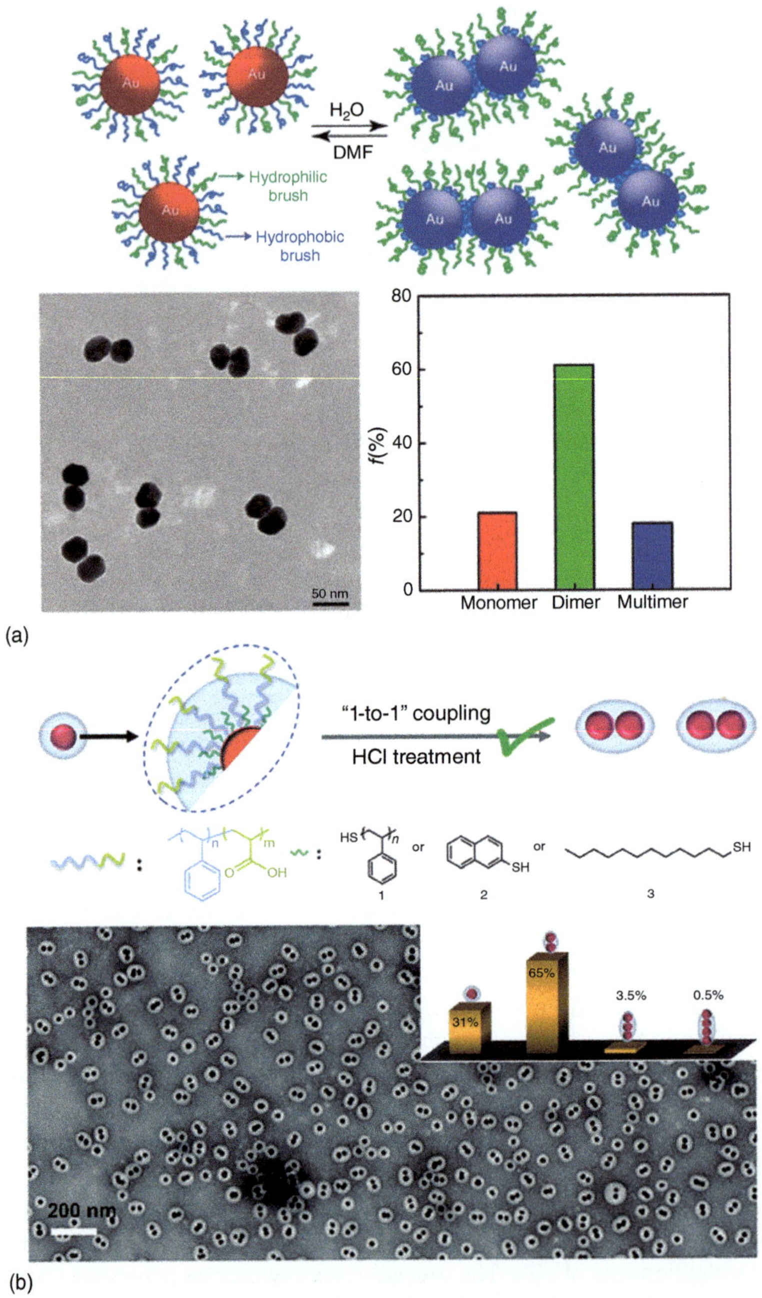

Figure 5.3 Directly self-assembly of nanospheres grafted with different polymers into dimers. (a) Self-assembly of NPs functionalized with mixed hydrophobic and hydrophilic polymers; top: schematic illustration of the self-assembly of AuNPs@PEG/PMMA; bottom left: TEM image of AuNPs dimers; bottom right: the statistical fractions of monomer, dimer, and multimeric structures. Source: Cheng et al. [25], Reproduced with permission from American Chemical Society. (b) Self-assembly of NPs functionalized with BCPs; top: schematics illustration of the one-step dimerization of AuNPs@PS-*b*-PAA by exploiting the structural transformation of polymer shells; bottom: TEM image and statistical fractions of AuNPs dimers. Source: Cheng et al. [26], Reproduced with permission from Royal Society of Chemistry.

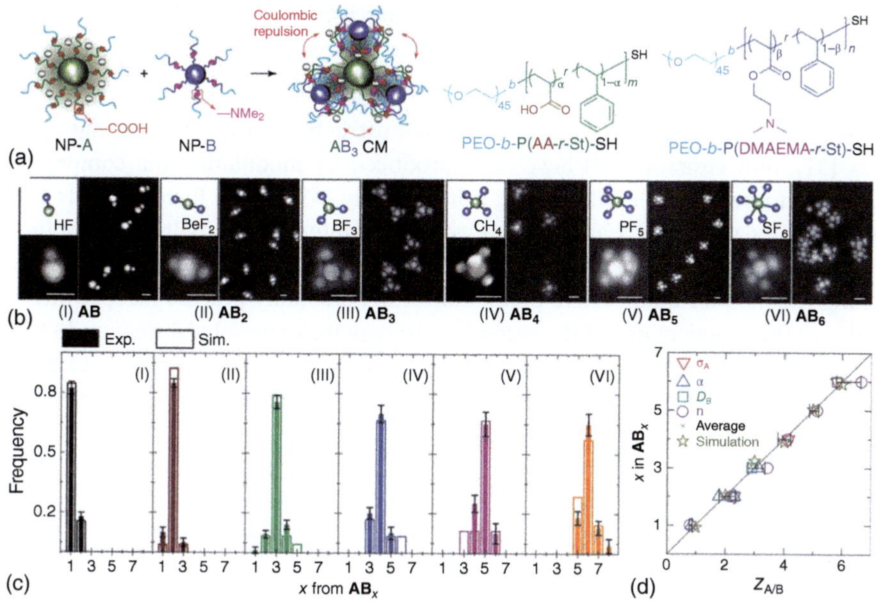

Figure 5.4 Self-assembly of binary PGNPs into CMs. (a) Schematic illustration of directional bonding of NP-A and NP-B, each coated with distinct block copolymer ligands, to form representative AB_3 CMs of NPs. (b) SEM images of AB_x CMs. (c) Distribution of CMs with different structures in simulation and experiments. (d) Variation in x in AB_x structures with increasing $Z_{A/B}$ (the ratio of the total number of acid groups on a single NP-A to the total number of base groups on single NP-B). Source: Yi et al. [28], Reproduced with permission from American Association for the Advancement of Science.

to the generation of CMs from NP cores of different compositions, such as Fe_3O_4 nanospheres, Ag nanospheres, and nanoplates [29].

5.3 Self-Assembly of PGNPs Into One-Dimensional (1-D) Structures

In this section, we review recent progress in the self-assembly of PGNPs into anisotropic 1-D structures in solution and in a condensed state. The key to the anisotropic self-assembly of PGNPs is the generation of directional interactions between NPs by designing the surrounding medium (e.g. solvent, polymer matrix) or engineering the surface chemistry of the PGNPs. Here we summarize existing strategies of assembling 1-D nanostructures into three main categories: (i) self-assembly of PGNPs in solution via various interactions such as hydrophobic interaction, neutralization reaction, and dipolar interaction (in Section 5.3.1), (ii) templated self-assembly of PGNPs in confined spaces (in Section 5.3.2), and (iii) the self-assembly of PGNPs in thin films (in Section 5.3.3).

5.3.1 Self-Assembly of PGNPs in Solution Guided by Various Molecular Interactions

5.3.1.1 Self-Assembly Driven by Neutralization Reaction

In the past decades, a variety of assembled nanostructures (e.g. linear strings [30], helix [31], and vesicles [32]) have been produced by modulating the competing nanoscale forces between NPs during self-assembly [33]. These forces include but are not limited to van der Waals forces, electrostatic interactions, hydrogen bonding, steric repulsion, and molecular dipole interactions [34]. In spite of great progress, lacking theories for quantitative prediction of the assembly process has become a barrier to the practical use of the assembly strategy for materials fabrication [33b]. Learning from polymerization framework in a molecular system proves to be a useful approach to predicting the self-assembly of NPs into polymer-like chains (i.e. so called colloidal polymers) [14, 35].

Inspired by the polycondensation of Nylon-66, Yi et al. reported a general paradigm for the copolymerization of binary inorganic NPs into linear nanostructures with periodic sequence, which expands the horizon of alternating copolymerization at the molecular level to nanoscale colloidal systems [36]. Binary inorganic NPs uniformly tethered with poly(styrene)-*b*-poly(acrylic acid-*r*-styrene)-SH [PS_x-*b*-$P(AA_{\alpha}$-*r*-$S_{1-\alpha})_y$] and poly(styrene)-*b*-poly(*N,N*-dimethylaminoethyl methacrylate-*r*-styrene)-SH [PS_m-*b*-$P(DMAEMA_{\beta}$-*r*-$S_{1-\beta})_n$] bearing either Lewis acid (i.e. –COOH) or Lewis base (i.e. $-NMe_2$) groups were used as nanoscale colloidal comonomers A and B for copolymerization, respectively. (Figure 5.5a) In the presence of acid catalyst, the neutralization between binary nanomers generated molecular dipole-like AB dimers that are further oriented and joined together to produce linear alternating nanoscale copolymers with alternating arrangements of NPs. The fast dimerization of conanomers is crucial to their sequential organization into alternating copolymers, resembling the preferential formation of dimers or charge-transfer complexes from comonomers in molecular systems.

Furthermore, shape-isotropic NPs such as triangular nanoplates modified with similar BCPs, PEO_n-*b*-$P(AA_{\alpha}$-*r*-$St_{1-\alpha})_m$, and PEO_k-*b*-$P(DMAEMA_{\beta}$-*r*-$St_{1-\beta})_l$ can also be used as colloidal monomers, namely NP-As, NP-Bs [29]. Depending on the length of the nonreactive block (i.e. PEO), the NP-B is preferentially attached to the edge or the face of a central disk-shape NP-A to produce edge-type or face-type clusters (Figure 5.5b). The regioselectivity is dictated by the different steric hindrances between planar and curved polymer brushes on the disc-shaped NP-As. When the PEO blocks were short such as $n = 45$ for [PEO_n-*b*-$P(AA_{\alpha}$-*r*-$St_{1-\alpha})_m$] ligands on NP-As, the steric shielding of short PEO blocks was negligible, so the neutralization between acid and base caused the bonding of NP-Bs onto the top and bottom faces of a central disc-shaped NP-A to form face-type AB_2 structures. Linear chains with alternating sequence of NP-As and NP-Bs were generated by varying the number of reactive groups (αm and βl) on the polymers, the length of polymer chains, and the feeding number ratios of NPs, $\psi_{B/A}$. Specifically, when the αm of PEO-*b*-P(AA-r-St) copolymers was 56 and the $\psi_{B/A}$ was ~1:1, linear chains with alternating sequence of NP-As and NP-Bs were produced (Figures 5.5b). In this case, the short polymer

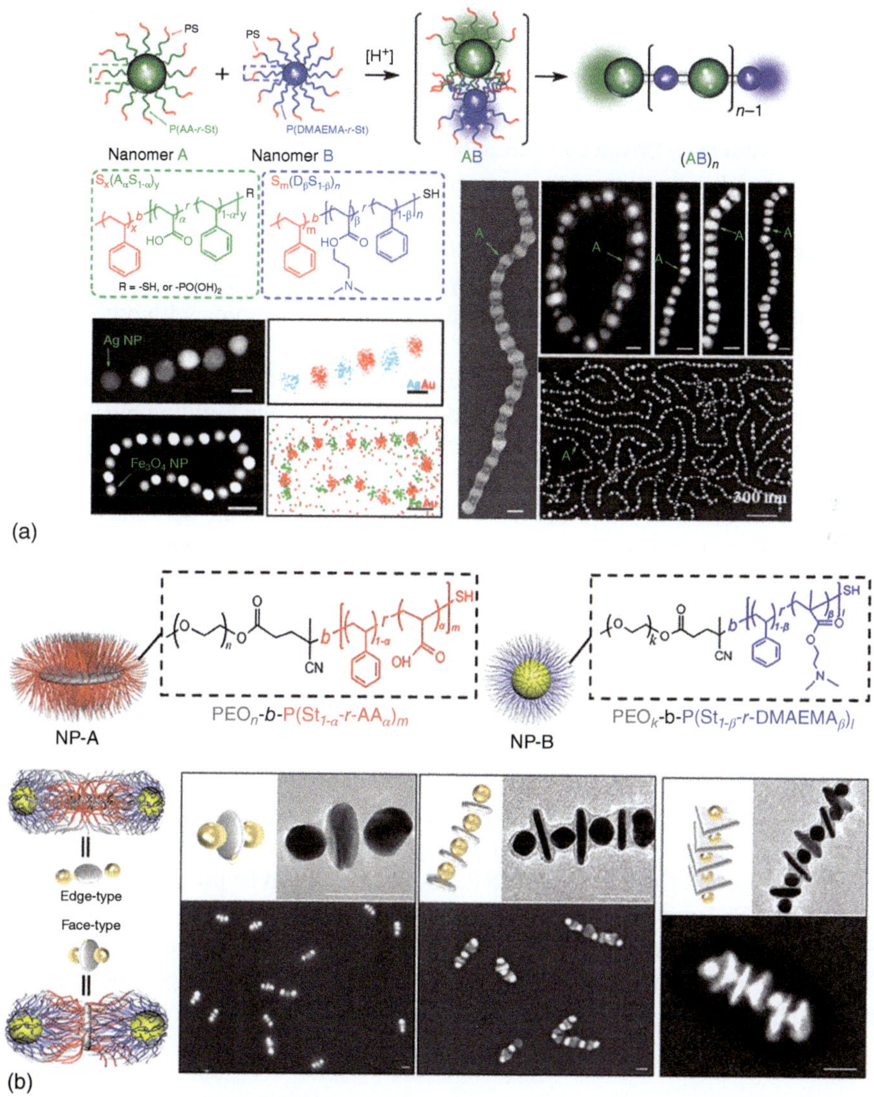

Figure 5.5 Alternating polymerization of nanocopolymers into linear nanostructures with periodic sequence. (a) Schematics of copolymerization of binary nanomers into ANCPs of $(AB)_n$ through interparticle neutralization and complexation; bottom: SEM images of ANCPs made from conanomers with different NP core or size combinations. Source: Yi et al. [36], Reproduced with permission from American Chemical Society. (b) Schematics of Ag nanoplates and AuNPs grafted with BCP ligands and their copolymerization to produce alternating chains governed by face-to-face assembly mode. Source: Lin et al. [29], Reproduced with permission from American Chemical Society. Scale bars are 20 nm in bottom left of (a), 50 nm in bottom right of (a), and 50 nm in (b) unless otherwise noted.

on the flat faces of NP-As was only able to consume base groups on one side of NP-Bs, thus resulting in base groups on the other side, which could further react with another NP-A.

5.3.1.2 Self-Assembly Driven by Hydrophobic Interaction

The differential binding strength between ligands and distinct crystal facets of shaped inorganic NPs enables the localized functionalization of NPs with polymers to construct PGNPs with anisotropic surface properties. Nanoscale hydrophobic attraction between these regiospecifically-grafted polymer brushes governs the self-assembly of PGNPs into desired nanostructures by combining with other types of repulsion forces [34].

Kumacheva et al. selectively modified two ends of CTAB stabilized AuNRs with hydrophobic thiol-terminated PS and used them as assembly building blocks that are structurally analogous to amphiphilic "A(polymer)-B(metal)-A(polymer)" type pom-pom tri-BCPs [2b, 22]. When water, a poor solvent for PS block, was added into a solution of PGNPs in DMF or THF, the variation in the solubility of the grafted PS chains triggered the self-assembly of PGNPs into various well-defined architectures (Figure 5.6a). In water/DMF mixture, the solvents were poor for the PS blocks but good for the hydrophilic CTAB-stabilized metal blocks. As a result, the metal–polymer triblocks assembled end-to-end to form plasmonic rings or linear chains to minimize the interfacial energy at both ends of NRs. When THF, a poor solvent for CTAB-coated metal blocks, was introduced in water/DMF mixture with appropriate water content (C_w), the metal–polymer triblocks associated in both side-by-side and end-to-end modes to form bundled chains in water/DMF/THF tertiary mixture. The emerging plasmonic coupling between AuNRs within chains (i.e. colloidal nanopolymers) is mainly dependent on the end-to-end distance between AuNRs and the length of AuNR chains, which can be precisely controlled by varying the PS length, incubation time, and solvent property (e.g. water content).

Furthermore, Kumacheva et al. reported the similarity between the self-assembly of polymer-grafted AuNRs and reaction-controlled step-growth polymerization [14, 39]. The kinetics and statistics of step-growth polymerization in molecular

Figure 5.6 Self-assembly of polymer-tethered NRs in selective solvents. (a) Self-assembly of CTAB-stabilized AuNRs tethered with thiol-terminated PS at two ends into chain-like structures in selective solvent. Source: Nie et al. [2b], Reproduced with permission from Nature Publishing Group. (b) TEM images of branched NR chains and cyclic structure assembled from AuNRs (top), variation in X_n with time t (bottom left), and variation in the polydispersity index (PDI) of the chains with X_n (bottom right). Source: Liu et al. [14], Reproduced with permission from American Association for the Advancement of Science. (c) SEM images of homopolymers of palladium NRs, and copolymers of palladium and gold NRs. Insets: high-magnification TEM image of fragments of a Pd NR chain and a Pd–Au chain. $C_w = 15$ wt %. Source: Liu et al. [37], Reproduced with permission from Wiley-VCH. (d) Schematic illustration of photocrosslinking of NR chain assembled from PS-r-PI end-tethered AuNRs and TEM images of the fragments of NR chains before and after photocrosslinking. Source: Lukach et al. [38], Reproduced with permission from American Chemical Society. Scale bars are 100 nm in (a), 50 nm in (b), 500 nm in (c), and 25 nm in (d).

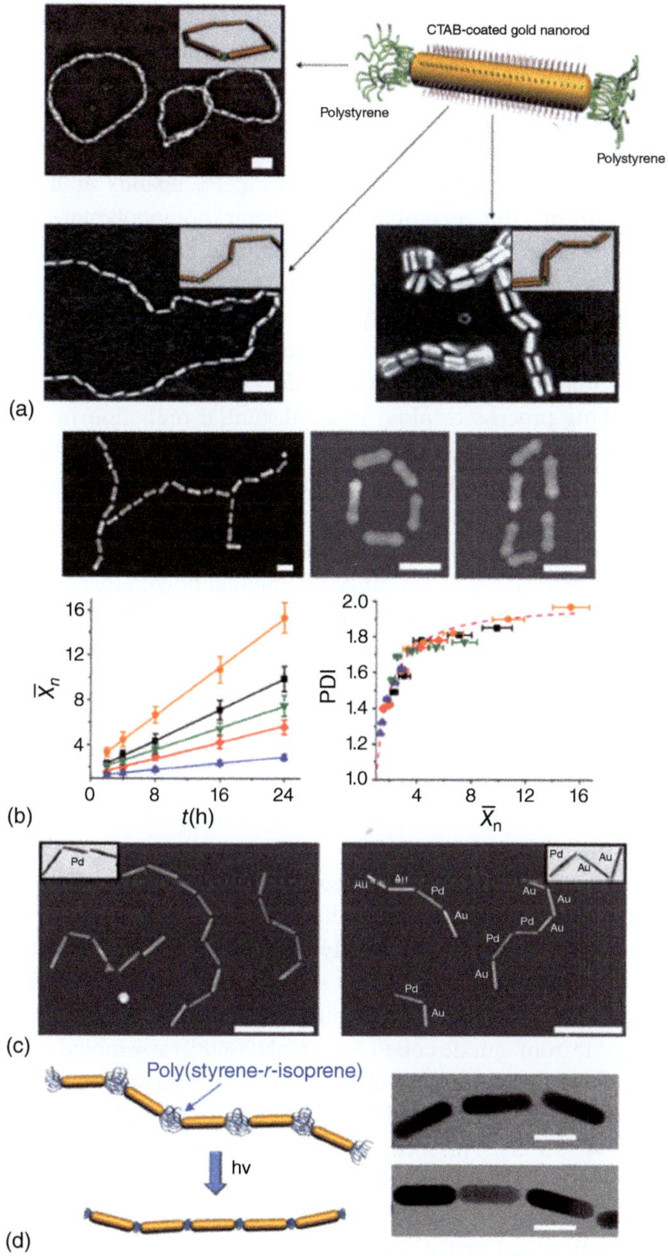

systems can be used to quantitatively predict the architecture of linear, branched, and cyclic assemblies of AuNRs, and the evolution in the aggregation number and size distribution of the linear colloidal nanopolymers over time (Figure 5.6b). The "polymerization" of polymer-grafted AuNRs into colloidal nanopolymers via non-covalent bonding could be terminated by introducing unique chain stoppers (e.g. Janus Au/Fe_3O_4 NPs with PS-tethered Au part) to control the length of resulting nanopolymers. This process resembles the use of monovalent small molecules as chain stoppers in typical polymerization reactions [18b].

This assembly strategy is applicable to metallic nanorods of other compositions or polymer ligands of different functionalities. For instance, palladium NRs exhibited facet-dependent affinity to thiols for preferentially grafting PS ligands at both ends. The PS-grafted palladium NRs assembled into linear homopolymer-like chains or co-assembled with gold NRs to form copolymer-like chains with random organization of two types of NRs in water/DMF mixture (Figure 5.6c) [37]. When photosensitive polymers such as poly(styrene-*r*-isoprene) (PS-*r*-PI) were used as ligands, the assembled colloidal chains could be photocrosslinked by UV irradiation after the photoinitiator azo-bis(isobutyronitrile) (AIBN) was added [38] (Figure 5.6d). The crosslinking process resulted in a substantial reduction in the interparticle distance and a notable increase in angles between adjacent NRs. These changes were adjustable by varying the time of irradiation, that is, the extent of crosslinking.

The "step-growth polymerization" mechanism lays solid foundation for understanding the formation and kinetics of a broad range of colloidal polymers, regardless of the shape and composition of NPs. The resemblance between colloidal and molecular polymerization reactions is useful for fundamental studies of polymerization reactions, as well as in the design of new nanoscale systems with desired properties. The heteroassembly of NPs with different dimensions, shapes, and compositions is promising for designing functional NP ensembles with a high degree of complexity and functionality.

5.3.1.3 Self-Assembly Driven by Dipolar Interaction

Another assembly method for preparing colloidal polymers is assisted by dipolar interactions between ferromagnetic metal or metallic oxide NPs (e.g. Fe, Co, Ni, γ-Fe_2O_3) [40]. The magnetic spin dipole intrinsically embedded in ferromagnetic NPs allowed for strong, directional associations to form colloidal polymers. For example, Pyun et al. reported a novel strategy for preparing hollow cobalt oxide nanochains [40a]. PS-coated ferromagnetic cobalt NPs (CoNPs@PS) assembled into NP chains via dipolar interactions in 1,2-dichlorobenzene (DCB). The preorganized cobalt NPs were subsequently oxidized by bubbling oxygen in their DCB solution at elevated temperature ($T = 175\,°C$). The formation of hollow cavities within the cobalt oxide NP shell was attributed to a nonuniform diffusion and reaction of O_2 with Co atoms throughout the metallic NP. During the process, oxidation of the NP was confined to the outer shell, depleting Co atoms from the colloidal core to satisfy the valency of O atoms in the growing cobalt oxide phase (Figure 5.7a).

The dipolar assembly is often inherently limited to magnetic colloids and is not directly amenable to organize materials of low magnetic susceptibility, such as AuNPs. To tackle this problem, Pyun et al. synthesized ferromagnetic core–shell gold–cobalt NPs (Au–Co NPs) using oleylamine capped AuNPs as seeds in the thermolysis of $Co_2(CO)_8$ in DCB. This method enables the dipolar assembly of a noble metal (e.g. Au, Pt) NP [41]. The fast development in colloidal chemistry offers

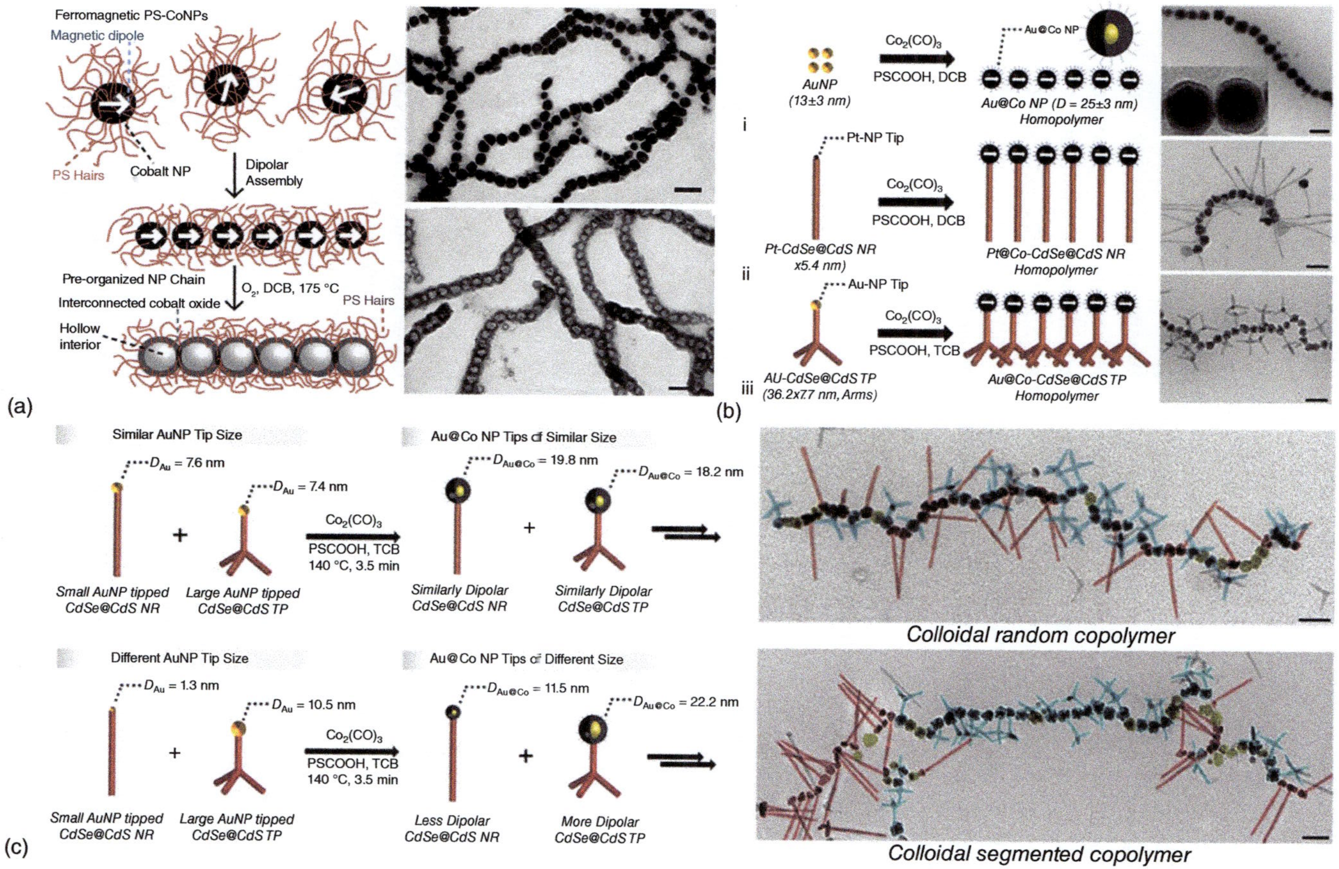

Figure 5.7 Colloidal polymer chains driven by dipolar interactions. (a) Schematic illustration of hollow cobalt oxide nanowires from colloidal polymerization of PS-coated CoNPs and TEM images before and after oxidation treatment. Source: Keng et al. [40a], Reproduced with permission from American Chemical Society. (b) Colloidal polymers based upon dipolar cobalt NP association using different building blocks. Top: Au@Co dipolar colloidal homopolymers; Source: Kim et al. [41], Reproduced with permission from Royal Society of Chemistry. Middle: dipolar "bottlebrush" homopolymers bearing CdSe@CdS NR pendant groups. Source: Hill et al. [40b], Reproduced with permission from Royal Society of Chemistry. Bottom: dipolar homopolymers with "giant t-butyl" groups. Source: Pavlopoulos et al. [40d], Reproduced with permission from Royal Society of Chemistry. (c) Random copolymers formed from with AuNP tips of near identical size (top) and segmented copolymers formed from monomers with significantly differently sized tips (bottom). Source: Pavlopoulos et al. [40c], Reproduced with permission from American Chemical Society. Scale bars are 50 nm.

a library of complex magnetic NPs composed of isotropic parts for use as assembly building blocks. For example, Pyun and coworkers developed a seven-step method to prepare gold@cobalt (Au@Co)-tipped CdSe@CdS NR or tetrapod (TP) [40b]. The hetero-structured semiconductor NRs spontaneously assembled into chains via magnetic dipolar associations of the deposited cobalt domains at elevated temperature (140 °C) in 1,2,4-tricholorobenzene (TCB) using carboxylic acid-terminated polystyrene (PS-COOH) as ligands. In these assemblies, CdSe@CdS NRs were presented as side chains densely grafted along the dipolar NP backbone to form bottle brush-type colloidal polymers (Figure 5.7b). When CdSe@CdS NR and TP-based building blocks served as comonomers in self-assembly, the geometry of the NRs/TPs and the size of the Au@Co tips governed the composition and sequence of the colloidal copolymers (Figure 5.7c) [40c]. Specifically, comparably sized AuNPs resulting in random copolymers, and highly differently sized tips resulting in segmented polymers is analogous to control of comonomer reactivity ratios in classical copolymerization processes.

5.3.2 Templated Self-Assembly of PGNPs into 1-D Structures

Templated assembly of neat PGNPs under confinement has proved to be an effective strategy for fabricating novel structures. Typical structurally defined templates include hard inorganic templates (e.g. CNTs, anodic aluminum oxide [AAO] channels), and soft templates (e.g. periodic nanostructures derived from the BCPs, homopolymer matrix) [42]. In the assembly process, pre-synthesized NPs interact with selected regions on templates through complementary interactions. For instance, NPs are usually located preferentially in the corona composed of complementary components within polymer assemblies. In this part, we present several representative examples of hard and soft template-assisted assembly of PGNPs in confined spaces.

5.3.2.1 Hard Template-Assisted Assembly of PGNPs

Porous AAO channels and CNTs have been widely adopted as hard templates in which cylindrical nanochannels provide 2-D confinement. Unlike traditional hard particles (e.g. crosslinked polymer beads), PGNPs can enter nanochannels whose diameter is even smaller than the overall diameter of PGNPs due to the realignment and stretching of the outer polymer chains [10b].

For example, Zhu et al. investigated the assembly of PS-tethered AuNPs in AAO cylindrical nanopores. The packing and assembly morphology of PGNPs is determined by the confinement strength, which is defined as the ratio of the channel diameter (D_C) to overall diameter of Au@PS NPs. Therefore, by varying D_C and molecular weight (MW) of PS ligands, the templated assembly generated a variety of structures such as linear chain, zigzag, two-NP layer, three-NP layer, and hexagonally packed NP structures (Figure 5.8a) [43].

Hard template confinement can be combined with external forces (e.g. electric field [EF], magnetic field) to guide the assembly of isotropic or anisotropic PGNPs, (e.g. PS-tethered AuNRs). Zhu et al. demonstrated the assembly of PS-grafted AuNRs

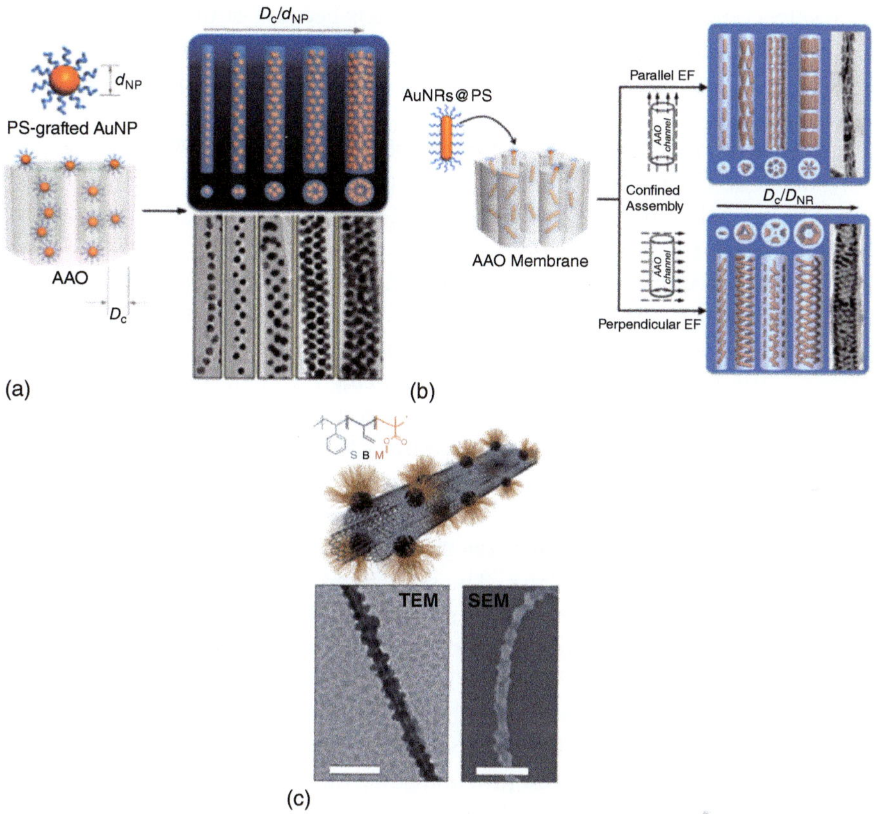

Figure 5.8 Confined self-assembly of PGNPs assisted by hard template. (a) Confined assembly of PS-tethered AuNPs in AAO cylindrical nanopores. Source: Liang et al. [43], Reproduced with permission from American Chemical Society. (b) Confined assembly of PS-tethered AuNRs in AAO cylindrical nanopores combined with controlled electric field. Source: Wang et al. [44], Reproduced with permission from American Chemical Society. (c) Carbon nanotubes templated assembly of PS-PB-PMMA Janus micelles. Source: Groschel et al. [45], Reproduced with permission from John Wiley & Sons.

into some more complicated packed arrays of AuNRs under an external EF [44]. Without EF, disordered structures with random orientation of AuNRs were obtained as a result of the anisotropic shape of NRs. EF provided a driving force for AuNRs to overcome the steric hindrance and to rotate along the EF line. D_C was referred to as the pore size of the AAO channel and $D_{NR\text{-}S}$ and $D_{NR\text{-}L}$ were defined as the overall effective diameters of AuNRs@PS when EF was parallel or perpendicular to the long axis of AAO channel, respectively, which can be adjusted by changing the MW of PS. When EF is parallel to the AAO channel, AuNRs@PS in different packing structures orientated along the long axis of the channel. The packing structures changed from single linear chain-like to triple-helix and finally to complex structures such as hexagonally packed structures, as $D_C/D_{NR\text{-}S}$ increased. Switching EF orientation from parallel to perpendicular led to different assembly structures of AuNRs@PS, in which $D_{NR\text{-}L}$ acted as the effective diameter. It is interesting that even at $D_c < D_{NR\text{-}S}$,

AuNRs@PS could still assemble into the channels because of the flexibility of PS grafts. Arrays of AuNRs@PS circulated along the long axis to form structures such as oblique single linear chain-like and triple-helix structures with $D_C/D_{NR\text{-}S}$ increasing (Figure 5.8b).

Muller et al. reported the selective assembly of Janus micelles of BCPs polystyrene-*b*-polybutadiene-*b*-poly(methyl methacrylate) (PS-PB-PMMA; SBM) with a cross-linked PB core, a dominant PS, and a minor PMMA patch onto multi-walled carbon nanotubes (MWNTs) in acetone [45].

In this system, acetone is a good solvent for PMMA, a near-theta solvent for PS, and a non-solvent for the MWNTs. When a mixture of the MWNTs and Janus micelles was sonicated in acetone, the Janus micelles densely adsorbed onto the MWNT's surface to form chain-like structures by minimizing the energetically unfavorable interactions between solvent and solvophobic interfaces (Figure 5.8c). The PMMA patches provided steric repulsion to stabilize the MWNTs as individuals, rather than forming bundles.

5.3.2.2 Self-Assembly of PGNPs Assisted by Soft Templates

Self-Assembly Using BCP as Templates The self-assembly of NPs can also be guided by "soft template" such as assemblies of supramolecular copolymers. The simplest BCPs are linear di-block copolymers composed of two distinct polymer chains covalently connected at their endpoints. The AB di-block copolymers can segregate to form periodic lamellar, cylindrical, cubic spherical, and interconnected network morphologies, depending on their composition and relative length of the blocks. These microdomain structures can host and guide the assembly of NPs into desired nanostructures [46]. In the assembly process, the enthalpic interactions between NPs and BCPs are largely dependent on the surface chemistry of NPs and polymer matrices, while the entropic interactions are determined by the size of NPs relative to the BCP domains [47].

Microphase-separated domains of BCPs can act as a scaffold to direct both the position and orientation of the NPs in self-assembly [47]. For example, cylindrical phase domains can host and guide the assembly of NPs into 1-D structures with different NP arrangements, when interactions between neighboring PGNPs, and between PGNPs and polymer matrices are modulated. One interesting supramolecular assembly system is constructed via hydrogen bonding between 3-pentadecylphenol (PDP) and pyridine groups of polystyrene-*b*-poly(4-vinylpyridine) (PS-*b*-P4VP). Zhu et al. reported the encapsulation of polymer-grafted AuNPs within cylindrical phases of PS-*b*-P4VP(PDP) assemblies and the release of linear NP arrays from films by breakdown of hydrogen bonds in ethanol (Figure 5.9a) [48]. Upon the solvent annealing, uniformly PS-coated AuNPs were selectively localized in the center of cylindrical PS phase resulting from their preferential interaction with the PS domain and unfavorable interaction with the P4VP(PDP) domain. For AuNPs tethered with a binary mixture of PS-and P4VP (AuNPs@PS/P4VP), the AuNPs were preferably localized at the interface between PS and P4VP(PDP)$_x$ domains, due to the redistribution of PS and P4VP ligands on the AuNP surface. Isolated cylindrical micellar structures embedded with organized AuNPs were obtained by

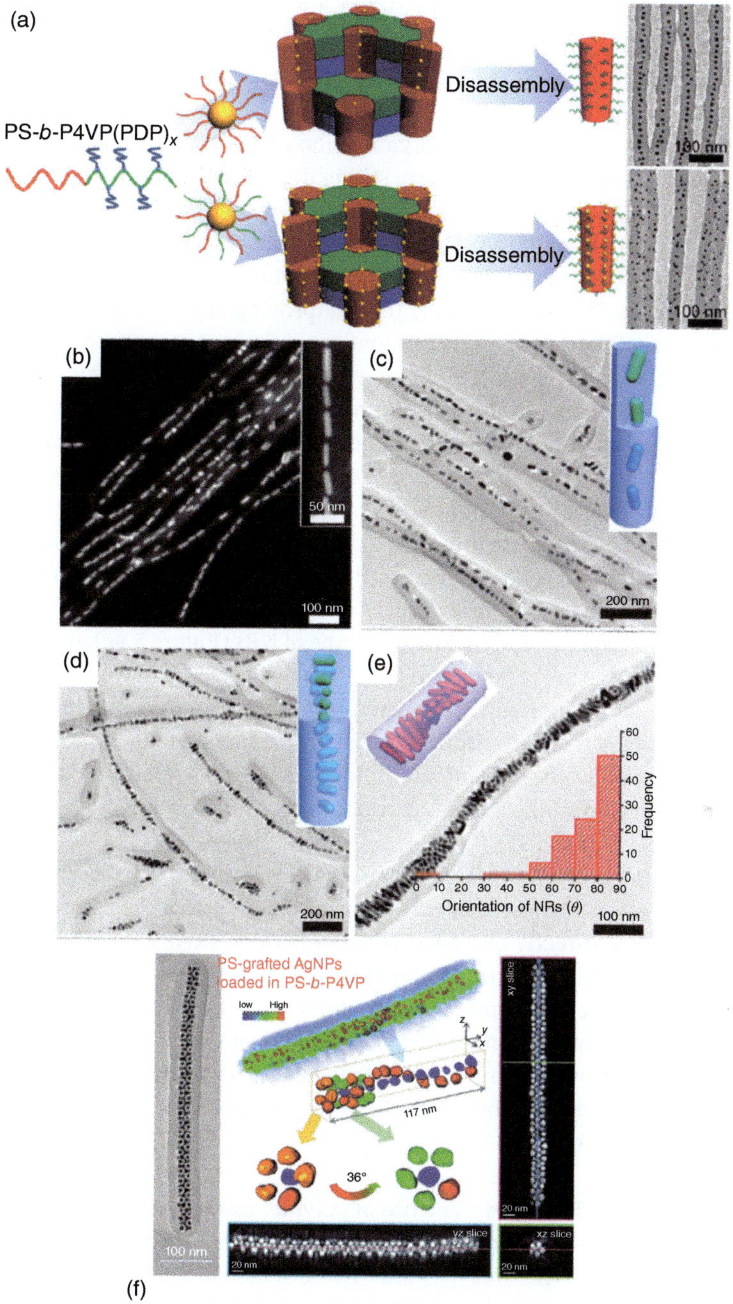

Figure 5.9 Self-assembly of PGNPs templated by supramolecular copolymer assemblies and the subsequent release of arrays from the template. (a) Generation of discrete AuNP arrays with different NP locations using PS-*b*-P4VP(PDP) assemblies. Source: Liang et al. [48a], Reproduced with permission from American Chemical Society. (b–e) TEM images of discrete AuNR arrays with different spatial arrangements using PS-*b*-P4VP(PDP) assemblies. Source: Li et al. [49], Reproduced with permission from American Chemical Society. (f) SEM image (left) and 3-D reconstruction image (right) of helically packed PS-tethered AgNPs confined in cylindrical PS domain of PS-*b*-P4VP matrix. Source: Sanwaria et al. [50], Reproduced with permission from Wiley-VCH.

breaking up the hydrogen bonding between P4VP and PDP and disassembling the supramolecular assemblies in ethanol. The morphology of the resulting NP-loaded micellar nanostructures (e.g. spheres, cylinders) can be effectively tailored by changing the composition of the blocks.

The same approach was applicable to the organization of PS-tethered anisotropic AuNRs in micellar structures [49]. In this case, strong depletion attractions arising from autophobic dewetting drove the PS-tethered AuNRs to assemble side-by-side to form an ordered smectic B phase in the cylindrical PS domains of PS-*b*-P4VP(PDP) assemblies. It was found that grafting PS brushes with different lengths on AuNRs suppressed the autophobic dewetting effect, thus improving the wettability and dispersion of NRs within PS domains. The orientation of the AuNRs can be tuned to be parallel or perpendicular to the PS cylinders by controlling the length and volume fraction of the AuNRs as well as the diameter of the cylindrical domains. Long AuNRs preferentially aligned parallel to PS domains to minimize the entropic penalty associated with the deformation of the surrounding PS matrix, whereas short rods aligned perpendicular to PS domains at a relatively high volume fraction of NRs (Figure 5.9b–d). For cylindrical micelles of PS(110 kDa)-*b*-P4VP(107 kDa) (PDP) with a large diameter ($D_{micelles}$ = 86 nm), the PS-tethered NRs hexagonally packed in a side-by-side mode within the cylinders and twisted along the cylindrical axis to relieve the entropic penalty and excluded volume of NRs (Figure 5.9e). A high level of complexity in stacking and orientation of PGNPs can also be achieved within BCP matrices. Nandan et al. reported that PS-modified AgNPs self-assembled into ordered superstructures (e.g. helixes) within cylindrical PS domain of the PS-*b*-P4VP film during the evaporation of the solvent, chloroform [50]. The subsequent disassembly of the film produced discrete helical structures when dispersed in methanol, a P4VP-selective solvent (Figure 5.9f).

In addition to the templating effect of microphase separation of BCP matrices, the wettability of PGNPs is crucial to the directed organization of PGNPs. The enthalpy penalty is significant when directly encapsulating the bare NPs within cylindrical micelles, as a result of the depletion attraction between NPs. However, grafting polymer brushes on NPs can largely suppress the autophobic dewetting effect, thus significantly improving the wettability and uniform dispersion of NRs within polymer domains. The enthalpic and entropic interactions between PGNP brushes and matrices can be modulated by tailoring the properties (e.g. architecture, composition, MW) of grafted polymer brushes.

Self-Assembly Using Other Soft Templates The use of polymer assemblies as skeletons for the deposition of inorganic components in solutions is a straightforward strategy for generating novel structures [51] and hybrid nanocomposites [52]. This synthetic approach relies on strong cooperative interactions [53] (e.g. electrostatic attraction forces [54], hydrogen bonding [55], covalent bonding [e.g. Au–S], etc.) between polymers and NPs or their precursors. Nnucleophilic moieties, such as carboxylic acid groups (–COOH) [52a, 56], or amine groups ($-NH_2$, $-NMe_2$, or pyridine) [57] can be integrated into solvophilic blocks or segments of amphiphilic copolymers for coordinating with various metal cations (e.g. $Fe_3{}^+/Fe_2{}^+$, $Cu_2{}^+$,

Pd_2^+, Ag^+, Cd_2^+, etc.). This method, however, usually does not produce inorganic NPs with high uniformity, high crystallinity, and defined shapes [58]. Compared with polymer assemblies, unimolecular micelles of such as star-like and bottle brush-like copolymers have precisely engineered yet robust architectures and relatively straightforward structure–property relationships, which makes them ideal templates for the synthesis and even assembly of inorganic NPs.

Lin et al. developed a general strategy to prepare a variety of 1-D necklace-like assemblies of inorganic nanodiscs [59]. These nanostructures can be regarded as organic–inorganic shish–kebabs, in which nanodisc kebabs were periodically located on a stretched polymer shish with regular spacing (Figure 5.10a). Polyrotaxane-based macroinitiators were synthesized by forming an inclusion complex between α-CD and linear polyethylene glycol (PEG). The macroinitiator was then used to polymerize amphiphilic poly(acrylic acid)-*b*-polystyrene (PAA-*b*-PS) and generate molecular complex structures as nanoreactors. The growth of inorganic components in nanoreactors involved the coordination between the PAA block of PAA-*b*-PS and desired metal precursors, and subsequent formation of disc-like nanocrystals (nanodiscs). This process generated a variety of PS-capped nanodiscs (e.g. semiconductor CdSe, magnetic Fe_3O_4, and ferroelectric $BaTiO_3$) with regular spaces. The nanodiscs were threaded by the flexible and stretched PEG chain (Figure 5.10b).

Water droplets can also serve as templates to guide the assembly of NPs into unique structures. For example, Zubarev et al. reported the spontaneous assembly of PS-coated AuNRs into defined rings [60]. When a carbon-coated grid was dipped in a CH_2Cl_2 solution of AuNRs and dried in air at room temperature, water droplets were condensed on the grid from humid air during the evaporation of CH_2Cl_2. The AuNRs were concentrated at the circumference of the water droplets and formed rings of assembled rods upon the evaporation of droplets, which is akin to the classical "breath figures" method. The polymer shell insured the high solubility and stability of the AuNR@PS in solution throughout the assembly process. There are very few, if any, rods inside the rings, which indicates a templated process.

5.3.3 The Self-Assembly of 1-D Structures in Polymer Films

The spontaneous diffusion of NPs in films of BCPs or polymer blends has long been observed, which can be interpreted by the equilibrium of enthalpic and entropic interactions between NPs and polymers [61]. As discussed in Section 5.3.2, the organization and localization of NPs within BCP domains are strongly dependent on the surface chemistry of NPs. In addition to BCPs, homopolymers can also serve as matrix to guide PGNPs to assemble into rich hybrid structures [62]. The assembly process is largely driven by the phase separation between the PGNPs and enthalpically incompatible polymer matrices.

For instance, poly(vinyl pyrrolidone) (PVP)-covered NP monolayers assembled at an air–water interface were transferred onto a PS thin film, and sank into the hydrophobic PS matrix upon exposure to chloroform vapor (Figure 5.11a). The subsequent solvent or thermal annealing swelled the PS matrix to a glassy, molten

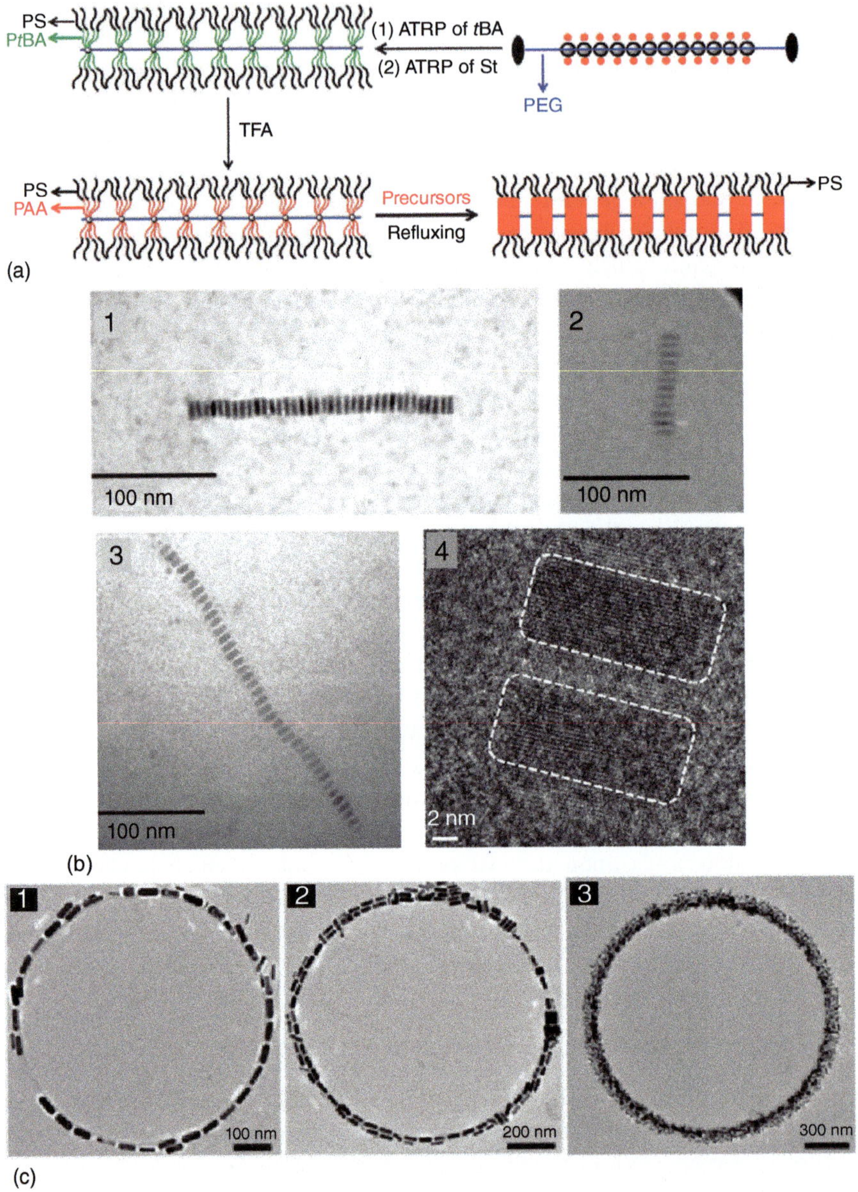

Figure 5.10 1-D assembled structures assisted by some special soft templates. (a) Schematic illustration of the synthesis of organic–inorganic shish-kebab-like nanohybrids composed of periodic nanodisc-like kebabs (red) on PEO shish (blue) by exploiting amphiphilic worm-like PAA-*b*-PS BCPs (lower left panel) as nanoreactor. (b) TEM images of semiconductor CdSe (1), magnetic Fe_3O_4 (2), and ferroelectric $BaTiO_3$ (3) nanonecklaces. HRTEM images of two semiconductor CdSe nanocrystal kebabs oriented perpendicularly to the substrate (4). White dashed rectangles are for guidance. Source: (a–b) Xu et al. [59]/AAAS/Public Domain. (c) TEM images of rings formed by AuNR@PS from a solution in CH_2Cl_2 templated by water droplets. Source: Khanal and Zubarev [60], Reproduced with permission from Wiley-VCH.

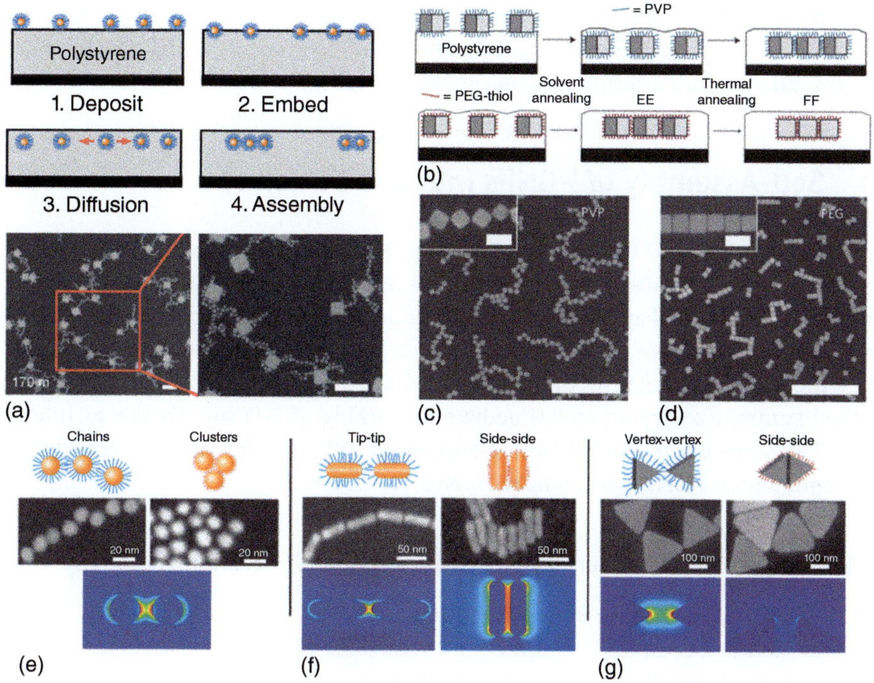

Figure 5.11 One-dimensional structures assembled in polymer thin-film through incompatible interactions between grafted brushes and matrix. (a) Schematic of the self-assembly process in the thin film and SEM images of mixed nanocube and nanosphere assemblies upon solvent annealing for 170 min, and a close-up view of the co-assembled structure. Source: Gao et al. [62a], Reproduced with permission from Royal Society of Chemistry. (b) Schematic of the assembly process of PVP or PEG-grafted nanocubes. (c–d) SEM images of nanocubes modified with PVP after solvent annealing (c) and PEG-alkanethiol ligands after both solvent annealing and thermal annealing (d). Source: (b–d) Gao et al. [63], Reproduced with permission from Nature publishing group. (e–g) Schematics, SEM images, and electrodynamic simulations of AuNP (e), AuNR (f), Ag triangular nanoprism (g) assemblies. Source: (e–g) Gao et al. [62b], Reproduced with permission from Royal Society of Chemistry. Scale bars are 200 nm in (a), 100 nm in (b).

state and enabled the migration, rotation, and redistribution of PVP-covered-NPs within the PS matrix [62a]. Tao et al. demonstrated the self-assembly of hydrophilic polymer-grafted silver nanocubes (AgNCs) into arrays of 1-D strings with controlled NC orientations in hydrophobic PS matrices [63]. The AgNCs exhibited two typical self-orientation models (i.e. face-to-face and edge-to-edge), depending on the steric interactions of grafted polymer ligands (Figure 5.11b). For cubes grafted with relatively long PVP ligands, the PVP chains were denser and stretched on the flat faces than the high curvature surfaces (e.g. edge or vertexes) of AgNCs. As a result, the steric hindrance overcame van der Waals forces between the faces of AgNCs, leading to the edge-to-edge orientation of AgNCs (Figure 5.11c). When short thiol-terminated PEG was used as ligands, the AgNCs assembled into chains with face-to-face orientations, because of the strong van der Waals interactions between the faces of interacting AgNCs (Figure 5.11d). This assembly approach was proven to be applicable to PGNPs with inorganic NP cores of various shapes (e.g. spheres, rods, and triangle plates) (Figure 5.11e–f), and binary mixtures of PGNPs possessing different diffusion rates within polymer matrices [62b].

5.4 Self-Assembly of PGNPs into 2-D Structures

Compared to 0-D or 1-D assemblies of PGNPs, 2-D structures of assembled PGNPs are mostly formed on substrates or at interfaces rather than in solution, which makes them useful in practical applications in such as sensing, nanophotonics, and energy fields [64]. In this section, we first introduce the confined self-assembly of PGNPs within the assembly structures of BCPs and on hard templates such as silicon wafers. We then summarize drying effect-mediated assembly of 2-D superlattice at interfaces. Finally, we discuss the substrate-supported self-assembly of PGNP/polymer, single component PGNPs, and binary PGNPs.

5.4.1 Templated Self-Assembly of PGNPs into 2-D Structures

5.4.1.1 Self-Assembly Using BCPs as Templates

As mentioned in Section 3.2, assembly structures of BCPs offer a rich variety of templates for guiding the organization of NPs with controlled particle location and patterns [65]. The selective incorporation of PGNPs into target microdomains of BCP assemblies is determined by the property of polymer ligands, that is, their compatibility with surrounding polymer matrix [66]. For example, Kramer et al. studied the assembly behavior of AuNPs grafted with PS or a mixture of PS and PVP in the lamellar phases of PS-*b*-PVP thin film [66c]. The PS-coated AuNPs were segregated into the center of the PS domains (light regions), whereas AuNPs coated with mixed polymers were preferentially localized at the interface of the PS and PVP blocks (Figure 5.12a,b).

Hawker et al. demonstrated the precise control over the spatial distribution of AuNPs functionalized with hydroxylated polyisoprene-*b*-poly(styrene) (PIOH-*b*-PS) in the matrix of poly(styrene-*b*-2-vinylpyridine) (PS-*b*-P2VP) films [67]. The hydrogen bonding between hydroxyl group of the NP-adjacent PIOH block of PGNPs and pyridine group on P2VP block of PS-*b*-P2VP BCPs in the matrix drove the segregation of NPs to locate in the lamellar P2VP domains, which counterbalances the enthalpy penalty of mixing of the outer PS layer on PGNPs with the P2VP domains (Figure 5.12c). By varying the number of hydroxyl groups (N_{OH}) and styrene repeating units (N_{PS}) per copolymer ligand and σ of ligands, the NPs could be positioned either in the PS or P2VP domains or at the PS/P2VP interface. Furthermore, when competing molecule 4-phenylpyridine (4PPy) was added, its interference with the hydrogen-bond formation induced the morphological transition of the hybrid assemblies. With increasing the amount of added 4PPy, the location of AuNPs was gradually shifted from the interface to the PS domain, and subsequently, the aggregation of NPs in the PS domain resulted in the formation of "peapod" structures (Figure 5.12d,e).

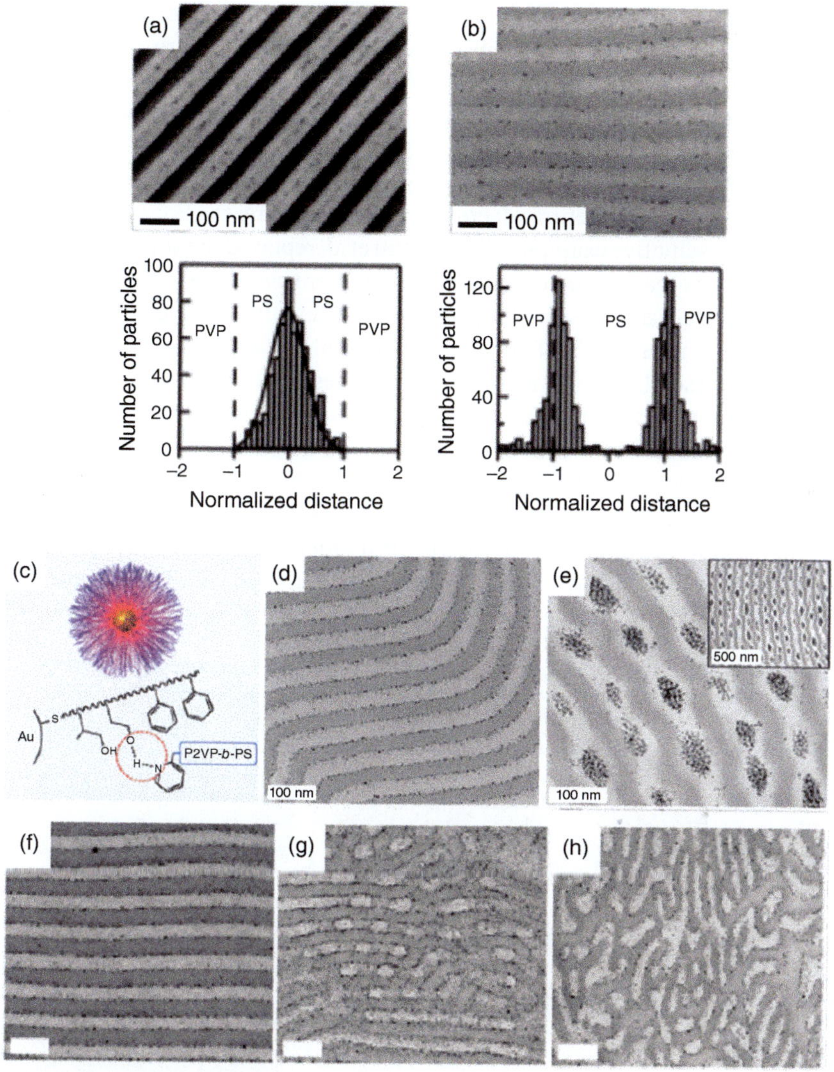

Figure 5.12 Self-assembly of PGNPs using BCPs as templates. (a-b) TEM images of PGNP/PS-PVP diblock copolymer composite films using AuNPs coated with (a) 100% PS-SH and (b) a 1:1 mixture of thiols that produces 20% PVP grafting. Graphs under them show the corresponding histograms of particle locations. Source: Chiu et al. [66c], Reproduced with permission from American Chemical Society. (c) Schematic illustration of hydrogen-bonding formation between hydroxyl groups in the PIOH inner shell and P2VP chains in the matrix. (d–e) Cross-sectional TEM micrographs of PS-*b*-P2VP and AuNPs tethered with hydroxylated PS_{34}-*b*-IOH_9 without (d) and with 5 M equiv. (e) 4PPy relative to total number of hydroxyl groups on the NPs. Source: (c–e) Jang et al. [67], Reproduced with permission from American Chemical Society. (f–h) Cross-sectional TEM images of PS-*b*-P2VP BCP containing AuNPs@PS at various $Ø_P$: (f) 0.04, (g) 0.07, (h) 0.09. Source: (f–h) Kim et al. [68], Reproduced with permission from American Chemical Society. The scale bars are 100 nm.

Kramer et al. investigated the effect of NP volume fractions ($Ø_P$) on the assembly morphology of PS-*b*-P2VP and AuNPs@PS mixtures [68]. The structure of the hybrid assemblies changed with increasing the $Ø_P$ of AuNPs@PS. At $Ø_P = 0.04$, the mixture assembled into well-ordered lamellar structures with the AuNPs (dark dots) segregated at the interfaces between the PS and P2VP domains. At $Ø_P \geq 0.09$, the hybrid assemblies adopted a bi-continuous morphology (Figure 5.12f–h). The domain spacings of the bi-continuous structures decreased with increasing $Ø_P$.

In another representative example, Nie and Zhu et al. reported the supramolecular assembly-mediated organization of AuNPs with different shapes (e.g. spheres, rods, and cubes) into large-area, free-standing 2-D superlattices with tunable spacing. This approach involves the assembly of polymer-capped AuNPs within the assemblies of BCPs and the disassembly of the assembled structures to release free-standing NP superlattices [69]. As shown in Figure 5.13a–b, lamellar structures of P4VP $(PDP)_{1.0}$ (the subscript 1.0 represents the ratio of PDP to PVP units) assemblies were used as sacrificial templates for assembling PS-capped AuNRs. The PDP molecules served as plasticizers to enhance the mobility of PGNPs in the assembly process, thus enabling the efficient packing and orientation of NPs to form superlattices. More recently, Zhu et al. used the same approach to assemble AuNRs end-tethered with PS into ordered

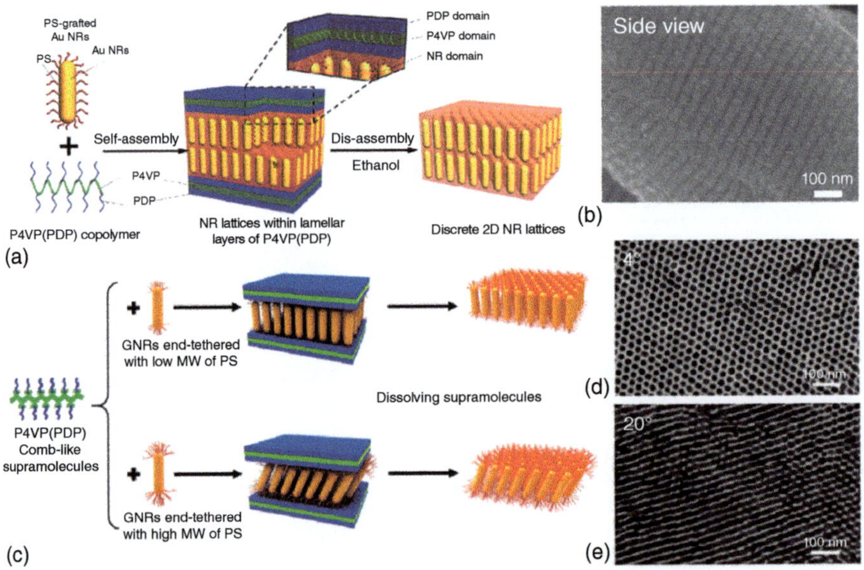

Figure 5.13 (a) Schematic illustration of the assembly of PS-grafted AuNRs into superlattices within confined space of lamellar supramolecule assemblies of $P4VP(PDP)_{1.0}$. (b) Representative SEM images of side view of AuNR superlattices assembled in $P4VP_{60\text{-kDa}}$ $(PDP)_{1.0}$ supramolecular structures after the removal of P4VP and PDP. Source: (a–b) Li et al. [69], Reproduced with permission from Wiley-VCH. (c) Schematic illustration of the assembly of AuNRs end-tethered with PS with different MW into ordered arrays assisted by supramolecules. (d–e) Representative TEM images of side-by-side arrays with different tilt: (d) $AuNRs@PS_{5\text{-kDa}}$ array with a tilt of 4°, (e) $AuNRs@PS_{20\text{-kDa}}$ array with a tilt of 20°. Source: (c–e) Li et al. [70], Reproduced with permission from American Chemical Society.

2-D arrays with tunable tilting angles [70]. The packing and tilting of the AuNR arrays strongly depend on the MW of the polymer ligands. By increasing the MW of PS, tilted arrays can be obtained. The average tilt angle (the angle between AuNRs in the arrays and the vertical direction of the substrate) changed from 4 to 37° when the MW of PS ligands increased from 5 kDa to 50 kDa (Figure 5.13c,e). The resulting AuNR arrays with different tilting exhibited tilt angle-dependent surface-enhanced Raman scattering (SERS) performance, which makes them attractive for use in photonics, electronics, plasmonics, etc.

5.4.1.2 Hard Template-Assisted Self-Assembly

The confinement of hard templates (e.g. patterned polymer film, AAO channels, CNTs) can be used to guide the assembly of PGNPs to fabricate composite nano or microstructures. For instance, Karim et al. reported the phase segregation of a blend of PS-grafted AuNPs and PS confined in polydimethylsiloxane (PDMS) microchannels with mesa-trench shape cross section (Figure 5.14a) [71]. Under the confinement of the channels, conformational entropy penalty arising from local perturbations of grafted and matrix chains drove the segregation of PGNPs from PS matrix to form spatially organized sub-micrometer structures within topographically patterned thin films. The strength of entropic confinement can be adjusted by tuning the MW of PS tethers and PS matrix, and the dimensions of the microchannels.

Lin et al. developed a flow-enabled self-assembly (FESA) strategy to control the drying of PS latex colloidal suspension for fabricating large-area periodic thread with tunable spacing [72]. The spacing between two threads of PS latex can act as "microchannel." The scheme in Figure 5.14b illustrates the formation process of uniform microchannels. The upper blade with uniform microchannels was fixed while the lower substrate was moved at a constant velocity during the drying of the solution of AuNPs@PS in toluene. During the gradual evaporation of toluene, the solution of AuNPs@PS was dewetted from the hydrophilic silicon wafer and segregated to the corners on the bottom of microchannels. The microchannel-guided deposition of AuNPs at the corners yielded two parallel threads inside the microchannel. This simple strategy opens the possibility for large-scale manufacturing of 2-D assemblies and materials for applications in such as optical and electronic devices.

5.4.2 Interfacial Assembly

Interfacial assembly is an effective and widely-used bottom-up strategy for preparing highly ordered 2-D or 3-D structures for applications in such as sensing, nanophotonics, and energy fields [73]. The capillary forces and reduction of interfacial energy are the dominant driving forces for the interfacial assembly of PGNPs. In the past two decades, this strategy has been used to assemble different types of NPs (e.g. magnetic particles, quantum dots, and semiconductors) into functional superlattice structures [74]. The properties of the resulting superlattice can be readily tuned by varying the size, shape, composition of the NP building blocks, and hence the spatial arrangement of the NPs [75]. Among others, 2-D plasmonic superlattices (so-called

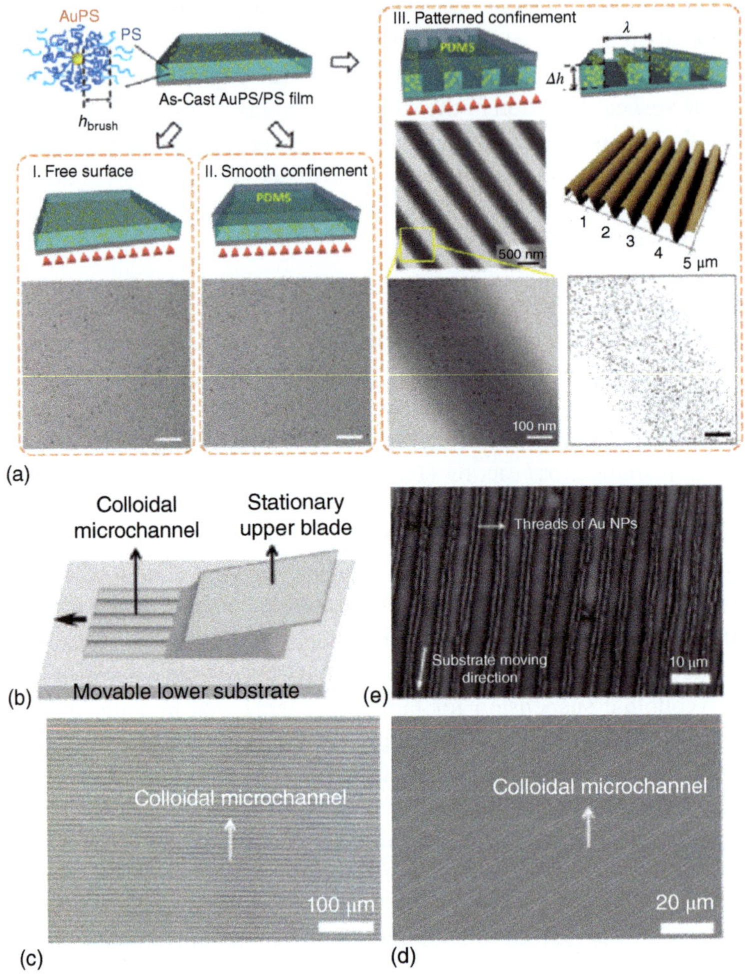

Figure 5.14 2-D structures assembled on substrate assisted by hard template. (a) Schematic representations for thermally annealing Au@PS/PS blend films with free air surface (I), smooth confinement (II), and channel-patterned confinement (III). The corresponding top-view TEM micrographs and 3-D AFM height image for 30% Au@PS/$PS_{3\text{-kDa}}$ films are shown below the schematics. Source: Zhang et al. [71], Reproduced with permission from National Academy of Sciences. (b) Schematic illustration of the formation of uniform microscopic cracks (i.e. microchannels) by FESA of the PS latex particle suspension. (c–d) Representative optical micrograph and SEM images of microchannels. (e) SEM image of aligned threads of AuNPs produced by using colloidal cracks as a template. Source: (b–e) Li et al. [72], Reproduced with permission from Wiley-VCH.

plasmene nanosheets) are appealing for applications in such as SERS and optoelectronic devices, owing to the collective properties arising from the plasmonic coupling between assembled NPs [76]. Especially, plasmene nanosheets of anisotropic NPs (e.g. nanorods, nanocubes, nanostars, nanooctahedra) have attracted great attention, because the strength and models of interparticle coupling can be tuned by controlling the particle shape and orientation [77].

Zhu et al. demonstrated the interfacial assembly of large-area monolayer superlattices of PS-grafted AuNPs for use in high-performance memory devices [78]. The ultrahigh concentration of AuNPs@PS was found to be the determinant factor determining the formation of large-area (centimeter-scale) superlattices at the air–water interface. During the assembly process, the high concentration of NPs promotes the spontaneous spreading of NPs to ensure the assembly of large-scale single-layered structures, and to suppress the formation of isolated islands in the monolayer superlattices (Figure 5.15a).

The interfacial assembly strategy is applicable to plasmonic NPs of various shapes for fabricating nanosheets that are useful for applications in photonic and sensing devices. Cheng's group pioneered the fabrication of freestanding sheets of PGNPs via evaporation-mediated interfacial assembly [75, 76d, 79]. In one interesting example, Cheng et al. demonstrated the entropy-driven interfacial assembly of freestanding, monolayered superlattices of PS-tethered AuNRs with controlled rod orientations [79a]. The authors spread a drop of PS-AuNR chloroform solution onto the water surface with an area of ~10 cm^2 to allow for slow evaporation of chloroform. As chloroform evaporated, the concentration of PS-tethered AuNRs increased, restricting the translational and orientational freedoms of the NRs. The rods sacrificed some of their orientational freedom in exchange for translation freedom by aligning parallel to each other, thereby minimizing the excluded volume and free energy. The PS-tethered AuNRs were assembled into horizontally aligned sheets (H-sheets) or vertically aligned sheets (V-sheets) at the air–water interface, depending on the assembly temperature (Figure 5.15b). The major products were H-sheets at ambient temperatures, and the yield of V-sheets increased with increasing the temperature or annealing time under a chloroform atmosphere.

More recently, Cheng et al. used the same approach to prepare 2-D plasmonic superlattices of PS-covered gold trisoctahedron (TOH) NPs with 24 facets and 14 vertices [79c]. Compared with nanosheets based on similar-sized AuNPs, gold TOH nanosheets showed a much higher Raman intensity due to hot spots created by the sharp edges and corners of the TOH NPs. The enhancement in the SERS intensity of the TOH superlattices could be optimized by controlling the ligand length and particle size. The superlattices were demonstrated as uniform SERS immunosubstrates for sensing rabbit IgG with a detection limit down to 1 pg ml^{-1} and a range from 1 pg ml^{-1} to 100 ng ml^{-1} (Figure 5.15c). The log-linear relationship between SERS intensity and protein concentration indicated the potential of TOH superlattices as robust, sensitive, and quantitative immunosubstrates. Furthermore, the same group fabricated large-area plasmene nanosheets of bimetallic Au@Ag NCs and shaped the nanosheets into 1-D nanoribbons and 3-D origamis using focused ion beam (FIB) technique (Figure 5.15d) [79b]. The 1-D nanoribbons exhibited width-dependent plasmonic properties. The three main characteristic resonance peaks, which corresponded to edge–edge, corner–corner, and edge–corner coupling modes, redshifted with increasing ribbon width but at different dispersion rates.

As we discussed above, interfacial assembly of PGNPs provides a powerful tool for the preparation of single-component 2-D superlattices. Moreover, this approach has also been demonstrated for the fabrication of 2-D or 3-D superlattices of

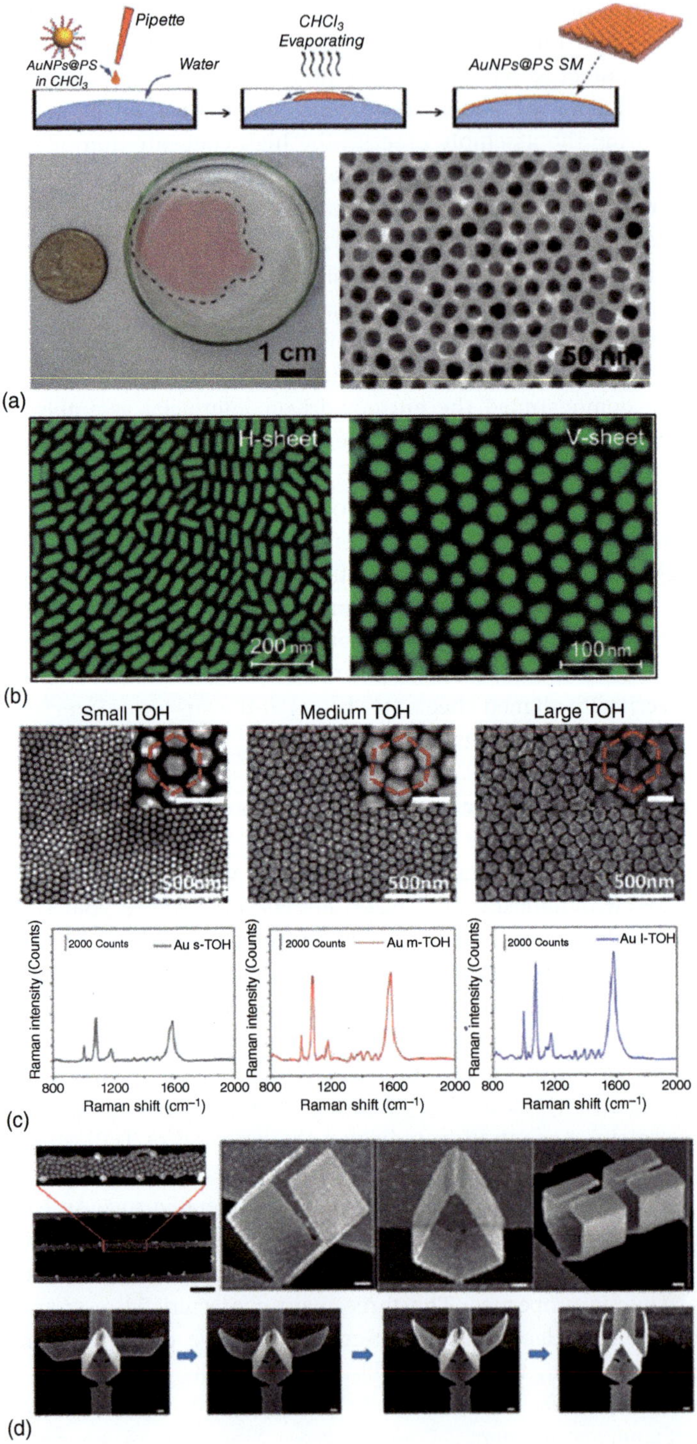
Pipette
AuNPs@PS in $CHCl_3$
Water
$CHCl_3$ Evaporating
AuNPs@PS SM
1 cm
50 nm
(a)
H-sheet
V-sheet
200 nm
100 nm
(b)
Small TOH
Medium TOH
Large TOH
500nm
2000 Counts
Au s-TOH
Au m-TOH
Au l-TOH
Raman intensity (Counts)
Raman shift (cm^{-1})
800
1200
1600
2000
(c)
(d)

Figure 5.15 2-D structures from interfacial assembly. (a) Illustration showing the strategy for fabricating AuNPs@PS monolayer through rapid liquid–liquid interface assembly on convex water surface and the photograph and TEM images of the single monolayer organized from $AuNPs_{16\text{-}nm}$@$PS_{5\text{-}kDa}$. Source: Wang et al. [78], Reproduced with permission from Wiley-VCH. (b) SEM images of circular H- and V-sheets of AuNRs@PS self-assembled at the air–water interface. Source: Ng et al. [79a], Reproduced with permission from American Chemical Society. (c) SEM images of superlattice nanosheets composed of PS-coated TOH different sizes (Insets display high magnification images showing hexagonal close-packed arrangement. Inset scale bar is 100 nm.) and corresponding SERS spectra of 4-ATP recorded at a laser wavelength of 633 nm. Source: Dong et al. [79c], Reproduced with permission from Royal Society of Chemistry. (d) SEM images of plasmene nanoribbons with width of 4 NCs and different origami structures: cube, diamond, hearts, and a series of images depicting the flapping motion of a bird's wings. Source: Si et al. [79b], Reproduced with permission from American Chemical Society.

binary PGNPs. For the sake of easy layout of figures, this content will be discussed later in the Section 5.4.3.3.

5.4.3 2-D Assemblies Within Thin Film

This section focuses on 2-D assembled structures of PGNPs in the thin films of polymer matrix and the 2-D ensembles composed of neat PGNPs. We discuss three typical categories: (i) PGNPs/homopolymer system, (ii) neat PGNP system, and (iii) binary blends of PGNPs.

5.4.3.1 PGNPs/Homopolymer System

Controlling the arrangement or dispersion of NPs within homopolymer matrix is crucial to fabricating composite materials with desired properties (e.g. mechanical strength, conductivity, and optical and magnetic properties) [80]. Tethering polymers, including homopolymers, mixed polymer brushes, or BCPs onto NPs has emerged as an important and effective strategy to control interactions between NPs and homopolymer matrices, thus fine-tuning the arrangement of NPs in the composites [81]. The energetic contributions (i.e. conformational entropy and interaction enthalpy) of grafted polymer chains in the system are dictated by their chemical nature, architecture, length, and σ.

Composto et al. investigated the morphological evolution and optical properties of nanocomposite films of P2VP and AuNRs@P2VP blends. The morphology of the composites was found to be affected by the rod volume fraction ($\varnothing_{AuNRs}$) and film thickness [82]. At a constant film thickness of 40 nm, the arrangements of AuNRs evolved from well-dispersed states to side-by-side aggregates of nanorods, and eventually to percolated networks with end-to-end nanorod linkages, when $\varnothing_{AuNRs}$ increased from 0.4% to 1.6% to 2.7% (Figure 5.16a). At a constant $\varnothing_{AuNRs}$ of 2.7%, thicker films (40 and 70 nm) showed percolated networks of end-to-end linked AuNRs, whereas thinner films (20 nm) exhibited mainly isolated AuNRs. Optically, the longitudinal surface plasmon resonance (L-SPR) peaks of the assemblies were correlated with the local orientation of the AuNRs. In the P2VP films, the L-SPR

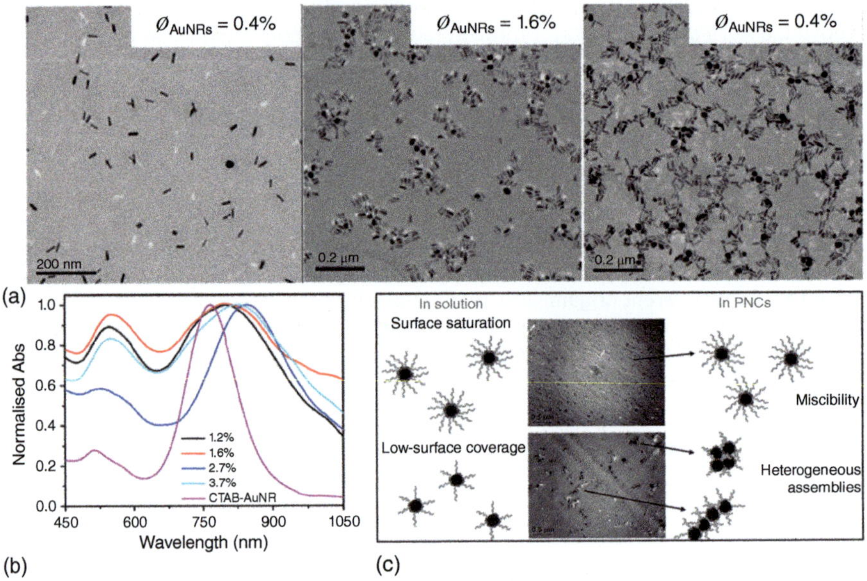

Figure 5.16 (a) TEM images of P2VP-AuNRs in P2VP films as a function of nanorod volume fraction. The $Ø_{AuNRs}$ are 0.4%, 1.6%, and 2.7%, from left to right. (b) UV–vis spectra of CTAB-coated AuNRs in water and PVP-grafted AuNRs in P2VP films at $Ø_{AuNRs}$ of 1.2%, 1.6%, 2.7%, and 3.7%. Source: Jiang et al. [82], Reproduced with permission from American Chemical Society. (c) P2VP-*b*-PS coated NPs assemblies in PS matrix: The top TEM image shows good NPs miscibility at surface saturation (the BCP concentration is 0.0015 g ml^{-1}). The bottom TEM image shows the formation of different NPs assemblies (strings, small clusters) for lower heterogeneous surface coverage (here the BCP concentration is 3.55×10^{-5} g ml^{-1}). The MW of PS matrix is 61.8 K. Source: Jouault et al. [81c], Reproduced with permission from American Chemical Society.

redshifted from 801 to 846 nm, as $Ø_{AuNRs}$ increased from 1.2% to 2.7%; and then blueshifted with further increasing $Ø_{AuNRs}$ to 3.7% (Figure 5.16b).

The elastic modulus and yield stress of an amorphous glassy polymer matrix can be improved by adding spherical NPs, but generally at the price of a lower ductility. In an interesting work by Kumar et al., the filling of PS-grafted SiO_2 NPs in a PS matrix enhanced all the physical properties (i.e. ductility, elastic modulus, and yield stress) of resulting nanocomposite films [80b]. This enhancement was attributed to the uniform dispersion of PS-grafted SiO_2 NPs in the PS matrix and the increased entanglement between the grafted PS chains and the PS matrix.

When BCPs are used as the ligands, the dispersion state or the assembly structure of the BCP grafted NPs can be tuned by designing the chemical nature and varying the composition of two blocks. As an example, Kumar et al. modified SiO_2 NP with PS-*b*-P2VP through the strong affinity of P2VP block with SiO_2 NP surface [81c]. The adsorbed P2VP block increased the interparticle distance, while the non-adsorbed PS block had favorable entropy of mixing with the PS matrix. The BCP coverage on NPs can be adjusted by tuning the BCP concentration for adsorption. At a BCP concentration of 0.0015 g ml^{-1}, the NPs dispersed well in the matrix due to high surface coverage of polymers on the NPs. In contrast, at a BCP concentration of 3.55×10^{-5} g ml^{-1}, a lower surface coverage yielded anisotropic assemblies such as strings and small clusters, presumably because of the formation of bald patches on the NP surfaces (Figure 5.16c).

5.4.3.2 Self-Assembly of Single-Component Neat PGNPs

Neat PGNP system refers to a system only containing one or multiple types of PGNPs and without any free polymers as matrix. We discussed the interfacial assembly of monolayered 2-D superlattice from one-component PGNPs in 5.4.2 section. In this section, we emphasize interfacial assembly of thin films or composites containing more than one layer of PGNPs.

For single-component PGNP systems, homopolymers or BCPs can be used as ligands for the assembly [83]. When homopolymers are used as ligands, the neat PGNPs can either form homogeneous nanocomposites or self-assemble into ordered structures through the microphase separation of the NP core and polymer grafts. As an example, Alivisatos et al. reported self-assembly of 5.2 ± 0.4 nm Au nanocrystals grafted with PS in the toluene into superlattices through controlled drying on an ethylene glycol interface in a Teflon trough [84]. The films were obtained under four different drying rates (5 hours, 4 minutes, 2 minutes, and 1 minute) and had millimeter to centimeter lateral dimensions and thicknesses of a single monolayer to hundreds of nanometers. When the $PS_{3\text{-kDa}}$ was used, the highest elastic modulus corresponds to a fairly disordered superlattice self-assembled within 2 minutes at 43 °C-uncovered conditions. The authors proved that residual stress can lead to elevated moduli in fast-drying superlattices, which indicated that increased structural disorder may actually correlate with increased superlattice elastic modulus. Besides, the MW of the PS had a significant effect on the mechanical properties of the PGNP films. The elastic modulus and hardness of the PGNP films for multilayer films and ranged from 6.1 ± 3.9 GPa to 19.5 ± 6.5 GPa and 116 ± 51 MPa to 171 ± 86 MPa as MW of the PS ranged from 1.1 to 20 kDa.

Composites composed of PGNPs are mechanically weak unless the grafted chains are long enough to form entanglements between PGNPs. Macfarlane et al. crosslinked silica NPs tethered with short reactive polymers to fabricate highly filled and mechanically robust nanocomposites [85]. Poly(glycidyl methacrylate) (PGMA) tethered SiO_2 NPs, together with diaminodiphenyl sulfone (DDS), were dissolved in mixtures of 2-butanone and 2-pentanoneand drop-casted into films and crosslinked polymer tethers with amine groups of DDS. The crosslinked PGNP films showed a significant enhancement in the hardness, modulus, and scratch resistance, compared to non-crosslinked PGNPs and crosslinked polymer films without NP reinforcement. Specifically, the crosslinked 116 nm-SiO_2@$PGMA_{32\text{-kDa}}$ films exhibited increased hardness of ~600 MPa and reduced modulus of ~11 Gpa, which is about 140% and 81% of the corresponding polymer-only materials, respectively.

Moreover, Koberstein et al. prepared NP assemblies embedded in silicon oxide with controllable interparticle distance using PDMS-modified iron oxide (γ-Fe_2O_3) NPs as building blocks [86]. The PDMS-grafted γ-Fe_2O_3 NPs self-assembled into superlattices with tunable interparticle distance dictated by MW of PDMS ligands (Figure 5.17a–c). When the films were treated with oxygen plasma or UV/ozone,

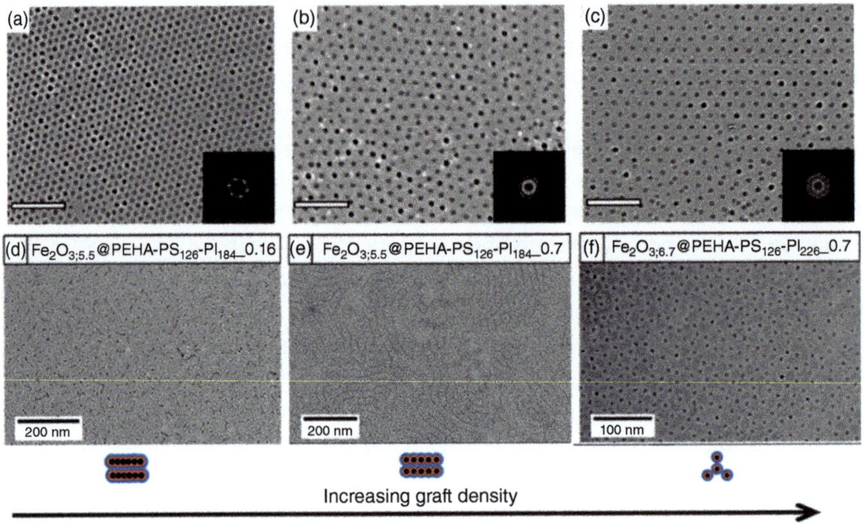

Figure 5.17 (a–c) TEM images of two-dimensional PDMS-modified 9.5 nm-Fe_2O_3 NP assemblies: (a) MW = 1 kDa; (b) MW = 5 kDa; (c) MW = 10 kDa; (Insets) 2D FFT power spectra. Scale bars indicate 100 nm. Source: Wu et al. [86], Reproduced with permission from American Chemical Society. (d–f) TEM images of a series of different types of nanoparticle arrays that can be realized by variation of σ. Source: Leffler et al. [87], Reproduced with permission from Wiley-VCH.

PDMS was converted to silicon oxide. The silicon oxide matrix made the NP assemblies inherently more robust than those within organic or polymeric matrices, in terms of tolerance to temperature and organic solvents.

When BCPs are used as ligands, the incompatibility between the NP core and polymer ligands drives their phase separation to form ordered hybrid structures. For instance, Forster et al. designed and synthesized poly(styrene-*b*-isoprene) (PS-*b*-PI) with pentaethylene hexamine (PEHA) terminal groups that have strong binding affinity to the surface of different NPs such as lead (II), PbS, ZnO, and Fe_2O_3 [87]. The PEHA-PS-*b*-PI coated PbS NPs (7 nm in diameter) self-assembled into ordered structures on carbon-coated copper grids, depending on the MW and σ of the BCP ligands. For short PS_{80}-*b*-PI_{117}-grafted NPs, the PGNPs were arranged into a hexagonal lattice in which the PS domains formed a spherical shell around the NPs and were fully separated from adjacent PS domains of another NPs by the PI matrix. For relatively long PS_{225}-*b*-PI_{368}-grafted NPs, the assembly structures transited from single NPs within spherical domains, to single NPs in strip-like domains, and eventually to linear NP arrays in stripe-like domains. Furthermore, the σ of BCP tethers also influenced the morphology of assembly structures. For 5.5 nm-Fe_3O_4 @PEHA-PS_{126}-*b*-PI_{184} with $\sigma = 0.16$ chains nm^{-2}, the self-assembly process yielded non-branched linear chains of contacted NPs surrounded by an inner PS domain (Figure 5.17d). With increasing σ from 0.70 to 0.77 chains nm^{-2}, the assembly morphologies evolved from linear arrays of separated NPs surrounded by inner PS domain (Figure 5.17e) to five- or six-membered rings of NPs in PS domains (Figure 5.17f).

5.4.3.3 Self-Assembly of Binary PGNPs Blends

Binary PGNPs system contains two types of PGNPs having different combinations of central NP cores with controlled size, shape or composition, and polymer grafts. The co-assembly of binary PGNPs has led to a wealth of superlattice structures with controlled stoichiometries and lattice parameters (e.g. lattice symmetry, interparticle spacing, etc.).

Bockstaller et al. studied binary mixtures of PS-coated AuNPs (or PI-coated AuNPs) and PS-grafted silica NPs and observed the formation of two distinct structures [88]. Mixed solutions of 20 nm-SiO_2@PS_{770}/$AuNPs_{3.5\text{-nm}}$@PS_{10} or 20 nm-SiO_2@PS_{770}/$AuNPs_{3.5\text{-nm}}$@PS_{15} in toluene were casted on the carbon-coated TEM grids. For 20 nm-SiO_2@PS_{770} and $AuNPs_{3.5\text{-nm}}$@PS_{10} mixture, their self-assembly generated structures with small AuNPs concentrated in the interstitial regions of SiO_2@PS_{770} (Figure 5.18a). However, for 20 nm-SiO_2@PS_{770} and $AuNPs_{3.5\text{-nm}}$@PI_{15} blends with incompatible polymer grafts, AuNPs segregated to the periphery of the clustered silica NPs (Figure 5.18b). Considering particle brush packing in a 2-D ordered array of neat 20 nm-SiO_2@PS_{770}, chain stretching is energetically favored

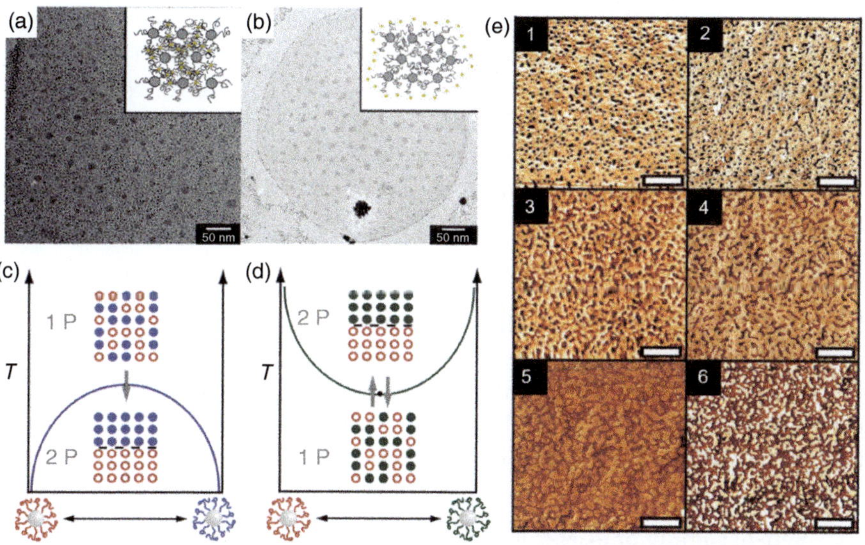

Figure 5.18 Self-assembly of binary PGNPs. (a–b) Structure formation in the neutral binary particle blend SiO_2@PS_{770}/AuNPs@PS_{10} (a) and incompatible SiO_2@PS_{770}/AuNPs@PI_{15} system (b). Source: Cheng et al. [74], Reproduced with permission from John Wiley & Sons. Illustration of ligand-induced UCST phase behavior (c) and LCST phase behavior (d) in particle brush blends. LCST blends allow for reversible cycling of blend through homogeneous one-phase (1 P) and phase-separated two-phase (2 P) states. (e) AFM phase images of phase separation in 8 nm-SiO_2@$PMMA_{194}$/8 nm-SiO_2@PS_{205} at varying composition and annealing temperature after thermal annealing for 48 h. Dark phase corresponds to PMMA. (1) PMMA/PS = 25:75; T = 200 °C. (2) PMMA/PS = 25:75; T = 140 °C. (3) PMMA/PS = 50:50; T = 200 °C. (4) PMMA/PS = 50:50; T = 160 °C. (5) PMMA/PS = 75:25; T = 200 °C. (6) PMMA/PS = 75:25; T = 140 °C. Source: Schmitt et al. [89]/AAAS/Public Domain.

over the formation of interstitials. In the 20 nm-SiO_2@PS_{770} and $AuNPs_{3.5\text{-nm}}$@PS_{10} mixture, the enthalpically neutral and sufficiently small AuNPs@PS redistributed to release stresses for the grafted polymer chains within the interstitial regions. In contrast, for the incompatible 20 nm-SiO_2@PS_{770} and $AuNPs_{3.5\text{-nm}}$@PI_{15} system, the interfacial energy between dissimilar polymer components exceeded the energy for stretching the grafted brush of SiO_2@PS_{770}, resulting in the exclusion of small incompatible NPs.

More recently, Bockstaller et al. showed that distinct polymer grafts on mixed PGNPs can induce upper or lower critical solution temperature (UCST/LCST)–type phase transition, similar to the phase behavior of the corresponding linear polymer blends (Figure 5.18c, d). Binary blends of PS-tethered 8 nm-SiO_2 NPs and PMMA-tethered 8-nm SiO_2 NPs were used as a model system because the corresponding linear polymers were widely studied for UCST polymer blends [90]. A solution of the mixed SiO_2@PS_{205} and SiO_2@$PMMA_{194}$ in THF was spin-casted into films with a thickness of 110 to 140 nm on silicon substrates. Cooling the binary PGNP films below the UCST temperature of polymers resulted in the organization of individual PGNPs into monotype microdomain structures. For the system with 50% of SiO_2@$PMMA_{194}$, when the annealing temperature decreased from 200 to 140 or 160 °C, the structures transformed from discrete island-type morphology (one-phase) to bi-continuous network-type structures (two phase) (Figure 5.18e). Furthermore, the SiO_2@$PMMA_{257}$/SiO_2@ $(PS\text{-}r\text{-}PAN)_{262}$ blends were observed to display a LCST behavior similar to that of PMMA/PS-*r*-PAN blends. The reversible phase separation of PGNPs offers new opportunities to fabricate composite materials with dynamically controlled structures and physical properties [89].

5.5 Self-Assembly of PGNPs into 3-D Structures

In this section, we summarize recent advances in the 3-D self-assembly of PGNPs. The discussions will be elaborated based on the morphology of the 3-D assembly structures: (i) clusters; (ii) vesicles; (iii) superlattices and crystals.

5.5.1 Self-Assembly of PGNPs into Clusters

Different from the CMs we discussed in Section 5.2, NP clusters described in this section are, in most cases, spherical aggregates without precise orientation of NPs. The clusters of inorganic NPs show application in high-performance catalysis, sensors, optics, and electronic devices [91]. Liquid-phase assembly of NPs into 3-D clusters can be achieved by implementing various interactions, such as hydrophobic interactions, electrostatic interactions, hydrogen bonding, or covalent bond.

Hydrophobic interactions are one of the most important nonspecific interactions used in solvent-mediated self-assembly of PGNPs. As discussed in Section 5.2 and 5.3, hydrophobic interactions can trigger the self-assembly of AuNRs@PS into

CMs and 1-D assemblies. Similarly, hydrophobic effect can induce the self-assembly of AuNRs@PS to form spherical clusters in selective solvent [2b, 22]. The application development of PGNP clusters is determined by our ability to control NP number per cluster, interparticle distance, and overall optical response. Liz-Marzán et al. developed an assembly strategy to fabricate clusters of PS-stabilized spherical AuNPs by using hydrophobic interactions (Figure 5.19) [2a]. In a solution of PGNPs

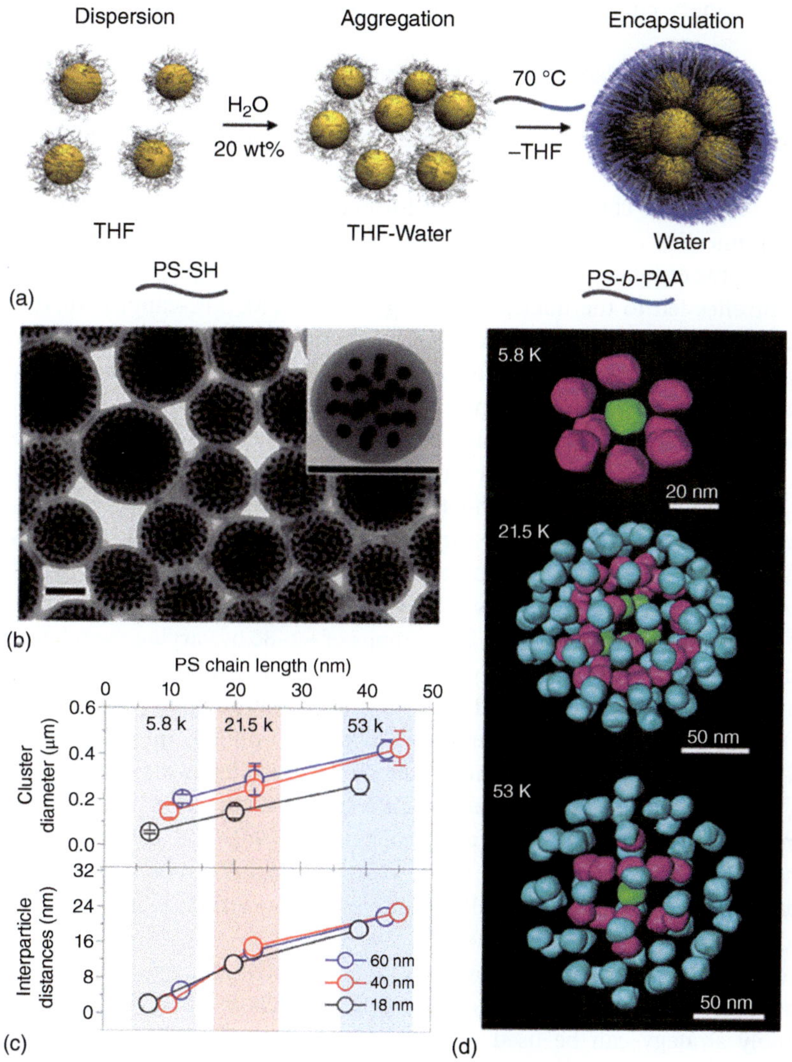

Figure 5.19 Clustering of PGNPs based on hydrophobic interaction. (a) Schematic illustration of the clustering of AuNPs@PS. (b) TEM image (scale bar = 200 nm) of clusters. (c) The effects of PS chain length and NP size on diameter and interparticle distance of clusters. (d) 3-D electron tomography reconstructions of clusters comprising 18 nm AuNPs with different lengths of the PS chains. Source: Sánchez-Iglesias et al. [2a], Reproduced with permission from American Chemical Society.

in THF, the addition of water, a poor solvent for PS, induced the clustering of NPs that was quenched by adding PS_{403}-*b*-PAA_{62} to encapsulate the clusters in the micelles. Thermal treatment of quenched clusters at 70 °C expelled solvent from the hydrophobic cores, yielding rigid NP clusters that are stable in water (Figure 5.19a,b). The size and interparticle distance of spherical clusters increased with increasing the MW of polymers (Figure 5.19c,d). This approach can be used to prepare unary clusters of NPs with different shapes (e.g. NRs and nanostars) [92], and binary clusters consisting of NPs with different sizes and relative ratio [93].

In addition to hydrophobic interactions, covalent bonding can also be used to control the clustering of PGNPs. Lin et al. synthesized stable PGNPs capable of reversibly clustering in response to light by using nanoreactors of amphiphilic star-like BCPs of poly(acrylic acid)-*b*-poly(7-methylacryloyloxy-4-methylcoumarin) (PAA-*b*-PMAMC) [94]. The star-like BCPs formed structurally stable spherical unimolecular micelles in the DMF/benzyl alcohol (V/V = 7 : 3) mixture. The coordination between gold precursors and the inner PAA compartment of unimolecular micelles led to the nucleation and growth of PMAMC-capped AuNPs. Due to the reversible photodimerization of the coumarin groups in the polymer chains, the PGNPs can be reversibly clustered and disassembled upon irradiation with 365-nm and 254-nm UV light, respectively.

The assemblies of BCPs in solutions or thin films can be used to guide the clustering of NPs to generate defined structures. For instance, the co-assembly of $AuNPs_{7.5\text{-nm}}$@$PS_{2\text{-kDa}}$ and $PS_{20\text{-kDa}}$-*b*-$P4VP_{17\text{-kDa}}$(PDP) supramolecules in films yielded hybrid sphere-within-lamellae structures with AuNPs located in the PS domains [95]. The breakup of hydrogen bonding between PDP and P4VP triggered the disassembly of the hybrid structures to release isolated NP clusters. The number of NPs per cluster was effectively tuned in the range of ~1–80 by varying the relative amount of NP and PDP. In another work, Weller et al. assembled a mixture of iron oxide NPs (IONPs) grafted with diethylenetriamine-terminated polyisoprene (PI-DETA) and poly(isoprene)-*b*-poly(ethylene glycol) (PI-*b*-PEG) into clusters in water. Quick injection of the mixture into water yielded micelles encapsulated with $IONPs_{8\text{-nm}}$ in the hydrophobic PI cores, which were subsequently crosslinked by thermal treatment in the presence of a radical initiator [96]. The size of the NP clusters increased with decreasing the molar ratio of PI-*b*-PEG to NPs. This clustering approach is also applicable to NPs of different compositions (e.g. CdSe/CdS quantum dots, AuNPs, and mixtures thereof).

Confined assembly of PGNPs is proven to be an effective strategy to fabricate hybrid structures of different morphologies (e.g. 1-D chains and 2-D sheets). This assembly strategy can be used to generate 3-D assemblies by using hard templates and soft emulsion droplets. For example, Grzelczak et al. demonstrated the reversible assembly of AuNPs@PS confined in the hollow cavity of permeable silica nanocapsules (Figure 5.20a) [97]. Permeable capsules loaded with AuNPs@PS were fabricated through steps including the clustering of AuNPs@PS in water/THF mixtures, encapsulation of clusters within BCP micelles, coating of the micelles with mesoporous silica shells, and removal of BCPs. When the water content in the

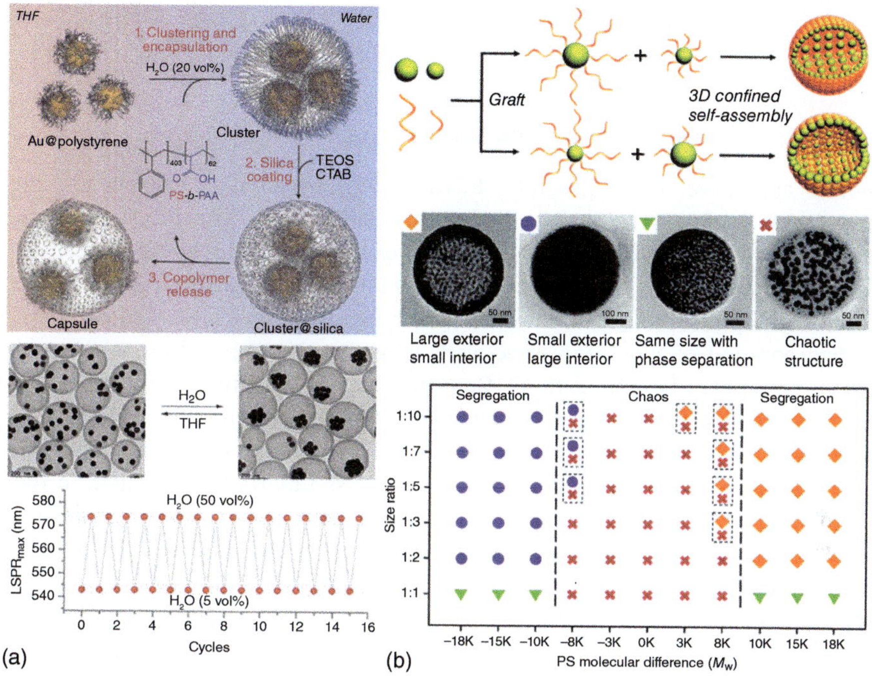

Figure 5.20 Clustering of PGNPs under confinement t. (a) Top: schematic illustration of fabrication of silica capsules containing clusters of AuNP@PS. Middle: TEM images of capsules with low (left) and high (right) water content. Bottom: evolution of the localized SPR maximum over 15 assembly–disassembly cycles. Source: Sanchez-Iglesias et al. [97], Reproduced with permission from Wiley-VCH. (b) Top: Schematic illustration of co-assembly of large and small AuNPs with varied polymer chain lengths into size-segregated assemblies upon solvent evaporation of the emulsion droplets. Middle: TEM images of the representative hybrid clusters. Bottom: the effect of size ratio of AuNPs and MW difference between PS ligands on the structure of the hybrid assemblies. Source: Yang et al. [32], Reproduced with permission from Royal Society of Chemistry.

THF/water mixture was repeatedly varied, the NP clusters in capsules reversibly assembled and disassembled, which was accompanied with the reversible blue and redshift of their plasmon bands. The cyclic clustering and associated optical response were highly reproducible, due to the confinement of a constant number of NPs in the cavity. This feature makes the system different from the solvent-induced NP self-assembly in open systems. In addition to hard templates, emulsion droplets can also serve as soft templates to confine the assembly of PGNPs. Zhu et al. co-assembled binary AuNPs@PS with distinct NP size and PS length into core–shell clusters in emulsion droplets (Figure 5.20b) [32]. The location of NPs in the clusters was controlled by varying the MW of PS grafts and hence the hydrophilicity of PGNPs. For binary PGNPs grafted with PS ligands of close MW (MW difference < 8 k), they co-assembled into chaotic structures due to the suppressed phase separation of the binary PGNPs. When the MW difference of the PS ligands was larger than 8 K, the AuNPs tethered with short PS chains always preferentially segregated into the outer layers of core–shell structures. When high MW PS ligands were used (e.g. $AuNPs_{8\text{-nm}}@PS_{20\text{-kDa}}$ and $AuNPs_{15\text{-nm}}@PS_{50\text{-kDa}}$), the PGNPs did not segregate in the assemblies, because of similar hydrophilicity of the binary PGNPs.

5.5.2 Self-Assembly of PGNPs into Vesicles

Vesicular assemblies of amphiphilic BCPs possess hydrophilic cavities for the effective encapsulation of various hydrophilic substances. The properties (e.g. thickness, permeability, rigidity, and stability) of the vesicular membranes can be readily tuned by tailoring the molecular structure of BCPs. Thus, these vesicles have shown broad applications in the fields of pharmaceutics, medicine, and catalysis. The incorporation of NPs into vesicle walls may give rise to novel properties and open up new potential applications. Conventionally, NPs-containing vesicles are usually constructed by embedding small ligand-stabilized NPs into the walls of BCP vesicles [98], or by *in situ* deposition of NPs on the surface of BCP vesicles [99]. In most of these cases, the particles are more or less randomly distributed in the wall. In contrast, the vesicular assembly of amphiphilic PGNPs allows for precise control over the location of NPs in the wall. The amphiphilic PGNPs consist of a hydrophobic inorganic core and a brush layer of hydrophilic or amphiphilic polymer chains, resembling amphiphilic molecules. The hydrophobic and hydrophilic parts can be tuned to balance the interfacial and repulsive energy and hence the morphology of assembled structures by varying the core size, and the composition, length, and σ of polymer brushes.

5.5.2.1 Self-Assembly of Hydrophilic Homopolymer-Grafted NPs

Hydrophilic homopolymer-grafted NPs can act as building blocks for assembling hybrid vesicles. For instance, Förster et al. self-assembled single-component PEO-coated CdSe/CdS core–shell NPs into hybrid vesicles by dispersing the PGNPs in water [100]. The structures assembled from $CdSe/CdS_{3.4\text{-nm}}@PEO_{2\text{-kDa}}$ was found to be dependent on the σ of PEO on NP surfaces: a structural transition of single PGNPs → cylinders → networks → vesicles was observed with decreasing σ. This observation suggests that the segregation of hydrophilic PEO grafts and hydrophobic NPs induces the anisotropic interactions between NPs, due to the minimization of interfacial energy of the system.

5.5.2.2 Self-Assembly of Mixed Homopolymer-Grafted NPs (M-PGNPs)

Tailoring the architecture of polymer ligands provides an efficient strategy to control the assembly of PGNPs. When a mixture of immiscible homopolymers is used as ligands, the lateral phase separation of the binary polymers on NPs can generate surface anisotropy to drive the self-assembly of M-PGNPs into defined nanocomposites in selective solvents [101]. The relative size ratio of brushes to NP core, $2R_g/d$, plays an important role in determining the assembly structures of M-PGNPs. At $2R_g/d \ll 1$, amphiphilic M-PGNPs cannot organize into vesicles, due to reduced flexibility and deformability of the polymer brushes. When $2R_g/d \gg 1$, the lateral phase separation of binary brushes cause M-PGNPs to form "AB" type hybrid amphiphiles where hydrophilic A and hydrophobic B components are covalently linked by a NP core. In this case, M-PGNPs prefer to assemble into cylindrical core–shell micelles or vesicles consisting of bilayer arrangement of PGNPs [101, 102]. At $2R_g/d \approx 1$, M-PGNPs tend to form a "ABA" type structures where solvophobic B compartment encapsulates inorganic NP core and solvophilic A compartments serve as both outer and inner stabilizing layer of vesicles [103]. In this case, the NPs form a monolayer array encapsulated within hydrophobic vesicular membranes.

Duan et al. functionalized AuNPs with two types of chemically distinct polymers such as, hydrophilic PEG and hydrophobic PMMA (Figure 5.21a). These M-PGNPs formed vesicular structures in an amphiphilicity-driven self-assembly process and the destruction of vesicles could be triggered by the reduction of pH or irradiation of light [103a]. In this study, the assembly of NP vesicles was achieved by film rehydration, a method commonly used to prepare polymersomes or liposomes. A two-step approach consisting of sequentially conducted "grafting to" and "grafting from" reactions was used to integrate binary polymer ligands on AuNPs with controlled structural parameters (molar ratio, MW, and σ) of mixed brushes [103b]. AuNPs were modified with a mixture of initiator and hydrophilic polymer chains. Subsequently, the surface-grafted initiators induced polymerization of hydrophobic or stimuli-responsive monomers on the surface of NPs [103c]. This approach was also applied to modify the surface of AuNRs with binary polymer brushes of PEO and polylactide (PLA) (Figure 5.21b) [103d]. Self-assembly of the PEO/PLA-grafted AuNRs by film rehydration yielded plasmonic vesicles with a monolayer of rods in the membranes. The formed AuNR vesicles with biodegradable polymer brushes can be destructed by either enzymatic degradation or near-infrared photothermal heating. Furthermore, Chen et al. fabricated ultrasmall plasmonic vesicles with diameters down to 60–65 nm from PEO/polylactide-tethered AuNRs (8×2 nm or 9×2 nm) using emulsions as templates. In this case, oil-in-water (e.g. chloroform-in-water) emulsions [105] or water-in-oil-in-water (e.g. water-in-chloroform-in-water) double emulsions [106] were used as soft templates, and the evaporation of chloroform containing AuNRs resulted in the interfacial assembly of AuNRs to form vesicles.

Recently, Song et al. prepared Janus M-PGNPs (Au@PS-Fe_3O_4@PEG and Au@PEG-Fe_3O_4@PS NPs) by grafting polymer brushes of different hydrophilicity on Au and Fe_3O_4 surfaces of Janus Au–Fe_3O_4 NPs [104]. These amphiphilic Janus NPs self-assembled into double-layered magneto-plasmonic vesicles (DL-MPVe) precisely controlled the orientation of the Janus NPs. The position of the Au and Fe_3O_4 in the vesicular membrane was determined by the amphiphilicity of the polymer brushes coated on Au or Fe_3O_4 surface. The Au@PEG-Fe_3O_4@PS M-PGNPs self-assembled into double-layered vesicles (DL-Ve 1) with Au located on the outside of vesicular shell and Fe_3O_4 located on the inside of the shell, whereas the vesicles assembled from Au@PS-Fe_3O_4@PEO (DL-Ve 2) featured a reversed orientation of Au and Fe_3O_4 parts. In contrast, Au–Fe_3O_4@PEG/PS (Au-Fe_3O_4 NPs functionalized with mixed PEG/PS on both Au and Fe_3O_4 surfaces) assembled into monolayered vesicles (ML-Ve 3) with randomly paralleled orientation of Au and Fe_3O_4 in the shell.

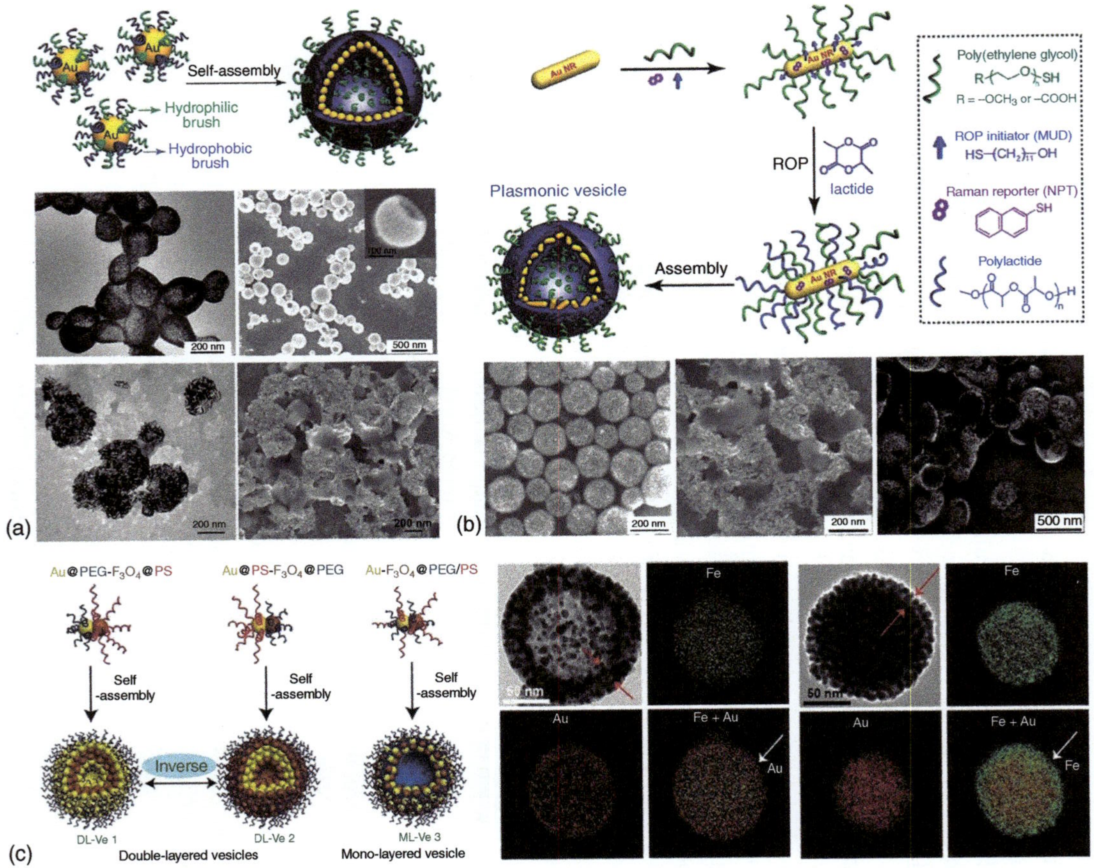

Figure 5.21 Self-assembly of M-PGNPs into vesicular structures. (a) Schematic illustration of self-assembly of AuNPs functionalized with mixed brushes of PEG and PMMA (top), TEM (middle left) and SEM images (middle right) of vesicles assembled from 14 nm AuNPs, and TEM (bottom left) and SEM images (bottom right) of vesicles irradiated with a 785 nm laser for 2 minutes. Source: Song et al. [103a], Reproduced with permission from American Chemical Society. (b) Schematic illustration of the functionalization of AuNRs with mixed polymer brushes (PEG and PLA) by using a sequentially combined "grafting to" and "grafting from" approach and self-assembly of functionalized AuNRs into vesicles (top), and SEM images of vesicles before treatment (bottom left), after 6 minutes laser irradiation at 2.5 W cm^{-2} (bottom middle) and after treatment by the enzyme for 15 hours (bottom right). Source: Song et al. [103d], Reproduced with permission from American Chemical Society. (3) Schematic illustration of different Janus Au-Fe_3O_4@PEG/PS NPs and the hierarchical self-assembly of Janus Au–Fe_3O_4@PEG/PS NPs into vesicles (left), and TEM-element mapping images of the DL-Ve 1 (middle) and DL-Ve 2 (right). Source: Song et al. [104a], Reproduced with permission from Wiley-VCH.

5.5.2.3 Self-Assembly of BCP-Grafted NPs (B-PGNPs)

When a mixture of homopolymers is used as ligands, it is difficult to quantitatively control or predict the relative density of each type of brushes on NP surfaces, because the absorption kinetics of polymers is strongly dependent on the length, composition, and architecture of polymers and the binding strength of terminal groups [107]. The use of BCPs in B-PGNPs allows for better control over the architectural complexity, chemical functionality, and composition of polymer chains on NP surfaces. Owing to the colloidal amphiphilicity originated by the conformational rearrangement of grafted brushes, B-PGNPs can self-assemble to form a variety of hybrid nanostructures (e.g. vesicles, tubules, and discs). For instance, Nie et al. prepared a class of B-PGNPs composed of AuNP cores and amphiphilic linear BCPs tethers, such as PEO-*b*-PS or poly(2-[2-methoxyethoxy]ethyl methacrylate)-*b*-polystyrene ($PMEO_2MA$-*b*-PS) [108]. The B-PGNPs self-assembled into well-defined vesicular and tubular nanostructures in selective solvents by using the film rehydration method (Figure 5.22a,b). The morphologies of assemblies were strongly dependent on the relative size of AuNPs (d_{Au}) and PS block in BCPs (R_0), characterized by the ratio of R_0/d_{Au} (Figure 5.22c). Specifically, the rehydration of B-PGNPs in water was suppressed at $R_0/d_{Au} > 0.5$. At $R_0/d_{Au} \approx 0.5$, the B-PGNPs favorably assembled into

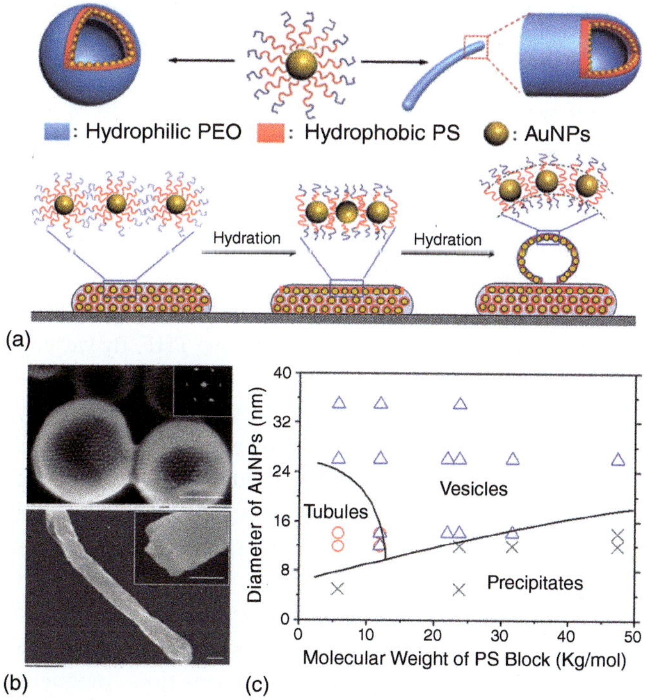

Figure 5.22 Self-assembly of B-PGNPs into vesicular and tubular structures. (a) Schematic illustration of self-assembly of AuNPs@PS-*b*-PEO into vesicles and tubules by rehydration. (b) SEM images of vesicles (top) and tubules (bottom). (c) Self-assembly "product diagram" of AuNPs@PS-*b*-PEO. Source: He et al. [108], Reproduced with permission from American Chemical Society.

tubules with a monolayer of AuNPs in the membranes. At $R_0/d_{Au} < 0.5$, vesicles were formed in self-assembly. The assembly behaviors of B-PGNPs were also influenced by the σ of BCPs on the NP surfaces. When the σ of PS-*b*-PEO was low (*c.* 0.03 chains nm^{-2}), the B-PGNPs assembled into chain vesicles with a monolayer of AuNP strings in the vesicular membranes. At higher σ (≥ 0.05 chains nm^{-2}), the assembly of B-PGNPs yielded non-chain vesicles with the uniform distribution of NPs in the vesicular membranes [109].

In addition to film rehydration [110], several other methods can be used to assemble B-PGNPs into vesicular structure; and they include: solvent selectivity-driven self-assembly, and microfluidic-assisted self-assembly. The former approach enables the full exploration of the key kinetic and thermodynamic factors in the assembly process [111]. The latter allows for continuous, stable generation of high-quality vesicles and other assembly structures from microfluidic devices, and for precisely controlling the size and morphology of assembly structures by manipulating hydrodynamics of fluidic flows [112].

5.5.2.4 Co-Assembly of Binary B-PGNPs or B-PGNPs/BCPs

Co-assembly of B-PGNPs and amphiphilic BCPs has proved to be an efficient approach for the fabrication of complex NP vesicles. Eisenberg et al. designed two types of B-PGNPs: one consists of crosslinked Pb acrylate cores surrounded by the PS_{155}-*b*-PEO_{45} corona, and another consists of AuNPs grafted with PS_{270}-*b*-PAA_{15} copolymers [113]. When water was added into solutions of mixed B-PGNPs and free BCPs of the same or similar composition as the surface-grafted polymers in dioxane, the mixed building blocks assembled into vesicular structures in which the NPs were located in the center portion of the vesicle walls. More recently, Nie et al. fabricated hybrid plasmonic vesicles by co-assembling PS_{x1}-*b*-PEO_{y1}-grafted AuNPs (~19 nm and 40 nm in diameter) and free BCPs of PS_{x2}-*b*-PEO_{y2} (Figure 5.23a) [114]. The co-assembly process was induced by adding water into solutions of the mixed amphiphiles in THF, followed by dialysis against water to remove THF. By varying the size of NP core and the block length of free and/or tethered BCPs, hybrid vesicles with various morphologies were obtained, such as Janus-like vesicles (JVs), patchy vesicles (PVs), and heterogeneous vesicles (HVs), as shown in Figure 5.23b–g. The binary mixture of amphiphilic BCPs and B-PGNPs could also co-assemble to generate Janus-like vesicles containing PEO_{45}-*b*-PS_{455}, AuNRs@PEO_{45}-*b*-PS_{455}, and PtNPs@PEO_{45}-*b*-PS_{455} for use as motors by using the laminar flow in a microfluidic device [115].

The assembly morphologies of B-PGNPs can be tuned by tailoring the architecture of BCP ligands on the NP surfaces. Furthermore, the use of inorganic NP cores with different compositions (e.g. CdSe and CdS quantum dots, magnetic Fe_3O_4 NPs) can impart the vesicular assemblies with new physical properties and functionalities, thus broadening their potential biomedical applications in controlled release of therapeutic drugs, and multimodality imaging and therapy of cancers [102a, 116]. For example, when magnetic Fe_3O_4 NPs were co-assembled with binary amphiphiles of free PS_{107}-*b*-PAA_4 BCPs and PS_{490}-*b*-PEO_{45}-grafted

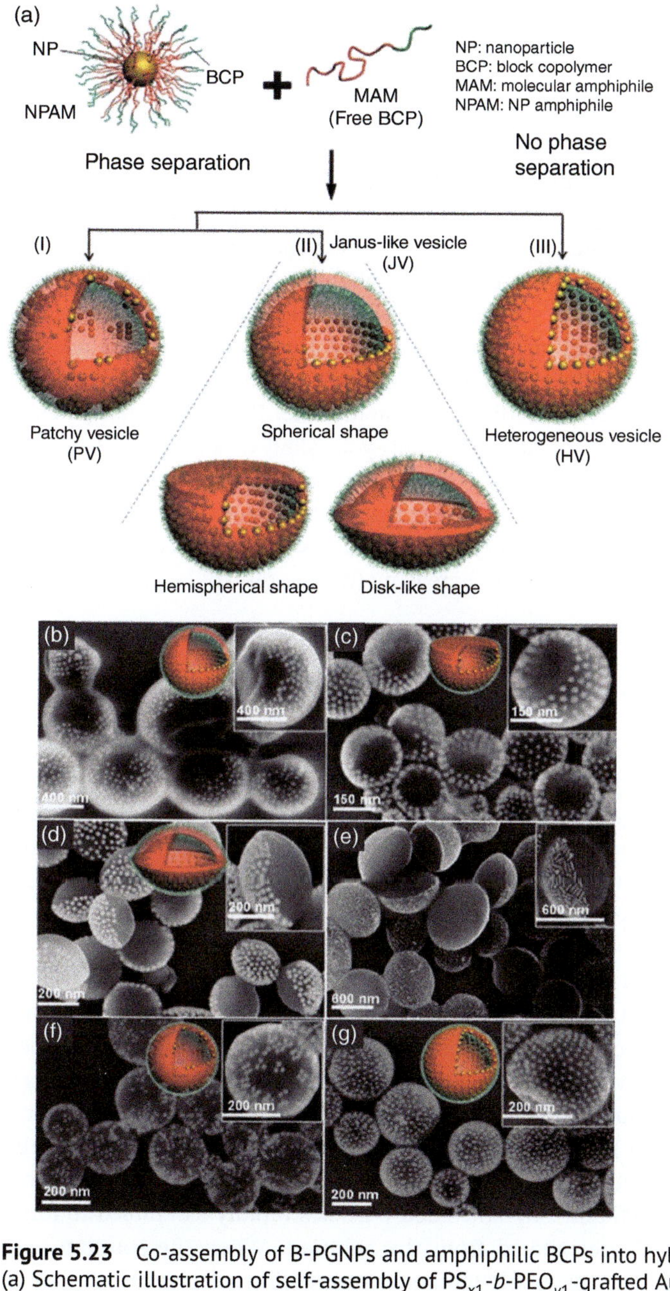

Figure 5.23 Co-assembly of B-PGNPs and amphiphilic BCPs into hybrid vesicles. (a) Schematic illustration of self-assembly of PS_{x1}-*b*-PEO_{y1}-grafted AuNPs and amphiphilic PS_{x2}-*b*-PEO_{y2} into hybrid vesicles with defined shape, morphology, and surface pattern. (b–e) Representative SEM images of Janus-like vesicles of spherical shape (b), hemispherical shape (c), disk-like shape (d), disk-like shape co-assembled form BCPs grafted nanorods and free BCPs (e). (f) SEM image of patchy vesicles. (g) SEM images of heterogeneous vesicles. Source: Liu et al. [114], Reproduced with permission from American Chemical Society.

AuNPs, the assembly process yielded hybrid magneto-plasmonic vesicles with controlled organization of binary inorganic NPs [117]. The morphology of assembled vesicles was determined by the size and mass fraction (ω) of NPs in the mixture. A large size of AuNPs (e.g. 30 and 50 nm) and low ω ($\leq$5.8%) of Fe_3O_4 NPs led to the formation of spherical JVs. Hemispherical JVs were obtained at small size of AuNPs (e.g. 20 nm) and medium ω (e.g. 2.5% and 5.8%), or large size of AuNPs (e.g. 30 and 50 nm) and high ω (e.g. 11%) of Fe_3O_4 NPs. The magnetic and optical responses of the hybrid vesicles were found to be dependent on the organization of NPs in the membranes. More recently, Nie et al. co-assembled of binary AuNPs@PS-*b*-PEO with distinct sizes into hybrid vesicles by dialyzing a solution of the binary B-PGNPs in THF against water [118]. Depending on their relative size and concentration, the assembly process yielded different structures: HVs with random distribution of B-PGNPs in the membrane, PVs with small patches of large B-PGNPs within the membranes, multi-yolk/shell vesicles (m-YSVs) with several large B-PGNPs segregated into the hollow cavity, and single-yolk/shell (s-YSVs) with single yolk in each shell. For instance, at constant relative molar concentration of B-PGNPs in the binary mixture, the assembly structures transited as HVs → JVs → m-YSVs → s-YSV, with increasing the size ratio of large B-PGNPs to small B-PGNPs.

5.5.3 Self-Assembly of PGNPs into 3-D Superlattices and Crystals

Self-assembly of DNA-grafted NPs has proven to be effective for fabricating colloidal superlattices and crystals with precise control over particle composition, lattice parameters, crystal symmetry, and crystal habits. Compared to DNA ligands, synthetic polymers have advantages such as ease of synthesis, high mechanical and chemical stability, rich functionalities, and being inexpensive. However, research on engineering of PGNP-based superlattices and crystals is still in its infancy.

5.5.3.1 Superlattices and Crystals Assembled in Solution

Recently, Macfarlane et al. proposed a class of PGNPs called nanocomposite tectons (NCTs) [119]. The molecular recognition between terminal groups in polymer brushes of NCTs can drive the assembly of binary NPs into defined superlattices (Figure 5.24). In one example, the hydrogen bonding of terminal diaminopyridine (DAP) and thymine (Thy) motifs in complementary PS ligands triggered the self-assembly of AuNP-based NCTs to form superlattices (Figure 5.24b) [119a]. After mixing, the DAP-PS and Thy-PS coated particles rapidly assembled to form random aggregates and precipitated from the solution. When the solution was heated to a characteristic disassembly temperature (melting temperature, T_m), the hydrogen bonding between NCTs became weaker, resulting in the redispersion of the aggregated NPs in the solution. The T_m of NCTs is affected by the collective interactions of DAP- and Thy-terminated polymer chains. Increasing NP size or deceasing polymer length led to a higher T_m, due to the increased areal density of DAP and Thy groups at the NCT periphery. Thermal annealing at a temperature just

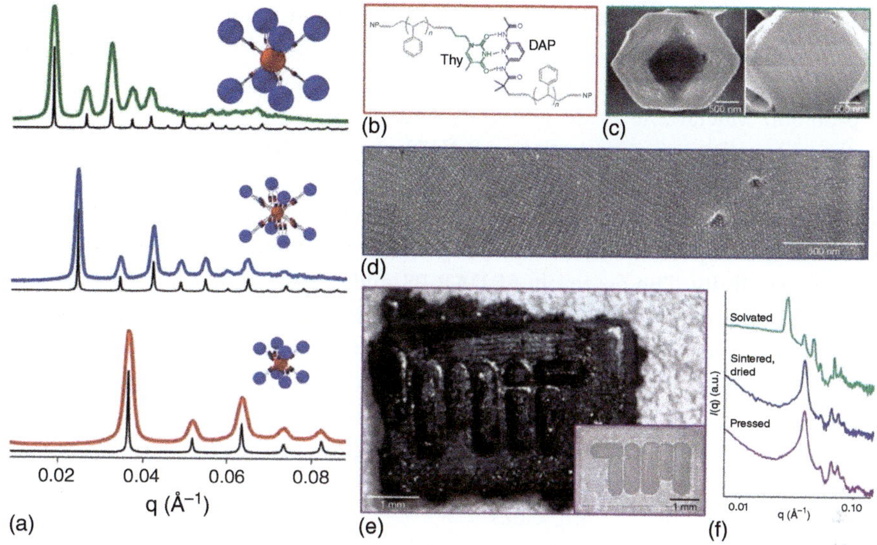

Figure 5.24 3-D superlattices and crystals assembled in solution. (a) SAXS data of bcc superlattices assembled by different NCTs: (green trace) $AuNPs_{20\text{-}nm}@PS_{11\text{-}kDa}$; (blue trace) $AuNPs_{10\text{-}nm}@PS_{11\text{-}kDa}$; (red trace) $AuNPs_{10\text{-}nm}@PS_{6\text{-}kDa}$; (black traces) predicted SAXS patterns for the corresponding perfect bcc lattices. Insets: unit cells for the lattices drawn to scale. Source: Zhang et al. [119a], Reproduced with permission from American Chemical Society. (b) The hydrogen bonding between Thy and DAP group. (c) SEM images of the surface morphology (left) and cross section (right) of a crystallite. (d) SEM image of a cross section of the solid sintered by crystallites. (e) Optical image of a sintered solid after pressing with a mold (inset). (f) SAXS data of the samples at different stages in the process. Source: (b–f) Santos et al. [120], Reproduced with permission from Springer Nature.

below T_m drove the NCTs to arrange into body-centered cubic (bcc) superlattices, which was characterized by small-angle X-ray scattering (SAXS). The same bcc superlattices were obtained for NCTs made from other polymer lengths or particle sizes (e.g. $NP_{20\text{-}nm}@PS_{11\text{-}kDa}$, $NP_{10\text{-}nm}@PS_{11\text{-}kDa}$ and $NP_{10\text{-}nm}@PS_{6\text{-}kDa}$) (Figure 5.24a).

Traditional methods for colloidal crystallization usually require a low dispersity of NP building blocks to produce an ordered array of particles. However, for polymer-grafted NCTs, a high polymer dispersity and a large core size deviation do not have a significant impact on the superlattice crystallinity [121]. This is because the flexible polymer chains can deform to accommodate irregularities in the size and shape of NP building blocks. In addition to AuNP, IONPs can also be used as the core of NCTs in which magnetic dipole interactions promote the bonding of NPs [122]. Interestingly, the lattices composed of only IONP or AuNP cores have bcc crystal symmetry, while the combinations of IONP and AuNP result in CsCl symmetry.

Notably, the superlattices of crystal constructed by Thy- and DAP-NCTs were determined by SAXS in assembly solvent. Removing the solvent from crystals is likely to cause the lattices disordered, as the polymer chains rapidly contract during drying. Thus, the formation of free-standing crystals requires methods to stabilize them against the loss of ordering during solvent removal. To solve this problem, Macfarlane et al. gradually added a non-solvent for the polymer into the dispersion of superlattices of NCTs (e.g. $AuNPs_{15.4\text{-nm}}$@$DAP\text{-}PS_{14\text{-kDa}}$ and $AuNPs_{15.4\text{-nm}}$@$Thy\text{-}PS_{14\text{-kDa}}$), which made polymer chains to adopt denser conformations and shrunk the lattices while preserving ordering [120]. The lattices can be subsequently dried for characterizing their mesoscopic shapes without a loss of their crystallinity. The assembly of DAP-PS and Thy-PS coated AuNPs results in rhombic dodecahedral crystals which are the thermodynamically favored Wulff constructions for bcc lattices (Figure 5.24c,f). Furthermore, NCT crystals can be sintered into bulk solids or molded into macroscopic structures with defined shapes by using laser-cut polyoxymethylene molds (Figure 5.24d,e). The microstructure remained ordered after sintering and pressing (Figure 5.24f). This work provides a versatile method to simultaneously control structural organization across the molecular to macroscopic length scales.

5.5.3.2 Binary Superlattice Assembled at Interfaces

As mentioned in Section 5.4.3.3, various 2-D superlattices of PGNPs can be obtained at the air–liquid interfaces. By controlling experimental factors, 3-D superlattice can also be assembled by this method. Alivisatos et al. prepared 2-D and 3-D binary nanocrystal superlattices (BNSLs) by slow drying of a toluene solution of 3.8 nm-Au@$PS_{3.0\text{-kDa}}$ and 13.4-nm-Fe_3O_4@$PS_{5.3\text{-kDa}}$ nanocrystals on top of an immiscible liquid, diethylene glycol (DEG) [123]. In most cases, when the total concentration of binary PS-grafted NP solution was not too low (when the total concentration of mixed NP spreading solution was reduced by 10–15 times compared with the original concentration to form 3-D superlattice [about 3 mg ml^{-1}], 2-D superlattice formed), diverse 3-D BNSLs with different order types could be obtained by adjusting the NP ratio, core size, and MW of grafted PS ligands on the two NP building blocks. For instance, when the NP ratio decreased from ~15:1 to ~7:1, the structure of BNLSs transformed from $NaZn_{13}$-type (Figure 5.25a–e) to bcc-AB_6-type ($C_{60}K_6$) (Figure 5.25f–j). Representative TEM images showed a [001]-oriented domain of the bcc-AB_6-type BNSLs and SEM image showed that the surface structure on the top layers matched the structural model of the (001) plane (Figure 5.25g). Further decreases in NP ratio yielded phase-pure BNSLs of lower stoichiometries, including $AuCu_3$ (Figure 5.25k–m), AlB_2 (Figure 5.25n–p) and NaCl (Figure 5.25q–s). Although these structures have been observed previously, it is unique that all the five BNSL structures were accessible from NPs with a single size ratio ($\gamma = 0.49$) only by changing the NP mixing ratio. The structure of BNSL assemblies can be adjusted by tuning the size ratio γ of binary NPs and MW of the grafted PS. NaCl-type BNSLs were the preferred structures when $AuNPs_{4.2\text{-nm}}$@$PS_{1.1\text{-kDa}}$ were used as small NPs for $\gamma = 0.4$, whereas the formation of $MgZn_2$-type and $CaCu_5$-type BNSLs was preferred at $\gamma = 0.69$ when $AuNPs_{6.1\text{-nm}}$@$PS_{5.3\text{-kDa}}$ were used as small NPs. The diversity of assembled structures highlights the versatility of using spherical polymer-grafted NPs as modular building blocks for constructing multicomponent assemblies.

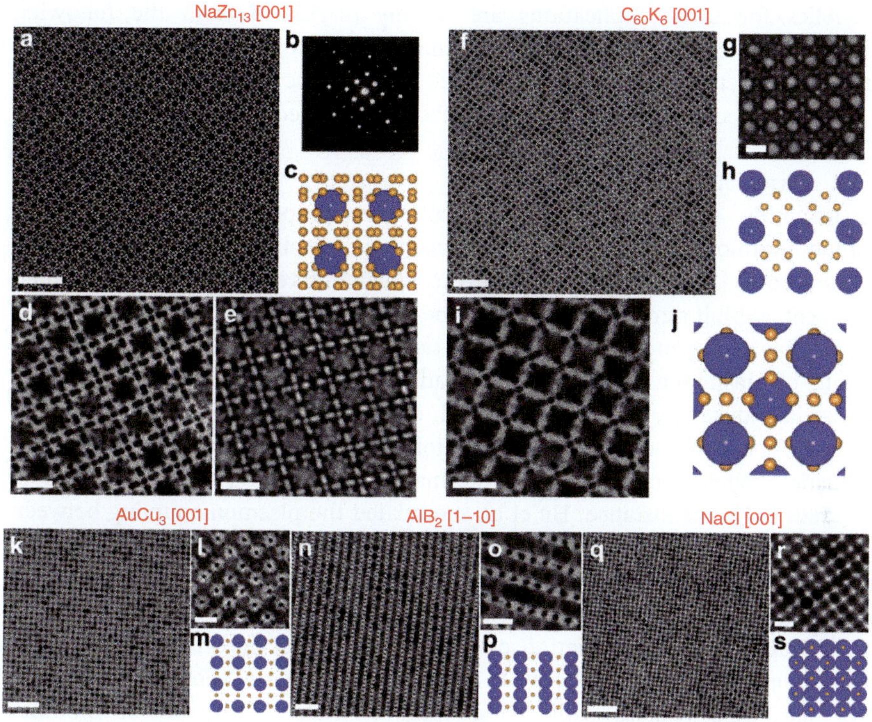

Figure 5.25 Structural diversity in 3-D BNSLs self-assembled from $AuNPs_{3.8\text{-}nm}@PS_{3.0\text{-}kDa}$ and 13.4 nm $Fe_3O_4@PS_{5.3\text{-}kDa}$ NPs. (a) Low-magnification TEM image and (b) corresponding SAED pattern, (d) high-magnification TEM image, and (e) HAADF-STEM image of $NaZn_{13}$-type BNSLs. (c) Structural model of the [001] projection of $NaZn_{13}$-type BNSLs. (f) Low-magnification TEM image, (i) high-magnification TEM image, and (g) SEM image of bcc-AB_6-type (isostructural with the $C_{60}K_6$ phase) BNSLs. (h, j) Structural models of the [001] projection (j) and the (001) surface (h) of bcc-AB_6-type BNSLs. (k, n, q) Low-magnification TEM images, (l, o, r) high-magnification TEM images, and (m, p, s) structural models of $AuCu_3$-type (k–m), AlB_2-type (n–p) and NaCl-type (q–s) BNSLs. Scale bars are (a) 100 nm; (d) 20 nm; (e) 20 nm; (f) 100 nm; (g) 20 nm; (i) 20 nm; (k) 100 nm; (l) 20 nm; (n) 50 nm; (o) 20 nm; (q) 100 nm; (r) 20 nm. Source: Ye et al. [123]/Springer Nature/Licensed under CC BY 4.0.

5.6 Representative Applications of Assembled PGNPs

Hybrid assemblies of PGNPs have shown a wide range of applications in such as cancer imaging and therapy, energy storage and conversion, information encryption, and optical/electronic devices [116d, 124]. Given that plenty of review articles have discussed the applications of hybrid assemblies, this section will only discuss several applications that benefit from the coupling of plasmons, excitons, magnetic moments between PGNPs.

5.6.1 Biological Applications: Imaging, Therapy, and Drug Delivery

In past decade, nanocomposites assembled form PGNPs have been widely studied for bio-imaging, therapy, and drug delivery [116d, 125]. The merits of PGNPs assemblies for these applications are mainly originated from the following three facts: (i) collective properties originated from couped NPs. (ii) combination or synergistic effect of multiple types of inorganic or organic components. (iii) stimuli-responsiveness of inorganic NP core or grafted polymer brushes.

5.6.1.1 Assemblies of Plasmonic PGNPs

The assembly of PGNPs may show collective properties arising from the plasmon–plasmon, exciton–plasmon, or magnetic coupling interactions between inorganic cores [126]. In the case of AuNPs, plasmonic vesicles of polymer-grafted AuNPs can exhibit strong absorption in the NIR range (~600–1100 nm) in which the light has deep penetration depth into tissues. The absorption wavelength and strength of plasmonic vesicles can be readily controlled by varying the shapes, interparticle distance, organization, and orientation of NPs within vesicular membranes, which enables their applications in such as photothermal and photoacoustic (PA) imaging, and photothermal ablation of cancers *in vivo* [103d, 111, 116d, 127]. For instance, He et al. modulated the plasmon coupling between AuNPs in vesicular assemblies by controlling NP size, aggregation number, and interparticle distance. Due to the strong absorption of formed vesicles in NIR range, the temperature of tumor injected with vesicles rapidly increased to 55–60 °C within 1 minute under irradiation of NIR laser (1 W cm^{-2}), thus facilitating the photothermal therapy of cancer [111]. In another work, Nie et al. prepared novel plasmonic vesicles with string-organized NPs in vesicular membranes (Figure 5.26a) [109]. Due to the close interparticle distance in the chains, these plasmonic vesicles exhibited a strong absorption at the NIR wavelength (*c.* 760 nm), thus enhancing their performance in PA imaging. Compared to non-chain vesicles with uniform distribution of NPs in the membranes, the chain vesicles exhibited eightfold signal enhancement in the PA imaging of tumors *in vivo*, while preserving their ability for the encapsulation and light-triggered release of therapeutic agents.

5.6.1.2 Assemblies of Magnetic PGNPs

The assembly of magnetic PGNPs into vesicles increases the stability and biocompatibility of NPs in a physiological environment, and the magnetic responsiveness and transverse relaxivity (r_2) of NPs for enhanced magnetic resonance (MR) imaging and drug delivery *in vivo*. For instance, Yang et al. fabricated magnetic vesicles with monolayer (MoMVs), double layer (DoMVs), and multilayer (MuMVs) by co-assembling a mixture of $IONPs_{9.2\text{-nm}}$@PS_{260}-*b*-PEO_{45} and free PS_{106}-*b*-PAA_4 BCPs (Figure 5.26b) [116a]. Among these vesicles and individual NPs, MuMVs exhibited the highest magnetization (6.98×10^{-13} emu/vesicle) and r_2 (293.6 mM^{-1} s^{-1}) due to high packing density of NPs. In contrast, individual IONPs showed a magnetic moment of 8.28×10^{-17} emu/particle and a r_2 of 108.7 mM^{-1} s^{-1}. Owing to the synergistic effect of external magnetic field-assisted and active tumor targeting, the RGD-conjugated MuMVs showed over 10-fold increase in the delivery of antitumor drugs in tumors *in vivo* for enhanced treatment, compared with control groups without RGD and magnetic field.

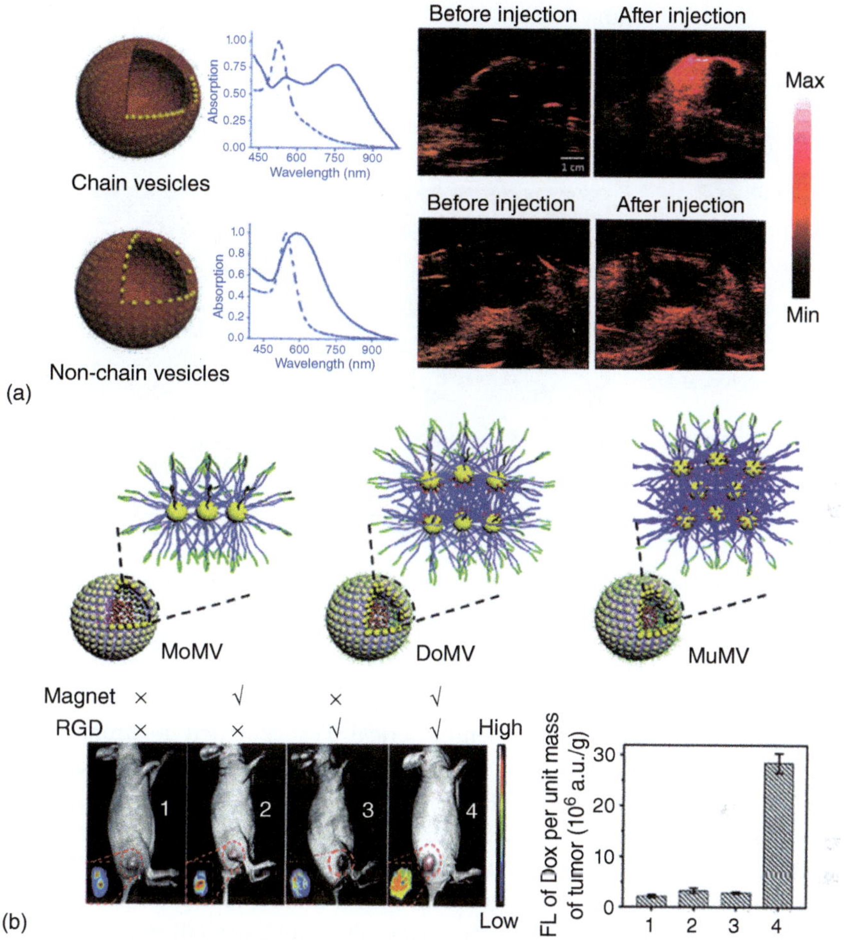

Figure 5.26 The biological application of assemblies fabricated form different PGNPs. (a) Plasmonic vesicles (chain and non-chain) assembled from AuNPs@PS-*b*-PEO as contrast agents in photoacoustic imaging of tumors *in vivo*. Source: Liu et al. [109], Reproduced with permission from Wiley-VCH. (b) Magnetic vesicles with controlled layers of NPs, co-assembled form IONPs@PS-*b*-PEO and free PS-*b*-PAA, for effective delivery of antitumor drugs in tumors. Source: Yang et al. [116a], Reproduced with permission from American Chemical Society.

5.6.1.3 Assemblies of Plasmonic-Magnetic PGNPs

Multi-modality imaging and combination therapy have proven to be effective for enhancing the outcomes of cancer imaging, diagnosis, and therapy. Self-assembly of PGNPs allows for readily integrating binary or multiple NPs into one system with well-controlled morphology and organization of NPs. The hybrid assemblies containing plasmonic and magnetic NPs can be used as contrast agents for enhanced multimodal imaging and vehicles for localized drug delivery and release. For instance, Nie et al. fabricated Janus-like MPVe with different shapes by co-assembly of Fe_3O_4 NPs (15 nm), free PS_{106}-*b*-PAA_4 and PS_{490}-*b*-PEO_{45}-grafted

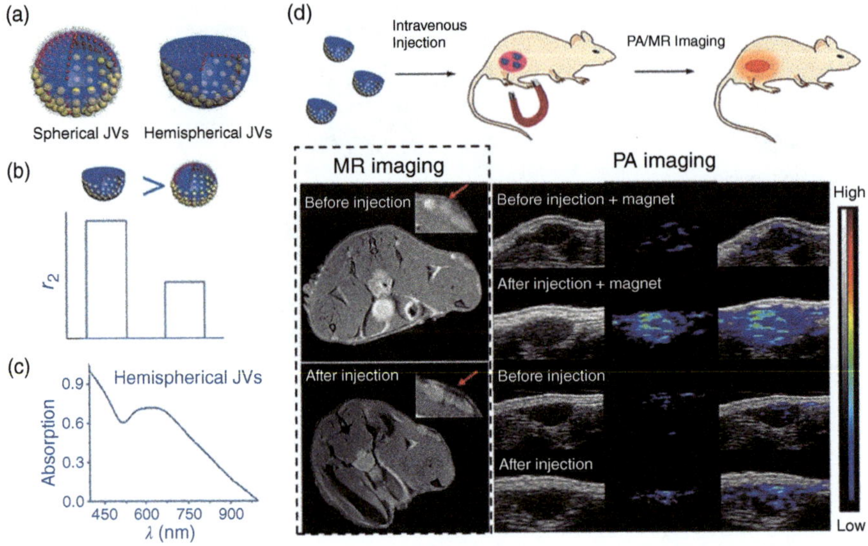

Figure 5.27 The biological application of hybrid MPVe assembled from AuNPs@PS-*b*-PEO, free PS-*b*-PAA, and Fe_3O_4 NPs. (a) Illustration and (b) r_2 of spherical JVs and hemispherical JVs. (c) Absorption of hemispherical JVs. (d) Magnetic-assisted bimodal PA and MR imaging of tumors *in vivo*. Source: Liu et al. [117], Reproduced with permission from Wiley-VCH.

AuNPs (Figure 5.27a) [117]. The hemispherical JVs exhibited a strong NIR absorption and higher r_2 than spherical JVs and individual Fe_3O_4 NPs, due to the dense and ordered packing of NPs in vesicular membranes (Figure 5.27b,c). When the magneto-plasmonic JVs were used as delivery vehicles, the release rate of encapsuled drugs could be modulated by NIR light and magnetic field. The JVs were also demonstrated as imaging agents for *in vivo* bimodal PA and MR imaging of tumors via intravenous injection. Moreover, when an external magnetic field was applied, the increased accumulation of the JVs in tumors drastically enhanced the signal of PA and MR imaging (Figure 5.27d). In another work, Chen et al. demonstrated the application of DL-PMVe assembled from PEG/PS grafted Janus Au–Fe_3O_4 NPs in biological PA and MR imaging [104a], as well as reactive oxygen species (ROS) enhanced chemotherapy [104b].

5.6.2 Dielectric Materials

Dielectric materials based on polymer nanocomposites show important applications in such as optoelectronics, pulsed power systems, energy harvesting, transistors, and inverters, owing to their high breakdown strengths, and great reliability and processability. Traditional nanocomposite dielectrics are usually prepared by mixing dielectric polymer and nanosized inorganic/organic fillers to enhance dielectric properties of materials. However, the blending process inevitably causes dielectric heterogeneities and hence lower breakdown strength, especially when the content of NP fillers is high. Nanocomposites assembled from PGNPs show uniform distribution and highly tunable content of NPs in the materials, which makes them ideal for dielectric applications. For instance, Vaia et al. constructed dielectric nanocomposites by direct assembly of TiO_2@$PS_{100\text{-kDa}}$ NPs without the need of adding free polymers as the matrix [128]. The formed nanocomposite containing 27 vol% of TiO_2 NPs exhibited a dielectric loss of 0.04 at 1 kHz and a dielectric constant of 6.4, compared to 2.7 for the PS film (MW = 10 kDa). In another work, Vaia et al. assembled polymer (PMMA or PS)-grafted silica NPs into nanocomposites featuring a much higher power density and charge/discharge efficiency than the amorphous polymer/NP blends [129]. For instance, the assemblies of PMMA-tethered silica showed an energy density of 1.58 J cm^{-3} at 320 V μm^{-1}, whereas the power density of blend film of PMMA and silica NPs was 1.06 J cm^{-3}. Furthermore, the charge/discharge efficiency (90%) of assemblies at an applied field of 320 V μm^{-1} was significantly higher than that of blends (67%). For practical use, NPs functionalized with bimodal (or multimodal) brush can be more valuable, owing to their additional functionalities and relatively independent control of enthalpic and entropic interactions between the NPs. Single-component nanocomposites assembled from multiple polymer brushes grafted NPs may show great promise for designing next-generation dielectric materials with broad applications.

5.7 Summary and Outlook

In this chapter, we summarize the self-assembly of PGNPs into 1-D, 2-D, and 3-D structures mainly from an experimental perspective. Despite much exciting progress, studies at this frontier are still in the early stage and there remain several challenges that call for further investigation.

(1) To improve complexity of assemblies by effectively manipulating different interactions in PGNPs systems. The PGNP assembly systems reported so far are largely based on simple interactions (e.g. hydrophobic effect), leading to the limited complexity in assembly architectures [34, 124c]. Integration of multiple interactions (e.g. electrostatic interactions, host–guest interactions, hydrogen bonding, halogen bonding, metal chelating, and even Lewis acid and base neutralization) into PGNP systems offers new possibilities for fabricating complex assembly structures. For instance, the use of binary interactions in PGNPs system makes it possible to co-assemble three different types of PGNPs into CMs with dozens of configurations. In principle, integration of diverse interactions in one system can be achieved by appropriate design of polymer ligands and tailored surface functionalization of NPs. However, there remains a grand challenge to precisely control the complementary interaction at nanoscale [33a]. Therefore, a better understanding of nonadditive interactions between PGNPs at different length scales is highly demanded.

(2) To establish a predictive framework for fabricating materials with desired architectures and synergistic properties. As an emerging class of hybrid building blocks, the nanoscale interactions between PGNPs and the underlying mechanisms of PGNP self-assembly remain largely unclear, due to the involvement of

multiscale forces and species. The structural–property relationships of PGNP assemblies are still not fully explored [83, 130]. The upcoming advances in computational modeling (e.g. better models, algorithms, and computational power) and *in situ* characterization tools (e.g. liquid–cell TEM) may offer an opportunity to tackle this challenge [131].

(3) To promote the discovery of novel or advanced applications of resulting materials. In the Section 5.6, the representative biological and dielectric applications of PGNP nanocomposites were briefly discussed. However, the applications of PGNP assemblies in many other fields have not been investigated. To promote the practical use of PGNP composites, the assemblies should be fabricated to feature novel macroscopic properties, which often require the organization of two or more functional NPs into large superstructures with desired architecture [132]. Moreover, as aforementioned, the quantitative prediction of the collective properties of assemblies with different architectures or diverse PGNPs is still lacking, which has become a barrier in facing the application development of assembled nanocomposites. Therefore, hybrid assemblies with hierarchically organized inorganic NPs should be further exploited. Therefore, more efforts should be devoted to the establishment of structural–property relationships and the development of killing applications of these assembled materials.

References

1 (a) Kalsin, A.M., Fialkowski, M., Paszewski, M. et al. (2006). *Science* 312: 420–424. (b) Hueckel, T., Hocky, G.M., Palacci, J., and Sacanna, S. (2020). *Nature* 580: 487–490. (c) Kim, Y.-J., Kim, J.-H., Jo, I.-S. et al. (2021). *J. Am. Chem. Soc.* 143: 13175–13183.

2 (a) Sánchez-Iglesias, A., Grzelczak, M., Altantzis, T. et al. (2012). *ACS Nano* 6: 11059–11065. (b) Nie, Z., Fava, D., Kumacheva, E. et al. (2007). *Nat. Mater.* 6: 609–614. (c) Chen, Q., Diesel, E., Whitmer, J.K. et al. (2011). *J. Am. Chem. Soc.* 133: 7725–7727.

3 (a) De Greef, T.F.A. and Meijer, E.W. (2008). *Nature* 453: 171–173. (b) Hirschberg, J.H.K.K., Brunsveld, L., Ramzi, A. et al. (2000). *Nature* 407: 167–170. (c) Wang, Y., Wang, Y., Breed, D.R. et al. (2012). *Nature* 491: 51–55.

4 (a) Gong, J., Li, G., and Tang, Z. (2012). *Nano Today* 7: 564–585. (b) He, J., Liu, Y., Hood, T.C. et al. (2013). *Nanoscale* 5: 5151–5166. (c) Xia, Y., Xiong, Y., Lim, B., and Skrabalak, S.E. (2009). *Angew. Chem. Int. Ed.* 48: 60–103.

5 Cho, E.S., Kim, J., Tejerina, B. et al. (2012). *Nat. Mater.* 11: 978–985.

6 Xia, Y., Yang, H., and Campbell, C.T. (2013). *Acc. Chem. Res.* 46: 1671–1672.

7 Zhang, H., Liu, Y., Yao, D., and Yang, B. (2012). *Chem. Soc. Rev.* 41: 6066.

8 Kramer, I.J. and Sargent, E.H. (2014). *Chem. Rev.* 114: 863–882.

9 Sung, J., Jo, P.S., Shin, H. et al. (2008). *Adv. Mater.* 20: 1505–1510.

10 (a) Grzelczak, M., Vermant, J., Furst, E.M., and Liz-Marzán, L.M. (2010). *ACS Nano* 4: 3591–3605. (b) Yi, C., Yang, Y., Liu, B. et al. (2020). *Chem. Soc. Rev.* 49: 465–508.

11 Fernandes, N.J., Koerner, H., Giannelis, E.P., and Vaia, R.A. (2013). *MRS Commun.* 3: 13–29.

12 (a) Liu, Y., Liu, Y., Yin, J.-J., and Nie, Z. (2015). *Macromol. Rapid Commun.* 36: 711–725. (b) Yi, C., Zhang, S., Webb, K.T., and Nie, Z. (2017). *Acc. Chem. Res.* 50: 12–21.

13 (a) A. v. (2003). Blaaderen. *Science* 301: 470–471. (b) Blaaderen, A.V. (2006). *Nature* 439: 545–546.

14 Liu, K., Nie, Z., Zhao, N. et al. (2010). *Science* 329: 197–200.

15 (a) Gröschel, A.H., Walther, A., Löbling, T.I. et al. (2013). *Nature* 503: 247–251. (b) Glotzer, S.C. and Solomon, M.J. (2007). *Nat. Mater.* 6: 557–562. (c) Liu, W., Halverson, J., Tian, Y. et al. (2016). *Nat. Chem.* 8: 867–873. (d) Chen, G., Gibson, K.J., Liu, D. et al. (2019). *Nat. Mater.* 18: 169–174.

16 (a) Li, N., Wu, F., Han, Z. et al. (2020). *J. Am. Chem. Soc.* 142: 11680–11684. (b) Edwardson, T.G.W., Lau, K.L., Bousmail, D. et al. (2016). *Nat. Chem.* 8: 162–170. (c) Liu, W., Tagawa, M., Xin, H.L. et al. (2016). *Science* 351: 582–586.

17 Bianchi, E., Capone, B., Coluzza, I. et al. (2017). *Phys. Chem. Chem. Phys.* 19: 19847–19868.

18 (a) Groschel, A.H., Schacher, F.H., Schmalz, H. et al. (2012). *Nat. Commun.* 3: 710. (b) Klinkova, A., Therien-Aubin, H., Choueiri, R.M. et al. (2013). *Proc. Natl. Acad. Sci. U.S.A.* 110: 18775–18779.

19 Chen, T., Yang, M., Wang, X. et al. (2008). *J. Am. Chem. Soc.* 130: 11858–11859.

20 Choueiri, R.M., Galati, E., Thérien-Aubin, H. et al. (2016). *Nature* 538: 79–83.

21 Rossner, C., Zhulina, E.B., and Kumacheva, E. (2019). *Angew. Chem. Int. Ed.* 58: 9269–9274.

22 Nie, Z., Fava, D., Rubinstein, M., and Kumacheva, E. (2008). *J. Am. Chem. Soc.* 130: 3683–3689.

23 Grzelczak, M., Sanchez-Iglesias, A., Mezerji, H.H. et al. (2012). *Nano Lett.* 12: 4380–4384.

24 Kim, A., Zhou, S., Yao, L. et al. (2019). *J. Am. Chem. Soc.* 141: 11796–11800.

25 Cheng, L., Song, J., Yin, J., and Duan, H. (2011). *J. Phys. Chem. Lett.* 2: 2258–2262.

26 Cheng, X., Zhao, G., Lu, Y. et al. (2018). *Chem. Commun.* 54: 7746–7749.

27 Urban, A.S., Shen, X., Wang, Y. et al. (2013). *Nano Lett.* 13: 4399–4403.

28 Yi, C., Liu, H., Zhang, S. et al. (2020). *Science* 369: 1369–1374.

29 Lin, X., Ye, S., Kong, C. et al. (2020). *J. Am. Chem. Soc.* 142: 17282–17286.

30 (a) Zhang, X., Lv, L., Ji, L. et al. (2016). *J. Am. Chem. Soc.* 138: 3290–3293. (b) Wang, H., Song, X., Liu, C. et al. (2014). *ACS Nano* 8: 8063–8073.

31 (a) Singh, G., Chan, H., Baskin, A. et al. (2014). *Science* 345: 1149–1153. (b) Kuzyk, A., Schreiber, R., Fan, Z. et al. (2012). *Nature* 483: 311–314.

32 Yang, Y., Wang, Y., Jin, S.-M. et al. (2019). *Mater. Chem. Front.* 3: 209–215.

33 (a) Batista, C.A., Larson, R.G., and Kotov, N.A. (2015). *Science* 350: 1242477. (b) Bishop, K.J.M., Wilmer, C.E., Soh, S., and Grzybowski, B.A. (2009). *Small* 5: 1600–1630.

34 Li, X., Liu, X., and Liu, X. (2021). *Chem. Soc. Rev.* 50: 2074–2101.

35 Ma, X., Gu, M., Zhang, L. et al. (2019). *ACS Nano* .

36 Yi, C., Yang, Y., and Nie, Z. (2019). *J. Am. Chem. Soc.* 141: 7917–7925.

37 Liu, K., Lukach, A., Sugikawa, K. et al. (2014). *Angew. Chem. Int. Ed.* 53: 2648–2653.

38 Lukach, A., Liu, K., Therien-Aubin, H., and Kumacheva, E. (2012). *J. Am. Chem. Soc.* 134: 18853–18859.

39 (a) Liu, K., Ahmed, A., Chung, S. et al. (2013). *ACS Nano* 7: 5901–5910. (b) Lee, A., Andrade, G.F., Ahmed, A. et al. (2011). *J. Am. Chem. Soc.* 133: 7563–7570.

40 (a) Keng, P.Y., Kim, B.Y., Shim, I.-B. et al. (2009). *ACS Nano* 3: 3143–3157. (b) Hill, L.J., Richey, N.E., Sung, Y. et al. (2014). *ACS Nano* 8: 3272–3284. (c) Pavlopoulos, N.G., Dubose, J.T., Hartnett, E.D. et al. (2016). *ACS Macro Lett.* 5: 950–954. (d) Pavlopoulos, N.G., Dubose, J.T., Pinna, N. et al. (2016). *Angew. Chem. Int. Ed.* 55: 1787–1791.

41 Kim, B.Y., Shim, I.-B., Monti, O.L.A., and Pyun, J. (2011). *Chem. Commun.* 47: 890–892.

42 Correa-Duarte, M.A., Perez-Juste, J., Sanchez-Iglesias, A. et al. (2005). *Angew. Chem. Int. Ed.* 44: 4375–4378.

43 Liang, R., Xu, J., Deng, R. et al. (2014). *ACS Macro Lett.* 3: 486–490.

44 Wang, K., Jin, S.M., Xu, J. et al. (2016). *ACS Nano* 10: 4954–4960.

45 Groschel, A.H., Lobling, T.I., Petrov, P.D. et al. (2013). *Angew. Chem. Int. Ed.* 52: 3602–3606.

46 Bockstaller, M.R., Mickiewicz, R.A., and Thomas, E.L. (2005). *Adv. Mater.* 17: 1331–1349.

47 Balazs, A.C., Emrick, T., and Russell, T.P. (2006). *Science* 314.

48 (a) Liang, R., Xu, J., Li, W. et al. (2014). *Macromolecules* 48: 256–263. (b) Li, W., Liu, S., Deng, R., and Zhu, J. (2011). *Angew. Chem. Int. Ed.* 50: 5865–5868.

49 Li, W., Zhang, P., Dai, M. et al. (2013). *Macromolecules* 46: 2241–2248.

50 Sanwaria, S., Horechyy, A., Wolf, D. et al. (2014). *Angew. Chem. Int. Ed.* 53: 9090–9093.

51 Yan, N., Liu, X., Zhu, J. et al. (2019). *ACS Nano* 13: 6638–6646.

52 (a) Zhang, M., Drechsler, M., and Müller, A.H.E. (2004). *Chem. Mater.* 16: 537–543. (b) Zhang, M., Estournès, C., Bietsch, W., and Müller, A.H.E. (2004). *Adv. Funct. Mater.* 14: 871–882.

53 Yan, C. and Wang, T. (2017). *Chem. Soc. Rev.* 46: 1483–1509.

54 Gorelikov, I., Field, L.M., and Kumacheva, E. (2004). *J. Am. Chem. Soc.* 126: 15938–15939.

55 (a) Cui, H., Chen, Z., Zhong, S. et al. (2007). *Science* 317: 647–650. (b) Yang, S., Yu, X., Wang, L. et al. (2010). *Macromolecules* 43: 3018–3026.

56 Underhill, R.S. and Liu, G. (2000). *Chem. Mater.* 12: 3633–3641.

57 (a) Ishii, T., Otsuka, H., Kataoka, K., and Nagasaki, Y. (2004). *Langmuir* 20: 561–564. (b) Huang, H., Chung, B., Jung, J. et al. (2009). *Angew. Chem. Int. Ed.* 48: 4594–4597.

58 Qiu, H., Gao, Y., Boott Charlotte, E. et al. (2016). *Science* 352: 697–701.

59 Xu, H., Xu, Y., Pang, X. et al. (2015). *Sci. Adv.* 1: e1500025.

60 Khanal, B.P. and Zubarev, E.R. (2007). *Angew. Chem. Int. Ed.* 46: 2195–2198.

61 Gupta, S., Zhang, Q., Emrick, T. et al. (2006). *Nat. Mater.* 5: 229–233.

62 (a) Gao, B., Alvi, Y., Li, V., and Tao, A.R. (2014). *CrystEngComm* 16: 9434–9440. (b) Gao, B., Alvi, Y., Rosen, D. et al. (2013). *Chem. Commun.* 49: 4382–4384.

63 Gao, B., Arya, G., and Tao, A.R. (2012). *Nat. Nanotechnol.* 7: 433–437.

64 (a) Cheng, W. (2017). *EPL* 119: 48004. (b) Cai, Z., Li, Z., Ravaine, S. et al. (2021). *Chem. Soc. Rev.* 50: 5898–5951.

65 (a) Sarkar, B. and Alexandridis, P. (2015). *Prog. Polym. Sci.* 40: 33–62. (b) Hoheisel, T.N., Hur, K., and Wiesner, U.B. (2015). *Prog. Polym. Sci.* 40: 3–32.

66 (a) Kim, B.J., Fredrickson, G.H., and Kramer, E.J. (2008). *Macromolecules* 41: 436–447. (b) Kim, B.J., Bang, J., Hawker, C.J. et al. (2007). *Langmuir* 23: 12693–12703. (c) Chiu, J.J., Kim, B.J., Kramer, E.J., and Pin, D.J. (2005). *J. Am. Chem. Soc.* 127: 5036–5037.

67 Jang, S.G., Kramer, E.J., and Hawker, C.J. (2011). *J. Am. Chem. Soc.* 133: 16986–16996.

68 Kim, B.J., Fredrickson, G.H., Hawker, C.J., and Kramer, E.J. (2007). *Langmuir* 23: 7804–7809.

69 Li, W., Wang, K., Zhang, P. et al. (2016). *Small* 12: 499–505.

70 Li, F., Wang, K., Deng, N. et al. (2021). *Mater. Interfaces* 13: 6566–6574.

71 Zhang, R., Lee, B., Stafford, C.M. et al. (2017). *Proc. Natl. Acad. Sci. U.S.A.* 114: 2462–2467.

72 Li, B., Jiang, B., Han, W. et al. (2017). *Angew. Chem. Int. Ed.* 56: 4554–4559.

73 Rey, M., Law, A.D., Buzza, D.M.A., and Vogel, N. (2017). *J. Am. Chem. Soc.* 139: 17464–17473.

74 Cheng, W., Campolongo, M.J., Cha, J.J. et al. (2009). *Nat. Mater.* 8: 519–525.

75 Si, K.J., Sikdar, D., Yap, L.W. et al. (2015). *Adv. Opt. Mater.* 3: 1710–1717.

76 (a) Prodan, E., Radloff, C., Halas, N.J., and Nordlander, P. (2003). *Science* 302: 419–422. (b) Nordlander, P., Oubre, C., Prodan, E. et al. (2004). *Nano Lett.* 4: 899–903. (c) Chen, Y., Fu, J., Ng, K.C. et al. (2011). *Crystal Growth & Design* 11: 4742–4746. (d) Shi, Q., Si, K.J., Sikdar, D. et al. (2016). *ACS Nano* 10: 967–976.

77 Huang, M.H. and Thoka, S. (2015). *Nano Today* 10: 81–92.

78 Wang, K., Ling, H., Bao, Y. et al. (2018). *Adv. Mater.* 30: 1800595–1800602.

79 (a) Ng, K.C., Udagedara, I.B., Rukhlenko, I.D. et al. (2012). *ACS Nano* 6: 925–934. (b) Si, K.J. (2014). *ACS Nano* 8: 11086–11093. (c) Dong, D., Yap, L.W., Smilgies, D.M. et al. (2018). *Nanoscale* 10: 5065–5071.

80 (a) Kao, J., Thorkelsson, K., Bai, P. et al. (2013). *Chem. Soc. Rev.* 42: 2654–2678. (b) Maillard, D., Kumar, S.K., Fragneaud, B. et al. (2012). *Nano Lett.* 12: 3909–3914.

81 (a) Li, Y., Krentz, T.M., Wang, L. et al. (2014). *Mater. Interfaces* 6: 6005–6021. (b) Zhao, D., Di Nicola, M., Khani, M.M. et al. (2016). *ACS Macro Lett.* 5: 790–795. (c) Jouault, N., Lee, D., Zhao, D., and Kumar, S.K. (2014). *Adv. Mater.* 26: 4031–4036.

82 Jiang, G., Hore, M.J.A., Gam, S., and Composto, R.J. (2012). *ACS Nano* 6: 1578–1588.

83 Lenart, W.R. and Hore, M.J.A. (2018). *Nano-Struct. Nano-Objects* 16: 428–440.

84 Gu, X.W., Ye, X., Koshy, D.M. et al. (2017). *Proc. Natl. Acad. Sci. U.S.A.* 114: 2836–2841.

85 Kubiak, J.M. and Macfarlane, R.J. (2019). *Adv. Funct. Mater.* 29.

86 Wu, C.-K., O'Brien, S., and Koberstein, J.T. (2008). *J. Am. Chem. Soc.* 130: 3516–3520.

87 Leffler, V.B., Mayr, L., Paciok, P. et al. (2019). *Angew. Chem. Int. Ed.* 58: 8541–8545.

88 Ojha, S., Beppler, B., Dong, H. et al. (2010). *Langmuir* 26: 13210–13215.

89 Schmitt, M., Zhang, J., Lee, J. et al. (2016). *Sci. Adv.* 2: 1–10.

90 Ahn, D.U., Wang, Z., Campbell, I.P. et al. (2012). *Polymer* 53: 4187–4194.

91 Jones, M.R., Osberg, K.D., Macfarlane, R.J. et al. (2011). *Chem. Rev.* 111: 3736–3827.

92 Grzelczak, M., Sánchez-Iglesias, A., and Liz-Marzán, L.M. (2014). *CrystEng-Comm* 16: 9425–9429.

93 Grzelczak, M., Sánchez-Iglesias, A., and Liz-Marzán, L.M. (2013). *Soft Matter* 9: 9094–9098.

94 Chen, Y., Wang, Z., He, Y. et al. (2018). *Proc. Natl. Acad. Sci. U.S.A.* 115: E1391–E1400.

95 Li, W., Liu, S., Deng, R. et al. (2013). *Macromolecules* 46: 2282–2291.

96 Schmidtke, C., Eggers, R., Zierold, R. et al. (2014). *Langmuir* 30: 11190–11196.

97 Sanchez-Iglesias, A., Claes, N., Solis, D.M. et al. (2018). *Angew. Chem. Int. Ed.* 57: 3183–3186.

98 (a) Lecommandoux, S., Sandre, O., Chécot, F. et al. (2005). *Adv. Mater.* 17: 712–718. (b) Krack, M., Hohenberg, H., Kornowski, A. et al. (2008). *J. Am. Chem. Soc.* 130: 7315–7320.

99 (a) Du, J. and Armes, S.P. (2005). *J. Am. Chem. Soc.* 127: 12800–12801. (b) Du, J., Tang, Y., Lewis, A.L., and Armes, S.P. (2005). *J. Am. Chem. Soc.* 127: 17982–17983. (c) Li, Y., Smith, A.E., Lokitz, B.S., and McCormick, C.L. (2007). *Macromolecules* 40: 8524–8526.

100 Nikolic, M.S., Olsson, C., Salcher, A. et al. (2009). *Angew. Chem. Int. Ed.* 48: 2752–2754.

101 Zubarev, E.R., Xu, J., Sayyad, A., and Gibson, J.D. (2006). *J. Am. Chem. Soc.* 128: 15098–15099.

102 (a) Coleman, B.R. and Moffitt, M.G. (2018). *Chem. Mater.* 30: 2474–2482. (b) Guo, Y., Harirchian-Saei, S., Izumi, C.M.S., and Moffitt, M.G. (2011). *ACS Nano* 5: 3309–3318.

103 (a) Song, J., Cheng, L., Liu, A. et al. (2011). *J. Am. Chem. Soc.* 133: 10760–10763. (b) Cheng, L., Liu, A., Peng, S., and Duan, H. (2010).

ACS Nano 4: 6098–6104. (c) Song, J., Fang, Z., Wang, C. et al. (2013). *Nanoscale* 5: 5816–5824. (d) Song, J., Pu, L., Zhou, J. et al. (2013). *ACS Nano* 7: 9947–9960.
104 (a) Song, J., Wu, B., Zhou, Z. et al. (2017). *Angew. Chem. Int. Ed.* 56: 8110–8114. (b) Song, J., Lin, L., Yang, Z. et al. (2019). *J. Am. Chem. Soc.* 141: 8158–8170.
105 Song, J., Yang, X., Jacobson, O. et al. (2015). *Adv. Mater.* 27: 4910–4917.
106 Song, J., Yang, X., Jacobson, O. et al. (2015). *ACS Nano* 9: 9199–9209.
107 (a) Wang, B., Li, B., Dong, B. et al. (2010). *Macromolecules* 43: 9234–9238. (b) Andala, D.M., Shin, S.H.R., Lee, H.-Y., and Bishop, K.J.M. (2012). *ACS Nano* 6: 1044–1050.
108 He, J., Liu, Y., Babu, T. et al. (2012). *J. Am. Chem. Soc.* 134: 11342–11345.
109 Liu, Y., He, J., Yang, K. et al. (2015). *Angew. Chem. Int. Ed.* 54: 15809–15812.
110 Lin, J., Wang, S., Chen, X., and Nie, Z. (2013). *ACS Nano* 7: 5320–5329.
111 He, J., Huang, X., Li, Y.C. et al. (2013). *J. Am. Chem. Soc.* 135: 7974–7984.
112 (a) He, J., Wang, L., Wei, Z. et al. (2013). *ACS Appl. Mater. Interfaces* 5: 9746–9751. (b) He, J., Wei, Z., Wang, L. et al. (2013). *Angew. Chem. Int. Ed.* 52: 2463–2468.
113 Mai, Y. and Eisenberg, A. (2010). *J. Am. Chem. Soc.* 132: 10078–10084.
114 Liu, Y., Li, Y., He, J. et al. (2014). *J. Am. Chem. Soc.* 136: 2602–2610.
115 Wang, L., Liu, Y., He, J. et al. (2015). *Small* 11: 3762–3767.
116 (a) Yang, K., Liu, Y., Liu, Y. et al. (2018). *J. Am. Chem. Soc.* 140: 4666–4677. (b) Ye, H., Yang, K., Tao, J. et al. (2017). *ACS Nano* 11: 2052–2059. (c) Song, J., Niu, G., and Chen, X. (2017). *Bioconjugate Chem.* 28: 105–114. (d) Song, J., Huang, P., Duan, H., and Chen, X. (2015). *Acc. Chem. Res.* 48: 2506–2515.
117 Liu, Y., Yang, X., Huang, Z. et al. (2016). *Angew. Chem. Int. Ed.* 55: 15297–15300.
118 Lamar, C., Liu, Y., Yi, C., and Nie, Z. (2020). *Giant* 1: 100010.
119 (a) Zhang, J., Santos, P.J., Gabrys, P.A. et al. (2016). *J. Am. Chem. Soc.* 138: 16228–16231. (b) Santos, P.J., Cao, Z., Zhang, J. et al. (2019). *J. Am. Chem. Soc.* 141: 14624–14632.
120 Santos, P.J., Gabrys, P.A., Zornberg, L.Z. et al. (2021). *Nature* 591: 586–591.
121 Santos, P.J., Cheung, T.C., and Macfarlane, R.J. (2019). *Nano Lett.* 19: 5774–5780.
122 Santos, P.J. and Macfarlane, R.J. (2020). *J. Am. Chem. Soc.* 142: 1170–1174.
123 Ye, X., Zhu, C., Ercius, P. et al. (2015). *Nat. Commun.* 6: 10052.
124 (a) Li, F., Lu, J., Kong, X. et al. (2017). *Adv. Mater.* 29. (b) Eygeris, Y., White, E.V., Wang, Q. et al. (2019). *Mater. Interfaces* 11: 3407–3416. (c) Li, Z., Fan, Q., and Yin, Y. (2021). *Chem. Rev.* .
125 (a) Fan, W., Yung, B., Huang, P., and Chen, X. (2017). *Chem. Rev.* 117: 13566–13638. (b) Fu, Q., Zhu, R., Song, J. et al. (2019). *Adv. Mater.* 31: e1805875. (c) Zheng, D., Yang, K., and Nie, Z. (2021). *VIEW* 20200067. (d) Yang, K., Zhang, S., He, J., and Nie, Z. (2021). *Nano Today* 36: 101046.
126 Nie, Z., Petukhova, A., and Kumacheva, E. (2010). *Nat. Nanotechnol.* 5: 15–25.
127 Song, J., Zhou, J., and Duan, H. (2012). *J. Am. Chem. Soc.* 134: 13458–13469.

128 Tchoul, M.N., Fillery, S.P., Koerner, H. et al. (2010). *Chem. Mater.* 22: 1749–1759.

129 Grabowski, C.A., Koerner, H., Meth, J.S. et al. (2014). *Mater. Interfaces* 6: 21500–21509.

130 Wang, Y., Desroches, G.J., and Macfarlane, R.J. (2021). *Nanoscale* 13: 426–443.

131 (a) Zhou, G., Zhang, B., Tang, G. et al. (2021). *Adv. Phys.: X* 6: 1866668. (b) Dijkstra, M. and Luijten, E. (2021). *Nat. Mater.* 20: 762–773.

132 Lee, M.S., Yee, D.W., Ye, M., and Macfarlane, R.J. (2022). *J. Am. Chem. Soc.* 144: 3330–3346.

6

Interfacial Property of Hairy Nanoparticles

Yilan Ye and Zhenzhong Yang

Tsinghua University, Department of Chemical Engineering, Beijing 100084, China

6.1 Introduction

Interfaces appear when incompatible phases of solids, liquids, or gases are in contact. Hairy nanoparticles (NPs) with polymers as coronas are highly structured at interfaces. When the hairy NPs are introduced to polymer matrices or emulsions, interfacial behavior of the systems can be significantly varied upon immobilization of the hairy NPs, thereby achieving a huge family of novel structures and functional materials. Polymers at the NPs surface are tunable in composition, molecular weight, and grafting density, and the hairy NPs exhibit a wealth of phase behaviors to engineer interfaces. In good solvents, polymers at the hairy NPs are highly swollen exhibiting mushroom or brush-like conformation. In selective solvents, Janus or patchy geometries of hairy NPs are generated by microphase separation of the grafted multiple polymers. Superstructures are constructed by self-assembly of the anisotropic NPs. Structured interfaces easily result from the hairy NPs for Pickering emulsions, colloidosomes, and interfacial jamming. Numerous interfacial applications are envisaged, such as in constructing responsive emulsions, structured liquids, heterogeneous catalysis, and functional nanocomposites.

Understanding interfacial behavior of the hairy NPs is important to manipulate interfaces and to design novel materials. Constituent, molecule weight, and graft density of polymers at the hairy NPs are key variables to govern microdomain structure of the corona and interaction among the NPs. In this chapter, we firstly introduce hairy NPs as interfacial building blocks by addressing the three aspects: (i) conformation of the grafted polymers in good solvents; (ii) formation of patchy or Janus structures by microphase separation within the corona in selective solvents; and (iii) interfacial activity of the hairy NPs. In the second section, we will introduce assembled structures of the hairy NPs at various interfaces: (i) in hairy NPs filled multiphase polymer nanocomposites, polymer–polymer interfaces are involved; (ii) self-assembly of hairy NPs to construct colloidal molecules or superstructures,

Hairy Nanoparticles: From Synthesis to Applications, First Edition. Edited by Zhiqun Lin and Yijiang Liu.

in which the corona–corona interactions are dominant; (iii) self-assembly of the hairy NPs at liquid–liquid interfaces; (iv) self-assembly at air–solid interfaces toward functional coatings; and (v) self-assembly at air–liquid interfaces toward superlattice nanosheets. In the third section, we will introduce theoretic consideration of entropy-governed interfacial behavior by the hairy NPs. In the fourth section, we will summarize recent advances in structural evolution of the hairy NPs at interfaces and interfacial jamming. In the last part, we will introduce some representative developments in hairy NPs based on single-chain nanoparticles (SCNPs) with precise compartmentalization at nanoscales.

6.2 Hairy NPs as Interfacial Building Blocks

6.2.1 Conformation of Grafted Polymers in Good Solvents

Polymeric corona is responsible to bridge the core of a hairy NP with the surrounding matrix. Conformation of the polymer corona is crucial to govern the interaction between NPs with matrix. Conformation of grafted polymer in good solvents as a function of graft density (σ) and degree of polymerization (N) is shown in Figure 6.1a [1]. As graft density (σ) increases, the polymer conformation is changed from mushroom-like random coil (pink) to expanded conformation in semidilute polymer brush (SDPB) regime (green), and more extended conformations in concentrated polymer brush (CPB) regime (blue). For large degree of polymerization (N) at high σ, the polymer adopts an SDPB conformation at the exterior region of the corona, while the CPB regime (yellow) is dominant at the interior region. The entropic effect to maintain random walk conformation is competitive with the enthalpic effect to stretch the chain. In the SDPB and CPB regimes, the polymer brush height (h) versus N follows the scaling laws of $h \sim N^{0.8}$ and $h \sim N^{0.6}$, respectively.

To enhance dispersion of the hairy NPs in a polymer matrix in the athermal state where the Flory–Huggins parameter $\chi \approx 0$, the grafted polymers should share the similar composition with the matrix. Hore et al. systematically characterized conformation of densely grafted polymer poly(methyl methacrylate) (PMMA) by small-angle neutron scattering (SANS) and self-consistent field theory (SCFT) calculation (Figure 6.1b) [2]. For the PMMA-grafted Fe_3O_4 (r_{core} = 2.5 nm) incorporated in PMMA matrix, the CPB and SDPB regions of the grafted PMMA were distinguished by SANS measurements. Using the core–shell-chain model, a shell thickness of 3.9 nm in the CPB region was obtained with the chain size scaling as $N^{0.5}$ in the SDPB region, implying that the grafted chain behaves as the ideal random walk at exterior region of the corona.

Length and conformation of the grafted polymers are crucial to tailor interparticle spacing of self-assembled superlattices of the hairy NPs [3]. For polystyrene (PS)-grafted Au NP and Fe_3O_4 NP with a relatively high grafting density (σ), height of the solvated polymer brush (h) followed the scaling law as $h \sim N^{0.75}$ in good solvents. This implied that the grafted PS chain at Au NP behaves brush-like in the CPB regime. Upon solvent evaporation, co-organization of the PS-grafted Au NP

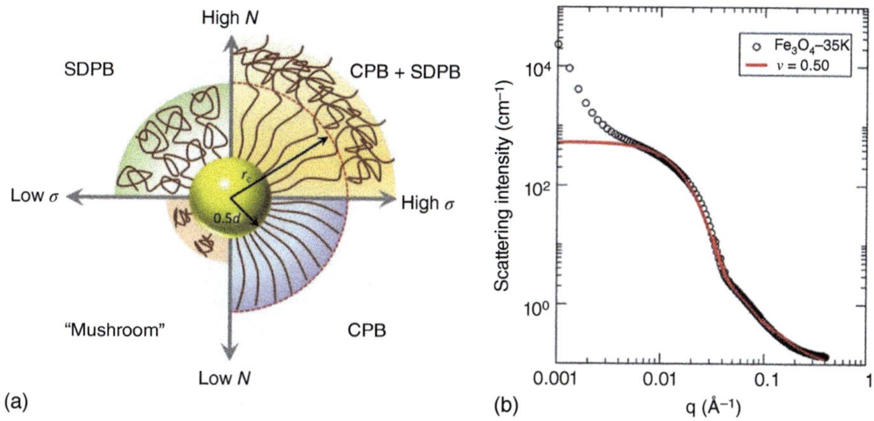

Figure 6.1 (a) Conformation of the grafted polymers at hairy NPs in good solvents. Source: Reprinted with permission from Yi et al. [1]. Copyright 2020, Royal Society of Chemistry. (b) SANS characterization of the grafted polymer conformation of PMMA-grafted Fe_3O_4 NP in PMMA. Source: Reprinted with permission from Hore et al. [2]. Copyright 2013, American Chemical Society.

and Fe_3O_4 NP gave rise to binary superlattices. Structure of the superlattices was tunable by varying number ratio of the Au NP to Fe_3O_4 NP. As shown in Figure 6.2a, various types of superlattices have been achieved, including NaCl type (1), $NaZn_{13}$ type (2), $C_{60}K_6$ type (3), and AlB_2 type (4). The grafted PS in the CPB regime is decisive to finely tune the interparticle spacing and stabilize the superlattice structure. The effects of molecular weight of the grafted polymers and the core size ratio of nanocrystals are orthogonal to tune the binary nanocrystal superlattices, implying the controllability is highly flexible.

Different from flexible linear polymers, bottlebrush polymers exhibit worm-like conformation, thus providing new opportunities to tail surface topology of the NPs. For the linear polyethylene glycol (PEG)-grafted NPs, conformation of PEG is transformed from mushroom-like to brush-like with increasing grafting density, and the adsorption of serum protein and uptake by macrophages of the hairy NPs are significantly reduced. In contrast, bottlebrush block copolymers (BBCPs) exhibit long persistence length, which is promising to tune NP topology and enhance medical performance [4]. At a polymer backbone, PEG and poly(lactic acid) (PLA) were sequentially polymerized forming the BBCPs (Figure 6.2c). By tuning the block ratio of PEG and PLA, both symmetric and asymmetric BBCPs were achieved. A triblock BBCP was derived after incorporation of another wide PEG brush at the middle region of the asymmetric BBCP. The BBCPs can be self-assembled into unique hairy NPs with a PLA core. The hairy NPs appear virus-like with the PEG brush corona. Microstructure of the PEG corona is highly dependent on PEG brush's width and block backbone length ratio of the BBCPs. The adsorption by proteins and uptake by cells are determined by microstructure and composition of the corona.

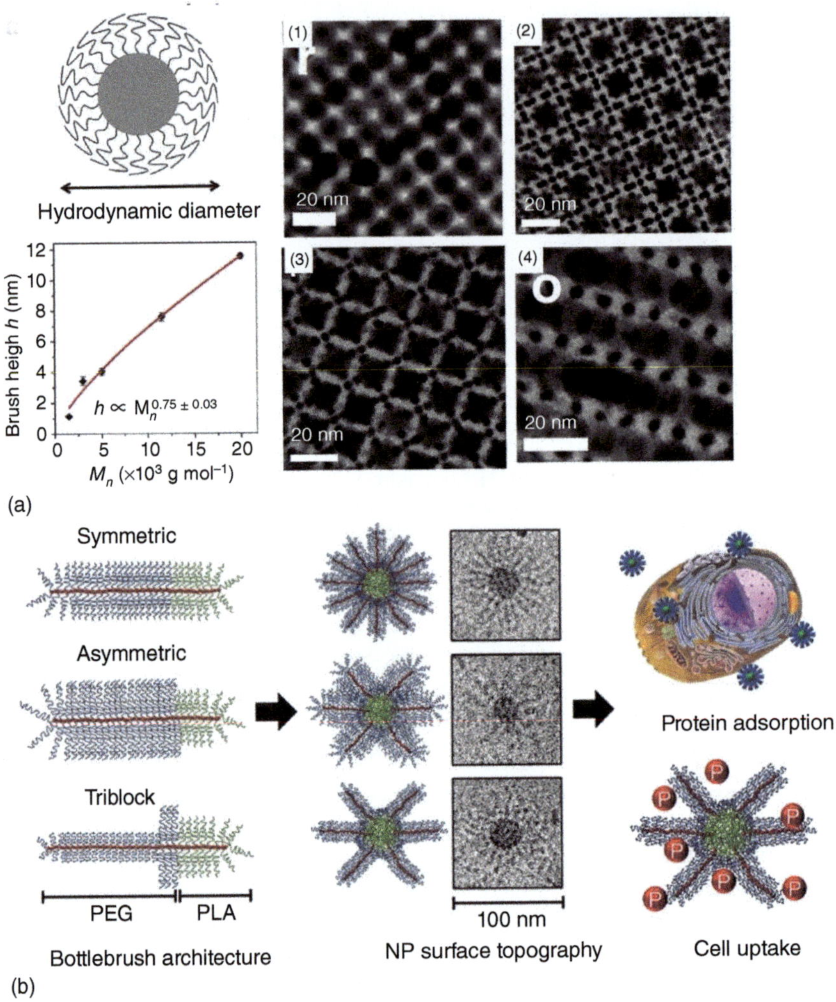

Figure 6.2 (a) Scaling law of polymer brush height (h) and molecular weight (M_n) for the PS-grafted at Au NP (left), and diverse coassembled superstructures of the PS-grafted Au NPs and Fe_3O_4 NPs (right). Source: Ye et al. [3]/Springer Nature/Licensed under CC BY 4.0. (b) Surface topographies of the polymeric bottlebrush-grafted NP and cell adsorption efficiency. Source: Adapted with permission from Grundler et al. [4]. Copyright 2021, American Chemical Society.

6.2.2 Patchy and Janus Geometry in Selective Solvents

Different from mushroom- and brush-like conformations in good solvents, microphase separation within the corona of hairy NPs occurs in poor solvents, giving asymmetric structures (Figure 6.3a) [5]. In general, patchy structures are favorable other than the core–shell structure with decrease in the NP diameter (D), while patch number (n) is increased with reduced polymer size (R) and lower molecular weight of the grafted polymers. As shown in the diagram, a patchy geometry becomes dominant over the core–shell structure at a low grafting density (σ) and small NP diameter (D). In the patchy region, the patch number (n) increases with NP diameter, which is less dependent on grafting density. The patches become small with the decrease in grafting density. In the case of three patches of NP diameter of 60 nm, the patch of 7 nm at $\sigma = 0.02$ chain nm^2 becomes smaller to 3 nm when the grafting density (σ) is decreased to 0.003 chain nm^2.

In order to avoid aggregation of NPs during formation of patches within the corona in poor solvents, it is necessary to perform the microphase separation in a dilute dispersion of the hairy NP. In the case of amphiphilic block copolymeric corona, formation of the patchy NPs can be achieved at a high solid content owing to the steric stabilization. Besides the variables of grafting length/density and the NP diameter, composition of the block copolymer corona and selective interactions with solvents are crucial to control number and spatial arrangement of the patches of the hairy NPs. The simulated phase diagram of the amphiphilic block copolymer-grafted hairy NPs in selective solvents is shown in Figure 6.3b [6]. At low chain number (f) and the solvophobic block fraction (α), only one patch is formed achieving a Janus structure. At high f and α, two patches are favored at the opposite poles. At higher f, three patches appear even at a low α. Therefore, control of number and composition of the tethered copolymers makes it easy to precisely control the valence of patchy NPs.

Patchy structures can also be achieved by surface segregation of binary polymers (Figure 6.3c) [7]. Poly(*tert*-butyl acrylate) (P*t*BA) and PS were sequentially grafted from the silica NP ($r = 10$ nm) through a difunctional initiator. At a given grafting density and molecular weight of P*t*BA ($M_n = 19.6$ kDa), molecular weight effect of PS on the microphase separation between PS and P*t*BA by solvent evaporation was examined. Multivalent patchy NPs were achieved. As molecular weight of the PS chain increases, the dark nanodomains become longer with a truncated wedge-like geometry. The compartmentalization becomes blurred with excessively increasing molecular weight of PS.

Patchy and Janus geometries featuring compartmentation can be achieved by self-assembly of block copolymers after structural evolution in selective solvents [8]. As shown in Figure 6.4a, corona-compartmentalized hairy NPs display varied microstructures of Janus (1), patchy (2), virus-like (3), and cylinders with patchy and Janus coronas (4, 5). Core-compartmentalized NPs as shown in Figure 6.4b, display richer microstructures from sphere (1, 2), disc (3), ellipsoid with a multilayer core (4), sphere with Janus and patchy cores (5–8), polymersome with compartmentalized membrane (9), and cylinder with a segregated core (10–12). Fine interplay between immiscibility of the blocks and stabilization of the soluble blocks results in the formation of various geometries and compartmentalized NPs. The sphere-on-sphere geometry (8) is taken as an example [9]. For polybutadiene-*block*-poly(2-vinylpyridine)-*block*-poly(*tert*-butyl methacrylate) (PB-*b*-P2VP-*b*-P*t*BMA), the microphase separation occurs upon feeding acetone as a poor solvent for both P2VP and PB. As P2VP and PB are strongly incompatible, PB is condensed as the core, while P2VP is divided into spherical patches at the PB core. P*t*BMA serves as the solvable brush to stabilize the sphere-on-sphere geometry.

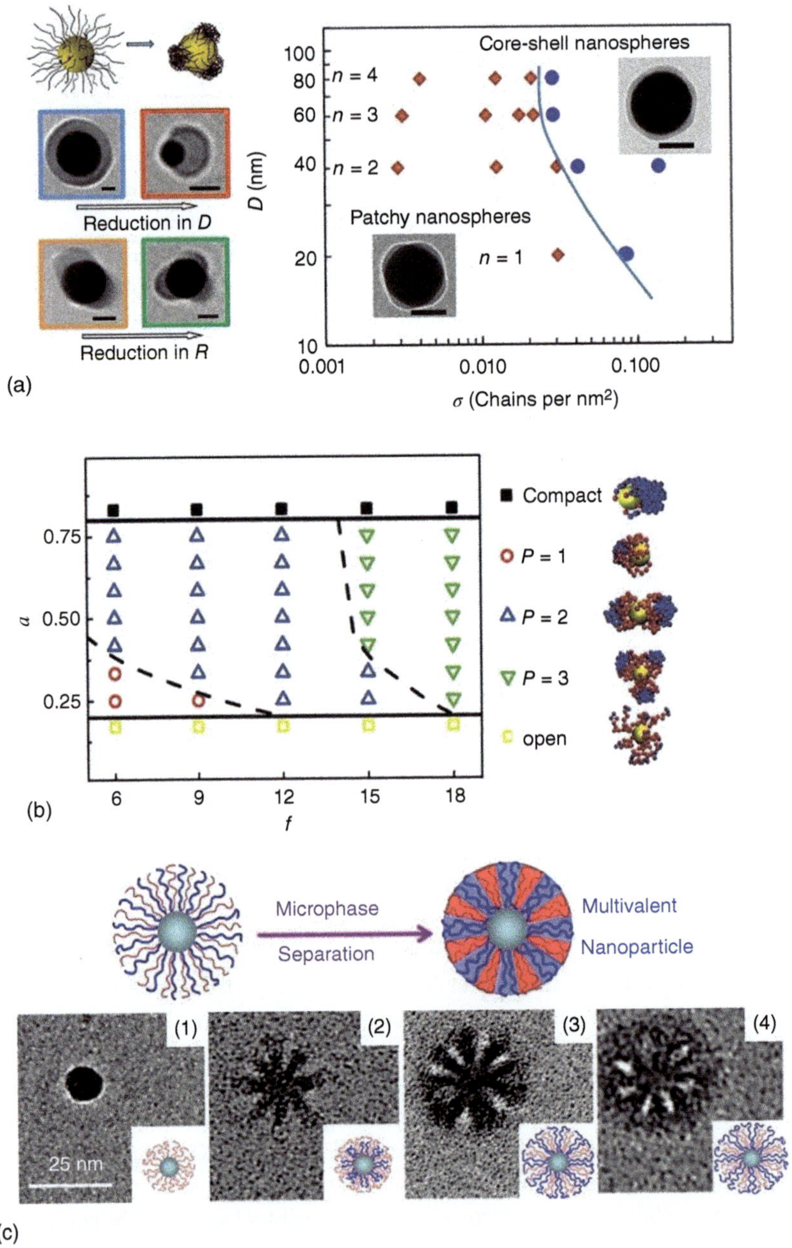

Figure 6.3 (a) Asymmetric geometries of hairy NPs in poor solvents. Scale bars: 20 nm. Source: Adapted with permission from Choueiri et al. [5]. Copyright 2016, Springer Nature. (b) Phase diagram of the amphiphilic block copolymer-grafted hairy NPs in selective solvents, and schematic structures. Source: Reprinted with permission from Zhou et al. [6]. Copyright 2017, Royal Society of Chemistry. (c) Schematic microphase separation, and TEM images of the representative patchy NPs grafting with binary polymers. Source: Adapted with permission from Bohannon et al. [7]. Copyright 2021, American Chemical Society.

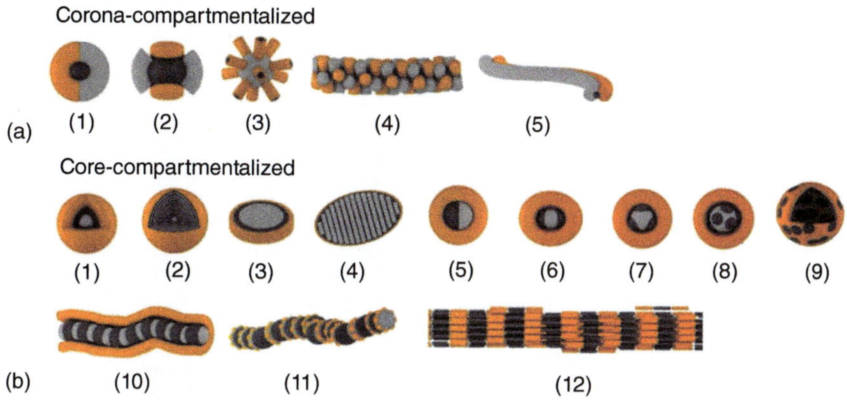

Figure 6.4 Patchy and Janus NPs derived from ABC triblock terpolymers in selective solvents. (a) Corona-compartmentalized, and (b) core-compartmentalized NPs. Source: Adapted with permission from Groschel et al. [8]. Copyright 2015, Royal Society of Chemistry.

6.2.3 Interfacial Activity as Colloids

Hairy NPs share a similar interfacial behavior as colloids with the Pickering effect to stabilize the interface between two immiscible fluids [10]. Lowered interfacial energy is responsible to resist coalescence and coarsening of the emulsion droplets. Contact angle θ is a key parameter to describe wettability of the NPs at interfaces. For a homogeneous colloid (Figure 6.5a): $\cos\theta = \frac{\gamma_{PO}-\gamma_{PW}}{\gamma_{OW}}$, where γ_{PO}, γ_{PW}, and γ_{OW} represent three interfacial tensions of colloid/oil, colloid/water, and oil/water, respectively. For $\theta < 90°$, the colloids are hydrophilic and preferentially wetted by water. For $\theta > 90°$, the colloid is hydrophobic and preferentially wetted by oil. Surface activity of the colloids is described by desorption energy. Maximum desorption energy is observed at $\theta = 90°$. Within the two windows at $0 < \theta < 20°$ and $160° < \theta < 180°$, the desorption energy is less than 10 $k_B T$, which is comparable with the characteristics of conventional surfactants [10]. Immobilization of the colloids at interfaces becomes weak, and the dynamic equilibrium of colloids between interface and aqueous phase (or oil phase) is remarkable.

In contrast to homogeneous colloids, Janus colloids feature two strictly segregated hemispheres corresponding to different polarities [11]. As shown in Figure 6.5b, the parameter of angle α is used to describe the position of the surface boundary between apolar and polar regions of the colloid, and β denotes the immersion angle of the colloid at the oil/water interface. Owing to the asymmetric characteristic, Janus colloids are amphiphilic. The Janus balance is determined by α and contact angles of polar area (θ_P) and apolar area (θ_A). The maximum amphiphilicity is attained at $\alpha = 90°$ and $\Delta\theta = \theta_P - \theta_A = 180°$. As the amphiphilicity is progressively tuned from $\Delta\theta = 0$ (in the case of homogeneous colloid) to 90°, the desorption energy increases significantly (Figure 6.5b). This implies that the Janus colloids are more effective to stabilize emulsions due to the high desorption energy [11].

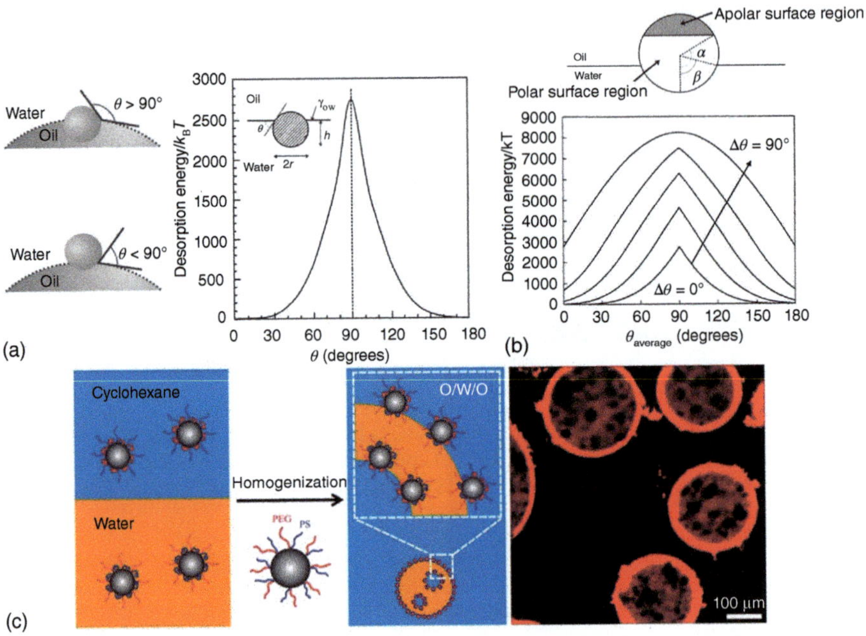

Figure 6.5 (a) Wettability of homogeneous NPs at interfaces (left), and dependence of the desorption energy at planar oil/water interfaces on contact angle θ (right). Source: Adapted with permission from Boeker et al. [10]. Copyright 2007, Royal Society of Chemistry. (b) Janus NPs at interfaces (left), and dependence of the desorption energy of Janus NPs at planar oil/water interfaces on contact angle θ (right). Source: Adapted with permission from Binks and Lumsdon [11] Copyright 2001, American Chemical Society. (c) Emulsification by the hairy NPs forming multiple emulsions. Water phase was dyed with Rhodamine B for easy observation. Source: Reproduced with permission from Liu et al. [12]. Copyright 2016, American Chemical Society.

Wettability of hairy NPs is essentially determined by composition, molecular weight, and architecture of the grafted polymers. As an example, hairy NPs grafting with amphiphilic polymers exhibit tunable surface wettability (Figure 6.5c) [12]. Specifically, ABC miktoarm star terpolymers consisting of PEG, PS, and poly[(3-triisopropyloxysilyl)propyl methacrylate] (μ-PEG-*b*-PS-*b*-PIPSMA) were grafted on the silica NP surface. The PS chains are freely flexible, while PEG chains are collapsed when the hairy NPs are initially dispersed in oil. A W/O emulsion is favored upon feeding water. An O/W emulsion forms upon feeding oil when the hairy NPs are initially dispersed in water. When using the cosolvents such as toluene for both PS and PEG, W/O emulsions are favored regardless existence history of the hairy NPs. Wettability of the silica NPs grafting with the amphiphilic miktoarm star polymers can be further tuned by host–guest complexation between PEG block and α-CD, leading to one-step formation of O/W/O multiple emulsions.

6.3 Hairy NPs Assembly at Various Interfaces

6.3.1 Dispersion in Polymer Nanocomposites

Hairy NPs are widely used as functional fillers in polymer nanocomposites. As the shielding layer, the grafted polymers usually bear the same chemical components as the matrix leading to athermal mixing ($\chi \approx 0$). However, this is insufficient to ensure single NP level dispersion since the unfavorable entropic surface tension always exists. In the athermal state, the entropic effect should be considered, which is closely related to graft density (σ) and molecular weight ratio (α) of brush polymer to matrix polymer. As shown in Figure 6.6, the diagram of PS-grafted SiO_2 NPs dispersed in PS matrix displays very rich morphologies covering spherical aggregates, connected sheets, strings, and dispersed states [13]. A good dispersion state of hairy NPs is achieved at a long graft length ($\alpha > 1$) with high graft density ($\sigma \sim 0.1$ chains nm^{-1} [2]). For a short graft length ($\alpha < 1$), aggregation occurs at low graft density σ. Interestingly, hairy NPs are interconnected into sheets or strings due to the Janus characteristics of hairy NPs at moderate σ and α. Specifically, the incompatibility between the SiO_2 core and the grafted PS chain drives the interfacial phase separation in analogy to microphase separation of block copolymers. As a result, polymeric corona is squeezed out from the NP core and allows threading to form the sheet-like or string-like structure. The assembled anisotropic structures of percolating particle sheets have been employed for improved mechanical properties of nanocomposites [14].

Different from melt state of nanocomposites with percolated structure of NPs preferred, single NP level dispersion in a glassy state is required for enhanced mechanical performances (Figure 6.7a) [13]. For glassy nanocomposites composed of PS-grafted silica NPs and PS matrix, sufficiently long chain (or large α) for the grafted polymer facilitates uniform dispersion and strong interfacial binding of the NPs with the matrix by entanglement. Both strength and toughness of glassy nanocomposites have been significantly enhanced. This unique performance is advantageous over traditional approaches with compromised toughness and modulus. Bimodal-grafted polymers of long and short lengths help excellent dispersion in the polymer matrix owing to delicate balance between enthalpic and entropic interfacial interactions [16]. Long chains are responsible for entanglement, while short chains are responsible for screening the attraction among particles. Tailoring graft density and molecular weight of the grafted polymers is highly effective to tune the decoupled entropic and enthalpic effects. Metal–organic framework (MOF) NPs were used as fillers to enhance both ductility and toughness of nanocomposites (Figure 6.7b) [15]. Poly(alkyl glycidyl ether)-grafted MOF NPs (~50 nm) can be well dispersed in PMMA. In the case of a longer grafting chain beyond the entanglement molecular weight of PMMA, fracture behavior of the PMMA/MOF nanocomposites is transformed from large crazes to massive fine crazing to shear banding. Ductility and modulus of nanocomposites are thus synchronously improved.

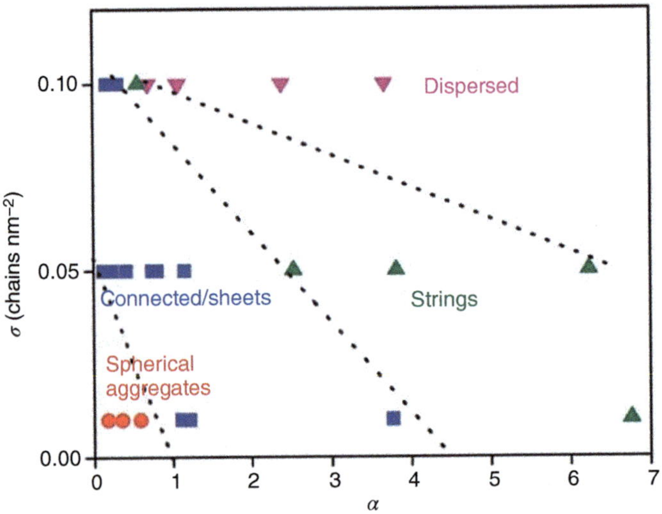

Figure 6.6 Morphology diagram of the PS-grafted SiO_2 NPs dispersed in PS matrix after annealing at 150 °C for 5 days, showing the effects of graft density (σ) and molecular weight ratio (α) of brush polymer to matrix polymer. Source: Reprinted with permission from Maillard et al. [13]. Copyright 2012, American Chemical Society.

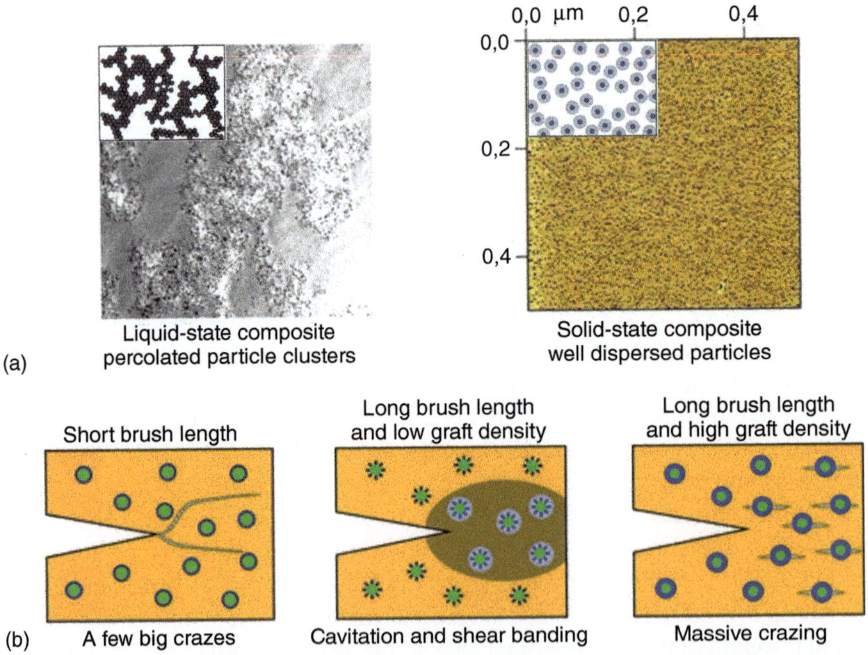

Figure 6.7 (a) Mechanical properties for nanocomposites: percolated clusters of NPs for melt state (left) and well-dispersed NPs for solid-state (right). Source: Adapted with permission from Maillard et al. [13]. Copyright 2012, American Chemical Society.
(b) Fracture behavior for polymers filled with hairy MOF as a toughening agent. Source: Reprinted with permission from Liu et al. [15]. Copyright 2020, American Chemical Society.

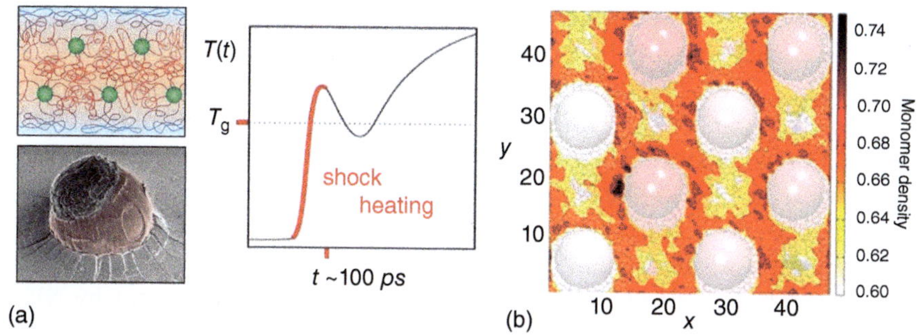

Figure 6.8 (a) Thin film of hairy NP nanocomposite under projectile impact. Source: Reprinted with permission from Hyon et al. [17]. Copyright 2021, American Chemical Society. (b) Heatmap of monomer density for matrix-free hairy NPs in PMA-grafted nanocomposites and neat PMA. Source: Reprinted with permission from Bilchak et al. [18]. Copyright 2017, American Chemical Society.

Self-assemblies of hairy NPs without matrix are promising for high energy absorption under projectile impact since load can be effectively transferred between canopy-entangled chains and from grafted polymers to the NPs (Figure 6.8a) [17]. Upon projectile impact, temperature at the impact site increases rapidly, leading to initial shock compression followed by longer-term viscoplastic heating of the deforming polymer melt. During melt drawing of the film, the distribution of entanglement density and entanglement number per chain of the grafted polymer effectively controls chain mobility near a free surface, which is crucial to acquire high energy absorption. Structurally, each NP acts as a giant crosslinking node to couple its surrounding nodes. The covalently anchored polymer chain at NP surface provides a mechanical network that supplements the entanglement network from the node–node NP interactions via entangled canopies. PS-grafted silica and Fe_3O_4 NPs with different grafting lengths were exemplified. For short PS chains, canopy interdigitation and entanglement are limited giving brittle films. A tougher film is achieved when long PS chains are highly entangled with grafted NP nodes.

Hairy NPs can form self-assemble into ordered arrays in the absence of matrices. Packing of the hairy NPs leaves a large amount of free volume, which is promising for gas separation [18]. As shown in Figure 6.8b, the free volume is indicated by the low monomer density in the simulated "heatmap." Compromise between NP ordering and chain entropy generates distortion of the hairy corona with increased free volume. The free volume can be greatly tuned by using the hairy NPs with varied grafting densities and chain lengths of polymers.

6.3.2 Anisotropic Assembly

Anisotropic directional assembly can be achieved by using either patchy/Janus NPs or homogeneous NPs. As discussed in Section 6.2.3, patchy and Janus geometries can be achieved by phase segregation of grafted polymers in selective solvents. Triblock terpolymers as grafts are advantageous to enhance colloidal stability and synthetic yield of anisotropic NPs [19]. The example terpolymer of poly(ethylene oxide)-*block*-poly(acrylic acid-*r*-styrene)-*block*-polystyrene (PEO_{45}-*b*-P[AA_{α}-*r*-$St_{1-\alpha}$]$_y$-*b*-PS_x-SH) contains a hydrophilic block and a hydrophobic block at the ends, and a random copolymer block in the middle. In the mixture solvent of DMF/H_2O, the random copolymer block of P(AA_{α}-*r*-$St_{1-\alpha}$)$_y$ serves as a buffering region between the hydrophilic PEO corona and hydrophobic PS core of the micelle (Figure 6.9b). Dynamic exchange of terpolymers at the NP surface and in the solution enables reorganization of terpolymers at the NP surface. Upon regulating fraction of PS block and water content in the mixture solvent, Janus/patchy NPs with varied patch numbers and morphology can be derived. The patchy/Janus NPs are capable to lead anisotropic assembly. Janus NPs can self-assemble into petal-like clusters, which were restricted by electrostatic repulsion and steric interaction between patches at neighboring NPs. Three-patch NPs can self-assemble into a chain-like structure, which is triggered by complexation between Co^{2+} and PAA segments.

For homogeneously grafted NPs, interplay of the attractive interaction between the NPs with different grafting polymers and local repulsion at the NP surface leads to anisotropy assembly of the NPs [20]. Poly(ethylene oxide)-*block*-(acrylic acid-*r*-styrene) [PEO-*b*-P(AA-*r*-St)] was grafted at a NP denoted as NP-A, and poly(ethylene oxide)-*block*-poly(*N*,*N* dimethylaminoethyl methacrylate-*r*-styrene) [PEO-*b*-P(DMAEMA-*r*-St)] was grafted at the NP denoted as NP-B. After protonation of the acid group at NP-A, the NP became negatively charged. After mixing NP-A and NP-B, the acid/base interaction from NP-A and NP-B is accessible owing to the chain flexibility. The Coulombic repulsion originating from charged groups allows the constrained assembly leading to directional bonding between the NPs. Number of bonds between NP-A and NP-B is regulated in a self-limiting manner owing to stoichiometry and reversibility of the reaction between the ligands. The anisotropic assembly of colloids can be used to mimic the construction of molecules through covalent bonding of atoms.

As discussed in Section 6.3.1, the PS-grafted silica NPs with moderate σ and α can assemble into strings in PS matrix [13, 14]. Incompatibility between the SiO_2 core and the grafted PS chain drives the interfacial phase separation and further anisotropy assembly. This phenomenon was exemplified by incorporation of homogeneously grafted NPs in incompatible polymers [21]. The unfavorable enthalpic interaction between the grafted NPs and the matrix drives the anisotropic assembly. Poly(vinyl pyrrolidone) (PVP) of different molecular weights and (11-mercaptoundecyl)tetra(ethylene glycol) (PEG) were grafted at the silver nanocubes with an edge length of 80 nm. The cubic shape is an interesting geometry to construct non-closely packed architectures via different coordination at facet, corner, and edge sites. When the hairy silver nanocubes are embedded in PS matrix, incompatibility between PS matrix and the grafted PVP or PEG drives the assembly of silver nanocubes. Molecular weight of the grafted polymers dictates the orientation of silver nanocubes due to steric effect (Figure 6.9c). Relatively long PVP chains prefer to graft at such surfaces (e.g. edge or vertex) with a high curvature. Interpenetration of polymers in the edge-to-edge junction mode is favored. Short

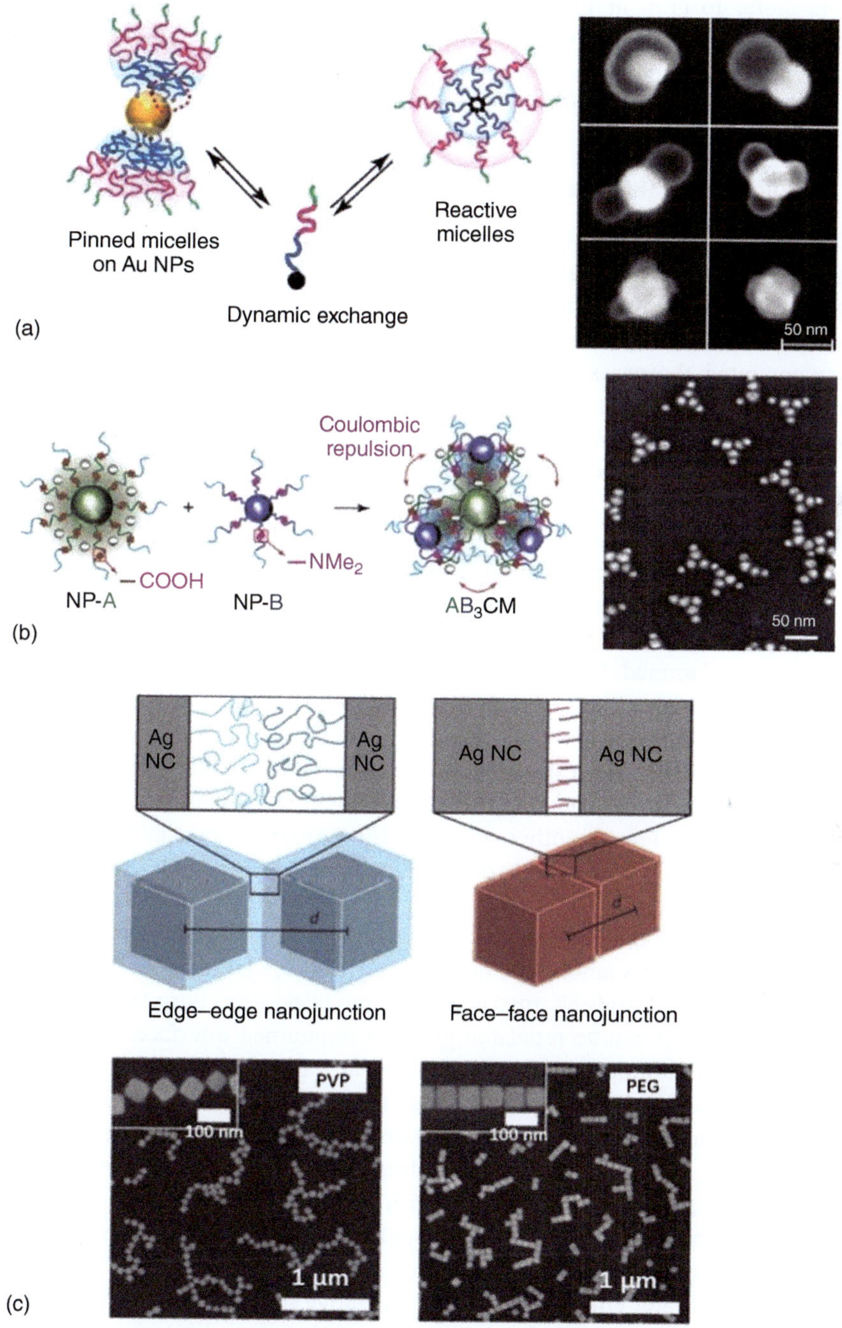

Figure 6.9 (a) Patchy and Janus hairy Au NPs. Source: Reprinted with permission from Yang et al. [19]. Copyright 2021, American Chemical Society. (b) Formation of AB_3 structure by directional assembly of NP-A and NP-B. Source: Reprinted with permission from Yi et al. [20]. Copyright 2020, The American Association for the Advancement of Science. (c) Assembly of hairy Ag nanocubes of different grafted polymers with varying lengths. Source: Adapted with permission from Gao et al. [21]. Copyright 2012, Springer Nature.

polymers prefer to graft at the plate surface of the nanocubes, which benefits face-to-face assembly. The orientation control of plasmonic nanocubes determines optical properties of the resultant superstructures. Similarly, face-to-face assembly of DNA-grafted nanocuboid can be tailored by multilinking bonds [22].

6.3.3 Liquid–Liquid Interfaces

Liquid–liquid interfaces are widely utilized to construct 2D and 3D confined spaces, enabling assembly toward superlattices and synthesis of Janus NPs [23], interfacial catalysis, and oil–water separation. 3D confined emulsion droplets are used to coassemble binary hairy NPs forming structured colloidosomes. Zhu et al. [24] demonstrated the coassembly of binary PS-grafted Au NPs into core–shell colloidosomes in emulsion droplets. The self-assembly of PS-grafted Au NPs occurs at chloroform/water emulsion interface. Interestingly, the NPs-grafted with a long chain PS preferentially locate in the core, while the NPs-grafted with a short PS are distributed in the shell. For the colloidosome as shown in Figure 6.10a, two Au NPs of 8 and 15 nm are distributed in the core and shell, respectively (i). On the other hand, 15 nm Au NPs are located in the core and 8 nm Au NP NPs are at the shell when they are grafted with long and short chains (ii), respectively.

Hairy nanoplates with amphiphilicity and zwitterionic characteristics are capable to form jammed layers at water/oil emulsion interfaces. Cheng et al. [25] developed pH-responsive interfacial jammed emulsion droplets by using zwitterionic Janus nanoplates (Figure 6.10b). The disk-like kaolinite NP was covalently grafted with pH-responsive polymer of poly[2-(dimethylamino)ethyl methacrylate] (PDMAEMA) on one side and hydrophobic polymer of poly(lauryl methacrylate) (PLMA) on the other side. Emulsion droplets are stabilized by the amphiphilic Janus nanoplates. The kaolinite NP is negatively charged, while grafted PDMAEMA becomes positively charged by protonation. At $pH = 7$, PDMAEMA is partially protonated and the nanoplates become zwitterionic, and a strong electrostatic attraction occurs among the nanoplates giving the jammed interface. The nanoplates become positively charged at $pH = 3$, and negatively charged at $pH = 12$. The interparticle electrostatic repulsion gives an unjammed interface. Therefore, the jammed and unjammed states of nanoplates at interfaces can be switched by changing surface charge, which is analogous to protein-based ion channels that can open and shut to control molecular transport. This design can be used to control interfacial molecular diffusion and drug release. Detailed introduction on interfacial jamming will be given in Section 6.5.

Coassembly of block copolymers (BCPs) and grafted NPs in emulsion droplets leads to diverse geometries with precise compartmentalization (Figure 6.10c) [26]. In the chloroform/water emulsion droplet, cooperative self-assembly of PS-*b*-P4VP and PS-grafted Au NPs occurs. After solvent evaporation, pupa-like, pinecone-like, bud-like, and Janus-like geometries with precise locations of Au NPs are obtained. The addition of PS chains to the mixture of BCPs and PS-grafted Au NPs can further regulate location of the Au NPs due to enthalpic interactions. As shown in Figure 6.10c, Au NPs are migrated to the surface of BCP aggregate and the

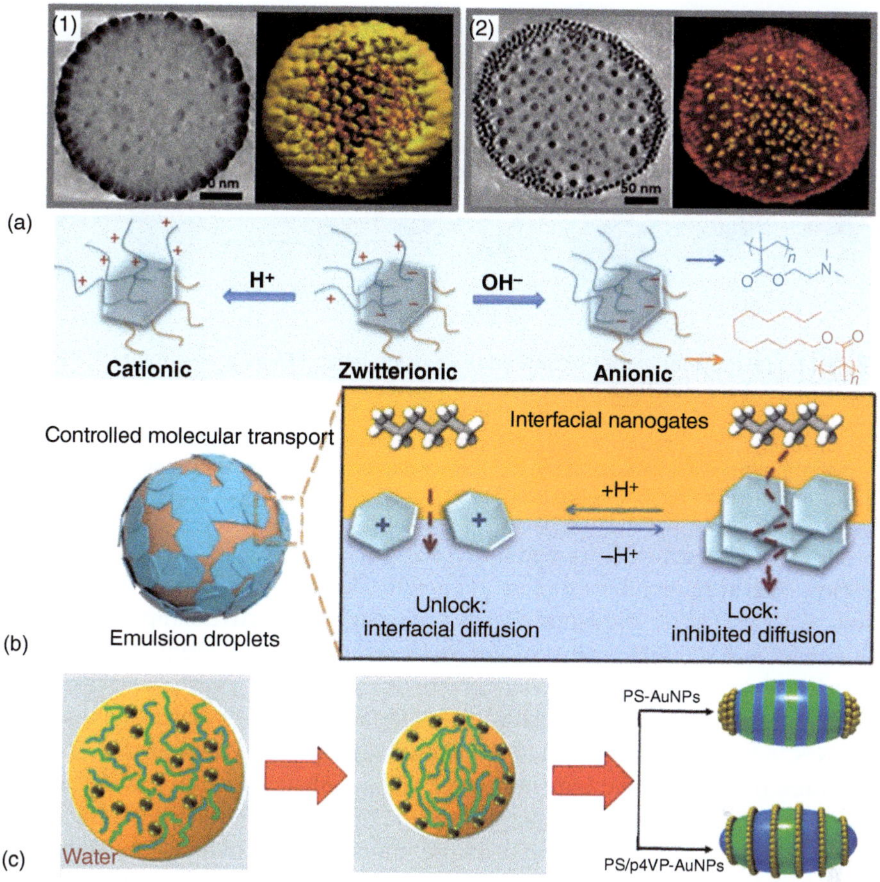

Figure 6.10 (a) Core–shell colloidosomes assembled by binary hairy Au NPs of 8 and 15 nm. Core of 8 nm NPs and shell of 15 nm NPs (1); core of 15 nm NPs and shell of 8 nm NPs (2). Source: Adapted with permission from Yang et al. [24]. Copyright 2019, Royal Society of Chemistry. (b) pH-responsive nanoplatelets and the jammed interfaces. Source: Adapted with permission from Luo et al. [25]. Copyright 2018, John Wiley and Sons. (c) Hybrid NPs by coassembly of BCPs and Au NPs in emulsion droplets. Source: Adapted with permission from Yan et al. [26]. Copyright 2017, American Chemical Society.

positioning can be precisely controlled by the entropic repulsion and the enthalpic attraction between the PS ligands on AuNPs and surfactants.

Incorporation of Janus hairy NPs in immiscible polymers can be achieved by melt extrusion processing. The hairy Janus NPs are promising as functional compatibilizers owing to combinational effects of solids and copolymers. An example Janus NP (40 nm) was used as a compatibilizer for the blend of poly(2,6-dimethyl-1,4-phenylene ether) (PPE) and poly(styrene-*co*-acrylonitrile) (SAN) (Figure 6.11a) [27]. The Janus SBM is composed of crosslinked PB as the core and flexible PS and PM as flexible brushes derived from polystyrene-*block*-polybutadiene-*block*-poly(methyl methacrylate) (SBM) triblock terpolymer. The Janus NP possesses a crosslinked PB core and equally sized PS/PMMA hemispheres.

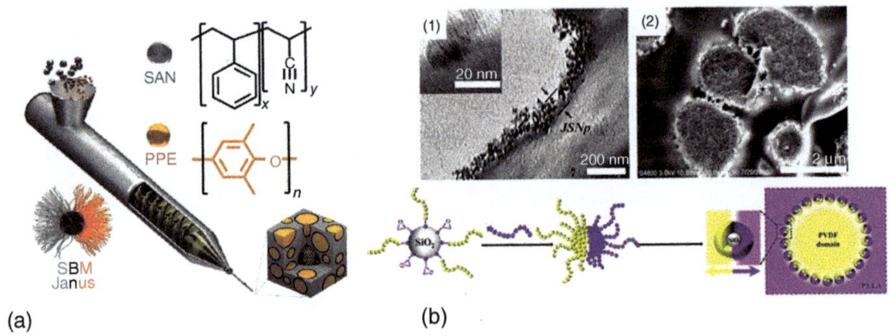

Figure 6.11 (a) Polymer blends of SAN and PPE using Janus NPs as compatibilizers. PPE droplets (yellow) are stabilized by the Janus NPs (black dots) within the SAN matrix (gray) during melt extrusion. Source: Adapted with permission from Bahrami et al. [27]. Copyright 2014, American Chemical Society. (b) TEM images of the PVDF/PLLA (50/50) blend demonstrating interfacial encapsulation. Source: Adapted with permission from Wang et al. [28]. Copyright 2017, American Chemical Society.

A raspberry-like structure is formed by densely packing the Janus NPs at the blend interface. The average melt droplets are greatly decreased to 300 nm by feeding 10 wt% of the Janus NP. The Janus NP has demonstrated effective compatibilization with exceptional stability against extensive shear and temperature profiles during industrial extrusion processing.

Positioning particulate compatibilizers at interfaces is achieved by using reactive Janus NPs toward processing immiscible melts [28]. The reactive Janus NPs should contain reactive groups and long molecular tails for ideal compatibilization effect. Epoxide group and PMMA are grafted onto the opposite sides of SiO_2 NP achieving the reactive Janus NP. In the example immiscible poly(vinylidene fluoride) (PVDF)/polylactic acid (PLLA) blend, the reactive Janus NPs are exclusively immobilized at the interface bridging the two immiscible phases after melt blending (Figure 6.11b). PLLA was in situ grafted onto the Janus SiO_2 NP by reaction of the residual carboxylic acid with epoxide group at the interface. Grafted PMMA at the Janus NP forms entanglement with PVDF matrix. Coalescence of PVDF domains is effectively suppressed by the reactive Janus NPs to enhance interfacial adhesion via chemical bonding. Morphology of the blend is stabilized by the reactive Janus NPs, which is preserved under strong shearing and annealing in the melt state. Reactive Janus NPs are important to attain high-performance multicomponent and multiphase materials by traditional processing methods.

Zubarev et al. [23] reported the breath-figure (BF) templated assembly of PS-grafted Au nanorods into rings from the Au $NR(PS)_n$ dispersion in CH_2Cl_2 at varied concentrations (Figure 6.12). At the chloroform/H_2O interface, evaporation of CH_2Cl_2 leads to condensation of water droplets from moisture, which reside on top of the PS-grafted Au nanorod dispersion. The water droplets serve as a template to induce formation of rings, while the nanorods are concentrated at the water droplet circumference. Rings were eventually formed after complete evaporation of the solvent. Diameter of the rings differs from 300 nm to microns but with a similar width of 50 nm. Orientation of the nanorods was random in most rings.

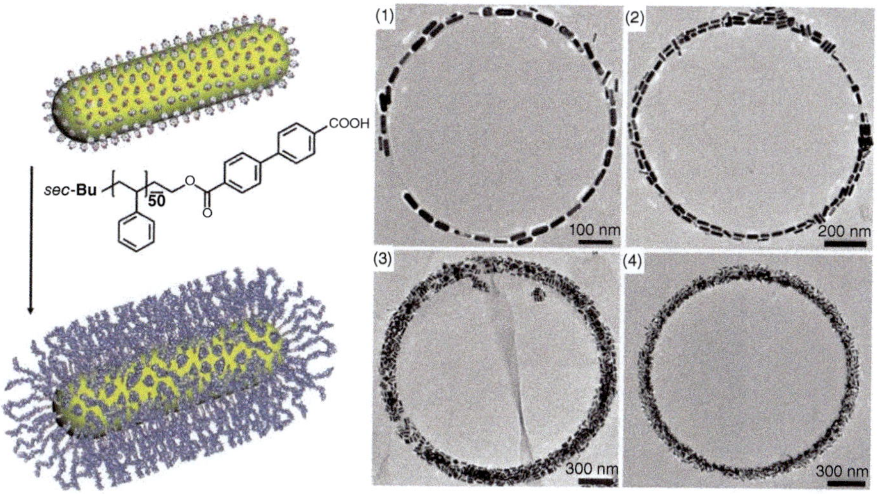

Figure 6.12 Schematic Au NR(PS)$_n$ and TEM images of different rings from the Au NR(PS)$_n$ dispersion in CH_2Cl_2 at varied concentrations. Source: Adapted with permission from Khanal et al. [29]. Copyright 2007, John Wiley and Sons.

A head-to-tail lining fashion was observed at low concentrations. Interparticle distance and alignment of the nanorods were also tailored by grafting PS with varied molecular weight and architecture.

6.3.4 Air–Solid Surfaces

Silica microcapsules consisting of corrosion inhibitors in the core and fluorescent moieties in the shell were used for coating onto copper substrates for anticorrosion protection and corrosion sensing. The shell displays emitting fluorescence at the damaged area, and the warning signal will last until a complete remedy is found. Upon alteration of pH, the coating becomes colored and the inhibitor is released for anticorrosion (Figure 6.13a) [30]. PVF nanofibers can be integrated to enhance the coating's strength and function. The rational design of functional microcapsules brings synergetic effects of good anticorrosion and sensing performance at air–solid interfaces.

Antifouling polymer coatings are crucial for medical devices and biosensors. On a prestressed PS film, Au NPs were sputtered to form a thin layer. Antifouling polymers (thiol-terminated poly(carboxybetaine) (PCB) were grafted on the Au NP layer for the antifouling coating. At high temperatures above the glass transition temperature of PS, the PS film shrunk, resulting in the nanoscale wrinkled structure with crowded PCB-grafted Au NPs. The wrinkling pattern is influenced by molecular weight of the polymers and hydration degree. Importantly, the nanoscale wrinkled structure brings highly sensitive localized surface plasmon resonance (LSPR) for detection of biotin–avidin and desthiobiotin–avidin complexations. Enhanced antifouling performance and generation of LSPR-active surface are thus simultaneously achieved (Figure 6.13b) [31].

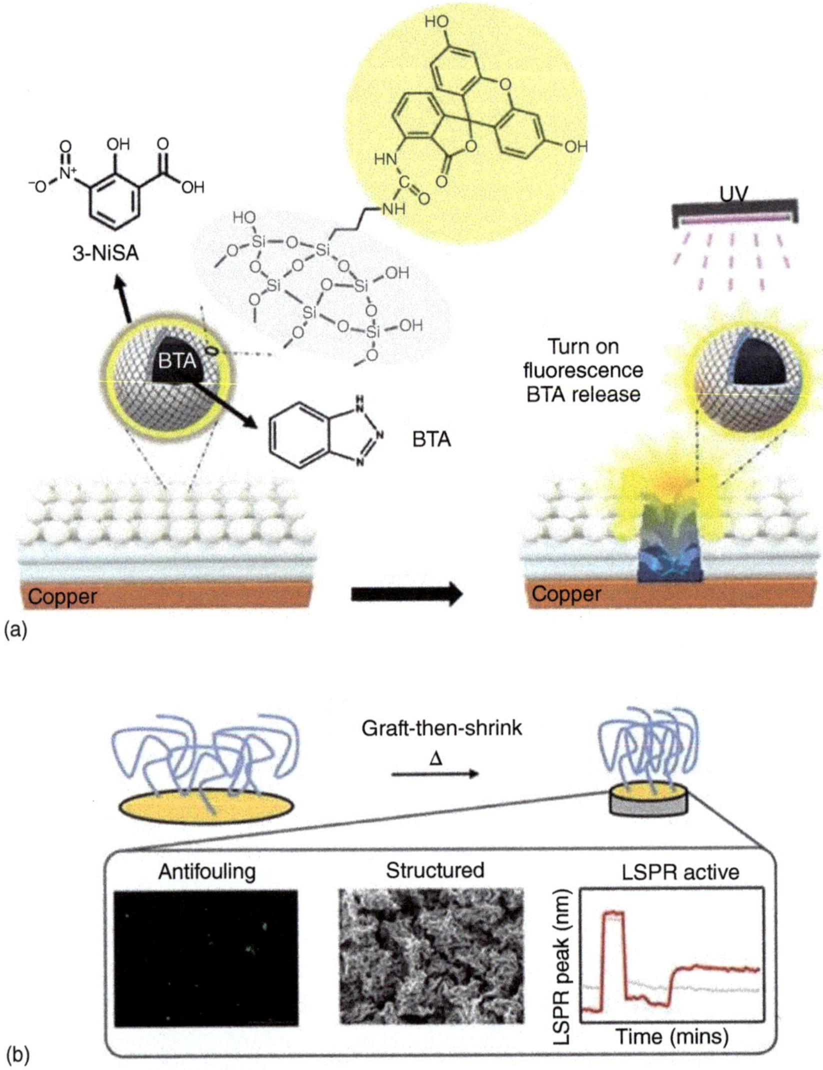

Figure 6.13 (a) Nanostructured coatings with sensing performance. Source: Adapted with permission from Salaluk et al. [30]. Copyright 2021, American Chemical Society. (b) Antifouling interfaces and localized surface plasmonic resonance biosensor. Source: Adapted with permission from Jesmer et al. [31]. Copyright 2021, American Chemical Society.

6.3.5 Air–Liquid Surfaces

Air–liquid interfaces are widely used to construct 2D assemblies, such as nanosheets and superlattices. For example, the lipid monolayer facilitates the adsorption of NPs at the liquid–air interface by electrostatic interaction. The interaction of DNA-grafted NPs with lipid membranes has gained growing interest for encapsulation of drugs and genes. Gang et al. [32] investigated the assembly of DNA-grafted

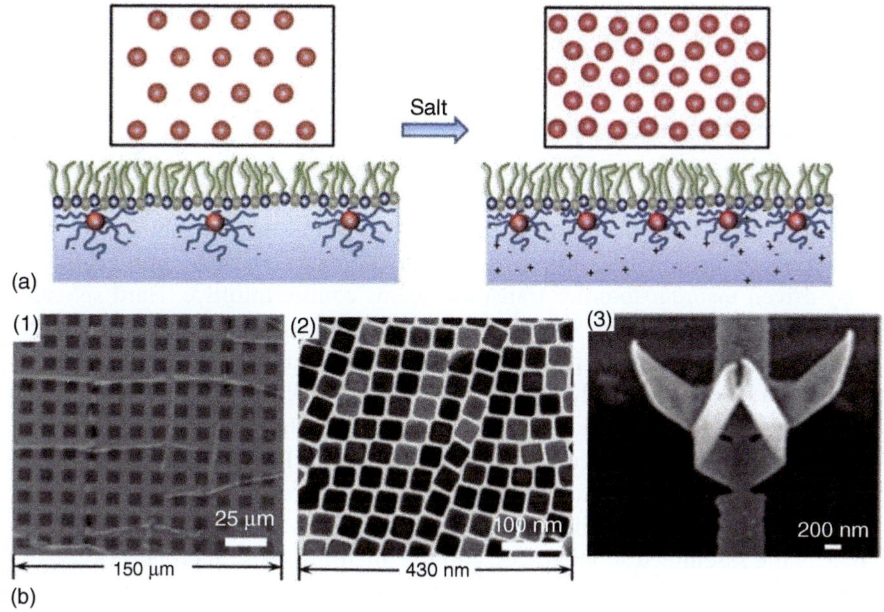

Figure 6.14 (a) DNA-grafted Au NPs, and assembly on a lipid layer at air–water interfaces with varied salt concentrations in the aqueous phase. Source: Reprinted with permission from Srivastava et al. [32]. Copyright 2014, American Chemical Society. (b) SEM (1) and TEM (2) images of the nanosheets, and SEM image (3) of a "flying bird" fabricated from the nanosheet. Source: Reprinted with permission from Yang et al. [24]. Copyright 2014, American Chemical Society.

Au NPs on the lipid layer at water–air interfaces. The grafted DNA chains are negatively charged, while the lipid layer is positively charged. The strong electrostatic interaction drives the formation of 2D hexagonally closely packed Au NP lattices at water–air interfaces (Figure 6.14a). Similar to the cocrystallization of hairy NPs at liquid–liquid interfaces as described in Section 6.2.1, the interparticle distance is tunable by changing conformation and size of the grafting polymers. The interparticle distance of DNA-grafted NPs at interfaces is sensitive to ionic strength in the aqueous phase. DNA conformation becomes less stretched with decreased persistence length at a high ionic strength by feeding salt in the aqueous phase, leading to a more densely packed superlattice.

2D superlattices of PS-grafted Au@Ag NPs were achieved by assembly at air–liquid interfaces (Figure 6.14b) [33]. The Au@Ag NPs of 20–30 nm contained the PS of a molecular weight of 50 kg mol^{-1}. Upon rapid evaporation of the solvent chloroform, patchy monolayers of the NPs were initially achieved, which can be further fused into a giant nanosheet with a lateral dimension of ~3 mm. Thickness of the monolayer and nanosheet kept the same at 40 nm. The free-standing nanosheets exhibit strong plasmonic resonance. Young's modulus of the nanosheets is high up to ~1 GPa. By processing with focused ion beam (FIB) lithography, the nanosheets can be folded into various 3D origami structures, for example, a "flying bird," which is promising for manipulating nanoscale objects.

6.4 Interfacial Entropy

Entropy (S) is defined as $S = k_B \ln W$, where W and k_B represent the number of states accessible to the system and Boltzmann's constant, respectively. Disordered phenomena are usually entropy favored due to increased state number (W). Entropy-driven disorder-to-order transition seems counterintuitive. Hard spheres exhibit random packing at low density. Ordering of the hard spheres occurs at high density since the hard spheres gain larger free volume and more entropy than their disordered states. Consequently, a disorder-to-order transition is driven when density of the hard sphere fluid exceeds a critical point. Similarly, entropic effect plays a vital role in determining interfacial behavior of hairy NPs. In particular, the conformational entropy of the grafted polymers is closely correlated with the formation of patchy geometries, spatial distribution of the NPs at interfaces, the interparticle distance of the assembled superlattices, and the micromechanical properties.

Hairy NPs embedded in thin films suffer a great entropic penalty due to the decreased number of states of grafted polymers in confined spaces. For PS ($M_n = 11.5\,kg\,mol^{-1}$) grafted Au NPs (12 nm) in a PS matrix ($2.8\,kg\,mol^{-1}$), the NPs were selectively distributed in the channel-like films with varied thicknesses (Figure 6.15a) [34]. The channel-like film consisted of alternating mesa with a thickness of $h_1 \approx 140$ and trench with a thickness of $h_2 \approx 20\,nm$. The PS-grafted Au NPs were exclusively segregated in mesas (dark strips in TEM image) of the patterned film. In contrast, the PS-grafted Au NPs were homogeneously distributed in PS matrix with smooth surface topography. Due to the conformational entropy loss of grafted chains at the Au NPs confined in ultrathin trenches, the NPs preferred to locate at the mesa areas with larger thicknesses. Such preferential segregation of hairy NPs in athermal blends is highly dependent on the confinement degree of (i) brush chains in trenches ($h_{brush}/h_{confine}$), (ii) matrix polymer chains trapped between gold cores and trench walls ($R_g/h_{confine}$), and (iii) matrix polymer chains confined between top and bottom trench walls (R_g/h_2). Variation of the matrix polymer chain size (R_g) and the trench thickness (h_2) allows to tune segregation of the PS-grafted Au NPs in the patterned mesa-trench regions. By varying molecular weight of the matrix PS from 2.8 to $360\,kg\,mol^{-1}$ and the film thickness from 85 nm to 140 nm, segregation of the NPs is greatly tunable under the channel-pattern confinement. The transition from weak to strong confinement is mediated by relative size of the grafted chain and matrix polymer ($h_{brush}/2R_g$).

Mechanical compression can bring entropic penalty for the hairy NPs in a polymer matrix [35]. In the binary mixtures of Janus NPs grafting with two ligands with varied lengths, two microdomains are formed at a fluid–fluid interface after a relatively strong phase separation (Figure 6.15b). Lateral compression can give rise to deformation of the grafted polymers at interfaces into planar brush-like, achieving a great loss in conformation entropy. Consequently, the transition is driven from random state of NP distribution to a long-ranged intercalation state.

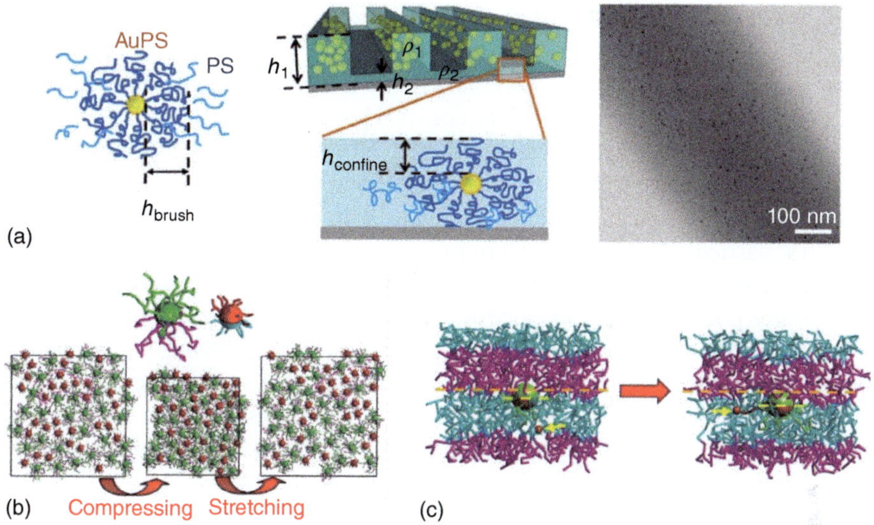

Figure 6.15 (a) Distribution of PS-grafted Au NPs in a polystyrene thin film under the channel-patterned confinement. Source: Reprinted with permission from Zhang et al. [34]. Copyright 2017, National Academy of Science. (b) Entropy-mediated mechanical response of Janus hairy NPs under compression and stretching. Source: Reprinted with permission from Liu et al. [35]. Copyright 2014, American Chemical Society. (c) Self-assembly of the Janus NPs at interfaces in the presence of block copolymers, which is controlled by entropic effect and reaction. Source: Adapted with permission from Chen et al. [36]. Copyright 2017, American Chemical Society.

Such transition provides extra space to adapt the deformation of long chains, and the steric repulsion among brushes is reduced with an increment of conformational entropy for the grafted polymers. In addition, the transition between random mixing and long-ranged intercalation states is reversible, which can be tuned by either compression or stretching. In the stretching process, mixing parameter (φ) turns to decrease as the interfacial area increases. After cessation of the stretching, φ returns to the original value.

Yan et al. have recently simulated the entropy effect in polymerization-induced interfacial self-assembly [36]. Janus NPs anchored at block copolymer microdomain interfaces will reorganize after polymerization with monomers in the preferential phase domains. Morphological evolution of the nanocomposites (Figure 6.15c) reveals the off-center distribution of Janus hairy NPs with respect to the phase interface during the polymerization. The off-center distribution causes an unfavorable enthalpic interaction, while a considerable fraction of the opposite site of the Janus NPs faces toward the chemically incompatible phase domain. However, the off-center distribution brings a significant entropy gain by disrupting the organization of block segments around the NPs. Therefore, Janus NPs are prone to migrate from the interfaces to the domains where the reaction may occur. The unique entropic effect mediated by the polymerization dominates over structural transition in the self-assembly of Janus NPs at the interface.

6.5 Interfacial Jamming

Jamming occurs when mobility of NPs is lost due to geometric constraints of neighboring NPs. Interfacial jamming of NPs occurs when assembled at the liquid–liquid interfaces if they are very crowded. Jammed NPs at interfaces are dynamically stable, which can be used to generate structured liquids with enhanced mechanical properties. This section highlights the advances in interfacial jamming of Janus hairy NPs, particularly those by in situ formed at interfaces. Typically, cooperative interfacial assembly between NPs and polymer ligands with complementary functional groups is driven by either electrostatic assembly or host–guest recognition. A wide range of NPs can be utilized covering Au, Fe_3O_4, POSS, MXene, and polymeric NPs with varied geometries, such as spheres, sheets, and cubes. The NPs render the structured novel materials with additional functions achieving reconfigurable ferromagnetic liquid droplets, MXene aerogels, 3D printable and biocompatible devices, and smart capsules. On-demand control of the jammed and unjammed states is highly desired for smart payloads and controlled release.

6.5.1 Electrostatic Assembly

Janus NPs can be generated in situ by electrostatic interaction when carboxylic acid-grafted PS NPs (PS NP-COOH) and amine-terminated poly(dimethylsiloxane) (PDMS-NH_2) meet at the interfaces of water/toluene emulsion droplets [37]. Within the window spanning the pKa of carboxylic acid (~4.2) and amine (~9), there exists a strong electrostatic attraction between deprotonated carboxyl group (COO^-) and protonated amine group (NH_3^+). Interfacial tension is greatly decreased, while the interface is fully covered by the Janus NPs. A jammed structure is constructed when more NPs enter the interface leading to a crowded state. Upon applying an electric field, shape of the droplets is deformed with increased surface area. Consequently, more Janus NPs are assembled at the interface to further lower interfacial energy. Upon removal of the electric field, the droplets fail to recover to their original shape due to the jammed structure of NPs at the interface (Figure 6.16a) [37]. The deformed water droplets are rather stable, with the deformed contour well preserved after one month. This interfacial jamming is pH-dependent. When an electric field is applied to the droplets at pH values above 5, the ellipsoidal shape is fixed after removal of the electric field. At a low pH below 4, the carboxyl group is fully protonated and the electrostatic attraction becomes weakening. The surface coverage of NPs is greatly decreased, and the ellipsoidal droplets are readily recovered to spherical upon removal of the electric field.

By constructing jamming monolayer of magnetic NPs, reconfigurable ferromagnetic liquid droplets are achieved exhibiting a reversible paramagnetic-to-ferromagnetic transformation of ferrofluid droplets (Figure 6.16b) [38]. An aqueous dispersion of carboxylated magnetic NP (Fe_3O_4-COOH) (22 nm in diameter) was mixed with a solution of amine-modified polyhedral oligomeric silsesquioxane (POSS-NH_2) in toluene. The POSS-NH_2 assemble at the interface forming strong electrostatic interaction with the Fe_3O_4-COOH NPs. A well-defined number of

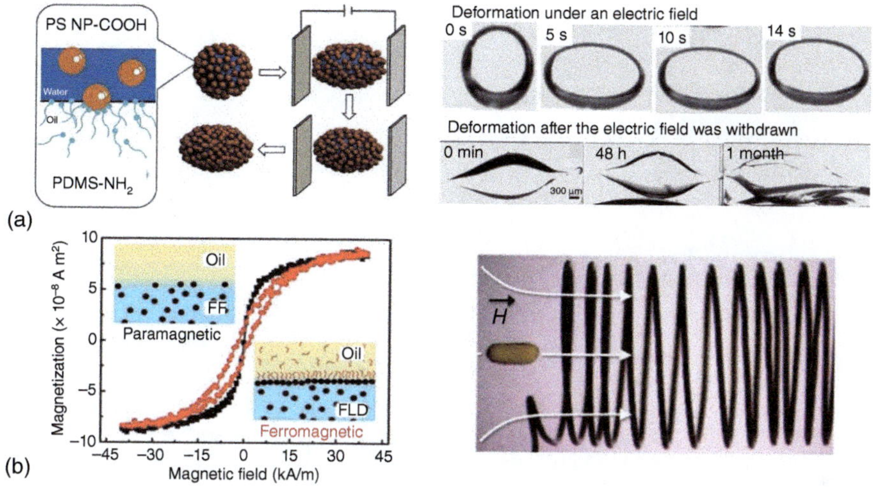

Figure 6.16 (a) Deformation and stability of the water droplets with a jammed interface in electric field. Source: Adapted with permission from Cui et al. [37]. Copyright 2013, The American Association for the Advancement of Science. (b) Magnetic hysteresis loops of the droplets with (red line) and without (black line) the interfacial layer of jammed NPs measured with a vibrating sample magnetometer (top), and a magnetized liquid cylinder is withdrawn under a magnetic field gradient generated by the aluminum solenoid (bottom). Source: Reprinted with permission from Liu et al. [38]. Copyright 2019, The American Association for the Advancement of Science.

POSS-NH_2 is conjugated at the magnetic NP forming the Janus NP surfactant. While the droplet shape is changing, the interfacial area increases and additional NP-surfactants assemble at the interface. Although the droplet is prone to recover the spherical shape to minimize the interfacial area and free energy of the system, the NP-surfactants are highly compressed forming a locked structure upon jamming in the deformed shape. The deformed droplet remains magnetized after removal of the external magnetic field. The ferromagnetic liquid droplets exhibit finite coercivity and permanent magnetization. The droplets can be easily reconfigured into different shapes while preserving the magnetic properties of solid ferromagnets with classic dipole interactions. The translational and rotational motions can be actuated by external magnetic field, which are promising in active matter, energy-dissipative assembles, and programmable liquid superstructures.

Bicontinuous jammed emulsions (or bijels) are interconnected with two immiscible liquids, which are promising for catalysis, energy storage, and molecular encapsulation. It is interesting to achieve bijels, particularly for the limit size below 5 μm. Carboxylic acid-functionalized hydrophilic PS NPs (PS NP-COOH) (~16 nm) and amine-functionalized hydrophobic polydimethylsiloxane (PDMS-NH_2, M_w = 1000 and 3000 g mol^{-1}) can coassemble into an elastic jammed monolayer of NPs at water–toluene interfaces. In the case of low molecular weight of PDMS-NH_2 (M_w = 1000 g mol^{-1}), double emulsions or water-in-oil emulsions are favorable. In contrast, water-in-oil emulsions are preferred in the case of high molecular weight of PDMS-NH_2 (M_w = 3000 g mol^{-1}). A bijel with submicrometre domains

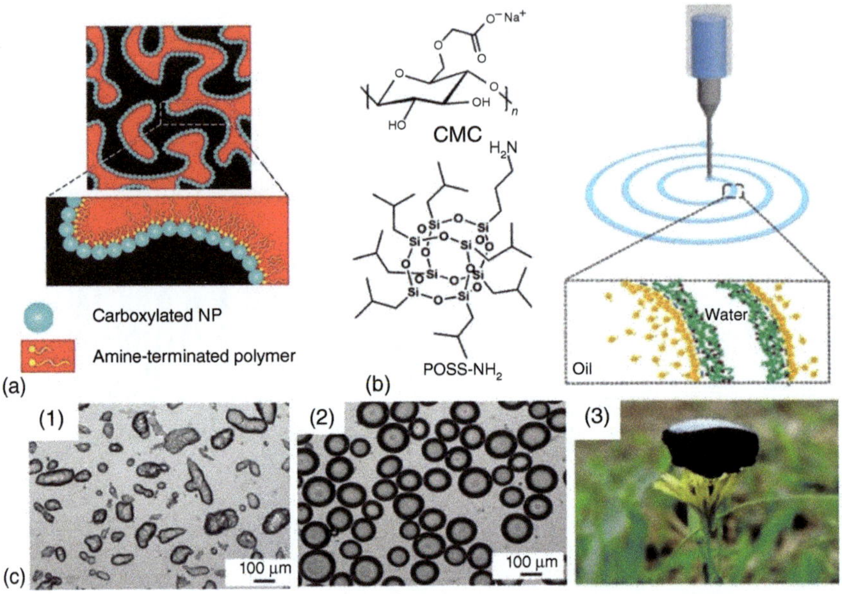

Figure 6.17 (a) Formation of a bijel by jammed Janus NPs in situ formed at oil–water interface. Source: Reprinted with permission from Huang et al. [39]. Copyright 2017, Springer Nature. (b) Assembly and jamming of CMCSs at the toluene–water interface. Source: Reprinted with permission from Xu et al. [40]. Copyright 2020, American Chemical Society. (c) Oil–water emulsions by using MXene surfactants. Source: Adapted with permission from Shi et al. [41]. Copyright 2019, John Wiley and Sons.

is accessible from a mixture of two polymers with different molecular weights in toluene (Figure 6.17a) [39]. The domain size is decreased from 10 to 1 μm at lower polymer concentration. The bicontinuous structure has arisen from the coarsening processes that the system undergoes after shearing. The coarsening involves varying local curvature imposed by the polymers with different molecular weights and a limited coalescence until a critical interfacial density of NPs.

A jammed structure at the toluene–water interface is also achieved by cooperative assembly of charged sodium carboxymethyl cellulose (CMC) and amine-functionalized polyhedral oligomeric silsesquioxane (POSS-NH_2) (Figure 6.17b) [38]. Specifically, the rigid and cubic POSS NPs endow the interface with enhanced mechanical strength. The jammed liquids exhibit solid-like behavior, which can be further processed by printing or molding into desired shapes with excellent structural stability. By 3D printing, threads of CMC/POSS-NH_2 liquid composites can be obtained with tunable diameters from 500 μm to 2.0 mm. The threads are stable for 1 month. By taking advantage of biocompatibility and negative charges of CMC, fluorescein isothiocyanate bovine serum albumin with positive charges can be adsorbed at inner wall of the tubule. This rational design is promising to fabricate biomedical devices.

2D platelet-like MXenes ($Ti_3C_2T_x$) are negatively charged and hydrophilic, which can be used to form jammed layers with POSS-NH_2 at water/toluene interfaces [41]. At increased concentrations of POSS-NH_2, the irregular emulsion droplets become smaller and more spherical (Figure 6.17c). Dense packing of the MXene at the interface endows the emulsions with good integrity, which is crucial to construct robust and conductive 3D aerogel frameworks. The material can be extremely light-weighted. This strategy enables the fabrication of functional MXene assemblies from mesoscale structured liquids to macroscale aerogels.

6.5.2 Host–Guest Molecular Recognition

Host–guest molecular recognition, which is sensitive to external triggers, has been utilized to drive interfacial jamming. Shi et al. [42] reported the photo-responsive jamming structure at the oil–water interface, which was formed by α-cyclodextrin (α-CD) modified Au NPs (~10 nm) and azobenzene terminated poly-L-lactide (Azo-PLLA) (Figure 6.18a). The host–guest molecular recognition between the α-CD and Azo moiety is responsible for the formation of the jamming interface achieving micron-sized emulsion droplets. Upon exposure to UV light, the *trans*-Azo is transformed to *cis*-Azo, which is excluded from the α-CD cavity. The interface becomes unjammed to release payloads. Such photo-responsive switching from jammed to unjammed states enables on-demand release of cargoes for the microcapsules. Similarly, a jammed layer forms by host–guest recognition between β-cyclodextrin (β-CD) modified Au NPs (~5 nm) and ferrocene (Fc) terminated poly-L-lactide (Azo-PLLA) at water/toluene interface (Figure 6.18b) [43]. The oil–water interface can be reversibly switched by redox process. The hydrophobic Fc group is transformed to hydrophilic Fc^+ upon feeding oxidants. The interface becomes unjammed after Fc escapes from the cavity of β-CD. The interface is recovered in the jammed state by feeding reductants after the uncharged Fc enters the cavity of β-CD. All-liquid "smart" superstructures including structured emulsions and programmable liquid devices are constructed accordingly, which are promising in responsive delivery and reaction systems.

6.6 Single-Chain NPs at Interfaces

6.6.1 Efficient Synthesis

Single-chain nanoparticles (SCNPs) have gained thriving interests with the development in synthetic chemistry. SCNPs with varied constituents, precise compartmentalization, and multifunction hybridization are promising in heterogeneous catalysis, oil–water emulsification and separation, and building blocks toward functional superstructures at nanoscales. Especially, SCNPs display diverse topologies ranging from colloid–chain, colloid–chain–colloid, and colloid–rod combinations (Figure 6.19a). We will introduce two effective approaches for synthesizing SCNPs and functional nanocomposites on a large scale.

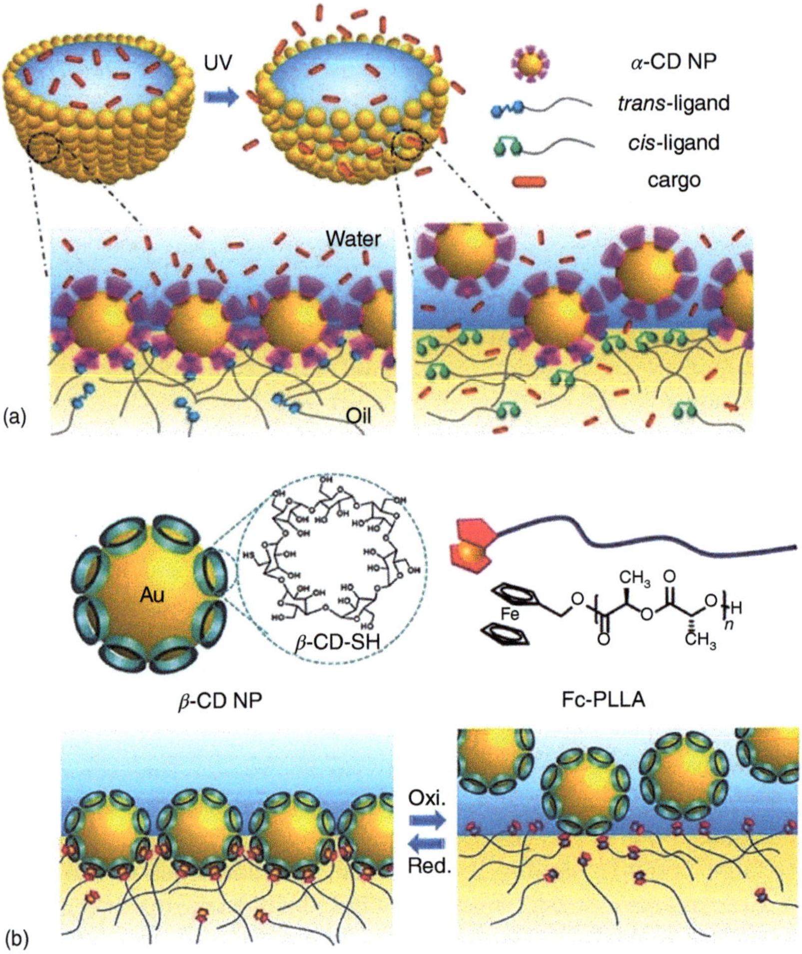

Figure 6.18 (a) Photo responsive interfacial jamming by host–guest recognition of α-CD modified Au NPs and Azo-PLLA. Source: Reprinted with permission from Li et al. [42]. Copyright 2021, John Wiley and Sons. (b) Chemical structures of the β-CD modified Au NP and Fc-PLLA, and reversible interfacial jamming through a redox process. Source: Reprinted with permission from Sun et al. [43]. Copyright 2021, American Chemical Society.

6.6.1.1 Electrostatic-Mediated Intramolecular Crosslinking Toward Large-Scale Synthesis of SCNPs

Traditional approaches for intramolecular crosslinking of polymer chains toward single-chain NPs are usually performed in extremely dilute solutions to avoid intermolecular crosslinking or gelation. Although some strategies, such as steric hindrance protection or continuous addition have been proposed to achieve intramolecular crosslinking at high polymer concentrations, efficient large-scale synthesis of SCNPs under facile conditions remains challenging. Yang et al. have recently proposed a new approach of long-ranged electrostatic-mediated intramolecular crosslinking of polymer chains at high

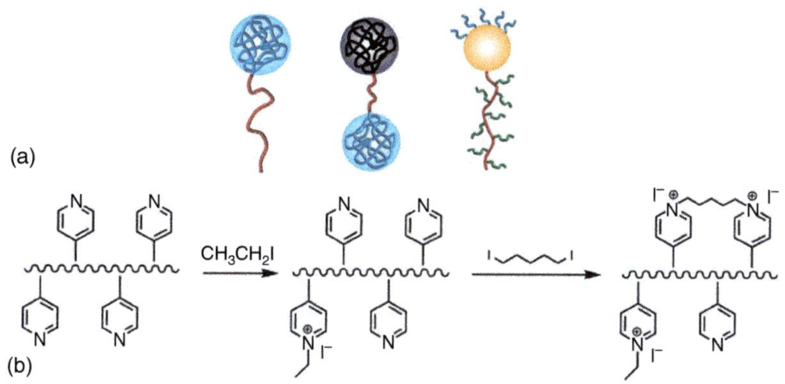

Figure 6.19 (a) SCNPs of diverse topologies covering colloid–chain, colloid–chain–colloid, and colloid-bottlebrush (rod); (b) Illustrative introduction of electrostatic interaction to the polymer by quaternization and intramolecular crosslinking by DIP. Source: Reprinted with permission from Xiang et al. [44] Copyright 2019, Chinese Chemical Society.

concentrations. As an example, electrostatic interaction was introduced by quaternization of P4VP with ethyl iodide. The intramolecular crosslinking was performed by feeding multivalent agents, such as diiodopentane (DIP) (Figure 6.19b). The SCNP was achieved by intramolecular crosslinking at a high concentration.

The SCNP was achieved by traditional intramolecular crosslinking of PS_{93k}-*b*-$P4VP_{35k}$ in an extremely dilute solution at 0.1 mg ml^{-1}, evidenced by decreased hydrodynamic diameter and apparent molecular weight (Figure 6.20a,b). The SCNP serves as the standard reference for intramolecular crosslinking under other conditions. In a concentrated solution at 100 mg ml^{-1}, PS_{93k}-*b*-$P4VP_{35k}$ appears larger than the polymer coil after crosslinking (Figure 6.20c), implying that intermolecular crosslinking occurs. Under electrostatic mediation at a charge ratio of 0.3 : 1, the intramolecular crosslinking was successfully performed at 100 mg ml^{-1}, evidenced by the same SCNP as the reference standard. However, as the charge ratio decreases below 0.2 : 1, intermolecular crosslinking takes place (Figure 6.20d). To verify the electrostatic-mediated intramolecular crosslinking, two charged P4VPs with different molecular weights were mixed for the crosslinking. Two narrow gel permeation chromatography (GPC) traces appear displaying the same as the corresponding GPC traces when the individual polymers are crosslinked (Figure 6.20e). A crosslinking diagram is provided regarding the relationship between charge ratio and critical polymer concentration for intramolecular crosslinking (Figure 6.20f). The critical polymer concentration increases substantially at a higher charge ratio. For the triblock terpolymer of PS_{33k}-*b*-$P4VP_{83k}$-*b*-PEO_{16k}, the electrostatic-mediated intramolecular crosslinking is achieved in the concentrated solution of 300 mg ml^{-1} at the charge ratio of 0.5 : 1. The electrostatic-mediated intramolecular crosslinking is related with dielectric permittivity (ε) of solvents. When the dielectric permittivity is progressively increased from 4.8 (chloroform) to 7.6 (tetrahydrofuran) to 37.6 (*N*,*N*-dimethylformamide), the critical polymer concentration is greatly increased.

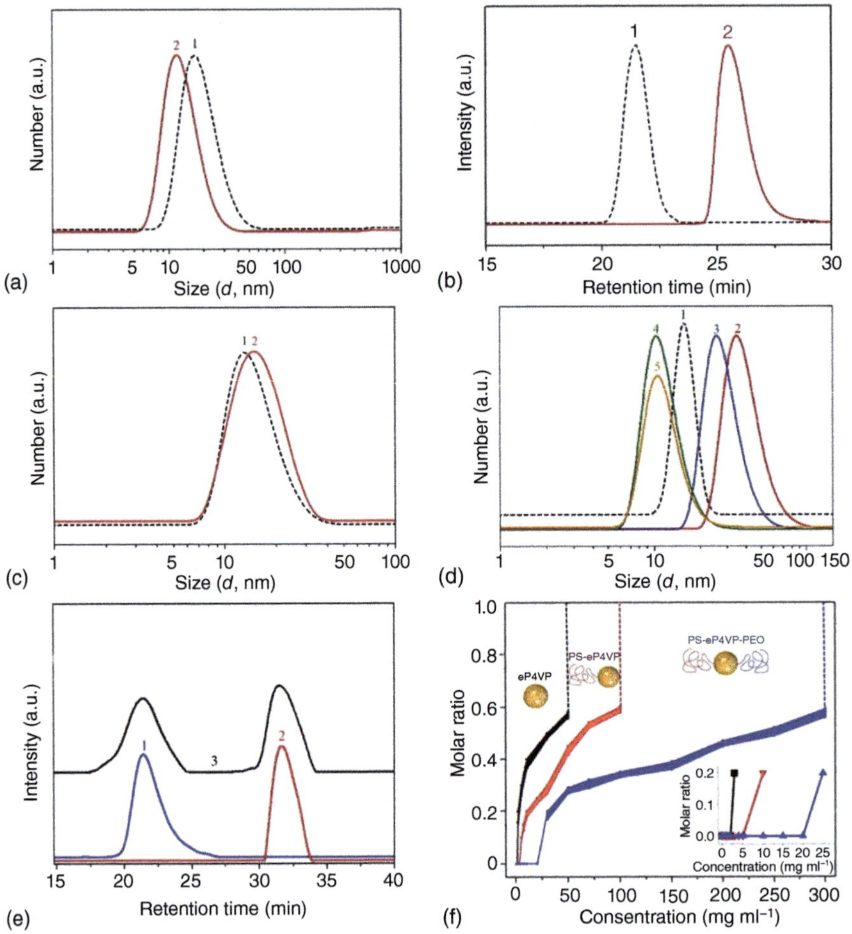

Figure 6.20 (a) DLS and (b) GPC traces of PS_{93k}-*b*-$P4VP_{35k}$ (1) and PS_{93k}-*c*$P4VP_{35k}$ by the intramolecular crosslinking at 0.1 mg ml^{-1} (2). (c) DLS traces of PS_{93k}-*b*-$P4VP_{35k}$ (1) and after modification with iodoethane at the iodoethane/4VP molar ratio of 0.3 : 1.0 (2). (d) DLS traces of PS_{93k}-*b*-$P4VP_{35k}$ (1) and after crosslinking, under the protection of varying iodoethane/4VP molar ratio of: 0 : 1.0 (2), 0.05 : 1.0 (3), 0.2 : 1.0 (4), and 0.3 : 1.0 (5). Crosslinking concentration was fixed at 100 mg mL^{-1}. (e) GPC traces of PS_{93k}-*c*$P4VP_{35k}$ (1), $PS_{9.8k}$-*c*$P4VP_{10k}$ (2), and the PS_{93k}-*b*-$P4VP_{35k}$/$PS_{9.8k}$-*b*-$P4VP_{10k}$ mixture (1 : 1) after the electrostatic-mediated intramolecular crosslinking (3). (f) Intramolecular crosslinking diagrams of the three representative PVP-contained polymers. Inset the illustrative SCNPs and diagrams at low-concentration regions. Source: Xiang et al. [44]/Chinese Chemical Society/Public Domain.

Feeding salt of potassium iodide (KI) results in intermolecular crosslinking rather than intramolecular one. This is understandable by screened electrostatic repulsion in the presence of salts.

The electrostatic-mediated intramolecular crosslinking is general to fabricate SCNPs with diverse geometry and function in many ways. In addition to covalent bonding to introduce charged groups either by copolymerization with ionic monomers or quaternization of vinyl pyridine segments, noncovalent bonding, such as protonation of vinyl pyridine segments is feasible in a fast and mild manner. The charge moiety can be easily deattached after crosslinking. For example, a polymeric colloidal dimer was synthesized at 100 mg ml^{-1} by orthogonal crosslinking of the two charged blocks of poly(isoprene)-*block*-poly(vinylpyridine) (PI-*b*-P4VP).

6.6.1.2 Grafting Single-Chain at NPs

Yang et al. have proposed fast grafting polymer single-chain onto NP surface via termination reaction to achieve inorganic/organic hybrid Janus SCNPs. Both living anionic polymerization and cationic polymerization are used to synthesize the living polymer chains. In order to ensure grafting single-chain, size of the polymer chain should be larger than the NP diameter. Once a single chain has been grafted onto the NP surface, the other chains are rejected from approaching due to a strong steric repulsion. A chloromethylphenyl-capped Fe_3O_4 NP is exemplified by grafting anionic PS single-chain. The Janus NP is uniform demonstrating asymmetric parachute shape (Figure 6.21a). The residual chloromethylphenyl groups at the NP surface are used to initiate atom transfer radical polymerization (ATRP) polymerization to conjugate functional polymers for example thermal responsive poly(*N*-isopropylacrylamide) (PNIPAM). The PNIPAM domain is visualized on one side of the Janus NP after selective staining with PTA, while the PS domain is visualized after another staining by RuO_4. The Janus NP is thermo-responsive and can be guided under a magnetic field. Onto amine-capped Fe_3O_4 NP, living cationic polymers could be easily conjugated by the fast termination reaction. Starting from the PVBC-Fe_3O_4 Janus NP, topology of the chain becomes bottlebrush after ATRP (Figure 6.21b). The bottlebrush is responsible for loading and controlled release of desired species, while the amine-capped Fe_3O_4 head is capable to recognize targets and steer under external magnetic field.

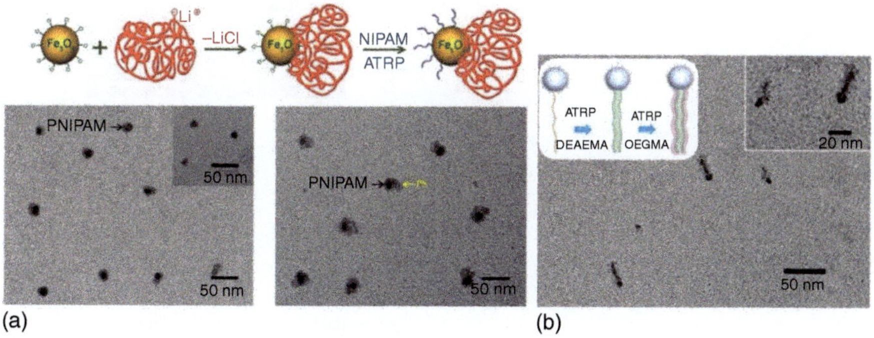

Figure 6.21 Single-chain-grafted NPs by fast termination of (a) anionic living polymers, (b) cationic living polymers and the derived bottlebrush conjugated NPs. Source: Reprinted with permission [45] Copyright 2016, American Chemical Society. Reprinted with permission from Jing et al. [46]. Copyright 2019, American Chemical Society.

6.6.2 Interfacial Applications

Janus SCNPs by electrostatic-mediated intramolecular crosslinking have been explored for interfacial catalysis. In the case of PS-*b*-P2VP-*b*-PEG, interchain electrostatic repulsion was introduced by forming the charged complex between P2VP with $Co_2(CO)_8$/THF. The intramolecular crosslinking was achieved by further complexation upon feeding $Co_2(CO)_8$ at a high solid content of 30 mg ml^{-1} (Figure 6.22a). Metallic Co was formed in situ at the crosslinked P2VP core by thermolytic reduction to derive the Janus composite SCNP of PS-P2VP@Co-PEG. The Janus composite SCNP displays highly efficient catalytic reduction of nitroaromatic compounds at emulsion interfaces at a SCNP/nitrobenzene ratio of 1 : 100. The catalytic activity is preserved at high levels after many cycles with a magnet (Figure 6.22b).

Janus composite SCNP of $PS_{5.2k}$-*c*PAA_{4k}@Fe_3O_4 was derived by in situ growth of Fe_3O_4 against the crosslinked *c*PAA core, which acts as a pH and magnetic dually responsive emulsifier. Similarly, Janus composite SCNP of PS-*c*P4VP@Au-PEO was derived from PS-*c*P4VP-PEO by in situ growth of Au within the *c*P4VP core. Silane moiety can be conjugated at the *c*P4VP@Au core surface, facilitating the construction of robust self-assembled monolayer or hollow structure by further sol–gel process. Favorable growth of Ni against the PS-*c*PI@PEO derives the magnetic responsive Janus composite SCNP of PS-*c*PI@Ni-PEO, which can be collected with a magnet (Figure 6.23). PS-*c*PI@Ni-PEO forms a monolayer at the emulsified paraffin sphere surface. A cyclohexane/water emulsion was achieved in the presence of PS-*c*PI@Ni-PEO, which can be collected with a magnet.

The head domain of tadpole-like Janus SCNP of PS-*c*PAA is easily converted as a colloidal initiator after crosslinking residual isocyanate with the free radical initiator 2,2′-azobis(2-methylpropionamidine) dihydrochloride (AIBA,

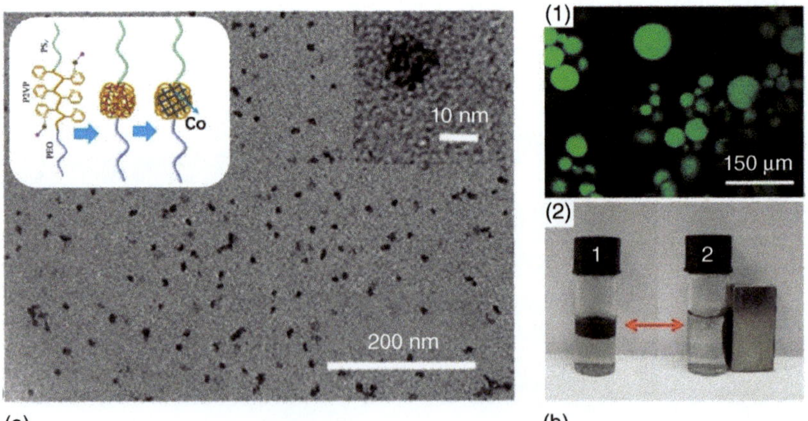

Figure 6.22 (a) TEM image of the Janus composite SCNP of PS-P2VP@Co-PEO by thermolysis. (b) LSCM image of the emulsion droplets stabilized with the Janus SCNP (1), magnetic collection of the emulsion droplets (2). Source: Adapted with permission from Xiang et al. [47]. Copyright 2020, American Chemical Society.

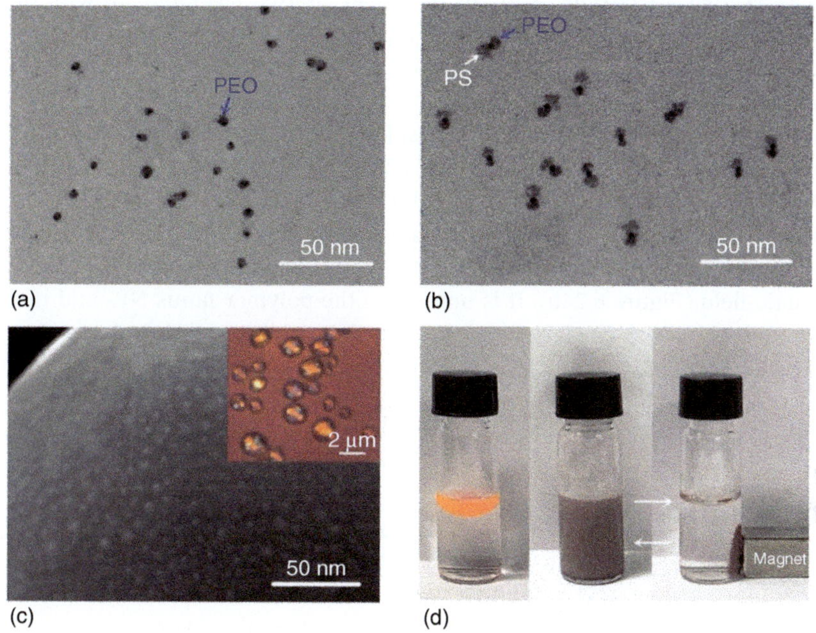

Figure 6.23 TEM images of the PS-*c*PI@Ni-PEO after staining with (a) PTA and (b) RuO_4. (c) SEM image of the paraffin sphere surface and inset polarizing optical image of the emulsion spheres stabilized with PS-*c*PI@Ni-PEO. (d) Cyclohexane/water mixture (left), cyclohexane/water emulsion stabilized with PS-*c*PI@Ni-PEO (middle), and collection of the emulsion droplets with a magnet (right). Oil-soluble dye of Coumarin 6 was added into cyclohexane for easy observation. Source: Reprinted with permission from Xiang et al. [44]. Copyright 2019, Chinese Chemical Society.

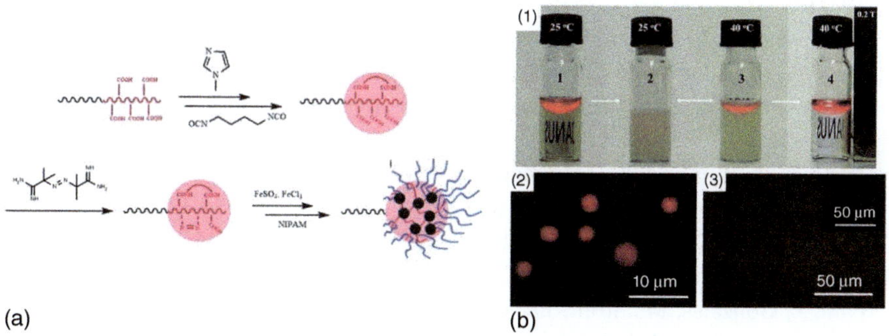

Figure 6.24 (a) Illustrative synthesis of the Janus composite NP of PS-*c*PAA@(Fe_3O_4-PNIPAM). (b) Feeding toluene to the PS-*c*PAA@(Fe_3O_4-NIPAM) aqueous dispersion at 25 °C, formation of the emulsion by ultrasonication at 25 °C, de-emulsification when heating the emulsion at 40 °C for 2 minutes, and magnetic field assisted de-emulsification at 40 °C (1), CLSM images of the emulsion droplets at 25 °C (2), de-emulsification showing no dark water domains in the top oil phase while the bottom aqueous phase contains no oil droplets (3). Source: Adapted with permission from Wang et al. [48]. Copyright 2020, Royal Society of Chemistry.

V50). The residual carboxylic acid at the head assists favorable growth of functional materials for example Fe_3O_4. Free radical polymerization of vinyl monomers and favorable growth of functional materials at the head allows for extension of composition and performance of the Janus NPs. The example Janus NP of PS-*c*PAA@(PINPAM-Fe_3O_4) displays magnetic and thermal dually responsive behavior (Figure 6.24a). At a low temperature below the critical value (~32 °C), a stable emulsion of toluene/water was formed in the presence of PS-*c*PAA@(PINPAM-Fe_3O_4). At a high temperature of 40 °C, the emulsion becomes unstable forming multiple emulsions. The de-emulsification becomes complete in a magnetic field (Figure 6.24b). It is noted that the polymer Janus NPs and the derived composite ones can be large-scale produced, providing a sufficient quantity of the materials for practical applications.

References

1 Yi, C., Yang, Y., Liu, B. et al. (2020). *Chem. Soc. Rev.* 49: 465–508.
2 Hore, M.J.A., Ford, J., Ohno, K. et al. (2013). *Macromolecules* 46: 9341–9348.
3 Ye, X., Zhu, C., Ercius, P. et al. (2015). *Nat. Commun.* 6: 10052.
4 Grundler, J., Shin, K., Suh, H.-W. et al. (2021). *ACS Nano* 15: 16118–16129.
5 Choueiri, R.M., Galati, E., Therien-Aubin, H. et al. (2016). *Nature* 538: 79–83.
6 Zhou, Y., Ma, X., Zhang, L., and Lin, J. (2017). *Phys. Chem. Chem. Phys.* 19: 18757–18766.
7 Bohannon, C.A., Chancellor, A.J., Kelly, M.T. et al. (2021). *J. Am. Chem. Soc.* 143: 16919–16924.
8 Groschel, A.H. and Mueller, A.H.E. (2015). *Nanoscale* 7: 11841–11876.
9 Groeschel, A.H., Schacher, F.H., Schmalz, H. et al. (2012). *Nat. Commun.* 3: 710–719.
10 Boeker, A., He, J., Emrick, T., and Russell, T.P. (2007). *Soft Matter* 3: 1231–1248.
11 Binks, B.P. and Lumsdon, S.O. (2000). *Langmuir* 16: 8622–8631.
12 Liu, M., Chen, X., Yang, Z. et al. (2016). *ACS Appl. Mater. & Inter.* 8: 32250–32258.
13 Maillard, D., Kumar, S.K., Fragneaud, B. et al. (2012). *Nano Lett.* 12: 3909–3914.
14 Akcora, P., Liu, H., Kumar, S.K. et al. (2009). *Nat. Mater.* 8: 354–U121.
15 Liu, C., Feng, S., Zhu, Z. et al. (2020). *Langmuir* 36: 11938–11947.
16 Rungta, A., Natarajan, B., Neely, T. et al. (2012). *Macromolecules* 45: 9303–9311.
17 Hyon, J., Gonzales, M., Streit, J.K. et al. (2021). *ACS Nano* 15: 2439–2446.
18 Bilchak, C.R., Buenning, E., Asai, M. et al. (2017). *Macromolecules* 50: 7111–7120.
19 Yang, Y., Yi, C., Duan, X. et al. (2021). *J. Am. Chem. Soc.* 143: 5060–5070.
20 Yi, C., Liu, H., Zhang, S. et al. (2020). *Science* 369: 1369–1374.
21 Gao, B., Arya, G., and Tao, A.R. (2012). *Nat. Nanotech.* 7: 433–437.
22 Y. Xiong, Z. Lin, D. Mostarac, B. Minevich, Q. Peng, G. Zhu, P. A. Sanchez, S. Kantorovich, Y. Ke, O. Gang, *Nano Lett.* 2021. DOI: https://doi.org/10.1021/acs.nanolett.1c03009.

23 Liu, Y., Wang, J., Shao, Y. et al. (2022). *Prog. Mater Sci.* 124: 100888.
24 Yang, Y., Wang, Y., Jin, S.-M. et al. (2019). *Mater. Chem. Front.* 3: 209–215.
25 Luo, J., Zeng, M., Peng, B. et al. (2018). *Angew. Chem. Int. Ed.* 57: 11752–11757.
26 Yan, N., Zhang, Y., He, Y. et al. (2017). *Macromolecules* 50: 6771–6778.
27 Bahrami, R., Loebling, T.I., Groeschel, A.H. et al. (2014). *ACS Nano* 8: 10048–10056.
28 Wang, H., Fu, Z., Zhao, X. et al. (2017). *ACS Appl. Mater. & Inter.* 9: 14358–14370.
29 Khanal, B.P. and Zubarev, E.R. (2007). *Angew. Chem. Int. Ed.* 46: 2195–2198.
30 Salaluk, S., Jiang, S., Viyanit, E. et al. (2021). *ACS Appl. Mater. & Inter.* 13: 53046–53054.
31 Jesmer, A.H., Huynh, V., Marple, A.S.T. et al. (2021). *ACS Appl. Mater. & Inter.* 13: 52362–52373.
32 Srivastava, S., Nykypanchuk, D., Fukuto, M., and Gang, O. (2014). *ACS Nano* 8: 9857–9866.
33 Si, K.J., Sikdar, D., Chen, Y. et al. (2014). *ACS Nano* 8: 11086–11093.
34 Zhang, R., Lee, B., Stafford, C.M. et al. (2017). *Proc. Natl. Acad. Sci. U S A* 114: 2462–2467.
35 Liu, Z., Guo, R., Xu, G. et al. (2014). *Nano Lett.* 14: 6910–6916.
36 Chen, P., Yang, Y., Dong, B. et al. (2017). *Macromolecules* 50: 2078–2091.
37 Cui, M., Emrick, T., and Russell, T.P. (2013). *Science* 342: 460–463.
38 Liu, X., Kent, N., Ceballos, A. et al. (2019). *Science* 365: 264–267.
39 Huang, C., Forth, J., Wang, W. et al. (2017). *Nat. Nanotech.* 12: 1060–1064.
40 Xu, R., Liu, T., Sun, H. et al. (2020). *Mater. & Inter.* 12: 18116–18122.
41 Shi, S., Qian, B., Wu, X. et al. (2019). *Angew. Chem. Int. Ed.* 58: 18171–18176.
42 Li, L., Sun, H., Li, M. et al. (2021). *Angew. Chem. Int. Ed.* 60: 17394–17397.
43 Sun, H., Li, M., Li, L. et al. (2021). *J. Am. Chem. Soc.* 143: 3719–3722.
44 Xiang, D., Chen, X., Tang, L. et al. (2019). *CCS Chem.* 1: 407–430.
45 Yao, X., Jing, J., Liang, F., and Yang, Z. (2016). *Macromolecules* 49: 9618–9625.
46 Jing, J., Jiang, B., Liang, F., and Yang, Z. (2019). *ACS Macro Lett.* 8: 737–742.
47 Xiang, D., Jiang, B., Liang, F. et al. (2020). *Macromolecules* 53: 1063–1069.
48 Wang, J., Chen, X., Lang, F. et al. (2020). *Chem. Commun.* 56: 3875–3878.

7

Hairy Hollow Nanoparticles

Huiqi Zhang

Nankai University, Collaborative Innovation Center of Chemical Science and Engineering (Tianjin), and College of Chemistry, State Key Laboratory of Medicinal Chemical Biology, Key Laboratory of Functional Polymer Materials (Ministry of Education), Tianjin 300071, China

7.1 Introduction

Hollow nanoparticles (NPs) are a special class of nanomaterials, typically with a void core inside a solid shell and a size scale between 1 and 1000 nm [1–12]. They have garnered tremendous interest from both the academic community and industry because of their distinct physicochemical properties, such as large surface areas, high loading and encapsulation capacities, low density, and dramatic swellability. These unique characteristics make them highly promising in many applications, such as nanoreactors [1–6, 8, 10], biomedical areas (e.g. drug delivery, bioimaging) [1–5, 7, 8, 10, 12], and various bioanalyses [4, 5, 8, 10, 11].

Over the past several decades, significant advances have been made in the development of hollow NPs in terms of their synthetic strategies and new characterization instruments and approaches, which have not only enabled the efficient fabrication of hollow NPs with various shell and internal structures and morphologies but also afforded deeper insight into their formation mechanism [1–12]. Such progress greatly facilitates the rational design and synthesis of well-defined, uniform hollow NPs with desired architectures, controlled dimensions, and customized physicochemical properties and largely expands their application areas.

Among the presently developed hollow NPs, hairy hollow NPs (i.e. hollow NPs with surface-grafted polymer brushes) have attracted rapidly increasing interest because the presence of functional polymer brushes on their surfaces can not only endow them with high surface chemistry tunability but also significantly modify their interfacial physicochemical properties (such as friction, adhesion, wettability, and biocompatibility) [13–15]. The polymer brushes can be grafted onto the hollow NP surfaces by using the "grafting-to" and "grafting-from" methods, with the former normally being accomplished via different coupling reactions and the latter being realized by using various surface-initiated polymerizations (SI-polymerizations or SI-Ps) [15–18]. In particular, the surface-initiated controlled/"living" radical

Hairy Nanoparticles: From Synthesis to Applications, First Edition. Edited by Zhiqun Lin and Yijiang Liu.

polymerizations (SI-CRPs) [15–18] and click chemistry-based coupling reactions [19, 20] have proven highly efficient for grafting various polymer brushes onto particle surfaces. According to the location of the polymer brushes on their shells, hairy hollow NPs can be divided into three types, i.e. those with polymer brushes on the exterior, interior, or both sides of the shells. The grafting of suitable polymer brushes onto the exterior surfaces of the hollow NPs can dramatically enhance their compatibility with the surrounding matrixes, thus largely improving their performance in different applications. For instance, the hairy hollow NPs with hydrophilic polymer brushes on their exterior surfaces not only show largely enhanced aqueous dispersion stability and antifouling ability in complex biological samples [14, 21], but also exhibit stimuli-responsivity when the polymer brushes are stimuli-sensitive [22, 23], which makes them highly promising in many real-world bioanalytical and biomedical applications (e.g. chemosensing, drug delivery, and bioimaging). On the other hand, the grafting of appropriate polymer brushes onto their interior surfaces might largely increase their loading capacities due to their interactions with the encapsulated species. In principle, all the hollow NPs can be transformed into hairy hollow NPs through first introducing some desired functional groups onto their shells and subsequent grafting of polymer brushes via either "grafting-to" or "grafting-from" method. So far, a great number of hairy hollow NPs with various morphologies as well as internal, shell, and hairy structures have been developed for different purposes. According to the chemical compositions of the solid shells in the hairy hollow NPs, they can be classified as hairy hollow polymer, inorganic, and organic/inorganic hybrid NPs, with their solid shell being chemically or physically cross-linked polymer, inorganic, and organic/inorganic hybrid layers, respectively. In Section 7.2, a detailed overview of the progress made in the design and synthesis of these hairy hollow NPs is presented. Their properties and applications are also described following the introduction of their synthesis. In Section 7.3, I present the pros and cons of each synthetic method and also highlight the future research direction and perspectives in this field. To my knowledge, although many review papers have summarized the research advances in the preparation of hollow NPs [1–12], a comprehensive overview of the progress in the design and synthesis of hairy hollow NPs has not been available yet.

7.2 Overview of the Progress in the Design and Synthesis of Hairy Hollow NPs

7.2.1 Synthetic Strategies for Hairy Hollow Polymer NPs

To date, several different kinds of synthetic strategies have been developed for the preparation of hairy hollow polymer NPs, which can be mainly divided into sacrificial template method, self-assembly (of block copolymers) method, and single-molecule templating (of bottlebrush copolymers) method. The introduction of these strategies and a detailed overview of the design and synthesis of hairy hollow polymer NPs using these strategies are presented below.

7.2.1.1 Sacrificial Template Method

The sacrificial template method has been most widely used for preparing hairy hollow polymer NPs owing to its general applicability, simple manipulation, and easy control of hollow particle internal void spaces and particle morphologies. It typically involves first the grafting of a cross-linked polymer shell and polymer brushes onto the sacrificial NP surfaces and subsequent removal of the template core, or the direct grafting of polymer brushes onto the hollow polymer NPs that are prepared in advance by using the sacrificial template method. The grafting of cross-linked polymer layers onto the sacrificial NP surfaces can be realized by using SI-Ps [15–18], precipitation polymerizations [2], and layer-by-layer (LbL) assembly method [24, 25]. The SI-P method is a typical "grafting-from" strategy that allows the formation of cross-linked polymer shells on NP surfaces either directly via the polymerization of monomers (including a cross-linker) from the initiating groups bound on the NPs or by first grafting polymer brushes bearing reactive groups onto NP surfaces via SI-Ps and their subsequent post-cross-linking. On the other hand, precipitation polymerization and LbL assembly methods belong to "grafting-to" strategies, with the former typically involving the formation of polymers in the reaction solutions and their deposition onto the NP surfaces and the latter involving the alternative adsorption of oppositely charged polymers onto the NP surfaces. The polymer brush-grafting methods mainly include various SI-Ps and coupling reactions. Significant progress has been made in this respect with the advent of SI-CRPs [15–18] and click chemistry methods [26–28]. The SI-CRPs enable the efficient synthesis of well-defined multilayer-structured NPs with controlled polymer shell and brush structures under mild conditions [15–18]. The click chemistry-based coupling methods (such as azide–alkyne and thiol-ene click reactions) have proven highly powerful for modifying NPs and fabricating functional polymer materials because of their quantitative yield, mild reaction conditions, and tolerance of a wide range of functional groups [26–28]. So far, many spherical nanomaterials, including nanosized SiO_2, polymer latex, and metal-based species, have been widely utilized as the sacrificial templates for fabricating hairy hollow NPs owing to their easy preparation in large amounts and high uniformity.

Surface-Initiated Polymerizations (SI-Ps) or Their Combination with Other Polymer Brush-Grafting Methods To date, several kinds of SI-Ps have been applied for preparing hairy hollow polymer NPs, including SI-CRPs [29–38], surface-initiated-photopolymerization (SI-PP) [39], and surface-initiated-ring opening polymerization (SI-ROP) [33, 40]. Among them, SI-CRPs have been mostly used because of their high controllability, mild reaction conditions, and "livingness" of the resulting polymers [15–18]. They include surface-initiated-atom transfer radical polymerization (SI-ATRP) [29–33, 35–37], surface-initiated-single electron transfer living radical polymerization (SI-SET-LRP) [34], and surface-initiated-reversible addition-fragmentation chain transfer (SI-RAFT) polymerization [38]. Silica NPs are mainly used as the hard template in these cases for preparing hairy hollow polymer NPs, mainly because they can be efficiently prepared via Stöber sol–gel reaction with high uniformity, readily tunable sizes, and high surface-modification

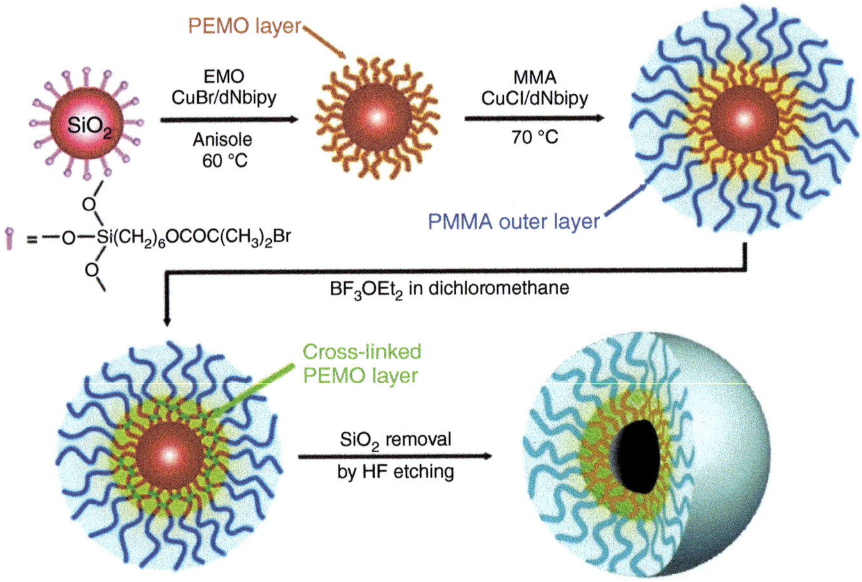

Figure 7.1 Schematic representation for the synthesis of polymeric hollow nanospheres. Source: Morinaga et al. [29], scheme 1 (p. 1161)/Reproduced with permission of American Chemical Society.

capability, and are easily etchable by using hydrofluoric acid (HF) or alkali solutions [29–37, 39, 40]. In addition, physically cross-linked poly(methacrylic acid) (poly(MAA) or PMAA) NPs have also been used recently as the sacrificial template for preparing hairy hollow polymer NPs via SI-CRP because of their easy and mild etching condition [38].

It was Fukuda and coworkers who reported the first fabrication of hairy hollow spherical polymer NPs with silica NP as the template by using SI-ATRP in 2007 (Figure 7.1) [29]. In this work, commercially available uniform silica NPs (diameter [D] = 740 nm) were first modified with (2-bromo-2-methyl)propionyloxyhexyltriethoxysilane (BHE) to obtain the silica NPs with surface-bound ATRP-initiating groups (alkyl bromide), which were then used as the immobilized ATRP initiator to induce the SI-ATRP of an oxetane group-carrying methacrylate (i.e. 3-ethyl-3-(methacryloyloxy)methyloxetane [EMO]). The polymerization proceeded in a living manner, leading to silica NPs with surface-grafted well-defined poly(EMO) (PEMO) of target molecular weight up to about 400 K and a graft density as high as 0.36 chains nm^{-2}. The resulting PEMO-grafted silica NPs were further used as the macroinitiator to induce the SI-ATRP of methyl methacrylate (MMA), which afforded silica NPs with surface-grafted block copolymer PEMO-*b*-PMMA (namely (PEMO-*b*-PMMA)-SiPs, PMMA refers to poly(MMA)). Uniform hairy hollow spherical polymer NPs with a size of 1.24 μm (by AFM) and good dispersibility in organic solvents were obtained through cross-linking the PEMO layer via the cationic ring-opening reaction of its oxetane groups and the selective removal of the silica core of the cross-linked (PEMO-*b*-PMMA)-SiPs by HF etching.

Following the above pioneering work, many hairy hollow polymer NPs with polymer brushes grafted on either the exterior [30, 31], interior [32, 33], or both sides [34] of the shells have been developed with silica NPs as the sacrificial template by using SI-CRPs alone [30–32, 34–37] or their combination with other polymer brush-grafting methods [33]. Based on the different cross-linking characters of their shell structures, they can be divided into chemically [30–34] and physically cross-linked ones [35–37]. In addition, most of these hairy hollow polymer NPs were endowed with various stimuli-responsivity [30–34, 36], mainly because stimuli-responsive hairy hollow NPs can control the release of their payloads in response to the external stimuli and show great potential in the fields of drug delivery and catalysis.

So far, some single or dual stimuli-responsive hairy hollow polymer NPs have been developed by simply introducing stimuli-responsive moieties into either the cross-linked polymer shells or polymer brushes or both of them. For instance, He and coworkers described the synthesis of hairy hollow poly(ionic liquid) (PIL) NPs with homogeneously embedded gold (Au) NPs inside both the PIL shell and polymer brushes and their use as thermoresponsive catalysts [30]. They were synthesized by first the preparation of "living" silica nanospheres with surface ATRP-initiating alkyl bromide groups, their subsequent grafting of poly(MIMC-*co*-MBA) shell (MIMC is the IL monomer (2-(1-methylimidazolium 3-yl)-ethyl methacrylate chloride) and MBA is *N*,*N*′-methylenebisacrylamide) and thermoresponsive poly(2-(dimethylamino)ethyl methacrylate) (poly(DMAEMA) or PDMAEMA) brushes via successive SI-ATRP, embedding Au NPs with an average diameter of 1.5 ± 0.2 nm onto both the PIL shell and polymer brushes, and final removal of the silica core. The hairy hollow PIL NPs proved to be highly active and stable thermoresponsive catalysts for the reduction of 4-nitrophenol.

Our group developed a facile, general, and efficient approach to prepare hydrophilic hollow molecularly imprinted polymer (MIP) NPs with photo- and thermo-responsive template binding and release behaviors in aqueous media (Figure 7.2) [31]. It involves first the preparation of uniform "living" silica nanospheres bearing surface ATRP-initiating groups (i.e. alkyl halide groups) via a one-pot sol–gel method, their subsequent grafting of azobenzene (azo)-containing MIP shell and poly(*N*-isopropylacrylamide)-*block*-poly(2-hydroxyethyl methacrylate) (poly(NIPAM or NIPAAm)-*b*-poly(HEMA) or PNIPAM-*b*-PHEMA) brushes via successive SI-ATRP, and final removal of the silica core. They proved to show apparently higher template binding capacities than the corresponding solid ones and obvious photo- and thermo-responsive template binding properties in aqueous solutions. Moreover, their pronounced light- and temperature-controlled template release in aqueous media was also demonstrated. In particular, the introduction of PNIPAM-*b*-PHEMA brushes onto hollow MIP NPs imparted them with high surface hydrophilicity both below and above the lower critical solution temperature (LCST) of PNIPAM. Note that although MIPs are synthetic receptors with tailor-made recognition sites and are promising substitutes for biological receptors (e.g. antibodies and enzymes) because of their good molecular affinity and selectivity, easy synthesis, high stability, and low cost, the development of useful strategies for

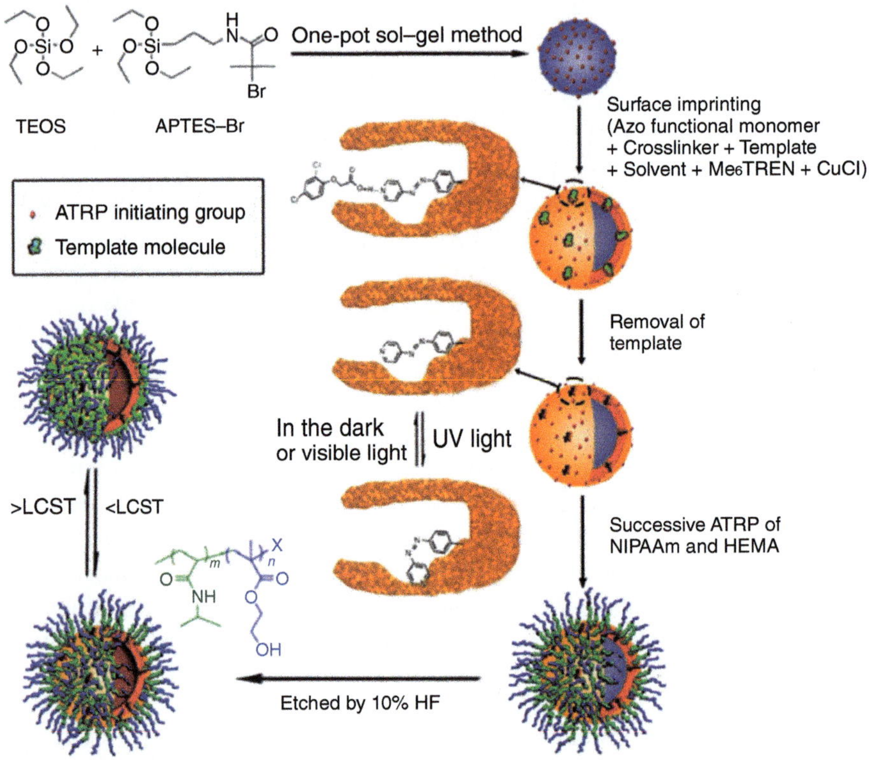

Figure 7.2 Schematic illustration for the preparation of photo- and thermo-responsive hydrophilic hollow azo-containing MIP NPs bearing PNIPAM-*b*-PHEMA brushes. Source: Li et al. [31], scheme 1 (p. 27341)/Reproduced with permission of American Chemical Society.

the preparation of water-compatible and (dual or multiple) stimuli-responsive MIP NPs with high template loadings has been a challenging task [21]. This work successfully addresses this issue and paves the way for the applications of such multifunctional MIPs in such areas as controlled drug/chemical delivery and smart bioanalysis.

Liu and coworkers reported the preparation of both temperature- and pH-responsive hairy hollow polymer NPs via SI-ATRP with silica NPs as the sacrificial template, which has temperature-responsive cross-linked polymer shells and pH-responsive polymer brushes on their interior walls [32]. Two-step SI-ATRPs were first carried out to successively graft poly(*tert*-butyl acrylate) (poly(*t*BA) or PBA) brushes and cross-linked poly(NIPAM-*co*-MBA) shells onto the silica NPs with surface ATRP-initiating groups. The subsequent removal of the silica core by etching with HF and transformation of the *tert*-butyl ester groups in polymer brushes into acrylic acid groups through hydrolysis with trifluoroacetic acid (TFA) eventually led to the production of the dually responsive hollow polymer NPs. The inner diameter, the cross-linking degree and the thickness of the shells, and the length of the functional brushes of the hairy hollow polymer NPs, could easily be

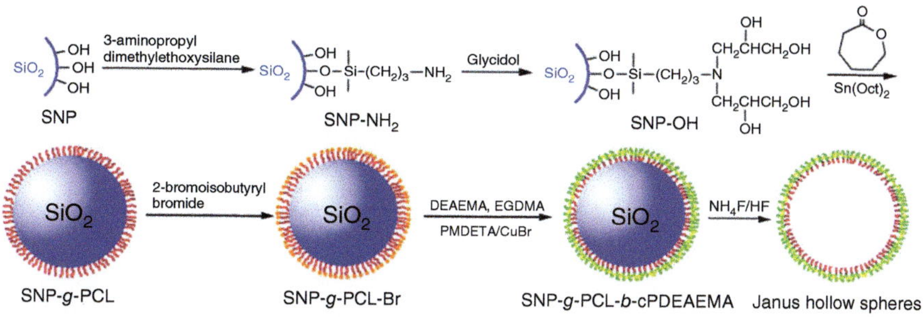

Figure 7.3 Schematic illustration for the preparation of pH-responsive hollow polymer NPs. Source: Zhao et al. [33], scheme 1 (p. 4147)/Reproduced with permission of Royal Society of Chemistry.

controlled by adjusting the preparation conditions. Such hairy hollow polymer NPs have potential applications in the controlled drug delivery area.

Zhang and Liang's groups developed pH-responsive hollow polymer NPs with hydrophobic poly(ε-caprolactone) (poly(ε-CL) or PCL) brushes on the interior wall of a pH-responsive shell for controlled drug delivery via the combined SI-ROP and SI-ATRP on the sacrificial silica NPs (Figure 7.3) [33]. Silica NPs with surface hydroxyl groups (as ROP-initiating groups) were first prepared through two-step surface modification of the silica NPs (prepared via Stöber method) with 3-aminopropyldimethylethoxysilane (APDMES) and glycidol successively. PCL brush-grafted silica NPs were readily obtained by using the above hydroxyl-functionalized silica NPs as the immobilized initiator to induce the SI-ROP of ε-CL, which were further transformed into silica NPs with surface alkyl bromide units (SiO_2-Br) by their reaction with 2-bromoisobutyryl bromide. Following the SI-ATRP of diethylaminoethyl methacrylate (DEAEMA) and ethyleneglycol dimethacrylate (EGDMA) (7 : 1 M ratio) with SiO_2-Br as the immobilized ATRP initiator and the removal of silica core, hollow polymer NPs with the biodegradable PCL brushes as the inner layer and the lightly cross-linked pH-responsive Poly(DEAEMA-*co*-EGDMA) as the outer layer were obtained. Such hollow polymer NPs can selectively load oil-soluble drug (Doxorubicin [DOX]) into their hydrophobic cavities and release it by adjusting the pH environment because of the pH-responsivity of the shell. The in vitro release exhibited a faster rate at acidic pH than at physiological pH. The intracellular uptake rate of the hollow polymer NPs at the pH of tumorous tissues was also much larger than that under the same conditions at the pH of the normal physiological environment.

Voit and coworkers demonstrated a facile, rapid, and one-pot approach for preparing both thermo- and pH-responsive hairy hollow nanocapsules (NCPs) with an ABA-type triblock copolymer membrane by using SI-SET-LRP technique with silica NPs as the sacrificial template (Figure 7.4) [34]. Silica NPs with surface-grafted ABA triblock copolymer brushes (where A block is a poly(NIPAM) chain and B block is a poly(DEAEMA-*co*-2-hydroxy-4-(methacryloyloxy) benzophenone) (poly(DEAEMA-*co*-BMA)) chain, $M_{n,\mathrm{GPC}} = 84\,000\,\mathrm{g\,mol^{-1}}$, molar mass dispersity $Đ = 1.40$) were successfully prepared via the successive SI-SET-LRP of NIPAM,

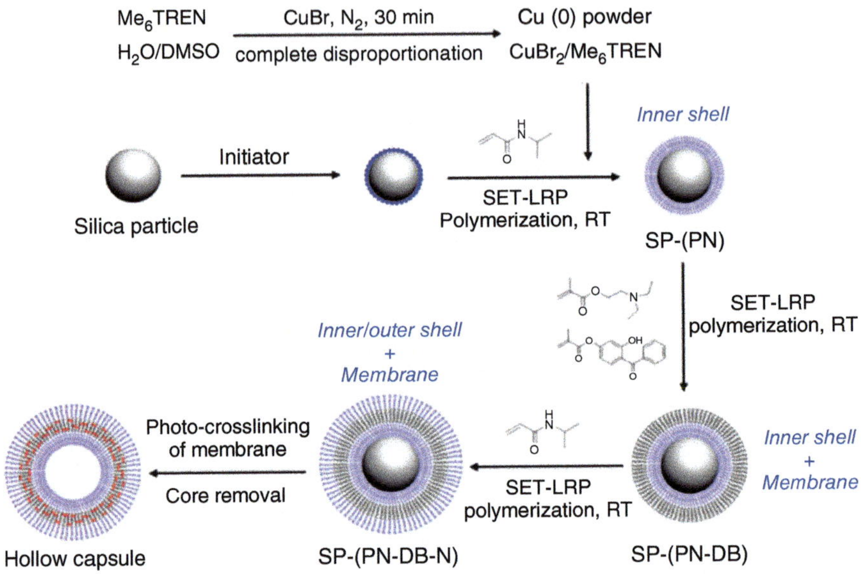

Figure 7.4 Formation of a pH and temperature dual-responsive hollow nanocapsule with ABA-type triblock copolymer brushes by sequential *in situ* chain extensions via SI-SET-LRP. Source: Liu et al. [34], scheme 1 (p. 16107)/Reproduced with permission of American Chemical Society.

that of DEAEMA and BMA, and that of NIPAM with silica NPs with surface alkyl bromide groups as the immobilized initiator and CuBr/Me_6TREN as the catalyst in DMSO/H_2O (15% v/v) at room temperature with the polymerization time being ≤30 minutes for each step. No purification is required between the successive block formation steps because of the almost full monomer conversion (>96% conversion) and the high-end group fidelity in each step. Following the photo-cross-linking of the B block of the grafted triblock copolymer brushes and removal of the silica core by using NH_4F/HF, hollow polymer NCPs with densely grafted PNIPAM brushes on both sides of the cross-linked polymer shell (diameter $D = {\sim}840$ nm, shell thickness = 64 nm, determined by cryo-TEM) were readily achieved. They showed both thermo- and pH-responsive swelling and shrinking owing to their presence of temperature-responsive outer and inner PNIPAM layers and a pH-responsive poly(DEAEMA-*co*-BMA) central membrane layer, which led to their easily tunable valve-like permeability (permeable, semipermeable, and impermeable). The hollow NCPs loaded with enzyme(s) could act as a dual-responsive enzymatic nanoreactor, where the enzymatic reactions could be reversibly switched "on" and "off" by using temperature as a main valve control and pH as a secondary sub-valve control. They represent the first nanoreactors with a tunable valve that can control their enzymatic reactions in a highly reproducible, efficient, specific, and successive manner over a broader enzyme-friendly environment (between pH 6 and 8). They are expected to have wide implications in the fields of synthetic biology and systems biology.

Besides the above hairy hollow polymer NPs with chemically cross-linked shells, some hairy hollow polymer NPs with physically cross-linked shells have also been developed via SI-CRPs with silica NPs as the sacrificial template by using different physical cross-linking strategies. Kang and coworkers reported on the synthesis of well-defined conductive hollow polyaniline (PANI) nanospheres via SI-ATRP [35]. Silica NPs with surface-bound ATRP-initiating groups were first prepared by immobilizing trichloro(4-chloromethylphenyl)silane onto silica NPs ($D = \sim 25$ nm). They were used as the immobilized ATRP initiator to induce the SI-ATRP of 4-vinylaniline (VAn), resulting in silica NPs with surface-grafted poly(VAn) (PVAn) (SiO_2-g-PVAn). Well-defined P(VAn-*graft*-PANI) hollow nanospheres with a conductive shell of about 15–40 nm in thickness and core void of about 25 nm in diameter were finally obtained through the surface oxidative graft copolymerization of aniline by using the aniline groups of PVAn on SiO_2-g-PVAn as the anchoring sites and the subsequent removal of the silica cores via HF etching. The resulting hollow nanospheres have good structural integrity because of chain entanglements in the dense comb-shaped P(VAn-g-PANI) brushes that constitute the shell.

Vamvakaki and coworkers reported the light-driven supramolecular engineering of water-dispersible, dual-responsive hairy polymer NCPs (Figure 7.5) [36]. Their synthetic route includes first the preparation of silica NPs with surface-bound ATRP-initiating alkyl bromide groups (hydrodynamic diameter $[D_h] = 267$ nm,

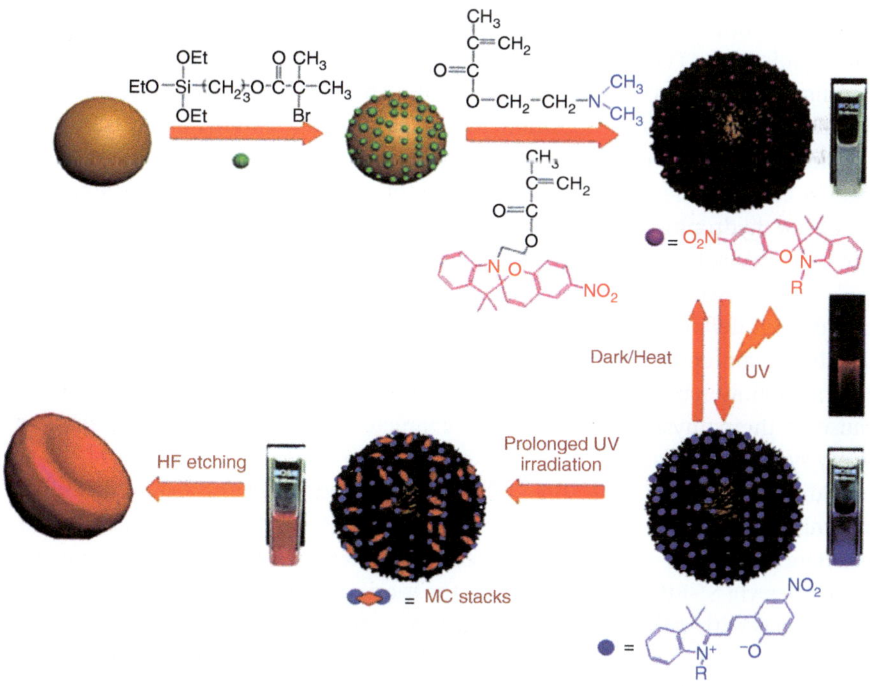

Figure 7.5 Synthetic procedure for fabricating the NCPs. Source: Achilleos et al. [36], scheme 1 (p. 5727)/Reproduced with permission from American Chemical Society.

determined by DLS), their subsequent controlled surface-grafting of a photoresponsive random copolymer PDMAEMA-*co*-PSPMA containing spiropyran (SP) units ($M_n = 66\,600\,\text{g mol}^{-1}$, $Đ = 1.29$; PSPMA refers to poly(1′-(2-methacryloxyethyl)-3′,3′-dimethyl-6-nitrospiro-(2H-1-benzopyran-2,2′-indoline)) via SI-ATRP (leading to core-brush hybrid NPs with D being $237 \pm 12\,\text{nm}$, determined by TEM), the prolonged UV irradiation of the hybrid particles to induce the isomerization of SP to merocyanine (MC) (to promote the formation of intraparticle H-type MC–MC stacks within the polymer layer, which act as noncovalent cross-linking points within the dense polymer brushes), and the final etching of silica template by using HF. The resulting physically cross-linked hollow NCPs can be disrupted remotely upon their exposure to visible light. The hydrophilic and pH-responsive nature of the DMAEMA unit allows the engineering of the vesicles in environmentally benign aqueous media and imparts the NCPs with pH-responsivity. Such light- and pH-responsive NCPs showed fluorescence in water, which can serve as addressable NCPs for biomedical applications.

Vamvakaki and coworkers developed a novel approach for preparing physically cross-linked rod-like hairy hollow polymer NPs of large size and aspect ratio with tunable shell thickness and flexibility via SI-ATRP [37]. They were synthesized by first grafting dense P(MMA-*co*-GlyMA)-*b*-PDMA block copolymer brushes (GlyMA refers to glycidyl methacrylate, PDMA = PDMAEMA; the grafting density of the polymer brushes is ~0.3 chains nm^{-2}) from the silica rods with surface-immobilized ATRP-initiating groups (with their average length and diameter being $4.5 \pm 2\,\mu\text{m}$ and $330 \pm 40\,\text{nm}$, respectively) via two-step SI-ATRP and subsequently etching the inorganic core. The solvent-incompatible inner block plays a decisive role in obtaining the rod-like polymer particles in the absence of chemical cross-linking, and the block copolymer composition affects the colloidal stability and flexibility of the hollow anisotropic colloids. An interior dense hydrophobic block of sufficient thickness enabled the polymer brush to retain the rod-like shape in water, and the exterior hydrophilic block allowed the stable dispersion of the particles in water. A combination of a relatively shorter hydrophobic block polymer chain ($M_{n,\text{GPC}} = 65\,000\,\text{g mol}^{-1}$) and a longer PDMA block ($M_{n,\text{GPC}} = 90\,000\,\text{g mol}^{-1}$) afforded well-defined and flexible hollow nanorods, whereas increasing the hydrophobic content of the copolymer resulted in rigid tube-like particles. Finally, the hairy hollow nanorods could be disrupted in a good solvent for both blocks because of their physically cross-linked character, which thus could be used in delivery systems or phase transformation applications.

In addition to the SI-CRPs, some other SI-P methods have also been utilized for preparing hairy hollow polymer NPs with silica NPs as the sacrificial template. For instance, Yin and coworkers reported the fabrication of hairy hollow polymer nanovesicles via SI-PP with silica NPs as the sacrificial template [39]. Hollow polymeric vesicles with well-defined PMMA brushes were prepared via first the surface-grafting of a cross-linked poly(DMAEMA-*co*-MBA) layer on the silica NPs with surface-functionalized thiol group ($D = 350\,\text{nm}$) via normal free radical reaction, their subsequent grafting of PMMA brushes through SI-PP of MMA in the presence of a hydrogen-abstraction photoinitiator 2-(2–3-epoxypropyloxy) thioxanthone (ETX), and final removal of the silica core by HF etching. It is the radical transfer addition polymerization taking place between the thiol groups on the silica NP surfaces and the propagating chains of PDMAEMA induced by AIBN that led to the grafting of the cross-linked poly(DMAEMA-*co*-MBA) layer onto the silica NP surfaces. Due to the presence of abundant tertiary amino groups in the cross-linked poly(DMAEMA-*co*-MBA) layer, it can generate radicals under UV irradiation to initiate the polymerization of MMA in the presence of ETX, thus leading to the successful grafting of PMMA brushes onto the poly(DMAEMA-*co*-MBA) layer. The hairy hollow polymer nanovesicles proved to be capable of encapsulating methyl orange in the apolar solvent toluene.

Savin and coworkers developed amphiphilic hairy hollow polymer NCPs through the sequential grafting of a cross-linked hydrophobic PCL layer onto the silica NPs with surface glyceryl groups (as ROP-initiating groups, $D = 70$ nm) via SI-ROP of ε-CL and a bis-CL monomer (as the cross-linker, 2.5 mol%) and hydrophilic poly(ethylene glycol) (PEG) brushes via the carbodiimide coupling chemistry, followed by removal of the silica core by HF etching (Figure 7.6) [40]. The resulting amphiphilic hairy polymer NCPs showed significant swelling in good, non-selective solvent conditions, and a collapsed hydrophobic core block in water. These amphiphilic NPs, both before and after core removal, could effectively stabilize hydrocarbons in water. In particular, the hairy hollow NCPs had c. 15 times greater uptake capacity in comparison with the solid ones.

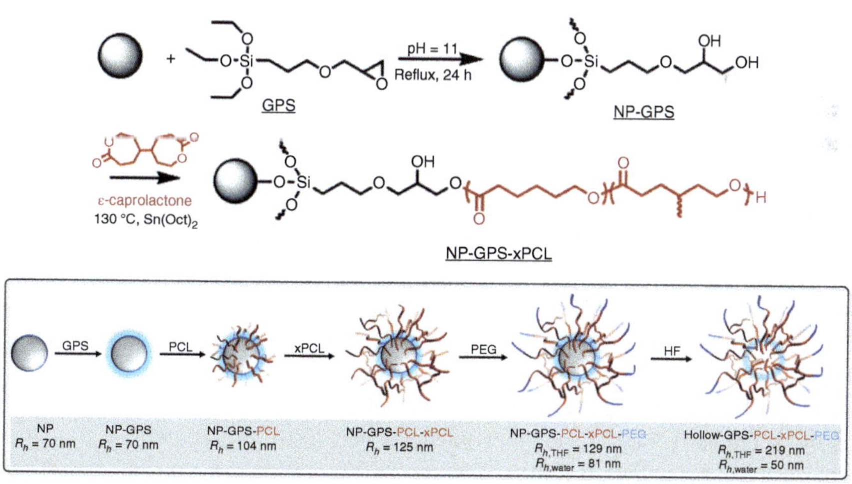

Figure 7.6 Reaction pathway to cross-linked PCL-functionalized silica NPs (NP-GPS-xPCL) (above). Schematic of the synthetic route to amphiphilic hairy hollow nanocapsules (below). Labels beneath each particle: NP, bare silica NPs; NP-GPS, NP after surface functionalization with (3-glycidyloxy)triethoxy silane; NP-GPS-PCL, NP-GPS after grafting linear PCL; NP-GPS-PCL-xPCL, NP-GPS-PCL after grafting cross-linked PCL; NP-GPS-PCL-xPCL-PEG, NP-GPS-PCL-xPCL after grafting PEG; Hollow-GPS-PCL-xPCL-PEG, NP-GPS-PCL-xPCL-PEG after silica core removal. Hydrodynamic radii as determined by DLS are given below each NP label. Source: Bentz et al. [40], scheme 2 & Fig. 1 (p. 5133)/Reproduced with permission of Royal Society of Chemistry.

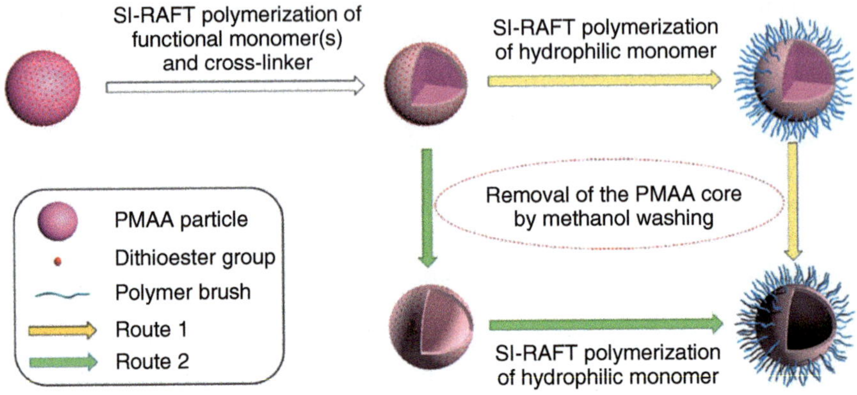

Figure 7.7 Schematic illustration for the preparation of well-defined uniform hydrophilic hairy hollow functional polymer micro- or nanoparticles by combining RAFT polymerization techniques and the sacrificial template method. Source: Zheng et al. [38], scheme 1 (p. 221)/Reproduced with permission of American Chemical Society.

As evidenced from the aforementioned literature, silica NPs were used as the sacrificial templates for preparing hairy hollow NPs in all the above cases. However, the removal of silica NPs requires rather harsh conditions (e.g. HF etching), which might not only be harmful to human bodies but also lead to significant environmental pollution. Therefore, the development of versatile approaches for the preparation of hairy hollow NPs by using sacrificial templates that can be easily etched under mild conditions is highly desirable. Very recently, our group developed a facile, general, and highly efficient strategy for preparing well-defined uniform hydrophilic hairy hollow functional polymer micro- and NPs by combining RAFT polymerization techniques and the sacrificial template method (with easily soluble PMAA micro/nanospheres as the template) (Figure 7.7) [38]. Two synthetic routes have been developed for this purpose. The first route involves one-pot synthesis of uniform "living" PMAA micro/nanospheres with surface-bound dithioester groups via reversible addition-fragmentation chain transfer precipitation polymerization (RAFTPP) [41], their successive grafting of cross-linked functional polymer layers and hydrophilic polymer brushes via SI-RAFT polymerization, and rapid etching of the PMAA core by simply washing the resulting core–shell-corona-structured particles with methanol under the assistance of sonication (3 × 5 minutes). Alternatively, such hairy hollow polymer particles can also be prepared by first the synthesis of uniform "living" hollow polymer particles (through controlled grafting of cross-linked functional polymer layers onto "living" PMAA particles prepared via RAFTPP and selective etching of the PMAA core by methanol washing) and their grafting of hydrophilic polymer brushes via SI-RAFT polymerization. Uniform hairy hollow polymer micro- or NPs with different hydrophilic polymer brushes and hydrodynamic diameters between 0.230 and 1.524 μm were successfully prepared via above routes. In particular, they could be easily imparted with pH–/thermoresponsivity, glutathione-induced degradability, and fluorescence by simply incorporating corresponding stimuli-responsive or functional units into their shells or polymer brushes, and well-defined uniform multiple stimuli-responsive fluorescent and degradable hydrophilic hairy hollow polymer NPs were readily obtained. Considering the high versatility of RAFTPP in one-pot and one-stage synthesis of uniform "living" PMAA micro/nanospheres, the general applicability of SI-RAFT polymerizations for the controlled grafting of various cross-linked polymer layers or hydrophilic polymer brushes, and easy, rapid, and environmental friendly etching of PMAA cores under mild conditions, we believe this strategy opens the new possibility for the development of various advanced hierarchical hollow polymer micro- and NPs with many potential applications, such as drug delivery and theranostics.

Precipitation Polymerization Combined with Different Polymer Brush-Grafting Methods Precipitation polymerization has proven to be a straightforward, facile, and efficient approach to afford uniform and clean micro- or nanosized polymer particles because of its easy operation and no need for any surfactant or stabilizer [2, 41]. So far, the combination of precipitation polymerization and different polymer brush-grafting methods (e.g. SI-CRPs and click chemistry) has also been demonstrated highly versatile for the synthesis of a wide range of well-defined hairy hollow polymer NPs by using silica NPs as the sacrificial template, which typically includes the grafting of various cross-linked polymer layers onto silica NPs via precipitation polymerization, introduction of certain reactive functionalities onto the resulting core–shell particle surfaces (this step can be omitted if the as-prepared core–shell particles have necessary surface functionalities for grafting polymer brushes), their grafting of polymer brushes via either "grafting-from" or "grafting-to" approach, and selective removal of the silica template core. A special kind of precipitation polymerization, named distillation precipitation polymerization (developed by Huang and coworkers in 2004 [42]), has been widely used for such a purpose. For instance, Kang and coworkers fabricated some well-defined hairy hollow polymer NPs by combining distillation precipitation polymerization and different click chemistries [43, 44]. In one example, they reported the synthesis of hairy hollow polymer nanospheres with a fluorescent shell and temperature-responsive polymer brushes via the combined use of distillation precipitation polymerization and thiol-ene click chemistry (Figure 7.8) [43]. Uniform silica nanospheres with surface-bound methacrylate groups ($D_n = 197$ nm) were first prepared by the one-pot sol–gel reaction of tetraethyl orthosilicate (TEOS) and 3-(trimethoxysilyl)propyl methacrylate (MPS). They were then used to induce the surface-initiated distillation precipitation copolymerization of *N*-vinylcarbazole (VCz) and divinylbenzene (DVB) to prepare nearly monodisperse silica core-poly(VCz-*co*-DVB) shell (briefly SiO_2@PVK) nanospheres with surface-bound vinyl groups (different shell thickness (14, 19, and 23 nm) were achieved by varying the weight ratio of (VCz + DVB) to SiO_2-MPS). Next, a thiol-terminated PNIPAM (PNIPAM-SH) was prepared via first the synthesis of PNIPAM with a dithioester end group ($M_{n,\mathrm{GPC}} = 11\,600\ \mathrm{g\,mol^{-1}}$, $Đ = 1.32$) via RAFT polymerization of NIPAM and the subsequent reduction of its dithioester end group into the thiol group by using $NaBH_4$. SiO_2@PVK nanospheres with PNIPAM brushes (i.e. SiO_2@PVK-PNIPAM) was obtained by the thiol-ene click

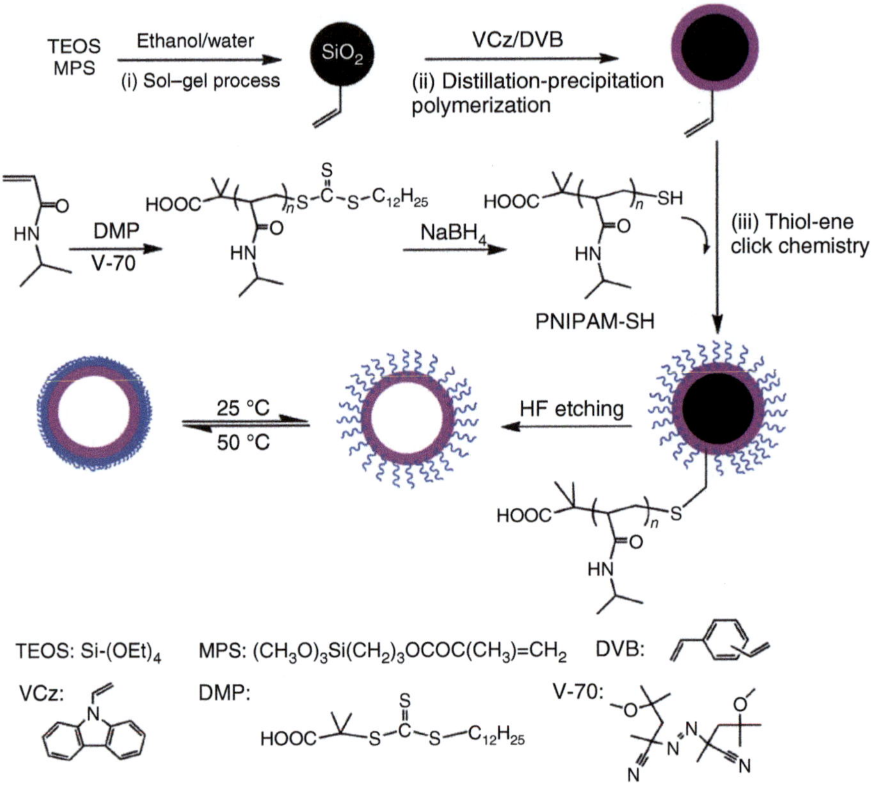

Figure 7.8 Schematic illustration of the fabrication of air@PVK-PNIPAM hairy hollow nanospheres by combined sol–gel reaction, distillation precipitation polymerization, and thiol-ene click chemistry. Source: Li et al. [43], scheme 1 (p. 5799)/Reproduced with permission of American Chemical Society.

reaction of PNIPAM-SH with the vinyl groups on the SiO_2@PVK nanospheres. The grafting density of PNIPAM brushes on the SiO_2@PVK nanospheres was about 0.1 chains nm^{-2}. Hairy hollow spherical NPs with a fluorescent and cross-linked PVK shell and temperature-responsive PNIPAM brushes were finally obtained after selective removal of the silica core from SiO_2@PVK-PNIPAM by HF etching. The combined use of distillation precipitation polymerization ("grafting from" technique) and thiol-ene click chemistry ("grafting to" technique) has provided a versatile tool for the synthesis of multifunctional hairy hollow polymer nanostructures. In another example, they further prepared hairy hollow nanospheres with surface-grafted binary polymer brushes via the combined use of distillation precipitation polymerization and dual "click" reactions (alkyne-azide and thiol-ene "click" chemistry) [44]. SiO_2@P(MAA-*co*-PMA-*co*-DVB) (PMA refers to propargyl methacrylate) core–shell hybrid nanospheres ($D = 179$ nm) were first prepared by surface-initiated distillation precipitation copolymerization of MAA, PMA, and DVB from silica nanospheres with surface vinyl groups ($D = 151$ nm) obtained from sol–gel reaction. The presence of propargyl groups from PMA and residual C=C double bonds from DVB on SiO_2@P(MAA-*co*-PMA-*co*-DVB) nanospheres enabled their respective alkyne-azide and thiol-ene reactions with the azido-terminated hydrophobic polystyrene (PS) (PSN_3, $M_n = 2550\,g\,mol^{-1}$) and PEG-SH ($M_n = 5000\,g\,mol^{-1}$). Selective removal of the silica core from the SiO_2@P(MAA-co-PMA-*co*-DVB)-PS/PEG nanospheres by HF etching eventually afforded the hairy hollow nanospheres with binary polymer brushes.

Some other hairy hollow polymer NPs have also been developed by others following the above strategy [45, 46]. For example, Chen and coworkers prepared hairy hollow polymer NPs with a pH-responsive poly(MAA-*co*-DVB) shell and temperature-responsive PNIPAM brushes by the combined use of distillation precipitation copolymerization and thiol-ene click chemistry (with SiO_2-MPS nanospheres as the sacrificial core material) and their use as stimuli-responsive drug carriers [45]. The resulting hairy hollow polymer NPs have tunable poly(MAA-*co*-DVB) shell thickness (by simply adjusting the initial MAA and DVB concentration during the distillation precipitation copolymerization) and grafting densities of PNIPAM brushes (decreased from 0.70 to 0.15 chains nm^{-2} with increasing the molecular weight of the grafted PNIPAm chains). The shell permeability of such hairy hollow NCPs toward the water-soluble Doxorubicin hydrochloride (DOX·HCl) (as the model drug) could be easily adjusted by controlling the thickness of the poly(MAA-*co*-DVB) shell, the grafting density of PNIPAm brushes, and the environmental pH and temperature. They also exhibited pH/temperature dual-responsive drug release behavior, which is promising for application as nanocontainers for controlled release of encapsulated guest molecules. Wang and coworkers also described the synthesis of thermoresponsive hairy hollow MIP nanospheres following the same strategy [46]. Their synthetic route includes first the synthesis of SiO_2@MIP NPs through grafting a cross-linked MIP layer (with (*S*)-(+)-ibuprofen [*S*-IBF], acrylamide [AM], and EGDMA as the template molecule, functional monomer, and cross-linker, respectively) onto SiO_2-MPS (180 nm) NPs via distillation precipitation polymerization, their surface-grafting of a thermosensitive PNIPAM brush via the thiol-ene click coupling reaction of PNIPAM-SH (obtained via RAFT polymerization of NIPAM and subsequent thiol-modification reaction: $M_n = 1.21 \times 10^4\,g\,mol^{-1}$, $Đ = 1.30$ for PNIPAM with a dithioester end group) and the vinyl groups on SiO_2@MIP NPs, and final removal of the silica core from the resulting SiO_2@MIP-PNIPAM nanospheres. The grafting density of PNIPAM brushes on the hollow nanospheres was about 0.18 chains nm^{-2}. Such hairy hollow NPs showed thermoresponsive size changes and drug (S-IBF) release properties (with about 50% of drug being released into acetonitrile-water solution through the specific holes of the imprinted shell at 25 °C under vibration and negligible drug being released at 45 °C owing to the collapse of the PNIPAM brushes).

To enhance the grafting density of the polymer brushes on the hairy hollow polymer NPs, Li et al. prepared hairy hollow polymer NPs with PEG brushes via the combined use of distillation precipitation polymerization and thiol-yne coupling reaction [47]. C=C double bond-modified silica NPs ($D_n = 223$ nm, determined by SEM) prepared via sol–gel reaction were used to induce the surface-initiated distillation precipitation polymerization to provide well-defined

SiO_2@poly(PMA-*co*-MAA-*co*-EGDMA) core–shell nanospheres. Following the grafting of PEG brushes onto the above core–shell nanospheres through the photoinitiated thiol-yne coupling reaction of PEG-SH ($M_n = 5000\,g\,mol^{-1}$) with the propargyl groups on the nanospheres and removal of the silica core by HF etching, hairy hollow polymer NPs (air@poly(PMA-*co*-MAA-*co*-EGDMA)-PEG) were obtained. A rather high grafting density of 0.95 chains nm^{-2} was achieved owing to the efficient coupling of two PEG-SH onto one propargyl unit. The combined precipitation polymerization and thiol-yne surface reaction provide a new and efficient synthetic strategy to construct hairy hollow polymer nanocontainers with highly functionalized surfaces.

Recent years have witnessed tremendous interest in the rattle-type (or yolk-shell) NPs because of their intriguing physicochemical properties and unique morphological features stemming from a movable core within a hollow shell [48, 49]. Kang and coworkers developed some hairy hollow polymer NPs with a movable metal nanocore (i.e. hairy rattle-type or yolk-shell polymer NPs) for different purposes through the combined use of distillation precipitation polymerization and different polymer brush-grafting methods. In their first such report, hairy rattle-type hybrid nanospheres with a metal nanocore (gold or silver) inside the cavity, a cross-linked poly(MAA-*co*-DVB) shell, and functional polymer brushes on the exterior surfaces were prepared (Figure 7.9) [50]. The incorporation of metal nanocore inside the cavity of the hairy hollow nanospheres was achieved by using the gold@silica or silver@silica core–shell NPs as the sacrificial template for the

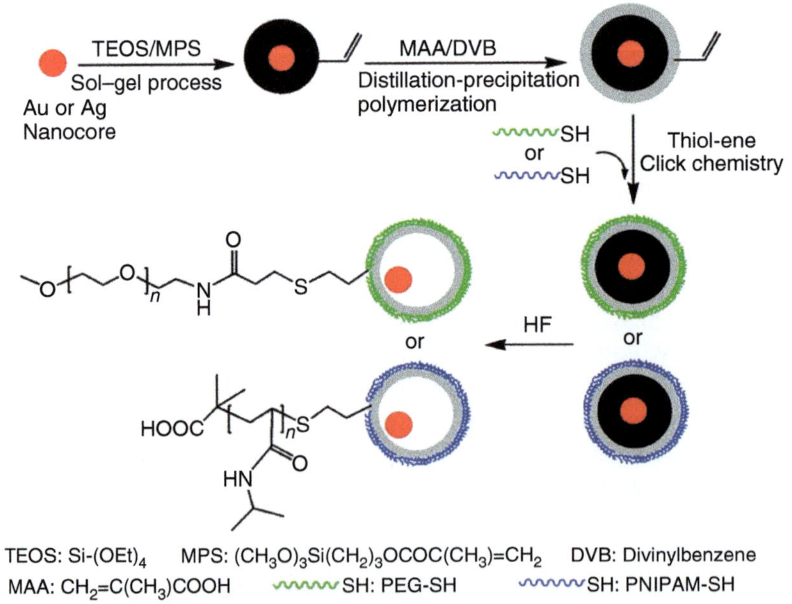

Figure 7.9 Schematic illustration of the synthesis of hairy metal@air@polymer hybrid microrattles with a metal nanocore, a cross-linked polymer shell, and functional polymer brushes on the exterior surface. Source: Li et al. [50], scheme 1 (p. 2365)/Reproduced with permission of American Chemical Society.

formation of metal@silica@polymer core-double shell nanospheres. PEG-SH or PNIPAM-SH were covalently grafted onto the residual C=C double bonds of the DVB unit on the nanosphere surfaces via the thiol-ene "click" reaction to afford a hairy surface of tailored functionality. These hairy rattle-type polymer nanospheres were successfully used for the confined heterogeneous catalytic reaction.

Cai and Kang's groups developed a general and effective strategy for the silica template-assisted synthesis of monodisperse hairy rattle-type polymer nanospheres by the combined use of surface distillation precipitation polymerization, click chemistry, supramolecular assembly, and their use as nanocatalysts. Such hairy nanorattles consist of a movable platinum (Pt) nanocluster and a temperature- and pH-responsive hairy polymer shell (i.e. Pt@air@P(MAA-*co*-(PMA-click-β-CD-guest-PVCL)). β-CD and PVCL refer to β-cyclodextrin and poly(*N*-vinylcaprolactam), respectively [51]. Their synthetic route includes the following five steps: (i) synthesis of Pt@SiO_2-MPS core–shell NPs by coating Pt nanoclusters with a uniform silica shell bearing surface vinyl groups via the sol–gel reaction of TEOS and MPS; (ii) preparing Pt@SiO_2@P(MAA-*co*-PMA) core-double shell NPs with surface alkyne groups by distillation precipitation copolymerization of MAA, PMA, and DVB from the Pt@SiO_2-MPS core–shell template; (iii) obtaining Pt@SiO_2@P(MAA-*co*-(PMA-click-β-CD)) NPs through the copper(I)-catalyzed alkyne-azide click coupling reaction of mono(6-azido-6-desoxy)-β-CD (N_3-β-CD) with the alkyne groups on the exterior surfaces of Pt@SiO_2@P(MAA-*co*-PMA) NPs; (iv) hairy Pt@SiO_2@P(MAA-*co*-(PMA-click-β-CD-guest-PVCL)) core-double shell NPs were generated via the supramolecular assembly of the adamantly-terminated PVCL (Ada-PVCL) chains with the surface-tethered β-CD units on the Pt@SiO_2@P(MAA-*co*-(PMA-click-β-CD)) NPs; (v) selective removal of the silica inner shell in the hairy Pt@SiO_2@P(MAA-*co*-(PMA-click-β-CD-guest-PVCL)) core-double shell NPs by HF etching led to the desired hairy Pt@air@P(MAA-*co*-(PMA-click-β-CD-guest-PVCL)) nanorattles. The protective and stimuli-responsive polymer shell on the hairy nanorattles not only promotes the efficient mass transfer to the encapsulated Pt nanoclusters but also functions as a physical barrier to prevent the coalescence of Pt nanocores. Furthermore, it affords a void space for the organic transformation to occur on the surface of the ligand-free Pt nanocluster in a controlled manner. Such hairy nanorattles were demonstrated to be a robust and reusable heterogeneous catalyst for catalytic reactions.

Chen and Kang's groups also described the silica template-assisted synthesis of some monodisperse hairy rattle-type nanospheres via the combined use of distillation precipitation polymerization and oxidation polymerization methods and their use as memory devices [52, 53]. In one example, they fabricated monodisperse hairy rattle-type polymer nanospheres with a movable Au nanocore inside the hollow cavity of an electroactive polymer shell (Au@air@PTEMA-g-P3HT hybrid nanorattles; PTEMA or poly(TEMA) refers to poly(2-(thiophen-3-yl)ethyl methacrylate); P3HT or poly(3HT) refers to poly(3-hexylthiophene)) [52]. Their synthetic route includes first the preparation of Au@SiO_2 core–shell NPs with surface methacrylate groups (Au@SiO_2-MPS) by performing Stöber sol–gel reaction on Au NP seeds, grafting a cross-linked poly(TEMA-*co*-DVB) layer onto

Au@SiO_2-MPS NPs via surface-initiated distillation precipitation copolymerization of TEMA and DVB (leading to Au@SiO_2@PTEMA core-double shell NPs), further grafting electroactive P3HT brushes onto Au@SiO_2@PTEMA NPs via the oxidative graft polymerization of 3-HT from the exterior surface of the Au@SiO_2@PTEMA core-double shell NPs (leading to Au@SiO_2@PTEMA-g-P3HT NPs), and final removal of the silica interlayer from Au@SiO_2@PTEMA-g-P3HT NPs by HF etching. A blend of the resulting hairy rattle-type polymer NPs and an electrically insulating PS was used to fabricate nonvolatile memory devices. The resulting Al/Au@air@PTEMA-g-P3HT + PS/ITO (ITO: indium-tin oxide) sandwich thin-film devices showed unique electrical behaviors, including insulator behavior, write-once-read-many-times and rewritable memory effects, and conductor behavior. In another example, they reported the synthesis of hairy rattle-type polymer nanospheres with a movable Au nanocore, a cross-linked poly(VAn-*co*-DVB) (VAn : 4-vinylaniline) shell, and PANI brushes (i.e. Au@air@PVAn-g-PANI nanorattles) following a similar synthetic strategy [53]. Such hairy nanorattles were prepared via first the synthesis of monodisperse Au@SiO_2-MPS core–shell nanospheres, their surface-grafting of a cross-linked poly(VAn-*co*-DVB) by distillation precipitation polymerization (leading to uniform Au@SiO_2@PVAn nanospheres), further grafting PANI brushes onto Au@SiO_2@PVAn nanospheres via surface-initiated oxidative graft copolymerization of aniline using the ANI moieties of PVAn as the anchoring site (providing Au@SiO_2@PVAn-g-PANI nanospheres), and final removal of the silica interlayer of the hairy Au@SiO_2@PVAn-g-PANI NPs by HF etching. An electrically insulating poly(vinyl alcohol) (PVA) matrix integrated with the above obtained nanorattles was used to fabricate the Al/Au@air@PVAn-g-PANI+PVA/ITO-coated poly(ethylene terephthalate) flexible device (with a turn-on voltage of −2.0 V and an ON/OFF state current ratio of more than 10^4), which can be switched to the ON state under a negative electrical sweep and reset to the initial OFF state by a reverse (positive) electrical sweep. The memory behaviors of the flexible device were almost unchanged under the bending test.

Some hairy hollow polymer NPs have also been prepared by the combined use of traditional precipitation polymerization and different polymer brush-grafting methods [54, 55]. For instance, Li and Cai's group described the silica template-assisted synthesis of hairy hollow nanocomposites with ruthenium-bipyridine complexes by applying traditional precipitation polymerization and click chemistry and their use for regulating the photoinduced electron/energy transfer-reversible addition-fragmentation chain transfer (PET-RAFT) polymerization upon multiple external stimuli and simplifying the procedures of post-polymerization purification [54]. Such hairy hollow polymer nanocatalysts had a cross-linked hyperbranched polyglycerol (HPG) shell and surface-conjugated binary polymers with polyfluorene-derived backbones, thermosensitive polymer brushes, and ruthenium(II)-bipyridine complex adducts. Their synthetic route includes the following five steps: (i) preparing narrowly dispersed SiO_2-MPS nanospheres with a diameter of 150 ± 12 nm via Stöber sol–gel reaction; (ii) grafting HPG to SiO_2-MPS nanospheres by surface traditional precipitation polymerization of methacryloyloxy-functionalized hyperbranched poly(glycerol-*co*-glycidyl propargyl ether) (HPGEm, $M_{n,\text{GPC}} = 6200\,\text{g mol}^{-1}$, $Đ = 1.19$), leading to SiO_2@HPGE core–shell NPs; (iii) obtaining hairy SiO_2@HPGE-PFPPN core–shell NPs via the click coupling reaction of azide group-containing poly[(9,9-bihexylfluorene)-*alt*-(5,5′-(2, 2′-bipyridine))]-*graft*-PNIPAM (PFPPN, $M_{n,\text{GPC}} = 29\,000\,\text{g mol}^{-1}$, $Đ = 1.58$) with the peripheral alkyne groups on the SiO_2@HPGE NPs; (iv) selective HF etching of the silica nanocore of the hairy SiO_2@HPGE-PFPPN NPs, followed by the incorporation of ruthenium(II) bipyridine moieties into a polyfluorene scaffold, led to the desired hairy hollow polymer nanocatalysts. Such hairy hollow nanocatalysts could regulate RAFT polymerization using light in the "ON/OFF" state and a temperature "BELOW LCST"/"ABOVE LCST." They have many advantages, such as high dispersity in various solvents and compatibility with many different monomers, as well as their heterogeneous nature and robust mechanical properties achieved through the cross-linked polymer shell. Their catalyzed RAFT photopolymerizations can proceed in both aqueous and organic media under low-energy light irradiation, at ppm-level doses of catalyst and in an oxygen-tolerant manner. In addition, they can be readily separated and reused via rapid centrifugation, and show inappreciable catalyst leakage and consistent catalytic performance even after multiple polymerization runs. In another example, You and Zhou's groups reported a novel silica template-assisted strategy for selectively grafting polymer brushes from the interior and/or exterior surfaces of bioreducible and temperature-responsive NCPs by the combined use of traditional precipitation polymerization and Ce^{4+}-initiated grafting polymerization [55]. Poly(NIPAM-*co*-PEG methacrylate) (P(NIPAM-*co*-PEGMA)) NCPs with hydroxyl groups at the interior, exterior, or both sides of the shells were prepared via first the surface precipitation copolymerization of NIPAM, PEGMA, and *N,N*′-cystaminebisacrylamide (CBA) on the SiO_2-MPS NPs ($D = \sim 100\,\text{nm}$) and the subsequent removal of the silica core by HF etching. The control of their hydroxyl group location was realized through tuning the addition time of PEGMA. The P(NIPAM-*co*-PEGMA) NCPs with poly(2-(dimethylamino)ethyl acrylate) (poly(DMAEA) or PDMAEA) brushes at the interior and/or exterior surfaces were then obtained via Ce^{4+}/mediated redox-initiated polymerization of DMAEA from the hydroxyl groups located on different places of P(NIPAM-*co*-PEGMA) NCPs. Such hairy polymer NCPs showed temperature-, pH-, and redox-responsivity.

In addition to the aforementioned synthesis of hairy hollow polymer NPs by the combined use of distillation or traditional precipitation polymerization of vinyl monomers and different polymer brush-grafting methods, some hairy hollow NPs with polydopamine (PDA) shells have also been prepared through first the surface precipitation polymerization (or oxidation polymerization) of dopamine (DA) on the sacrificial template, their subsequent grafting of polymer brushes, and final removal of the sacrificial template [56, 57]. For example, Kohri and coworkers described a facile method to prepare hairy hollow polymer NPs with a colorless PDA thin layer and PHEMA brushes by the combined use of oxidation polymerization of DA and SI-ATRP with polystyrene (PSt or PS) NPs ($D = 410$ and $188\,\text{nm}$) as the sacrificial template (Figure 7.10) [56]. The *in situ* oxidative copolymerization

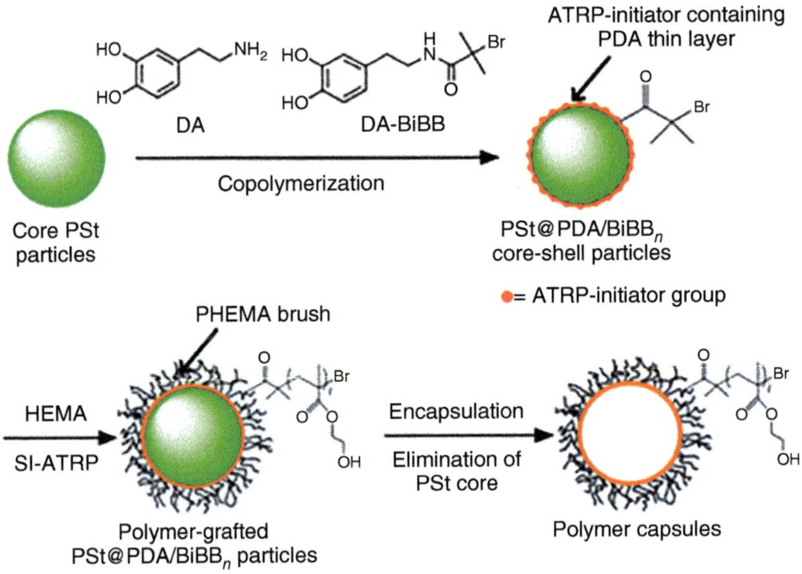

Figure 7.10 Schematic representation of the preparation of polymer capsules using a PDA thin layer as a basis for polymer brushes. Source: Kohri et al. [56], Fig. 1 (p. 2697)/ Reproduced with permission of Royal Society of Chemistry.

of DA and ATRP initiator (BiBB)-bearing DA (DA-BiBB) onto PSt core particles led to PSt@PDA/BiBB$_n$ ($n = 2$, the molar ratio of BiBB to DA) with surface-bound ATRP-initiating groups. Following the use of PSt@PDA/BiBB$_n$ as the immobilized ATRP initiator for the SI-ATRP of HEMA and removal of the PSt core from the obtained PSt@PDA/PHEMA core–shell-corona NPs by using THF, colorless hairy hollow PDA NPs with PHEMA brushes were obtained. Their hollow core sizes and capsule wall thicknesses could be tailored by choosing PSt sacrificial templates with different sizes and controlling the SI-ATRP reaction conditions. The successive preparation of functional PHEMA capsules by further grafting other polymer chains (e.g. PDMAEMA) and post-functionalization of hydroxyl groups of PHEMA chains with a fluorescent dye (dansyl chloride) were also demonstrated.

Jia and coworkers also reported the synthesis of hairy hollow PDA NPs with PDMAEMA brushes and their use as the thermo-controlled pesticide release system [57]. PDA-coated SiO_2 NPs were prepared by coating a PDA layer onto SiO_2 nanospheres ($D_n = 220$ nm) via the oxidation polymerization of DA. The thermoresponsive PDMAEMA brushes (the LCST of PDMAEMA is c. 40 °C) were then grafted on SiO_2@PDA particle surfaces via the metal-free SI-PP of DMAEMA without adding any photoinitiator or photosensitizer under UV light irradiation, following a previously reported method [58]. The subsequent HF etching of the silica core from the above-obtained SiO_2@PDA@PDMAEMA NPs led to the desired hairy hollow polymer NCPs ($D_h = 459$, 450, and 269 nm at 20, 30, and 50 °C, respectively [determined by DLS]). They not only showed a high loading ability of a pesticide avermectin (Av) (up to 52.7% [w/w]) but also exhibited temperature-controlled

Av release behavior. They hold much promise in various applications, such as in controlled drug release and agriculture-related fields.

Layer-by-Layer (LbL) Assembly Combined with Polymer Brush-Grafting Method LbL assembly method is a technique widely used for developing thin films with defined nanostructures on both planar surfaces and micro- or NPs, which can be readily implemented by the alternative adsorption of oppositely charged polymers from aqueous media onto the target surfaces [24, 25]. The combination of LbL assembly (on the sacrificial template particle surfaces) and polymer brush-grafting method has been used to develop hairy hollow polymer micro- or NPs, which involves first the alternative adsorption of oppositely charged polymers onto the sacrificial template surfaces, selective removal of the template core from the resulting core–shell particles, and final grafting of polymer brushes onto the interior or exterior of the hollow particle shells. For instance, Kim and coworkers described the preparation of polyelectrolyte hollow capsules with polymer brushes grafted on the interior of their shells via first the fabrication of polyelectrolyte hollow capsules with initiator bound on the interior of their shells and the subsequent grafting-from polymerization (Figure 7.11) [59]. Poly(allylamine hydrochloride) (PAH), poly(styrene sulfonate) (PSS), and potassium peroxodisulfate (KPS) were used as the positive polyelectrolyte, negative polyelectrolyte, and water-soluble initiator, respectively. The hollow capsules with initiator bound on the interior of the shells were fabricated through first the adsorption of KPS onto the commercially available weakly cross-linked melamine formaldehyde (MF) colloidal particle surfaces, the stepwise adsorption of PAH and PSS, and the final selective removal of the MF core by its decomposition in HCl solution. Different contents of a water-soluble monomer styrene sulfonate (SS) (25 and 40 wt%) were then dispersed in a suspension of initiator-bound hollow capsules to grow PSS brushes via the grafting-from polymerization method at 70 °C

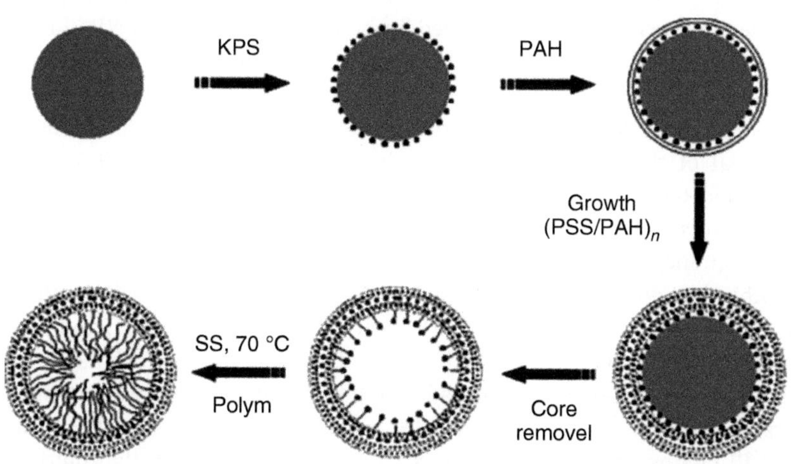

Figure 7.11 Schematic representation of the procedures used for fabricating initiator-coated hollow capsules and grafting-from polymerization inside the capsules. Source: Choi et al. [59], fig 1 (p. 1097)/Reproduced with permission of John Wiley & Sons.

for four hours. Since the initiator is present only on the interior of the shells of the hollow capsules, the PSS brushes could only be selectively grafted inside the capsules. The capsules were separated from the solution of the SS monomers by centrifugation after the polymerization, whose shapes were found to be dependent on the initial SS content. The hollow capsules obtained via polymerization in solutions containing 25 and 40 wt% of SS were "caved-in" capsules (as a result of insufficient amounts of polymer inside the capsules) and "semispherical" capsules (owing to the generation of sufficient polymers [PSS] to fill the capsules), respectively. The size of the "caved-in" and "semispherical" capsules was about 600 and 800 nm (determined by SEM), respectively. The influence of the confined interior of the hollow capsules on the detailed polymerization behavior was studied through polymerizing SS (with 40 wt% of SS monomer) by using four different approaches to form "floating-outside" polymers (obtained from free-floating initiators outside the hollow capsules), "floating-inside" polymers (obtained from free-floating initiators inside the hollow capsules), "immobilized-outside" polymers (grown from the outer surface of highly cross-linked MF particles), and "immobilized-inside" polymers (grown from the inner surface of the capsules). The floating-inside polymers have a molecular weight and molar mass dispersity that are about twice those of the floating-outside polymers, indicating the existence of some confined-space effects. In particular, the immobilized-inside polymers grown from the inner surface have a molecular weight that is an order of magnitude higher than both the floating-inside and the floating-outside polymers, which was considered to be likely due to both immobilization and confined-space effects.

7.2.1.2 Self-Assembly (of Block Copolymers) Method

The self-assembly of block copolymers in solutions has garnered tremendous interest over the past several decades because of the high diversity and complexity of the resulting nanoassemblies (e.g. spheres, cylinders, lamellae, vesicles, etc.) [60]. This versatile technique has also been utilized for preparing hairy hollow NPs. Liu's group pioneered this research direction. They developed a series of hairy hollow polymer NPs with polymer brushes grafted on different places of the shells by directly cross-linking the self-assembled block copolymers or by their combination with the removal of the cores through selective etching [61–65]. In 1997, Liu and Ding reported their first preparation of hairy hollow NPs through first the fabrication of nanovesicles with a poly(2-cinnamoylethyl methacrylate) (PCEMA) shell and polyisoprene (PI) chains on both the interior and exterior of the shell via the self-assembly of a diblock copolymer PI-*b*-PCEMA (with 160 units of isoprene (Ip) and CEMA, respectively) in hexanes/THF (with 40% THF by volume) and the subsequent photo-cross-linking of the PCEMA shell (Figure 7.12) [61]. Following the above work, they also prepared water-soluble hollow nanospheres as potential drug carriers by first the formation of vesicles in THF/hexanes with PCEMA as the shell and PI chains stretching into the solution phase from both the interior and exterior of the PCEMA shell (the hollow cavity diameter of the vesicles is c. 38 nm) via the self-assembly of a PI-*b*-PCEMA sample (with 88 units of isoprene and 230 units of CEMA), the cross-linking of the PCEMA shell, and the subsequent

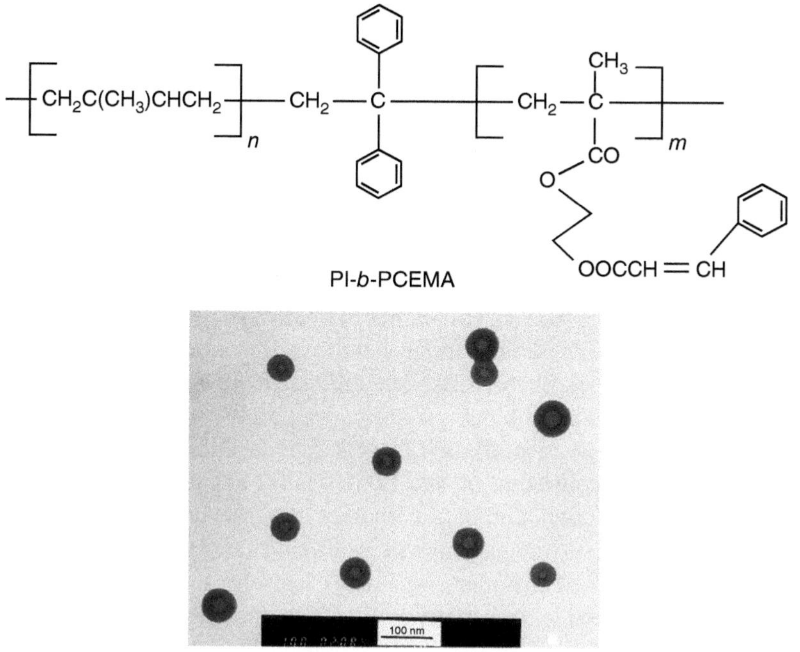

Figure 7.12 The chemical structure of the diblock copolymer PI-*b*-PCEMA (above) and its vesicles prepared in hexanes/THF with 40% THF (the TEM sample was prepared by spraying the vesicle solution on a carbon-coated copper grid) (below). Source: Ding et al. [61], fig 1 (p. 655)/Reproduced with permission from American Chemical Society.

conversion of PI chains into water-soluble poly(2,3-dihydroxyl-2-methyl-butane) chains [62]. The resulting water-soluble hairy hollow nanospheres could encapsulate a large amount of rhodamine B in methanol and released the compound into water at a tunable rate depending on the amount of ethanol added to the aqueous medium, which hold much promise as drug carriers in controlled drug release applications. Later on, the same group also prepared semi-shaved hollow nanospheres (i.e. hairy hollow nanospheres with PI brushes only on the interior shell) from PI-*b*-PCEMA by first fabricating hairy hollow nanospheres with PI on both sides of the shell and the subsequent treatment of the cross-linked vesicles with ozone for a short period (which could selectively degrade the outer PI chains to yield semi-shaved hollow nanospheres) [63]. Another hairy hollow polymer NPs with hydrophilic polymer brushes on the interior of the shells were also prepared by Liu's group through first the synthesis of a diblock copolymer poly(solketal methacrylate)-*b*-PDMAEMA (PSMA-*b*-PDMAEMA, with 105 SMA units and 122 DMAEMA units) via ATRP, the formation of spherical micelles with PDMAEMA coronas and PSMA cores in water with 5 vol% THF, cross-linking of the PDMAEMA coronas with 1,2-bis(2′-iodoethoxyl)ethane, and final hydrolysis of the acetonide groups of PSMA [65]. Furthermore, they also developed hairy hollow nanospheres with polymer brushes only on the exterior of the shells by first the synthesis of a triblock copolymer PI-*b*-PCEMA-*b*-P*t*BA (with 370 units of isoprene, 420 units of

CEMA, and 550 units of *t*BA), its formation of spherical micelles with a P*t*BA corona, PCEMA shell, and PI core in THF/methanol, UV cross-linking of the PCEMA shell, and degradation of the PI block by ozonolysis to yield nanospheres [64]. Such hairy hollow polymer NPs could be loaded with rhodamine B.

Following the above pioneering works, significant advances have been made in the development of hairy hollow polymer NPs via cross-linking the self-assembled block copolymers. In particular, some stimuli-responsive hairy hollow polymer NPs have been prepared by introducing stimuli-responsive components into their shells or polymer brushes via the self-assembly method [66, 67]. For example, Liu and Hu utilized the self-assembly of a triblock copolymer to prepare pH-responsive hollow NCPs with poly(ethylene oxide) (PEO) brushes and regularly sized nanochannels on the shell (Figure 7.13) [66]. Such NCPs were fabricated via first the formation of vesicles with the soluble PEO block as the corona and insoluble P*t*BA and PCEMA chains forming the vesicular shell from a pseudo miktoarm copolymer (μ-(P*t*BA)(PCEMA)(PEO)) consisting of one P*t*BA chain (100 repeat units), one PCEMA chain (130 repeat units), and approximately 1.14 PEO chains (114 repeat units) (Figure 7.13a) (in THF/water), photo-cross-linking the PCEMA wall with permeated P*t*BA chain-formed cylinders, and final hydrolysis of P*t*BA to PAA (poly(acrylic acid) or poly(AA)). Such hairy hollow NPs exhibited pH-responsive

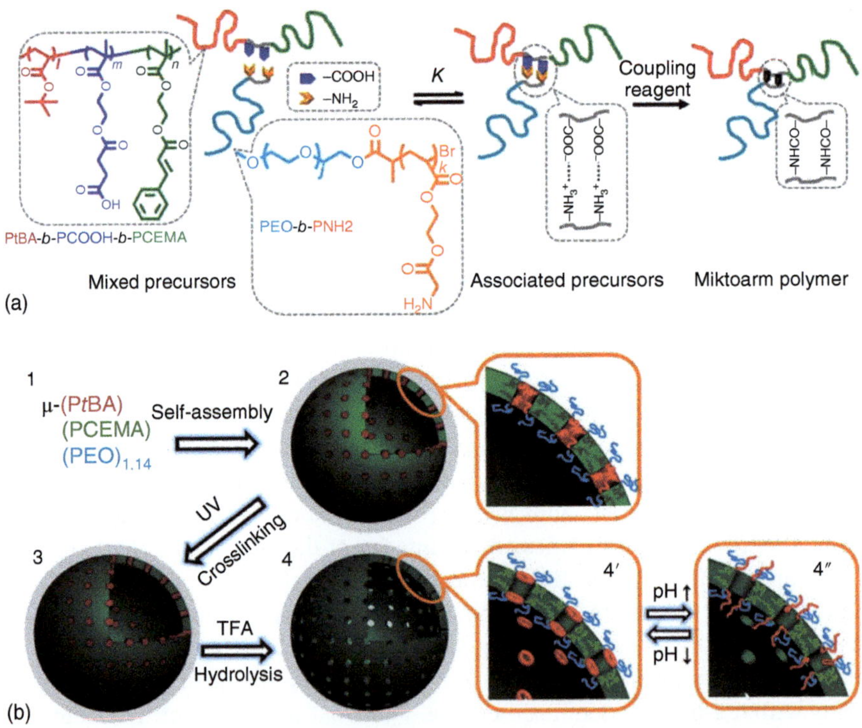

Figure 7.13 (a) Preparation of μ-(PtBA)(PCEMA)(PEO)$_{1.14}$ by the "association-and-reaction" strategy. (b) Preparation of vesicles permeated by pH-responsive nanochannels. Source: Hu and Liu [66], scheme 1 (p. 5097)/Reproduced with permission of American Chemical Society.

reagent release in aqueous media because of their regularly packed uniform PAA-gated nanochannels (in acidic media, these PAA chains are H-bonded and assume compact conformations [4′ in Figure 7.13b]). In basic media, the PAA chains expand and may escape from the nanochannels due to entropy gain (4″). Hu and coworkers also developed pH-responsive porous NCPs with surface-grafted PEG brushes ($D_h = 220$ nm at pH 7.4 [determined by DLS]; the shell thickness < 12.5 nm [determined by AFM]) through first the synthesis of a ternary graft copolymer (PGMA-g-(PCEMA-*r*-MPEG-*r*-PDEAEMA)) (PGMA refers to poly(glycidyl methacrylate) or poly(GMA), MPEG refers to PEG methyl ether) by grafting three types of polymer chains (each with a –C≡CH end group) onto the backbone polymer ($PGMA_{36}$-N_3, it was prepared via first the synthesis of $PGMA_{36}$ via ATRP of GMA and the subsequent epoxy ring-opening reaction of $PGMA_{36}$ with sodium azide) via click chemistry, its self-assembly in DMF/water into vesicles with PEG chains as the corona and a dominant PCEMA continuous phase that was interspersed by PDEAEMA domains as the wall, and photo-cross-linking of PCEMA in the wall [67]. The resulting hairy hollow NCPs showed a larger hydrodynamic size and enhanced release of the encapsulated pyrene at lower pH, mainly because the capsule walls became porous following the protonation of the PDEAEMA chains and their stretching into the water.

Besides the aforementioned covalently connected block copolymers, some supramolecular multiblock copolymers have also been used to prepare hairy hollow polymer NPs via the self-assembly method, which not only simplifies the synthetic process but also enables easy template etching under mild conditions [68, 69]. For example, O′Reilly and coworkers described the fabrication of hairy hollow polymer NPs with hydrophobic polymer brushes grafted on the interior shell surfaces from the self-assembled supramolecular metallo-triblock copolymers (Figure 7.14) [68]. A supramolecular amphiphilic metallo-triblock copolymer

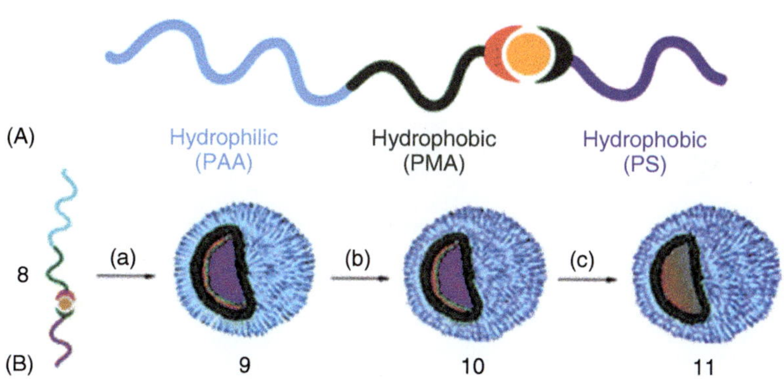

Figure 7.14 (A) Schematic representation of the triblock metallopolymer with a Pd metal center. (B) (a) Solution self-assembly of the amphiphilic triblock metallopolymer to afford uniform spherical micelles. (b) Selective cross-linking of the PAA shell of micelles to afford covalently cross-linked NPs. (c) Removal of the PS core to afford a hairy hollow PAA-*b*-PMA nanocage. Source: Moughton et al. [68], scheme 1 & 6 (p. 2364–2367)/Reproduced with permission of Royal Society of Chemistry.

(i.e. PAA-*b*-PMA-Pd-PS) (PMA refers to poly(methyl acrylate) or poly(MA)) (Figure 7.14A) was obtained by first the preparation of a Pd-functionalized diblock copolymer (P*t*BA$_{90}$-*b*-PMA$_{45}$-Pd) through the synthesis of a diblock copolymer P*t*BA$_{90}$-*b*-PMA$_{45}$ with a SCS "pincer" ligand end group ($M_{n,\text{GPC}} = 13\,500$ g/mol, $Đ = 1.15$) via RAFT polymerization by using a SCS "pincer" ligand- containing trithiocarbonate RAFT agent, its removal of trithiocarbonate end group (to avoid its coordinating with Pd), and subsequently complexing with Pd(II), cleavage of the *t*-butyl ester group in P*t*BA$_{90}$-*b*-PMA$_{45}$-Pd (leading to PAA-*b*-PMA-Pd) by TFA, and final supramolecular coordination with a pyridine end-functionalized PS homopolymer (prepared via NMP, $M_{n,\text{NMR}} = 3600$ g/mol, $Đ = 1.18$). Uniform spherical micelles ($D_\text{h} = 74$ nm, determined by DLS) were obtained through the solution self-assembly of the supramolecular amphiphilic triblock copolymer (Figure 7.14B). The subsequent cross-linking of the PAA shell of the micelles by using a diamine cross-linker and selective removal of the PS core by protonation of the pyridine moiety on the PS chain-end at low pH eventually led to hairy hollow polymer nanospheres with hydrophobic PMA brushes and functional Pd(II) complexes on the internal wall ($D_\text{h} = 168$ nm, determined by DLS).

Tian and coworkers also developed hairy hollow polymer NPs with thermoresponsive polymer brushes on the exterior of the shell for controlled drug delivery by the self-assembly of a supramolecular triblock copolymer (Figure 7.15) [69].

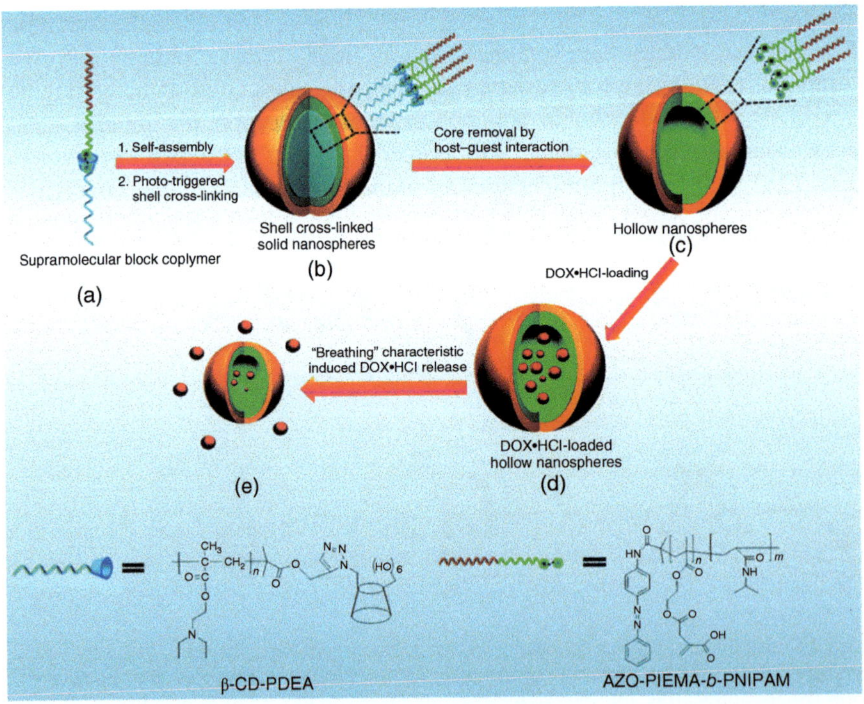

Figure 7.15 Preparation of PIEMA-*b*-PNIPAM-based hairy hollow polymer nanospheres and their controlled release of DOX·HCl. Source: Zhang et al. [69], scheme 1 (p. 8529)/ Reproduced with permission of Royal Society of Chemistry.

The synthetic route for such hairy hollow NPs involves three steps: (i) the synthesis of poly((itaconoyloxy)ethyl methacrylate)-*block*-PNIPAM with an azo end group (AZO-PIEMA-*b*-PNIPAM) (via first the successive ATRP of HEMA and NIPAM and the subsequent esterification of the hydroxyl groups in the block copolymer with itaconic anhydride [IA]) and β-CD-terminated PDEA (i.e. β-CD-PDEA, PDEA = PDEAEMA) (via ATRP using β-CD-Br as the initiator); (ii) construction of the supramolecular block copolymer via the host–guest interaction between AZO and β-CD of the above two polymers; (iii) self-assembly of the supramolecular triblock copolymer into spherical micelles with PDEA as the core and PIEMA-*b*-PNIPAM as the shell, photo-cross-linking of the middle PIEMA layer in the micelles, and the removal of PDEA core in the cross-linked micelles by dissociating the host–guest interaction under 365 nm UV irradiation at a pH of 6.8 (Figure 7.15a–c). The resulting hairy hollow nanospheres showed thermoresponsive "breathing" characteristics owing to their presence of PNIPAM brushes, which led to the temperature-controlled release of DOX·HCl (Figure 7.15d,e). They also show high biocompatibility, and the DOX-loaded hollow nanospheres showed a higher cytotoxicity against A2780 cells at 37 °C than that at 25 °C.

Despite the above progress, it is still highly challenging to prepare hairy hollow polymer NPs from the self-assembled block copolymers on a large scale because the self-assembly processes are generally performed at high dilution. Recently, aqueous polymerization-induced self-assembly (PISA) has been used to address this issue because it allows the single-step preparation of various block copolymer nanostructures (e.g. worm-like micelles and vesicles) at high solid concentrations (typically 10–30% w/w) in a reproducible manner [70, 71]. For example, O'Reilly and coworkers developed hairy hollow polymer NPs with tunable membrane permeability by aqueous emulsion PISA (Figure 7.16) [72]. Aqueous RAFT-mediated photoinitiated PISA (photo-PISA) of a mixture of water-miscible 2-hydroxypropyl methacrylate (HPMA) and water-immiscible GlyMA as the core-forming monomers was carried out under emulsion polymerization conditions with 405 nm visible light irradiation at 37 °C (using PEG_{113} macro-CTA as both the steric stabilizer block and the surfactant for stabilizing the heterogeneous monomer-in-water solution), leading to single-phase epoxy-functionalized PEG_{113}-*b*-P(HPMA-*co*-GlyMA) diblock copolymer vesicles with $DP_{PHPMA} = 320$ (80 mol%) and $DP_{PGlyMA} = 80$ (20 mol%)

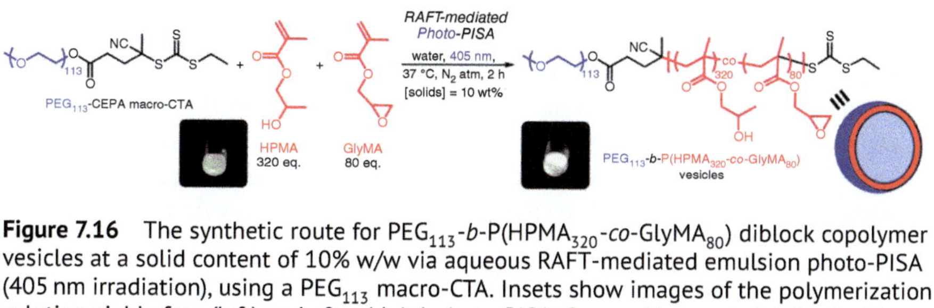

Figure 7.16 The synthetic route for PEG_{113}-*b*-P($HPMA_{320}$-*co*-$GlyMA_{80}$) diblock copolymer vesicles at a solid content of 10% w/w via aqueous RAFT-mediated emulsion photo-PISA (405 nm irradiation), using a PEG_{113} macro-CTA. Insets show images of the polymerization solution vial before (left) and after (right) photo-PISA. Source: Varlas et al. [72], fig 1A (p. 12645)/Reproduced with permission from Royal Society of Chemistry.

at 10% w/w solid content (Figure 7.16). Similarly, a catalytic enzyme (horseradish peroxide)-loaded PEG_{113}-*b*-P($HPMA_{320}$-*co*-$GlyMA_{80}$) polymersome nanoreactors were also prepared via one-pot aqueous emulsion photo-PISA under the above same reaction condition except adding the enzyme into the reaction solution. The epoxy-functionalized PEG_{113}-*b*-P(HPMA-*co*-GlyMA) diblock copolymer vesicles and enzyme-loaded ones had DLS-derived D_h of 188.3 ± 5.6 and 182.6 ± 2.5 nm, respectively, and average shell thickness of 28.0 ± 3.0 and 27.9 ± 2.6 nm (determined by cyto-TEM), respectively. Further functionalization of the epoxide groups in P(HPMA-*co*-GlyMA) shell of the vesicles with a series of primary amines and cross-linking diamines led to a distinct increase in the vesicular membrane thickness and as a consequence in less-hydrated nanoreactors with tunable permeability toward small molecule substrates, as determined by enzymatic assays.

Hong and Zhang's group also developed hairy hollow polymer nanospheres with adjustable pH-responsive drug release performance via PISA [73]. A series of stimuli-responsive vesicles with D_h ranging from 310 to 561 nm (determined by DLS) were fabricated at 15% solid content via the RAFT dispersion copolymerization of 2-(diisopropylamino)ethyl methacrylate (DIPEMA) and benzyl methacrylate (BzMA) with varying feed molar ratios and different target degree of polymerization (DP) of P(DIPEMA-*co*-BzMA) using PEG as solvophilic macro-chain-transfer agents. The vesicles had a P(DIPEMA-*co*-BzMA) shell and PEG brushes on both the interior and exterior of the shell. They showed pH-responsivity because the tertiary amino groups of DIPEMA (pK_a ~6.3) can be protonated in acidic solutions, which could enhance the hydrophilicity of the membrane-forming P(DIPEMA-*co*-BzMA) blocks. By tuning the molar ratio of DIPEMA/BzMA in the hydrophobic P(DIPEMA-*co*-BzMA) blocks and their DP, the pH-responsive kinetics of vesicles (or the release kinetics) of the payloads (rhodamine B) from the vesicles might change from an explosive release (for the vesicles with higher content of DIPEMA units) to much retarded release (upon reducing the content of DIPEMA units in the P(DIPEMA-*co*-BzMA) blocks and elongating the P(DIPEMA-*co*-BzMA) blocks with the same content of DIPEMA units).

7.2.1.3 Single-Molecule Templating (of Core–Shell Bottlebrush Polymers) Method

Bottlebrush polymers are a class of branched macromolecules with one or more polymeric side chains tethered to each repeating unit of a long linear polymer backbone [74–76]. They have emerged as versatile single-molecule templates for the preparation of well-defined (hairy) hollow polymer NPs with unique architectures and controlled dimensions [77–83]. Wooley and coworkers reported the first fabrication of hollow polymer nanospheres from a bottlebrush copolymer prepared via a tandem synthetic strategy, combining ring-opening metathesis polymerization (ROMP) and NMP (Figure 7.17) [77]. It involves first the synthesis of a NMP-ROMP inimer (1) containing an alkoxyamine functionality (as a universal initiator for NMP) and a norbornene (NB) functionality (as a monomer for ROMP), the synthesis of poly(1) (4, $M_{n,GPC}$ = 122 kDa, $Đ$ = 1.13) via ROMP of 1 with Grubbs' catalyst $RuCl_2(CHC_6H_5)[P(C_6H_{11})_3]_2$, the use of 4 as the polyfunctional NMP

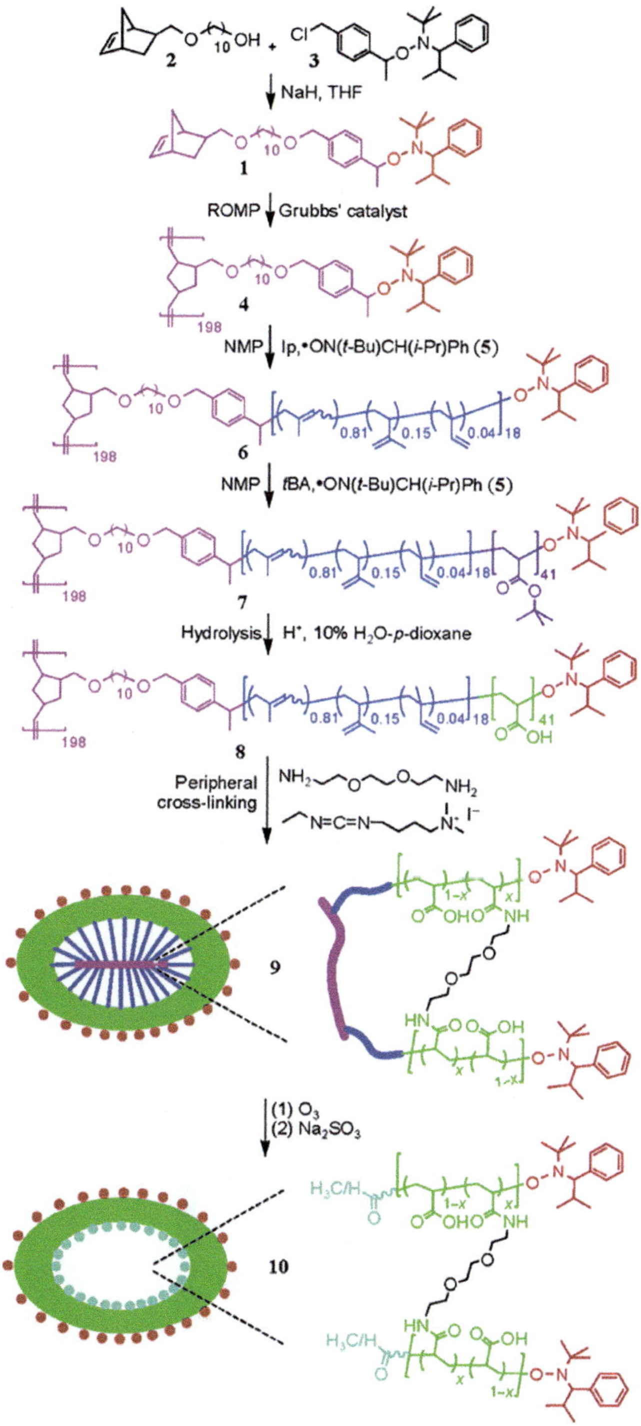

Figure 7.17 Fabrication of hollow nanospheres from a bottlebrush copolymer prepared via a tandem synthetic strategy. Source: Cheng et al. [77], scheme 1 (p. 6808)/Reproduced with permission of American Chemical Society.

macroinitiator for the sequential polymerizations of isoprene (Ip) and *t*BA in the presence of 5 to suppress biradical coupling (leading to core–shell brush copolymer 7 [$M_{n,\mathrm{GPC}} = 1410\,\mathrm{kDa}$, $Đ = 1.23$]), generation of the amphiphilic core–shell brush copolymer 8 through the complete conversion of the *t*BA units of 7 into AA units by hydrolysis, cross-linking of 8 with 2,2′-(ethylenedioxy)bis(ethylamine) in the presence of 1-[3′-(dimethylamino)propyl]-3-ethylcarbodiimide methiodide (a catalyst) (resulting in the peripherally cross-linked brush copolymer 9), and final selective degradation of the poly(isoprene) (PIp)-based core of 9 using ozone followed by reduction with Na_2SO_3. The resulting hollow nanospheres had an average diameter (D_{av}) of 39.7 ± 10.6 nm and an average height (H_{av}) of 1.98 ± 0.39 nm as determined by AFM. Since such hollow NPs still have alkoxyamine "living" groups on their surfaces, they can be further functionalized by grafting with various polymer brushes to afford hairy hollow NPs.

Rzayev and coworkers developed an efficient strategy for the preparation of hairy organic (or polymeric) nanotubes with controlled length and functionalities from core–shell bottlebrush copolymers with degradable cores and cross-linkable shells that adopt cylindrical conformations in solution [78–81]. In their first report, a core–shell bottlebrush copolymer with a polymethacrylate backbone and triblock copolymer branches (or side chains) was prepared by the combined use of RAFT polymerization and ROP. The triblock copolymer branches contained a degradable PLA inner block, a cross-linkable PS-*co*-poly(4-(3-butenyl)styrene) (PSB) middle block, and a hydrolyzable PMMA outer block in the side chains (Figure 7.18) [78].

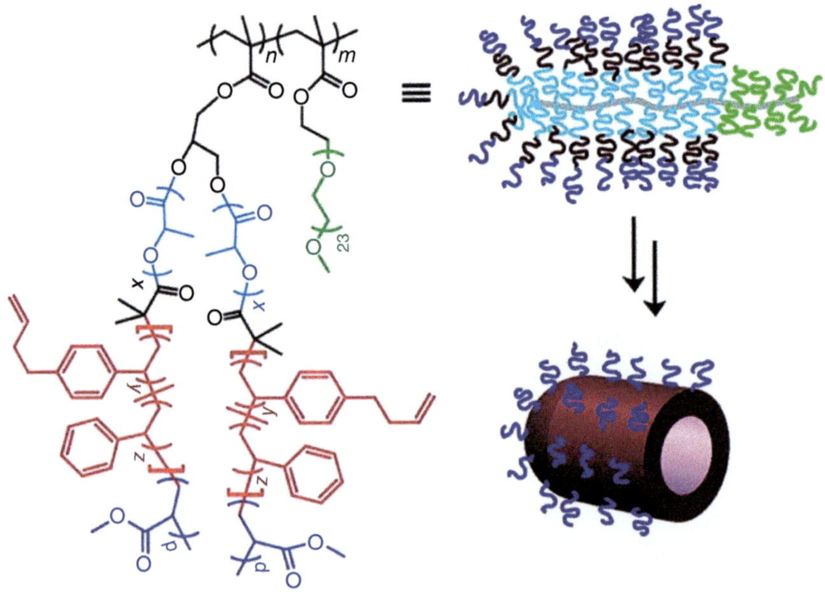

Figure 7.18 Chemical structure and schematic illustration of the multicomponent bottlebrush copolymer used for the preparation of well-defined hairy polymer nanotubes. Source: Huang and Rzayev [78], fig 6 (p. 6884)/Reproduced with permission of American Chemical Society.

Following the intramolecular cross-linking of the bottlebrush copolymer by treating its diluted solution with Grubbs generation I catalyst (for cross-metathesis of olefinic groups in the PSB block) and the subsequent etching of PLA chains from the core of the cross-linked bottlebrush copolymer with acid (hydrolysis of the PMMA brushes also occurred under the condition of PLA etching, leading to hydrophilic PAA brushes), water-soluble hairy polymer nanotubes with hydrophilic PAA brushes and hydrophobic pores were obtained. To create nanotubes with one open end, another block with PEO brushes was introduced into the bottlebrush copolymer to function as the sacrificial "stopper" (the green part in Figure 7.18). The hairy polymer nanotubes had an average length of 32 ± 6 nm, pores of 3–5 nm in diameter, and shell thicknesses of 5–7 nm (determined by TEM).

Following the above strategy, Rzayev and coworkers developed many other hairy polymer nanotubes for different purposes. In one example, they fabricated hairy polymer nanotubes with easily tunable surface functionality via first the synthesis of a core–shell bottlebrush copolymer with a PLA inner block, a cross-linkable PSB middle block, and a poly(styrene-*co*-maleic anhydride) (PSMA) reactive outer block in the side chains and subsequent cross-linking of PSB middle block with Grubbs' catalyst and removal of the PLA core [79]. Such hairy polymer nanotubes had an average length of 35 ± 5 nm and pore diameter of ~5 nm in diameter in the dry state (determined by TEM) and hydrodynamic diameters of 107–135 nm in a PBS solution (determined by DLS). Their surface-grafted PSMA brushes (containing anhydride groups) could react with water and amine-terminated oligoethylene glycols (OEGs) of varying lengths and afford precise control of the outer surface characteristics. The resulting negatively charged tubular nano-objects showed surface chemistry-dependent internalization by HeLa cells. The presence of hydrophobic groups on the nanotube surfaces was crucial for their association with the cellular membrane and subsequent endocytosis, whereas longer OEG chains on the nanotube surface completely inhibited the uptake. These hairy polymer nanotubes with well-defined cavities and easily tunable surface functionalities hold much promise as drug carriers. In another example, amphiphilic hairy polymer nanotubes with hydrophobic exterior surfaces and hydrophilic PAA brushes on the interior surfaces were also prepared through first the synthesis of a bottlebrush copolymer with triblock copolymer side chains (containing a degradable PLA inner block, a hydrolyzable P*t*BA middle block, and a cross-linkable PSB outer block), shell-cross-linking of the PSB block of the bottlebrush copolymer with Grubbs' catalyst, and subsequent removal of the PLA core and deprotection of *t*-butyl groups of P*t*BA by acidic hydrolysis [80]. Such hairy polymer nanotubes had 36 nm-long tubular nanostructures (with the average diameter of the internal cavity being 6 nm). They could effectively transport positively charged dyes from the aqueous phase to chloroform instead of negatively charged dyes. In addition, they also showed a remarkable size selectivity in the transport of larger organic guests, which can be attributed to the presence of a well-defined internal cavity. Rzayev and coworkers also developed organosoluble hairy poly(pyrrole) (PPy) nanotubes by the single-molecule templating of a bottlebrush copolymer with triblock copolymer side chains bearing a PLA inner block, a poly(4-(pyrrolylmethyl)styrene) (PMS)

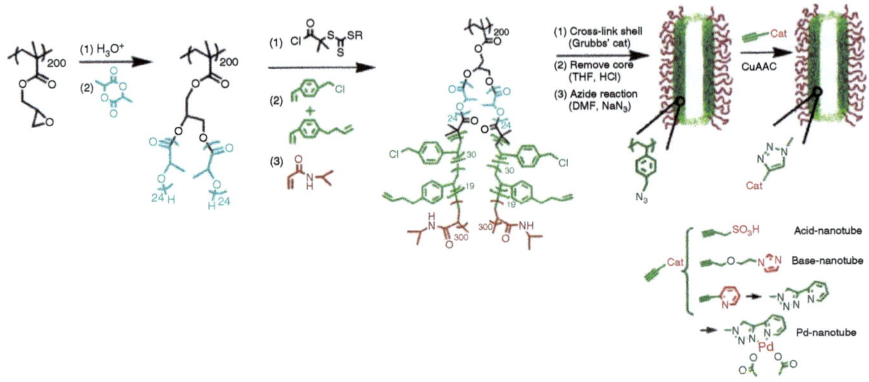

Figure 7.19 Synthetic route for the core–shell bottlebrush copolymer and schematic representation of the synthesis of polymer nanotube catalysts. Source: Xiong et al. [82], scheme 1 (p. 2)/Reproduced with permission of IOP Publishing Ltd.

middle block, and a PS exterior block [81]. The cross-linking of the middle PMS block through the oxidation copolymerization of its pendant pyrrole groups with a free pyrrole monomer and etching of the PLA core led to organosoluble hairy PPy nanotubes with a narrow length distribution (their length = 50 ± 6 nm; their average diameter = around 20 nm, determined by TEM) and good solubility in common organic solvents. They hold much promise in the field of optoelectronic nanodevices and biotechnology.

Recently, Huang and coworkers utilized the above single-molecule templating strategy to fabricate a soluble hairy polymer nanotube-supported catalyst system from a core–shell bottlebrush copolymer (Figure 7.19) [82]. A bottlebrush copolymer with a polymethacrylate backbone and triblock copolymer side chains (containing a degradable PLA inner block, a cross-linkable poly(4-vinylbenzyl chloride-*co*-4-(3-butenyl)styrene) (PVBC/BS) middle layer with functional PVBC segments, and a PNIPAM outer layer) was prepared via the combined RAFT polymerization and ROP. It was transformed into polymer nanotubes with surface-grafted PNIPAM brushes through cross-linking the PVBC/PS block with Grubbs' catalyst and subsequent removal of the PLA core by acidic hydrolysis. Further conversion of the benzyl chloride groups in the functional PVBC wall of the hairy polymer nanotubes by treatment with sodium azide led to azide-functionalized hairy polymer nanotubes with an average length of 35 ± 5 nm. Some organic or metal catalysts (e.g. sodium prop-2-yne-1-sulfonate (SPS), 1-(2-(prop-2-yn-1-yloxy)ethyl)-1H-imidazole (PEI), and $Pd(OAc)_2$) were anchored onto the tube walls to functionalize the organic nanotubes via copper-catalyzed azide–alkyne cycloaddition reaction. The resulting hairy polymer nanotube-supported catalysts showed high catalytic efficiency (for Knoevenagel condensation [PEI-functionalized base nanotube catalyst], one-pot cascade reactions including an acid-catalyzed acetal hydrolysis and the subsequent base-catalyzed Knoevenagel condensation [SPS-functionalized acid nanotube catalyst + PEI-functionalized base nanotube catalyst], and Suzuki–Miyaura reaction [Pd-functionalized nanotube catalyst]) and site-isolation features owing to the "confined effect" and the accessible cavity microenvironments of their tubular structures.

Later on, Huang's group also prepared acid–base bifunctional amphiphilic hairy polymer nanotubes containing PEI groups as the basic sites on the nanotube wall and benzenesulfonic acid (BSA) groups as acidic sites on the corona of nanotubes, following the same strategy as above except changing the PNIPAAM outer block in the above bottlebrush copolymer (as shown in Figure 7.19) to a copolymer of phenyl 4-vinylbenzene sulfonate (PVBS) and NIPAAM [83]. The acidic sites (i.e. BSA groups) were produced through the hydrolysis of the PVBS units into sulfonic acid sodium salt and the subsequent ion exchange reaction in an acidic aqueous solution by dialysis using a semipermeable membrane. The length and pore diameters of the resulting acid–base nanotubes were about 37 ± 4 and 4 ± 2 nm, respectively. They could be completely dissolved in water and showed strong organic molecular binding ability. In particular, they exhibited high catalytic performance for one-pot deacetalization-Knoevenagel cascade reactions in water, owing to their hydrophobic cavity microenvironments, the presence of a hydrophilic corona, and site-isolation features.

7.2.2 Synthetic Strategies for Hairy Hollow Inorganic NPs

A number of hairy hollow inorganic NPs have also been developed by grafting polymer brushes onto two kinds of hollow inorganic NPs, i.e. carbon nanotubes (CNTs) [84] and hollow silica nanoparticles (HSNPs) [85, 86]. The hairy CNTs and HSNPs have been prepared by the direct grafting of polymer brushes onto the commercially available CNTs [19, 87–110] and the sacrificial template strategy combined with sol–gel chemistry and various polymer brush-grafting methods [111–123], respectively. An overview of the progress made in their design and synthesis is summarized below.

7.2.2.1 Direct Grafting of Polymer Brushes onto Hollow Inorganic NPs

CNTs are hollow cylindrical tubes consisting of carbon (graphite) with a high aspect ratio (~1000) and sp^2 hybridization. According to the number of graphite layers, they can be single-walled nanotubes (SWNTs) or multi-walled nanotubes (MWNTs) [84]. CNTs have proven to be ideal candidates for electronic devices, chemosensors, transistors, electron field emitters, lithium-ion batteries, white light sources, hydrogen storage cells, cathode ray tubes, electrostatic discharge, and electrical-shielding applications. However, the large surface area and high aspect ratio of the CNTs, together with their attractive van der Waal interactions, lead to their significant aggregation, thus making them difficult to mix with organic materials and greatly limiting their broad applications. Therefore, tremendous efforts have been devoted to addressing this issue. One of the most efficient strategies for solving this problem is to modify the CNT surfaces with polymer brushes.

So far, many hairy CNTs have been fabricated by grafting polymer brushes onto the CNT surfaces through either "grafting-to" or "grafting-from" strategies [87]. Some "grafting-to" methods have been developed to prepare hairy CNTs with various

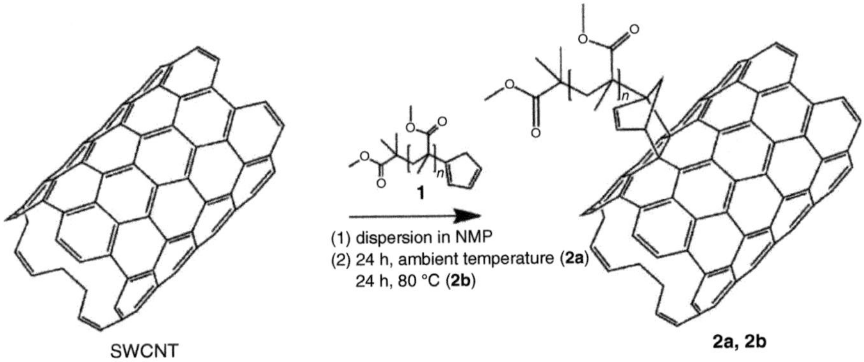

Figure 7.20 Synthetic pathway for the direct [4 + 2] functionalization of SWNTs with cyclopentadiene-capped PMMA. Source: Zydziak et al. [92], scheme 1 (p. 3375)/Reproduced with permission of American Chemical Society.

polymer brushes (e.g. PEO [88], PS [89], and poly(VCz) [90]) through the amidation or esterification reactions of amino- or hydroxyl-functionalized polymers with the carboxylic acid moieties at the ends and defect sites of shortened SWNTs. In addition, radical coupling [91] or click chemistry [19] has also been used to prepare hairy CNTs. For example, Jérôme and Kiriy's groups prepared hairy CNTs through reacting sp^2-hybridized carbons of CNTs with poly(2-vinylpyridine) (poly2VP) chains terminated with 2,2,6,6-tetramethylpiperidinyl-1-oxyl (TEMPO) radical-stabilizing group [91]. Poly2VP with controlled molecular weights (3–5 kg mol^{-1}) and narrow molar-mass dispersity (1.25) and end-capped by TEMPO were prepared via the CRP of 2VP in the presence of benzoyl peroxide and TEMPO. Barner-Kowollik and coworkers realized one-step functionalization of SWNTs via the Diels–Alder click chemistry (Figure 7.20) [92]. The unmodified SWNTs react as dienophiles in a single-step Diels–Alder [4 + 2] cycloaddition with cyclopentadienyl-capped PMMA (PMMA-Cp, $M_n = 2900\,g\,mol^{-1}$, $Đ = 1.2$) under mild conditions at both ambient temperature and 80 °C in the absence of any catalyst. The average grafting density of the polymer chains on SWCNTs was 0.029 and 0.039 chains nm^{-2} for samples obtained at ambient temperature and at 80 °C, respectively. Later on, the same group enhanced the grafting density of the polymer brushes on the obtained hairy SWNTs up to 0.0774 chains nm^{-2} through the hetero Diels–Alder (HDA) reaction of SWNTs pre-functionalized with a pyridinyl-based dithioester and PMMA-Cp ($M_n = 2700\,g\,mol^{-1}$, $Đ = 1.14$) at room temperature [19].

Over the years, a number of SI-polymerization-based "grafting-from" strategies have been developed for preparing hairy CNTs. In 2003, Ryu and Ajayan's groups reported the single-step *in situ* synthesis of polymer-grafted SWNTs via SI-anionic polymerization [93]. Dried pristine SWNTs were dispersed by sonication in purified cyclohexane, to which *sec*-butyllithium was added and sonicated in a bath for one hour. St was then added and polymerized at 48 °C for two hours under sonication. Carbanions were introduced on the SWNT surfaces by treatment with the anionic initiator, which not only allowed the negatively charged nanotube bundles into separate ones but also provided initiating sites for the polymerization of St.

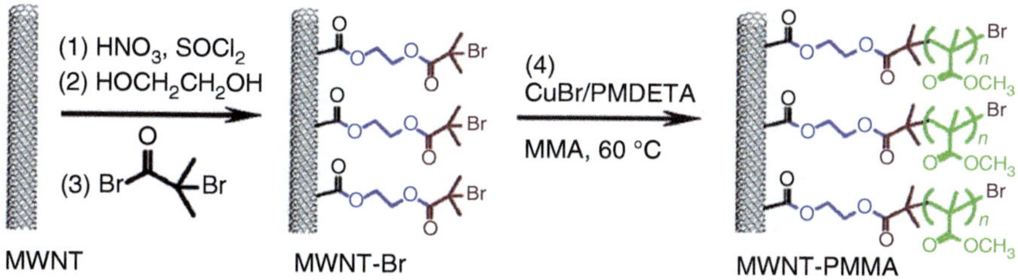

Figure 7.21 Schematic illustration of the preparation of hairy MWNTs via SI-ATRP. Source: Kong et al. [96], scheme 1 (p. 412)/Reproduced with permission of American Chemical Society.

The process requires no nanotube pretreatment and works well with as-produced SWNTs. Well-defined composites with a homogeneous dispersion of nanotubes were obtained. Nevertheless, the achieved degree of polymer functionalization was low (~10 wt%).

Many different SI-CRPs have also been applied to prepare hairy CNTs, including SI-ATRP [94–99], RAFT polymerization [100], and NMP [101, 102]. In these cases, CNTs functionalized with CRP-initiating or chain transfer groups were first prepared and then used as the immobilized CRP-initiators or chain transfer agents to induce the controlled polymerizations of monomers for grafting polymer brushes. For example, Yan and coworkers described a general strategy for grafting well-defined polymers onto MWNTs via SI-ATRP (Figure 7.21) [96], which involves first the synthesis of carbonyl chloride unit-functionalized MWNTs (MWNT-COCl) via reacting thionyl chloride with carboxyl-contained MWNT (MWNT-COOH) previously made by oxidation of the crude MWNT with 60% HNO_3, the formation of MWNTs with ATRP-initiating groups (MWNT-Br) through introducing hydroxyl groups onto the MWNT surfaces by reacting MWNT-COCl with glycol (generating MWNT-OH) and the subsequent reaction of MWNT-OH with 2-bromo-2-methylpropionyl bromide, and the final grafting of PMMA brushes from MWNT-Br via SI-ATRP of MMA (resulting in MWNT-PMMA). The thickness of the polymer layer on the functionalized MWNTs can be well controlled by the feed ratio (in weight) of MMA to MWNT-Br. In addition, the surface-grafting of PMMA-*block*-PHEMA brushes on MWNT surfaces was also realized via the ATRP of HEMA using MWNT-PMMA as the macroinitiator.

Hong et al. prepared water-soluble MWNTs with surface-grafted temperature-responsive shells by SI-RAFT polymerization [100]. It involves first the synthesis of MWNTs with surface-bound ATRP-initiating groups (MWNT-Br) through the formation of MWNT-COOH by HNO_3 oxidation and the reaction of MWNT-COOH with 2-hydroxyethyl-2′-bromoisobutyrate, the preparation of RAFT agent-functionalized MWNTs by reacting MWNT-Br with PhC(S)SMgBr, and the final controlled grafting of PNIPAM chains from MWNTs via SI-RAFT polymerization. The resulting MWNT-*g*-PNIPAM showed good solubility in water, chloroform, and THF and temperature-responsivity in the aqueous solution.

Dehonor and Terrones's groups prepared hairy N-doped MWNTs (briefly CN*x*) via SI-NMP [101]. It involves first the synthesis of CN*x* with surface-bound nitroxide groups through the free radical functionalization of CNx tubes with an initiator and a nitroxide controller, and the subsequent SI-NMP of St. The advantage of this method is the absence of any acid treatment for the CNx during the surface introduction of CRP-initiating groups.

In addition to SI-CRPs, some SI-ROPs have also been used for preparing hairy CNTs, including ROMP of norbornene [103, 104], cationic ROP of 3-ethyl-3-(hydroxymethyl)oxetane [105], coordination-insertion-based ROP of ε-CL [106], anionic ROP of glycidol [107], and ROP of 3-(benzyloxycarbonyl)-L-lysine *N*-carboxyanhydride [108]. The above strategies also involve first the synthesis of CNTs with surface ROP-initiating or catalyzing groups via the surface-modification of CNTs and their subsequent use as the immobilized initiators or catalyst for the ROPs of different monomers.

A "grafting-from" strategy based on both SI-CRP and ROP was also utilized for preparing hairy CNTs [109, 110]. For instance, Hadjichristidis and Sakellariou's group reported the synthesis of novel diblock copolymer-grafted MWNTs via a combination of ROP and SI-ATRP [109]. They first prepared MWNTs with surface-bound 2-hydroxyethylbenzocyclobutene via the Diels–Alder cycloaddition reaction and then used them as the immobilized initiator for SI-anionic ROP of CL. The OH-end groups of the resulting MWNT-g-PCL were then reacted with 2-bromoisobutyryl bromide to provide MWNT-PCL-Br, which was used for subsequent ATRP of St or DMAEMA to afford the MWNTs grafted with diblock copolymers. Mosnáčková et al. prepared MWNTs covalently modified with PHEMA-g-PCL brushes (MWCNTs-PHEMA-g-PCL) via the combined use of SI-RAFT and ROP [110]. MWNTs with surface-bound RAFT moieties were first prepared and used as the immobilized RAFT agent for the SI-RAFT polymerization of HEMA, leading to the MWNTs with PHEMA brushes. Subsequent ROP of CL from the pendant hydroxyl groups of PHEMA on MWNTs led to MWCNTs-PHEMA-g-PCL. MWCNTs-PHEMA-g-PCL containing 24 wt% MWCNTs was a thermoplastic elastomer and could retain its elastic properties at least up to 100 °C. It also exhibited an excellent, fully reversible, repeatable, and fast photomechanical actuation behavior.

7.2.2.2 Sacrificial Template Strategy Combined with Sol–Gel Chemistry and Polymer Brush-Grafting Methods

HSNPs have garnered enormous interest because of their well-defined hollow structure, thermal and mechanical stability, low density, large surface areas, facile surface functionalization, and biocompatibility [85, 86]. Many hairy HSNPs with surface-grafted polymer brushes have also been prepared to enhance the functions of HSNPs. They are typically prepared via the combined use of sacrificial template strategy, sol–gel chemistry, and various polymer brush-grafting methods (including "grafting-to" [e.g. coupling reaction] and "grafting-from" [i.e. SI-CRPs]). For an example of using the "grafting-to" coupling reaction, Lu and Liu's groups developed some hairy HSNPs with PEG brushes through first the preparation of HSNPs via the sol–gel reaction of TEOS on the sacrificial PS NP template and removal of the PS core by calcination at 550 °C, their surface-functionalization with amino groups, and final grafting of PEG brushes via the coupling reaction (using monomethoxy PEG [$M_n = 5$ kDa] 4-nitrophenyl carbonate) [111]. The average diameter and shell thickness of the resulting PEG-g-HSNPs are approximately 210 and 37 nm (determined by TEM), respectively. Such hairy hollow NPs showed good dispersity in aqueous solutions and very low in vitro cytotoxicity. In addition, PEG-g-HSNPs loaded with paclitaxel (PTX) exhibited a potent capacity to kill cancer cells.

To improve the controllability of the polymer brush-grafting process and achieve densely grafted polymer brushes, different SI-CRPs have been used to prepare hairy HSNPs, including SI-NMP [112], SI-ATRP [113–121, 124], and SI-RAFT polymerization [122]. For instance, Tsujii and coworkers developed hairy HSNPs with surface-grafted PS brushes via first the preparation of monodisperse ZnS NPs ($D_n = 120$–400 nm), their coating with a thin silica layer via sol–gel reaction, surface modification with an alkoxyamine derivative, surface-grafting of PS brushes ($M_n = 30\,000$–$114\,000$ g mol^{-1}, $Đ < 1.5$) with a grafting density of 0.9 chains nm^{-2} by SI-NMP, and final selective dissolution of the ZnS core from ZnS@SiO_2-PS NPs by using HCl [112].

Cai and Zhou's group developed charged polymer brush-grafted HSNPs (PSPMA-g-HSNPs) ($D_n = 422$ nm [determined by SEM and TEM] and 739.2 nm [determined by DLS]) via first the one-pot preparation of HSNPs ($D_n = 342$ nm) through the sol–gel reaction of TEOS (using ammonia as the catalyst) on PS NPs ($D_n = \sim 286$ nm, prepared via the emulsion polymerization) (PS core became dissolved by ammonia during the sol–gel reaction [123]), their surface functionalization with ATRP-initiating groups (alkyl bromide), and final surface-grafting of negatively charged poly(3-sulfopropyl methacrylatepotassium salt) (PSPMA) brushes via SI-ATRP (Figure 7.22) [124]. They exhibited a good swelling behavior in aqueous media and a good hydration-lubricating effect. In addition, they could also encapsulate a model drug (aspirin) inside the hollow NPs and showed a significant sustained drug release. The above results demonstrated that PSPMA-g-HSNPs are a promising material for joint lubrication and arthritis treatment.

Dai and coworkers described the synthesis of hairy hollow porous silica NPs with gas separation capability, which involves first the preparation of hollow porous silica NPs (average particle size ~60 nm) following a literature method [113], their surface-attachment of ATRP-initiating groups (alkyl chloride), and the subsequent grafting of different polymer brushes (including PMMA, poly(MMA-*random*-poly(ethylene glycol)methyl ether methacrylate) (PMMA-*r*-PEGMEMA), and PIL) by surface-initiated activators regenerated by electron transfer (ARGET) ATRP (SI-ARGET ATRP) [114]. As a control, hairy solid silica NPs were also prepared following the same procedure except changing the hollow porous silica NPs to 15 nm solid porous silica NPs. The resulting hairy hollow and solid silica NPs had hydrodynamic sizes of 250–500 and 90–150 nm, respectively (determined by DLS). The PIL nanocomposite films exhibited a much higher CO_2 permeance than PMMA and PMMA-*r*-PEGMEMA systems with a similar silica content. In addition, the hollow silica-mixed matrix membranes showed much higher CO_2 permeability than the 15 nm solid silica films, owing to

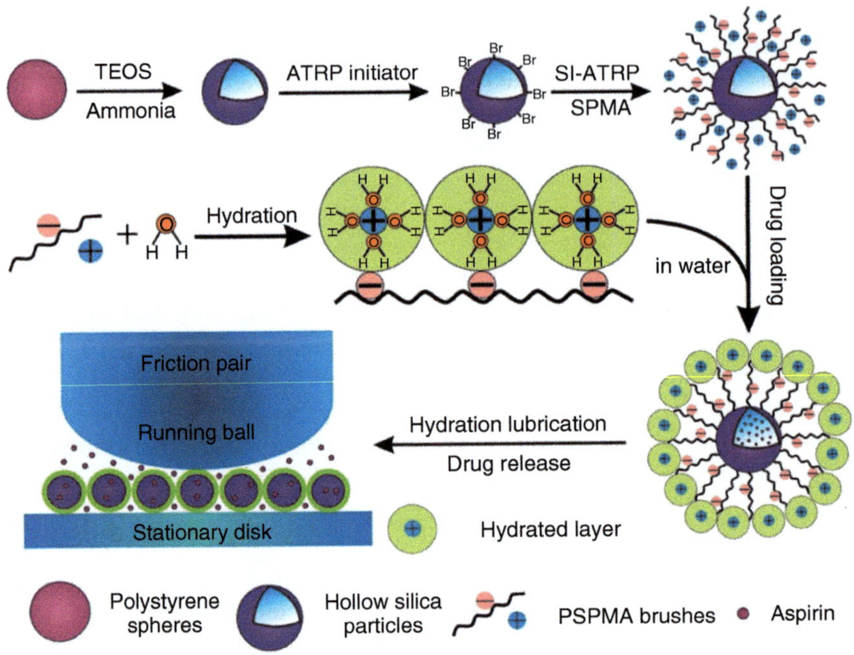

Figure 7.22 Fabrication of PSPMA-*g*-HSNPs, hydration mechanism of the negatively charged polymer brushes, drug loading and release, and hydration lubrication between steel and steel contacts. Source: Liu et al. [124], scheme 1 (p. 4922)/Reproduced with permission of American Chemical Society.

their hollow porous structures. The hollow silica particle brush films also exhibited the same mechanical properties as the 15 nm solid silica-derived ones.

Hairy hollow rod-shaped silica NPs were also developed by Ohno and coworkers via first the fabrication of silica-coated α-Fe_2O_3 nanorods with surface-bound ATRP-initiating groups (with an aspect ratio of 5.3 and an average long axis of 630 nm) through the sol–gel reaction on α-Fe_2O_3 nanorods and the subsequent surface-functionalization, their use as the immobilized ATRP initiator for surface-grafting of PMAA via SI-ATRP, and final selective removal of α-Fe_2O_3 core by etching with HCl [115]. The hairy hollow silica nanorods grafted with polymer chains of a proper length could form lyotropic liquid crystals.

A number of stimuli-responsive hairy HSNP-based drug delivery carriers have also been developed via the combined use of sacrificial template strategy, sol–gel chemistry, and SI-ATRP. For example, Lu and Liu's groups developed some hairy HSNPs with PMAA brushes via first the preparation of HSNPs (their average diameter and average shell thickness were about 248 and 31 nm [determined by TEM]) by the sol–gel reaction of TEOS on the PS template and removal of the PS core by calcination at 550 °C, their surface-functionalization with ATRP-initiating groups (alkyl bromide), and final grafting of PMAA brushes via SI-ATRP of sodium methacrylate [116]. The resulting hairy HSNPs with PMAA brushes (PMAA-g-HSNPs) showed pH-responsivity owing to their presence of pH-responsive PMAA brushes

(PMAA brushes had a lower ionization degree and low solubility in the acidic aqueous solution [e.g. pH 3.4], but a higher ionization degree and high solubility when the pH was higher than 7). Calcein blue and fluorescein isothiocyanate (FITC)-labeled dextran ($M_n = 10$ kDa) could be encapsulated in the cavities of PMAA-g-HSNPs with a negligible amount in PMAA brushes at pH 2, and their pH-triggered release from such vesicles was observed at pH 7.4.

Tang and coworkers reported the synthesis of PDMAEMA brush-coated hollow mesoporous silica nanoparticles (HMSNs) ($D_n = 500$ nm [determined by TEM]) and their pH-responsive controlled drug release. The hairy HMSNs were obtained via first the preparation of HMSNs via the sol–gel reaction of TEOS with docecylamine and poly(vinylpyrrolidone) (PVP-10) as the co-templates, their surface-attachment of ATRP-initiating groups (alkyl bromide), and subsequent surface-grafting of PDMAEMA brushes via SI-ATRP [117]. The formed ordered pore channels provided mass exchange between the internal and external of HMSNs, while the surface-grafted PDMAEMA brushes acted as a good gatekeeper to control access to the pores via a pH-dependent open–close mechanism. The resulting HMSN-g-PDMAEMA could efficiently load a drug (DOX) and showed well-controlled pH-responsive DOX release. They have potential applications in site-selected drug release and gene delivery. Zhao and coworkers developed the triple-responsive hairy HMSNs for multistimuli-responsive drug delivery (Figure 7.23) [118]. The synthetic route of the hairy HMSNs includes first the preparation of HMSNs ($D = {\sim}150$ nm [determined by TEM]) through the formation of silica core/silica mesoporous shell NPs via two-step sol–gel reactions and removal of the silica core by using Na_2CO_3, their successive surface-attachment of reduction-cleavable disulfide units, light-cleavable *o*-nitrobenzyl ester units,

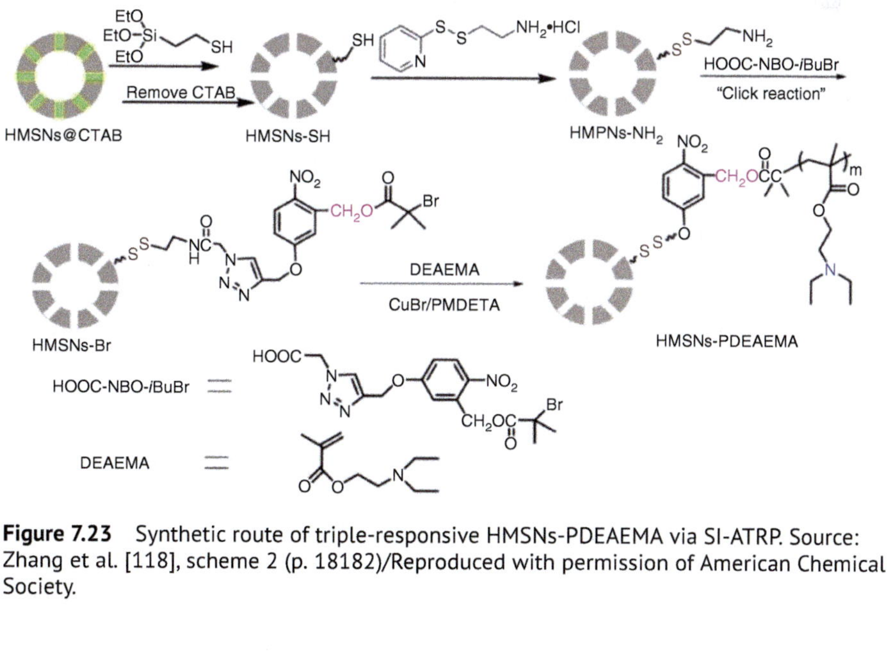

Figure 7.23 Synthetic route of triple-responsive HMSNs-PDEAEMA via SI-ATRP. Source: Zhang et al. [118], scheme 2 (p. 18182)/Reproduced with permission of American Chemical Society.

and ATRP-initiating groups (alkyl bromide), and the final surface-grafting of pH-responsive PDEAEMA brushes via SI-ATRP. The resulting hairy HMSNs could efficiently encapsulate DOX with a high loading capacity and showed controlled release in response to the stimuli of reducing agent, acid environment, or UV light irradiation. In addition, these hollow nanocarriers could be efficiently internalized by HeLa cancer cells. They showed several distinctive advantages including enhanced release of DOX into the cytoplasm under external UV light irradiation, better cytotoxicity against HeLa cells, and superior control over drug delivery and release. They are highly promising as a drug delivery platform for cancer therapy.

Beagan and Halwani's groups described the synthesis of pH-responsive hairy HMSNs with modified end group on the polymer brushes via first the preparation of HMSNs (D_n = ~340 nm, determined by SEM/TEM) through the formation of silica core/silica mesoporous shell NPs (D_n = 230–340 nm) via two-step sol–gel reactions and removal of the silica core by using Na_2CO_3, their surface-attachment of ATRP-initiating groups (alkyl bromide), surface-grafting of PDEAEMA brushes (dry thickness c. 30 nm) via SI-ATRP, and the final end group modification of the polymer brushes with glucosamine [119]. The resulting HMSN-g-PDEAMA-glucosamine had pH-dependent sizes due to the deprotonation of the tertiary amino groups in PDEAEMA chains (from c. 950 nm in acidic media to c. 500 nm in basic media). They showed a high loading capacity of DOX (c. 64%) and pH-responsive drug release capability. In addition, they also exhibited a strong ability to kill breast cancer cells (MCF-7 and MCF-7/ADR) at low drug concentrations in comparison to free DOX.

Xu and coworkers developed redox-triggered gatekeeper-enveloped star-like hollow silica nanoparticles (SHNPs) with surface-grafted polymer brushes as the intelligent delivery systems (Figure 7.24) [120]. Such hairy SHNP drug nanocarriers (SHNP-PGEA) were fabricated via first the preparation of SHNPs by using star-shaped PbS NPs (sizes ≈ 65 and 85 nm) as the sacrificial templates via sol–gel reaction, their successive attachment of epoxy, disulfide, and adamantine groups, and final grafting of polymer brushes via their supramolecular self-assembly with a CD-terminated polymer (CD-PGEA, it was synthesized via ATRP of GMA by using CD-containing ATRP-initiator and the subsequent ring-opening reaction with ethanolamine [EA]). They showed significantly higher gene transfection efficiency than CD-PGEA. CD-PGEA gatekeepers on the hollow NP surfaces could control the release of loaded drugs inside the silica cavities. Compared with their ordinary spherical hollow silica NP-based counterparts, SHNP-PGEA carriers with six sharp horns possessed better performances for cellular uptake, gene delivery, and antitumor effects.

Later on, Xu and coworkers further developed well-defined hairy hollow silica nanostars with encapsulated Au caps via first the formation of hollow silica nanostars with encapsulated Au caps (star-like Au@SiO_2) (with comparable oversize of ~120 nm and encapsulated Au nanocaps of ~40 nm) through the sol–gel reaction of TEOS on the PBS-Au star-like hetero-nanostructures and removal of the sacrificial PBS core by using HCl etching, their surface-attachment of ATRP-initiating groups (alkyl bromide), the surface-grafting of PGMA brushes via SI-ATRP, and their

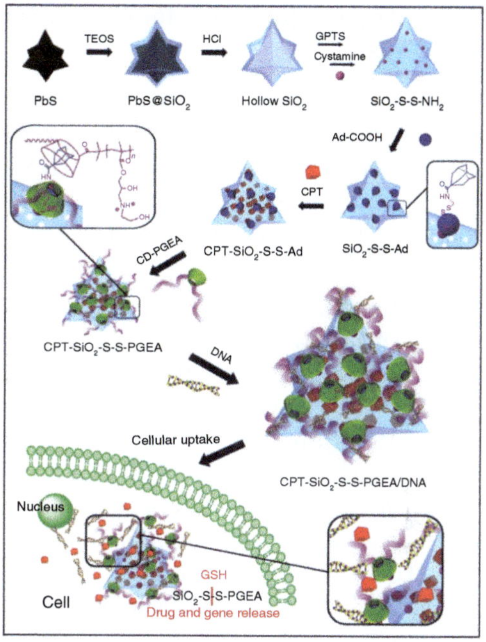

Figure 7.24 Schematic illustration of the preparation and surface functionalization of SHNPs and the resultant redox-responsive drug/gene co-delivery. Source: Zhao et al. [120], fig 1 (p. 6469)/Reproduced with permission of John Wiley & Sons.

surface-modification with EA to transform PGMA brushes to PGEA brushes [121]. The resulting star-like Au@SiO_2-PGEA NPs combined the photothermal effect for photothermal therapy (PTT) because of their Au caps, excellent drug loading capability for chemotherapy owing to the interior hollow feature of star-like Au@SiO_2 NPs, and good transfection performances for gene therapy due to the presence of polycationic PGEA brushes on the surface. These characteristics, together with their favorable endocytosis because of their sharp horns, led to their enhanced antitumor effectiveness.

Hong and coworkers developed a hairy silica nanotube (SNT)-based drug carrier with pH-sensitive polymer brushes via first the fabrication of well-defined SNTs and their subsequent surface-grafting of pH-responsive diblock copolymer brushes via SI-RAFT polymerization [122]. A block copolymer PDMAEMA-*b*-PS (the DP of PDMAEMA is 149 and that of PS is 2505)-based polymer nanorod was first synthesized in one-pot via PISA, which was utilized as the template for preparing hybrid silica nanorods through sol–gel reaction of TEOS (leading to the deposition of silica onto the weak polyelectrolyte PDMAEMA block of the polymer nanorods). Calcination of the resulting silica hybrid nanorods afforded hollow SNTs (the average diameter and wall thickness of the SNTs are 95 and 16 nm, respectively [determined by TEM]). Hairy SNTs with well-defined structure and micropores in the SNT wall were obtained by grafting diblock copolymer PDEAEMA-*b*-POEGMA brushes onto SNTs (premodified with RAFT agent) via SI-RAFT polymerization. They showed pH-responsive drug (DOX) release behavior owing to the presence of pH-sensitive PDEAEMA block in the polymer brushes. This, together with their good biocompatibility and excellent endocytosis property, makes them highly promising in drug and gene delivery.

7.2.3 Synthetic Strategies for Hairy Hollow Organic/Inorganic Hybrid NPs

So far, four strategies have been developed to fabricate hairy hollow organic/inorganic hybrid NPs, including direct deposition of polymer layers onto hollow inorganic NPs by SI-polymerizations [125], self-assembly method [126], single-molecule templating method [127–130], and sacrificial template method combined with polymer brush nanoreactors [131–134].

7.2.3.1 Direct Deposition of Polymer Layers onto Hollow Inorganic NPs by SI-Polymerizations

The combination of depositing a cross-linked polymer layer onto the hollow inorganic NPs and subsequent grafting of polymer brushes onto the resulting hollow hybrid NPs by SI-polymerizations has become a facile and straightforward approach for fabricating hairy hollow hybrid NPs. For instance, Zhou and Li's groups reported the preparation of hairy hollow magnetic hybrid molecularly imprinted polymer nanorods (HMMINs) for fast removal of sulfamethazine (SMZ) from aqueous media [125]. The synthetic route for HMMINs includes the following four steps: (i) preparing magnetic halloysite nanotubes (MHNTs) through introducing iron ions into the lumen and surface of HNTs using a wetness impregnation step and their subsequent reduction using ethylene glycol at high temperature under the protection of a nitrogen atmosphere; (ii) preparing MHNTs with surface-bound ATRP-initiating groups (alkyl bromide) (MHNTs@Br) by the surface modification of MHNTs via the sequential silanization and amidation reactions; (iii) preparing magnetic surface-imprinted core–shell nanorods (MMINs) via SI-ATRP with an antibiotic drug SMZ as the template, 4-vinylpyridine (4-VP) as the functional monomer, EGDMA as the cross-linker, and PMDETA/CuBr as the catalyst; and (iv) grafting PHEMA brushes onto the MMIN surfaces via the SI-ATRP. The resulting HMMINs had a core–shell structure with an ultrathin MIP film (12 nm) and PHEMA brushes (2–4 nm) and uniformly dispersed magnetic NPs on the HNT surfaces. They exhibited a high binding capacity (37.64 ± 1.36 micromol g^{-1}), fast binding kinetics, and good selectivity toward SMZ in aqueous media (because of their surface-grafting of hydrophilic polymer brushes), as well as good regeneration performance.

7.2.3.2 Self-Assembly Method

Duan and coworkers reported the preparation of hairy hollow hybrid plasmonic vesicles by the self-assembly of surface-enhanced Raman scattering (SERS)-encoded amphiphilic gold NPs for cancer cell targeting and traceable intracellular drug delivery (Figure 7.25) [126]. It involves first the fabrication of gold NPs with a Raman reporter by coating the citrate-stabilized gold NPs (14 nm in size) with the Raman dye BGLA (with a brilliant green backbone and a multivalent disulfide anchoring group), their subsequent grafting of the hydrophilic PEG brushes and hydrophobic copolymer (PMMAVP) of MMA and 4-VP using the tandem "grafting-to" (thiolated PEG and an ATRP initiator with a disulfide group [i.e.

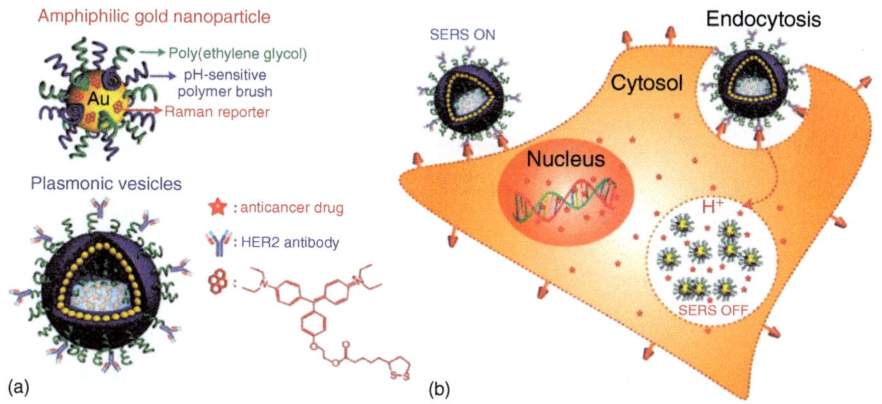

Figure 7.25 (a) Schematic illustration of the amphiphilic gold NP coated with Raman reporter BGLA and mixed polymer brushes of hydrophilic PEG and pH-sensitive hydrophobic PMMAVP grafts and the drug-loaded plasmonic vesicle tagged with HER2 antibody for cancer cell targeting. (b) The cellular binding, uptake, and intraorganelle disruption of the SERS-encoded pH-sensitive plasmonic vesicles. Source: Song et al. [126], fig 1 (p. 13459)/Reproduced with permission of American Chemical Society.

2,2′-dithiobis[1-(2-bromo-2-methylpropionyloxy)]ethane [DTBE]] were attached on gold NPs by forming covalent Au—S bonds) and "grafting-from" (PMMAVP brushes were grown from the above-functionalized gold NPs with ATRP-initiating groups through SI-ATRP) methods, and their self-assembly into vesicles via the film rehydration method commonly used for preparing polymersomes of amphiphilic block copolymers. The resultant amphiphilic NPs on average have 82 PEG ($M_n = 5$ kDa) grafts and 164 PMMAVP ($M_n = 24$ kDa) grafts and a grafting density of 0.1 chain nm^{-2}. Thirty five percentage of the thiolated PEG chains were heterofunctional PEG with a carboxylic acid end group, which were used for chemically conjugating a cancer-targeting ligand (a monoclonal antibody to HER2 protein). The 4-VP moieties (~10%) in the hydrophobic PMMAVP brushes have a pK_a of 5.4, which makes the copolymer chains water-soluble below pH 5.4 due to the protonation of pyridine groups. This hydrophobic-to-hydrophilic transition at pH 5.4 enabled the stimulated disruption of plasmonic vesicles in intracellular acidic compartments such as late endosomes/lysosomes (pH ~4.7–5.5), which led to the dramatic decrease of the scattering properties and SERS signal (the plasmonic vesicles showed greatly enhanced SERS intensity in comparison with single gold NPs due to strong interparticle plasmonic coupling). The as-prepared hairy hollow hybrid plasmonic vesicles have a diameter of about 200 nm and a shell thickness of less than 20 nm (composed of a monolayer of amphiphilic NPs, where the hydrophobic PMMAVP brushes collapse to form the shell matrix with gold NPs embedded and the extended hydrophilic PEG chains cover the exterior and interior surfaces). They showed efficient cancer cell-targeting ability (because of their having targeting units at the terminal of PEG brushes) and pH-responsive drug (DOX) release following the pH-responsive disassembly of the plasmonic vesicles in acidic intracellular compartments. More importantly, the strong fluorescence of DOX at 590 nm and the light scattering of the plasmonic vesicles allow for independently tracking the cargo and carriers using fluorescence and dark-field imaging, respectively. Particularly, the dramatic changes in scattering properties and SERS signals following the disassembly of the vesicles in intracellular acidic compartments can also serve as independent feedback mechanisms to signal cargo release from the vesicles. Such hairy hollow hybrid plasmonic vesicle can serve as a promising platform for targeted combination therapy and theranostic applications.

7.2.3.3 Single-Molecule Templating Method

Lin and coworkers developed a general single-molecule templating strategy for fabricating hairy hollow nanocrystal/polymer hybrid NPs with precisely controlled dimensions, compositions, and architectures by using star-like triblock copolymers as the templating nanoreactors (Figure 7.26) [127–130]. In one example, they prepared hairy hollow hybrid NPs through first the synthesis of a series of amphiphilic multi-arm star-like triblock copolymers that are covalently linked to a small core (i.e. β-CD-(PS-*b*-P*t*BA-*b*-PS)$_{21}$) via the sequential ATRP by using a β-CD-based ATRP initiator, the hydrolysis of the block P*t*BA in the above triblock copolymers into PAA, selectively sequestering precursors in the intermediate PAA block, and final formation of a nanocrystal/PAA hybrid layer through subsequent hydrolysis and condensation of the appropriate precursors in the mixed solvents of DMF and benzyl alcohol (9 : 1 v/v) [128]. Hairy hollow hybrid NPs with hydrophobic PS

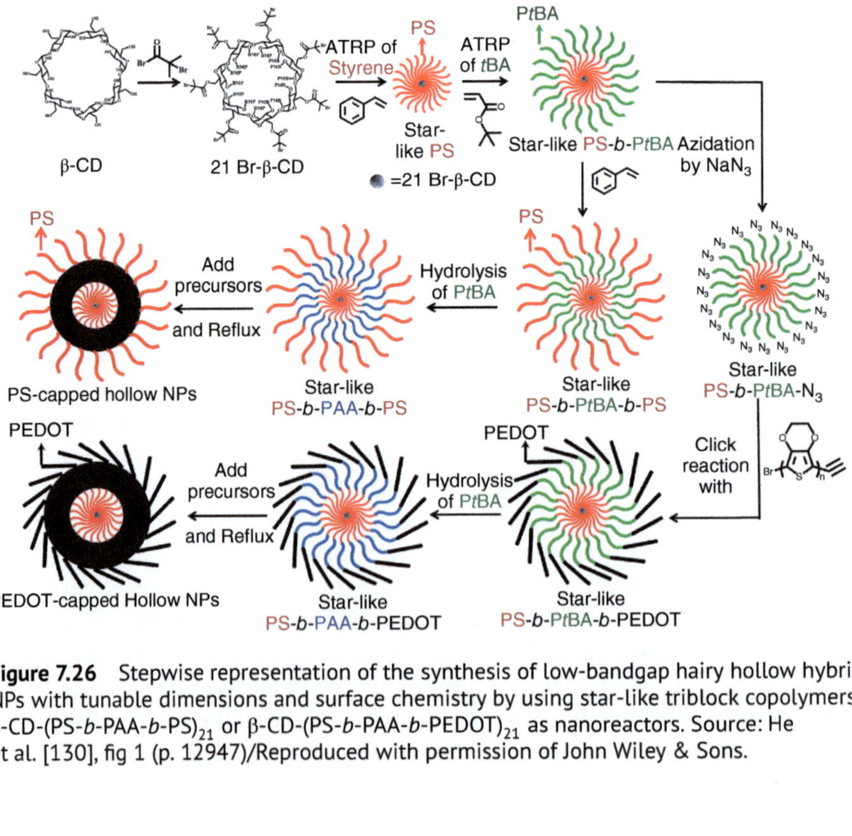

Figure 7.26 Stepwise representation of the synthesis of low-bandgap hairy hollow hybrid NPs with tunable dimensions and surface chemistry by using star-like triblock copolymers β-CD-(PS-*b*-PAA-*b*-PS)$_{21}$ or β-CD-(PS-*b*-PAA-*b*-PEDOT)$_{21}$ as nanoreactors. Source: He et al. [130], fig 1 (p. 12947)/Reproduced with permission of John Wiley & Sons.

blocks tethered on both the interior and exterior of the Au/PAA hybrid shell were readily obtained by confining the Au precursors in the intermediate PAA regime of a star-like triblock copolymer (β-CD-(PS-*b*-P*t*BA-*b*-PS)$_{21}$) template. The sizes of such hairy hollow hybrid NPs can be conveniently controlled by varying the length of the innermost PS block and intermediate P*t*BA block during ATRP. The strategy can also be extended to the production of a variety of other hairy hollow hybrid NPs with different sizes, including hairy hollow semiconductor Cu_2O NPs. Similarly, water-soluble hairy hollow Au/PAA hybrid NPs with hydrophilic PEO brushes were also prepared by using a star-like triblock copolymer β-CD-(PS-*b*-PAA-*b*-PEO)$_{21}$ as the templating nanoreactor.

Later on, Lin and coworkers also prepared some uniform permanently ligated hairy hollow hybrid NPs with precise dimension control and tunable optical properties following the above strategy [129]. A series of hairy hollow Au/PAA hybrid NPs with different diameters of hollow interior and shell thickness were obtained by tuning the molecular weights of the innermost PS block and intermediate P*t*BA blocks of the star-like triblock copolymers (β-CD-(PS-*b*-P*t*BA-*b*-PS)$_{21}$). Their characteristic surface plasmon resonance (SPR) properties could be readily adjusted by altering either the shell thickness or external diameter, which may greatly expand their applications in catalysis, SERS, imaging, sensors, optics, electronics, and optoelectronics. In addition, their presence of intimately and permanently ligated PS chains on the shell surfaces endows them with good solubility and long-term stability in nonpolar solvents, which may enable a homogeneous dispersion in a polymer matrix for use in functional nanocomposites. Furthermore, these hairy hollow hybrid NPs also proved to be a class of efficient nanocarriers. The same group further utilized the above strategy to prepare uniform hairy hollow hybrid semiconducting NPs (i.e. lead chalcogenide [PbS and PbTe] NPs) with tailorable dimensions, compositions, surface chemistry, and near-infrared absorption (Figure 7.26) [130], which are not easily accessible by conventional approaches. Amphiphilic star-like triblock copolymers [β-CD-(PS-*b*-PAA-*b*-PS)$_{21}$ and β-CD-(PS-*b*-PAA-*b*-PEDOT)$_{21}$, PEDOT refers to poly(3,4-ethylenedioxythiophene)] were used as nanoreactors for fabricating uniform hairy hollow hybrid PbS/PAA and PbTe/PAA NPs. The former triblock copolymer was synthesized as described above, and the latter one was prepared by the two-step sequential ATRP of St and *t*BA, their end group functionalization with azide unit, subsequent click reaction with a conjugated PEDOT bearing an alkyne end group, and final hydrolysis of the intermediate P*t*BA blocks into PAA. Compared to their solid counterparts, the near-IR absorption of the hollow NPs is blueshifted owing to the hollow interior. This strategy can be readily extended to other types of intriguing low-bandgap hairy hollow hybrid NPs for diverse applications.

7.2.3.4 Sacrificial Template Method Combined with Polymer Brush Nanoreactors

Guo and coworkers developed a new approach to prepare hairy hollow hybrid NPs with the spherical polyelectrolyte brushes grafted on the sacrificial PS NPs as nanoreactors (Figure 7.27) [131–134]. It involves first the preparation of PS NPs (average size being c. 100 nm) with surface-grafted PAA brushes

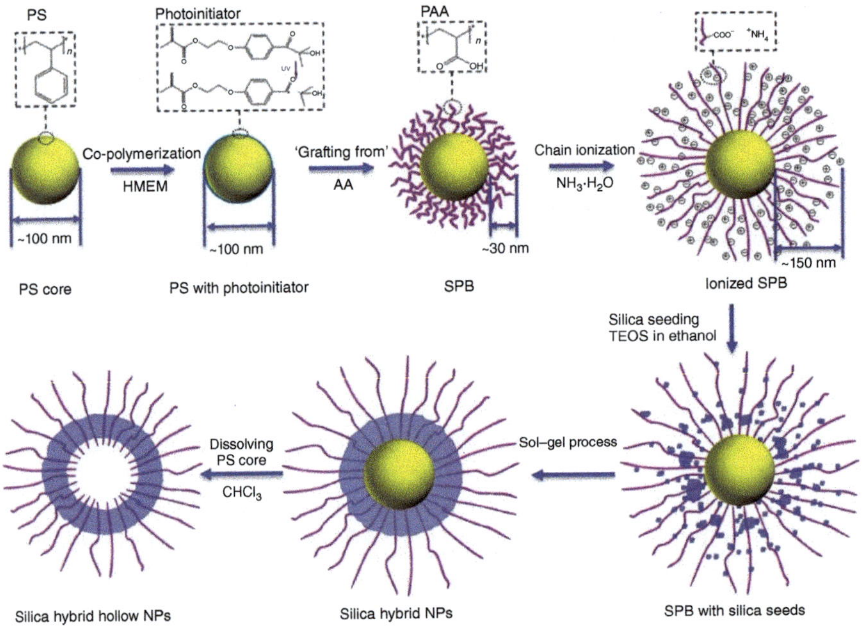

Figure 7.27 Schematic illustration of the fabrication of hairy silica hybrid NPs and hollow silica NPs. Source: Han et al. [132], fig 3 (p. 9860)/Reproduced with permission of American Chemical Society.

(0.03–0.06 chains nm^{-2}) via the two-step emulsion polymerization [135], formation of hybrid NPs through depositing a silica layer (via sol–gel reaction) onto the PS NP surfaces with the assistance of the ionized PAA brushes (with adsorbed positively charged ammonium ions), and final removal of the PS core by etching with chloroform. Their silica shells proved to have a unique inner-loose-outer-dense structure [132, 133], which seems to be ideal for their use as drug carrier. In addition, a small portion of PS-PAA copolymers should remain near the internal cavity surfaces of the hairy hollow hybrid NPs [133]. These hairy hollow hybrid NPs showed high (re)dispersity in aqueous media and obvious pH-responsive size changes owing to the change in the ionization degrees of PAA brushes following the pH change [131]. When TEOS and bis[3-(triethoxysilyl)propyl]tetrasulfide (TETS) were employed as the mixed precursors to introduce disulfide bonds into the silica framework during the sol–gel reaction process, the resulting hairy hollow hybrid NPs contained glutathione (GSH)-degradable disulfide units in the silica shell, which could show both pH- and GSH-responsive drug release capability [134].

7.3 Conclusions and Perspectives

This book chapter presents a detailed overview of the research progress made in the design and synthesis of hairy hollow NPs, as well as their properties and applications. As evidenced by the discussed literature, tremendous advances have been made in fabricating hairy hollow NPs with various morphologies, compositions, and interior, shell, and hairy structures following many breakthroughs in the fields of polymer chemistry, materials science, and nanotechnology. A range of useful synthetic strategies for the hairy hollow NPs has been developed, such as sacrificial template method (combined with other methods), self-assembly method, single-molecule templating method, and direct grafting of polymer layers (both cross-linked polymer layers and polymer brushes) method. In particular, the introduction of CRPs into this field has greatly revolutionized this area because of their excellent controllability, high compatibility with a wide range of monomers, and "livingness" of the resulting hollow NP surfaces, which paves the way for the rational design and highly efficient synthesis of various advanced functional hairy hollow NPs with well-defined internal, shell, and hairy structures and largely improved physicochemical properties under mild reaction conditions. In addition, the use of click chemistry in this field also significantly facilitates the development of hairy hollow NPs. So far, hairy hollow NPs have demonstrated their high potential as nanoreactors, biomedical materials (e.g. drug delivery carriers, bioimaging), and various bioanalytical devices.

Despite the above significant progress, the presently developed synthetic strategies for the hairy hollow NPs still have their pros and cons. For example, although the sacrificial template method has demonstrated its versatility for the preparation of hairy hollow NPs owing to its general applicability, simple manipulation, and easy control of hollow particle internal void spaces and particle morphologies, it normally can only provide hollow NPs larger than 50 nm because of the difficulty in obtaining smaller-sized templates. In addition, the procedure for the selective removal of hard templates typically involves very harsh conditions (e.g. by HF etching) and often introduces uncontrollable compositional and structural complexity (leading to NPs with partial coverage, inhomogeneous shell thickness, poorly defined composition, and so forth [129]). In comparison, the self-assembly and single-molecule templating methods allow for the preparation of rather small hairy hollow NPs through rationally designing the chemical structures of the block copolymers and bottlebrush copolymers. However, the self-assembly approach is usually performed in highly diluted solutions (leading to very low yields) and is rather time-consuming, and the core–shell bottlebrush copolymers and star multiblock copolymers used in the single-molecule templating methods normally require complicated multistep synthetic procedures, which are difficult to scale up and thus greatly hamper their broad applications. The direct grafting of polymer layers (both cross-linked polymer layers and polymer brushes) method can only be used in limited special cases where hollow inorganic NPs are commercially available. Recently, some versatile approaches have been developed to address some of the above existing problems. For example, the use of uniform physically cross-linked "living" PMAA nanospheres (prepared in one-pot via the RAFTPP) as the sacrificial template has provided a facile, general, and highly efficient platform for developing various advanced hierarchical hollow polymer NPs of tailor-made structures and desired properties (e.g. multiple stimuli-responsivity) under mild preparation and green etching conditions (in alcohol for 3×5 minutes) [38]. In addition, the aqueous PISA has proven to be an efficient self-assembly method for fabricating hairy nanovesicles at high solid concentrations (typically 10–30% w/w) [72, 73]. Nevertheless, it is still imperative to develop facile and convenient methods to be capable of synthesizing hairy hollow NPs with well-defined size, shape, internal shell, and hairy structures on a large scale for their real-world applications. In particular, the development of versatile synthetic strategies for hairy hollow NPs with complex multilevel hollow features that can better use the internal spaces and those with new anisotropic shapes should be paid more attention in future researches.

More future efforts should also be devoted to fabricating hairy hollow NPs fully compatible with their application purposes instead of just showing the proof-of-principle. For instance, fully biodegradable hairy hollow NPs with good biocompatibility and hemocompatibility, as well as stimuli-responsive drug release and imaging capabilities should be developed for their use as effective drug nanocarriers. Moreover, the development of more advanced multifunctional hairy hollow NPs is also highly desirable for their practical applications. Although much remains to be done to make hairy hollow NPs suitable for real-world uses, it can be envisioned that this situation will definitely improve with the rapid development of new powerful synthetic strategies and more insights gained from their in-depth structure–property studies. Considering the almost infinite diversity of hairy hollow NPs in terms of their structures, properties, and compositions, as well as their ease of functionalization and surface modification, we are confident that practically applicable hairy hollow NPs may not lie in the too-distant future.

Acknowledgment

The author thanks the financial supports from National Natural Science Foundation of China (Nos. 21574070, 21774063, 22071121) and Project supported by NCC Fund (NCC2020PY12).

References

1 Fu, G.-D., Li, G.L., Neoh, K.G., and Kang, E.T. (2011). *Prog. Polym. Sci.* 36: 127–167.
2 Li, G.L., Möhwald, H., and Shchukin, D.G. (2013). *Chem. Soc. Rev.* 42: 3628–3646.
3 Li, Y. and Shi, J. (2014). *Adv. Mater.* 26: 3176–3205.
4 Qi, J., Lai, X., Wang, J. et al. (2015). *Chem. Soc. Rev.* 44: 6749–6773.
5 Wang, X., Feng, J., Bai, Y. et al. (2016). *Chem. Rev.* 116: 10983–11060.
6 Prieto, G., Tüysüz, H., Duyckaerts, N. et al. (2016). *Chem. Rev.* 116: 14056–14119.
7 Wang, J. and Li, N. (2017). *J. Mater. Chem. B* 5: 8430–8445.
8 Bentz, K.C. and Savin, D.A. (2018). *Polym. Chem.* 9: 2059–2081.
9 Yu, L., Yu, X.Y., and Lou, X.W. (2018). *Adv. Mater.* 30: 1800939.

10 Soares, S.F., Fernandes, T., Daniel-da-Silva, A.L., and Trindade, T. (2019). *Proc. R. Soc. A* 475: 20180677.

11 Bhogal, S., Kaur, K., Malik, A.K. et al. (2020). *TrAC-Trends Anal. Chem.* 133: 116043.

12 Zhao, D., Yang, N., Xu, L. et al. (2022). *Nano Res.* 15: 739–757.

13 Arnold, R.M., Patton, D.L., Popik, V.V., and Locklin, J. (2014). *Acc. Chem. Res.* 47: 2999–3008.

14 Badoux, M., Billing, M., and Klok, H.-A. (2019). *Polym. Chem.* 10: 2925–2951.

15 Sun, W., Liu, W., Wu, Z., and Chen, H. (2020). *Macromol. Rapid Commun.* 41: 1900430.

16 Ma, S., Zhang, X., Yu, B., and Zhou, F. (2019). *NPG Asia Mater.* 11: 1–39.

17 Yan, J., Bockstaller, M.R., and Matyjaszewski, K. (2020). *Prog. Polym. Sci.* 25: 677–710.

18 Zoppe, J.O., Ataman, N.C., Mocny, P. et al. (2017). *Chem. Rev.* 117: 1105–1318.

19 Zydziak, N., Preuss, C.M., Winkler, V. et al. (2013). *Macromol. Rapid Commun.* 34: 672–680.

20 Ma, Y., Gao, J., Zheng, C., and Zhang, H. (2019). *J. Mater. Chem. B* 7: 2474–2483.

21 Zhang, H. (2014). *Polymer* 55: 699–714.

22 Weir, M.P. and Parnell, A.J. (2011). *Polymers* 3: 2107–2132.

23 Li, D., Xu, L., Wang, J., and Gautrot, J.E. (2021). *Adv. Healthcare Mater.* 10: 2000953.

24 Richardson, J.J., Bjornmalm, M., and Caruso, F. (2015). *Science* 348: aaa2491.

25 Correa, S., Dreaden, E.C., Gu, L., and Hammond, P.T. (2016). *J. Controlled Release* 240: 364–386.

26 Hoyle, C.E. and Bowman, C.N. (2010). *Angew. Chem. Int. Ed.* 49: 1540–1573.

27 He, H. and Gao, C. (2011). *Curr. Org. Chem.* 15: 3667–3691.

28 Geng, Z., Shin, J.J., Xi, Y., and Hawker, C.J. (2021). *J. Polym. Sci.* 59: 963–1042.

29 Morinaga, T., Ohkura, M., Ohno, K. et al. (2007). *Macromolecules* 40: 1159–1164.

30 He, X., Liu, Z., Fan, F. et al. (2015). *J. Nanopart. Res.* 17: 74.

31 Li, C., Ma, Y., Niu, H., and Zhang, H. (2015). *ACS Appl. Mater. Interfaces* 7: 27340–27350.

32 Mu, B., Liu, P., Tang, Z. et al. (2011). *Nanomed. Nanotechnol.* 7: 789–796.

33 Zhao, Z., Zhu, F., Qu, X. et al. (2015). *Polym. Chem.* 6: 4144–4153.

34 Liu, X., Appelhans, D., and Voit, B. (2018). *J. Am. Chem. Soc.* 140: 16106–16114.

35 Fu, G.D., Zhao, J.P., Sun, Y.M. et al. (2007). *Macromolecules* 40: 2271–2275.

36 Achilleos, D.S., Hatton, T.A., and Vamvakaki, M. (2012). *J. Am. Chem. Soc.* 134: 5726–5729.

37 Chambon, L. and Vamvakaki, M. (2020). *Soft Matter* 16: 833–841.

38 Zheng, C., Zhou, Y., and Zhang, H. (2020). *ACS Appl. Polym. Mater.* 2: 220–233.

39 Chen, F., Jiang, X., Liu, R., and Yin, J. (2011). *Polym. Chem.* 2: 614–618.

40 Bentz, K.C., Ejaz, M., Arencibia, S. et al. (2017). *Polym. Chem.* 8: 5129–5138.

41 Zhang, H. (2013). *Eur. Polym. J.* 49: 579–600.

42 Bai, F., Yang, X., and Huang, W. (2004). *Macromolecules* 37: 9746–9752.
43 Li, G.L., Xu, L.Q., Tang, X. et al. (2010). *Macromolecules* 43: 5797–5803.
44 Li, G.L., Wan, D., Neoh, K.G., and Kang, E.T. (2010). *Macromolecules* 43: 10275–10282.
45 Chen, L., Peng, Z., Zeng, Z. et al. (2014). *J. Polym. Sci., Part A: Polym. Chem.* 52: 2202–2216.
46 Guo, C., Wang, B., and Shan, J. (2015). *Chin. J. Chem.* 33: 225–234.
47 Li, G.L., Yu, R., Qi, T. et al. (2016). *Macromolecules* 49: 1127–1134.
48 Priebe, M. and Formm, K.M. (2015). *Chem. Eur. J.* 21: 3854–3874.
49 Moon, G.D. (2020). *Nanomaterials* 10: 675.
50 Li, G.L., Xu, L.Q., Neoh, K.G., and Kang, E.T. (2011). *Macromolecules* 44: 2365–2370.
51 Li, X., Cai, T., and Kang, E.-T. (2016). *Macromolecules* 49: 5649–5659.
52 Cai, T., Zhang, B., Chen, Y. et al. (2014). *Chem. Eur. J.* 20: 2723–2731.
53 Zhang, B., Cai, T., Li, S. et al. (2014). *J. Mater. Chem. C* 2: 5189–5197.
54 Huang, Y., Zhang, X.R., Ye, S. et al. (2019). *Nanoscale* 11: 13502–13510.
55 Yan, L., Wang, Z.-K., Yan, J.-J. et al. (2013). *Polym. Chem.* 4: 1243–1249.
56 Kohri, M., Kohma, H., Shinoda, Y. et al. (2013). *Polym. Chem.* 4: 2696–2702.
57 Sheng, W., Li, W., Li, B. et al. (2015). *Macromol. Rapid Commun.* 36: 1640–1645.
58 Sheng, W., Li, B., Wang, X. et al. (2015). *Chem. Sci.* 6: 2068–2073.
59 Choi, W.S., Park, J.-H., Koo, H.Y. et al. (2005). *Angew. Chem. Int. Ed.* 44: 1096–1101.
60 Rodríguez-Hernández, J., Chécot, F., Gnanou, Y., and Lecommandoux, S. (2005). *Prog. Polym. Sci.* 30: 691–724.
61 Ding, J. and Liu, G. (1997). *Macromolecules* 30: 655–657.
62 Ding, J. and Liu, G. (1998). *J. Phys. Chem. B* 102: 6107–6113.
63 Ding, J. and Liu, G. (1998). *Chem. Mater.* 10: 537–542.
64 Stewart, S. and Liu, G. (1999). *Chem. Mater.* 11: 1048–1054.
65 Zhang, Z., Liu, G., and Bell, S. (2000). *Macromolecules* 33: 7877–7883.
66 Hu, H. and Liu, G. (2014). *Macromolecules* 47: 5096–5103.
67 Song, J., Wei, Y., Hu, J. et al. (2018). *Chem. Eur. J.* 24: 212–221.
68 Moughton, A.O., Stubenrauch, K., and O'Reilly, R.K. (2009). *Soft Matter* 5: 2361–2370.
69 Zhang, H., Tian, W., Suo, R. et al. (2015). *J. Mater. Chem. B* 3: 8528–8536.
70 Canning, S.L., Smith, G.N., and Armes, S.P. (2016). *Macromolecules* 49: 1985–2001.
71 D'Agosto, F., Rieger, J., and Lansalot, M. (2020). *Angew. Chem. Int. Ed.* 59: 8368–8392.
72 Varlas, S., Foster, J.C., Georgiou, P.G. et al. (2019). *Nanoscale* 11: 12643–12654.
73 Xu, X.-F., Pan, C.-Y., Zhang, W.-J., and Hong, C.-Y. (2019). *Macromolecules* 52: 1965–1975.
74 Rzayev, J. (2012). *ACS Macro Lett.* 1: 1146–1149.
75 Xie, G., Martinez, M.R., Olszewski, M. et al. (2019). *Biomacromolecules* 20: 27–54.

76 Li, Z., Tang, M., Liang, S. et al. (2021). *Prog. Polym. Sci.* 116: 101387.
77 Cheng, C., Qi, K., Khoshdel, E., and Wooley, K.L. (2006). *J. Am. Chem. Soc.* 128: 6808–6809.
78 Huang, K. and Rzayev, J. (2009). *J. Am. Chem. Soc.* 131: 6880–6885.
79 Huang, K., Jacobs, A., and Rzayev, J. (2011). *Biomacromolecules* 12: 2327–2334.
80 Huang, K. and Rzayev, J. (2011). *J. Am. Chem. Soc.* 133: 16726–16729.
81 Huang, K., Canterbury, D.P., and Rzayev, J. (2010). *Chem. Commun.* 46: 6326–6328.
82 Xiong, L., Yang, K., Zhang, H. et al. (2016). *Nanotechnology* 27: 115603.
83 Xiong, L., Zhang, H., He, Z. et al. (2018). *New J. Chem.* 42: 1368–1372.
84 Kumar, S., Rani, R., Dilbaghi, N. et al. (2017). *Chem. Soc. Rev.* 46: 158–196.
85 Li, X., Yang, Y., and Yang, Q. (2013). *J. Mater. Chem. A* 1: 1525–1535.
86 Sharma, J. and Polizos, G. (2020). *Nanomaterials* 10: 1599.
87 Sakellariou, G., Priftis, D., and Baskaranw, D. (2013). *Chem. Soc. Rev.* 42: 677–704.
88 Sano, M., Kamino, A., Okamura, J., and Shinkai, S. (2001). *Langmuir* 17: 5125–5128.
89 Hill, D.E., Lin, Y., Rao, A.M. et al. (2002). *Macromolecules* 35: 9466–9471.
90 Wu, W., Zhang, S., Li, Y. et al. (2003). *Macromolecules* 36: 6286–6288.
91 Lou, X., Detrembleur, C., Pagnoulle, C. et al. (2004). *Adv. Mater.* 16: 2123–2127.
92 Zydziak, N., Hübner, C., Bruns, M., and Barner-Kowollik, C. (2011). *Macromolecules* 44: 3374–3380.
93 Viswanathan, G., Chakrapani, N., Yang, H. et al. (2003). *J. Am. Chem. Soc.* 125: 9258–9259.
94 Yao, Z., Braidy, N., Botton, G.A., and Adronov, A. (2003). *J. Am. Chem. Soc.* 125: 16015–16024.
95 Qin, S., Qin, D., Ford, W.T. et al. (2004). *J. Am. Chem. Soc.* 126: 170–176.
96 Kong, H., Gao, C., and Yan, D. (2004). *J. Am. Chem. Soc.* 126: 412–413.
97 Sun, T., Liu, H., Song, W. et al. (2004). *Angew. Chem. Int. Ed.* 43: 4663–4666.
98 Zhang, Y., He, H., and Gao, C. (2008). *Macromolecules* 41: 9581–9594.
99 Priftis, D., Sakellariou, G., Baskaran, D. et al. (2009). *Soft Matter* 5: 4272–4278.
100 Hong, C.-Y., You, Y.-Z., and Pan, C.-Y. (2005). *Chem. Mater.* 17: 2247–2254.
101 Dehonor, M., Masenelli-Varlot, K., González-Montiel, A. et al. (2005). *Chem. Commun.* 5349–5351.
102 Zhao, X.-D., Fan, X.-H., Chen, X.-F. et al. (2006). *J. Polym. Sci., Part A: Polym. Chem.* 4656–4667.
103 Gómez, F.J., Chen, R.J., Wang, D. et al. (2003). *Chem. Commun.* 190–191.
104 Liu, Y. and Adronov, A. (2004). *Macromolecules* 37: 4755–4760.
105 Xu, Y., Gao, C., Kong, H. et al. (2004). *Macromolecules* 37: 8846–8853.
106 Zeng, H., Gao, C., and Yan, D. (2006). *Adv. Funct. Mater.* 16: 812–818.
107 Zhou, L., Gao, C., and Xu, W. (2009). *Macromol. Chem. Phys.* 210: 1011–1018.
108 Li, J., He, W.-D., Yang, L.-P. et al. (2007). *Polymer* 48: 4352–4360.
109 Priftis, D., Sakellariou, G., Mays, J.W., and Hadjichristidis, N. (2010). *J. Polym. Sci., Part A: Polym. Chem.* 48: 1104–1112.

110 Mosnáčková, K., Mrlík, M., Mičušík, M. et al. (2021). *Macromolecules* 54: 2412–2426.
111 Lay, C.L., Liu, H.Q., Wu, D., and Liu, Y. (2010). *Chem. Eur. J.* 16: 3001–3004.
112 Ladmiral, V., Morinaga, T., Ohno, K. et al. (2009). *Eur. Polym. J.* 45: 2788–2796.
113 Li, B., Yang, X., Xia, L. et al. (2013). *Sci. Rep.* 3: 2128.
114 Wang, Z., Chen, H., Wang, Y. et al. (2021). *ACS Appl. Mater. Interfaces* 13: 27411–27418.
115 Huang, Y., Sasano, T., Tsujii, Y., and Ohno, K. (2016). *Macromolecules* 49: 8430–8439.
116 Lay, C.L., Tan, H.R., Lu, X., and Liu, Y. (2011). *Chem. Eur. J.* 17: 2504–2509.
117 Yu, F., Tang, X., and Pei, M. (2013). *Microporous Mesoporous Mater.* 173: 64–69.
118 Zhang, Y., Ang, C.Y., Li, M. et al. (2015). *ACS Appl. Mater. Interfaces* 7: 18179–18187.
119 Beagan, A., Lahmadi, S., Alghamdi, A. et al. (2020). *Polymers* 12: 2749.
120 Zhao, N., Lin, X., Zhang, Q. et al. (2015). *Small* 11: 6467–6479.
121 Wang, R., Zhao, N., and Xu, F.-J. (2017). *Adv. Funct. Mater.* 27: 1700256.
122 Zhou, J., Zhang, W., Hong, C., and Pan, C. (2015). *ACS Appl. Mater. Interfaces* 7: 3618–3625.
123 Chen, M., Wu, L., Zhou, S., and You, B. (2006). *Adv. Mater.* 18: 801–806.
124 Liu, G., Cai, M., Zhou, F., and Liu, W. (2014). *J. Phys. Chem. B* 118: 4920–4931.
125 Ma, P., Zhou, Z., Dai, J. et al. (2016). *RSC Adv.* 6: 9619–9630.
126 Song, J., Zhou, J., and Duan, H. (2012). *J. Am. Chem. Soc.* 134: 13458–13469.
127 Liu, Y., Wang, J., Zhang, M. et al. (2020). *ACS Nano* 14: 12491–12521.
128 Pang, X., Zhao, L., Han, W. et al. (2013). *Nat. Nanotechnol.* 8: 426–431.
129 Chen, Y., Yang, D., Yoon, Y.J. et al. (2017). *J. Am. Chem. Soc.* 139: 12956–12967.
130 He, Y., Pang, X., Jiang, B. et al. (2017). *Angew. Chem. Int. Ed.* 56: 12946–12951.
131 Huang, S., Yu, X., Dong, Y. et al. (2012). *Colloid. Surface A* 415: 22–30.
132 Han, H., Li, L., Wang, W. et al. (2017). *Langmuir* 33: 9857–9865.
133 Han, H., Li, L., Yang, Q. et al. (2018). *J. Mater. Sci.* 53: 3210–3224.
134 Yang, Q., Li, L., Zhao, F. et al. (2019). *J. Mater. Sci.* 54: 2552–2565.
135 Guo, X., Weiss, A., and Ballauff, M. (1999). *Macromolecules* 32: 6043–6046.

8

Self-Assembly of Binary Mixed Homopolymer Brush-Grafted Silica Nanoparticles

Bin Zhao[1], Ping Tang[2], Phoebe L. Stewart[3], Rong-Ming Ho[4], Christopher Y. Li[5], and Lei Zhu[6]

[1] *University of Tennessee, Department of Chemistry, Knoxville, Tennessee 37996, United States*
[2] *Fudan University, The State Key Laboratory of Molecular Engineering of Polymers, Key Laboratory of Computational Physical Sciences, Department of Macromolecular Science, Shanghai 200433, China*
[3] *Case Western Reserve University, Department of Pharmacology and Cleveland Center for Membrane and Structural Biology, Cleveland, Ohio 44106-4965, United States*
[4] *National Tsing Hua University, Department of Chemical Engineering, Hsinchu 30013, Taiwan*
[5] *Drexel University, Department of Materials Science and Engineering, Philadelphia, Pennsylvania 19104, United States*
[6] *Case Western Reserve University, Department of Macromolecular Science and Engineering, Cleveland, Ohio 44106-7202, United States*

8.1 Introduction

Hierarchical self-assembly of inorganic nanoparticles (NPs), including metallic NPs with strong surface plasmon resonances, magnetic NPs with magnetic properties, and quantum dots (e.g. CdSe and CdS) with tunable fluorescence characteristics, has attracted substantial research interest due to their collective properties [1–3]. However, it has been challenging to precisely control their self-assembly. One potential solution is to graft polymer chains onto inorganic NPs and combine the desired properties of both inorganic NPs and polymers. For example, grafting two polymers on the surface of NPs with a sufficiently high grafting density, in the form of either mixed homopolymer brushes or block copolymer brushes, gives bicomponent hairy NPs [4–11], which can respond to changes in the environment, such as solvent selectivity, temperature, and pH [1, 6–8, 12–19]. The response is manifested by the reorganization of the self-assembled nanostructures formed by the two polymers on the NP surface, which can in turn be used to control their hierarchical assembly. Hence, the structures and properties of these bicomponent hairy NPs and their assemblies can be reversibly tuned by external stimuli. In this sense, they can have a broad range of potential applications such as Pickering emulsion, antifouling, metamaterials, nanocomposites, optoelectronics, catalysis, sensing, and medical therapeutics and diagnostics [1, 6, 12–23]. For example, amphiphilic bicomponent polymer brush-grafted NPs belong to a special type of particulate surfactants, which are ideal for Pickering emulsions to stabilize oil–water mixtures at an extremely low concentration. These Pickering emulsions have been widely used in biomedicine, food,

Hairy Nanoparticles: From Synthesis to Applications, First Edition. Edited by Zhiqun Lin and Yijiang Liu.

fine chemical synthesis, and cosmetics. Also, the self-assembled superstructures of bicomponent hairy NPs can be considered as a unique metamaterial with photonic bandgap properties, which could be useful for optical filters, photonic sensors, and waveguides.

Binary mixed homopolymer brushes grafted on a flat surface (i.e. the limiting case of binary mixed homopolymer brush-grafted NPs (MBNPs) with an infinitely large particle size) have been studied theoretically since 1991 and experimentally since 1999 [6, 24–54]. Marko and Witten predicted through a theoretical study that under the equilibrium melt conditions, symmetric binary mixed homopolymer brushes on a flat substrate undergo lateral microphase separation, producing a "rippled" nanostructure, instead of vertical phase separation, forming a "layered" structure, and the feature size is on the order of polymer chain dimensions [26]. In solvents, mixed brushes exhibit even richer behavior than in the melt, because the two grafted polymers can have different interactions with the solvent and the solvent quality for either polymer can be tuned [29–33]. A variety of self-assembled nanostructures have been predicted, including "rippled," "layered," and "dimpled" phases. Experimentally, the surface-responsive properties of binary mixed polymer brushes have been demonstrated by many researchers [35–54]. For example, Sidorenko et al. reported the synthesis of mixed polystyrene (PS)/poly(2-vinylpyridine) brushes on flat substrates by a two-step surface-initiated conventional free radical polymerization process from a surface-immobilized azo initiator and showed the switching of the surface property of the obtained mixed brushes from hydrophobic to hydrophilic and vice versa by exposing the brushes to different solvents [35]. However, direct visualization of the self-assembled morphologies of mixed brushes on flat substrates was hindered due to the difficulty in using transmission electron microscopy (TEM) and small angle X-ray or neutron scattering (SAXS and SANS) to study the mixed brushes on silicon wafers or glass substrates.

Grafting binary mixed homopolymer brushes on NPs not only allows for the use of common polymer characterization techniques such as differential scanning calorimetry (DSC), TEM, and SAXS but also introduces a new parameter to the mixed-brush system – the curvature of the substrate [6]. For larger NPs with diameters significantly greater than the sizes of the grafted polymers, the brushes can be considered as grafted on a quasi-flat substrate. For smaller NPs, whose size is much smaller than that of the polymer chains, the hairy NPs can be considered as miktoarm star copolymers [55]. When the dimensions of the polymer brushes and the NP are similar, ordered patchy particle morphologies can be obtained, as predicted by simulation studies [56–60]. In this book chapter, we summarize our theoretical and experimental studies on the self-assembly of well-defined MBNPs with various polymer molecular parameters and core NP sizes. Section 8.2 focuses on computer simulations of individual MBNPs with similar sizes for the core NP and the grafted polymer chains, as well as the hierarchical structures of the MBNP assemblies. Section 8.3 presents the experimental studies of the morphologies of well-defined mixed homopolymer brushes grafted on silica particles of various diameters by TEM. Section 8.4 shows the in situ three-dimensional (3D) morphologies of MBNPs in solvents and polymer matrices using TEM tomography. The chapter is concluded with a summary and outlook in Section 8.5.

8.2 Computer Simulations of the Self-Assembled Morphology of MBNPs

In our initial study, a self-consistent field theory was used to predict the self-assembled nanostructures of individual MBNPs via numerical implementation [57]. This numerical method can successfully circumvent the "pole problem," which is caused by the spherical coordinate system in the conventional finite element grid. Figure 8.1a shows the self-assembled morphology of symmetric MBNP with the NP radius $R = 4$, the chain length $N_A = N_B = 20$, and the total grafting density (σ_{total}) = 0.6. Note that "symmetric" in this section is defined as identical chain lengths for the two grafted polymers, i.e. $N_A = N_B$. By varying the grafting density ratio σ_B/σ_A, different nanostructures are obtained. When $\sigma_B/\sigma_A = 1$, a ripple structure is observed. When $\sigma_B/\sigma_A = 2$, a structure of 10 isolated domains formed by the minority polymer A distributed symmetrically in the matrix of the majority polymer B (i.e. a Z10/A islanded structure) is obtained. When $\sigma_B/\sigma_A = 3$, a Z12/A islanded structure is obtained. Finally, when $\sigma_B/\sigma_A = 5$, a layered structure with the A species covering the B species (L_{BA}) is observed, where the loosely grafted A chains are pushed to stretch out by the densely grafted

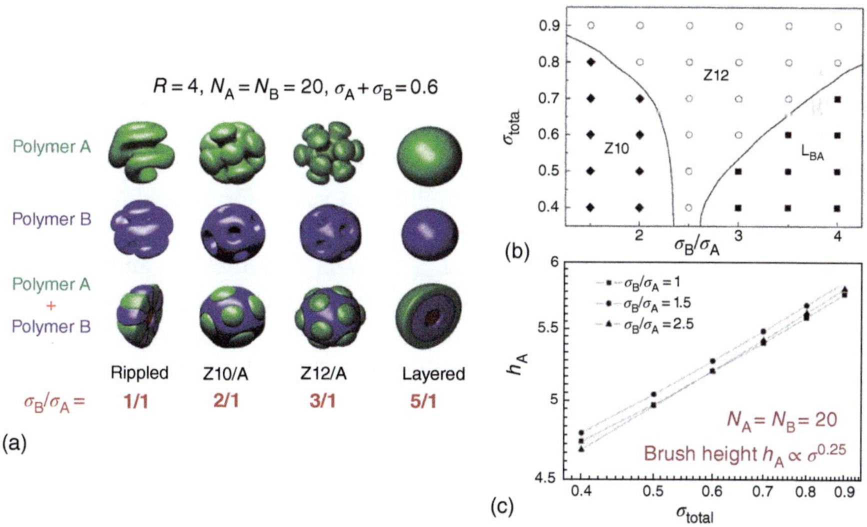

Figure 8.1 (a) Simulated morphology of individual MBNPs at different grafting density ratios, σ_B/σ_A, with the NP radius $R = 4$, degrees of polymerization $N_A = N_B = 20$, and total grafting density (σ_{total}) = 0.6. (b) Phase diagram of MBNPs as a function of σ_{total} and σ_B/σ_A with $R = 4$ and symmetric chain length $N_A = N_B = 20$. (c) Brush height of the islands formed by the A polymer (h_A) versus σ_{total} with $R = 4$, $N_A = N_B = 20$, and different σ_B/σ_A values. Source: Adapted from Ref. [57]. Copyright American Institute of Physics 2013.

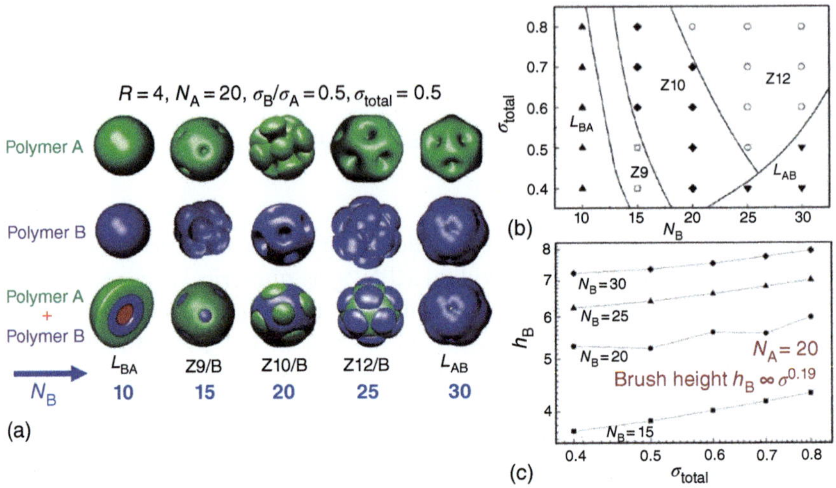

Figure 8.2 (a) Simulated morphology of individual MBNPs with asymmetric chain lengths at $N_A = 20$: $R = 4$, $\sigma_B/\sigma_A = 0.5$, and $\sigma_{total} = 0.5$. (b) Phase diagram of MBNPs as a function of σ_{total} and N_B with $R = 4$, $N_A = 20$, and $\sigma_B = \sigma_A = 0.25$. (c) Brush height of the islands formed by the B polymer (h_B) versus σ_{total} with $R = 4$, $N_A = 20$, $\sigma_B/\sigma_A = 0.5$, and different N_B values. Source: Adapted from Ref. [57]. Copyright American Institute of Physics 2013.

B chains and form the outer layer. Figure 8.1b shows the phase diagram for these symmetric mixed-brush-grafted NPs at different σ_{total} values. When the σ_{total} is relatively low, a low σ_B/σ_A ratio favors the Z10/A structure, and a high σ_B/σ_A favors the L_{BA} structure. At high σ_{total} values, Z12/A is favored. Because of the short chain lengths ($N_A = N_B = 20$), a relatively large number of domains resulted to avoid high stretching of the polymer chains. The brush height of the islands formed by the A polymer (h_A) has a scaling relationship with the σ_{total} under different σ_B/σ_A ratios (1, 1.5, and 2.5): $h_A \propto \sigma_{total}^x$, where $x = 0.24$–0.26 (averaging 0.25, see Figure 8.1c).

For the asymmetric mixed brushes with different values for N_A and N_B, the situation becomes different. When $R = 4$, $N_A = 20$, $\sigma_B/\sigma_A = 0.5$, and $\sigma_{total} = 0.5$, the self-assembled morphology evolves from L_{BA} to Z9/B, Z10/B, Z12/B, and finally to L_{AB} (the layered structure with the B polymer outside) as the N_B increases from 10 to 30 (Figure 8.2a). The phase diagram as a function of σ_{total} and N_B is shown in Figure 8.2b. Different from the symmetric brush case, the Z9 islanded structure exists, although it is found in a small region of the phase diagram. The Z12 structure prevails in the phase diagram with the largest region. The brush height of the B domains also scales as a function of σ_{total} with the exponent x ranging from 0.14 to 0.24 (averaging 0.19, see Figure 8.2c).

The self-assembled structure of MBNPs can also be influenced by the particle size. As shown in Figure 8.3a, by increasing R from 0.5 to 4, various islanded structures can be realized, with the number of islands ranging from 2 to 8. The Z2 islanded structure is a dumbbell-like structure with a D_∞ symmetry. The Z3

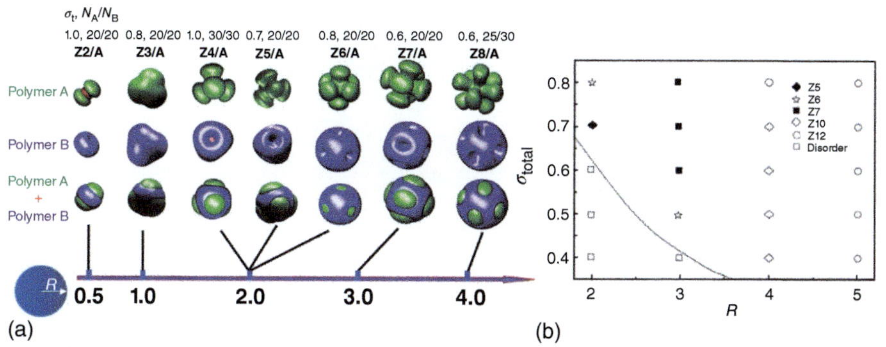

Figure 8.3 (a) Typical structures of individual MBNPs at $\sigma_B/\sigma_A = 2$ with different combinations of R, σ_{total}, and N_A and N_B values. The number of A islands varies from 2 to 8. (b) Phase diagram of MBNPs with different spherical radius R, $N_A = N_B = 20$, and $\sigma_B/\sigma_A = 2$. Lines are used to guide the eyes. Source: Adapted from Ref. [57]. Copyright American Institute of Physics 2013.

structure consists of a planar equilateral triangle with a D_{3h} symmetry, while the Z4 islanded structure consists of a tetrahedron with a T_d symmetry. The Z5, Z6, and Z7 islanded structures appear to be a triangular bipyramid, a regular octahedron, and a pentagonal bipyramid with D_{3h}, O_h, and D_{5h} symmetry, respectively. The Z8 islanded structure appears to be a square antiprism. Interestingly, these islanded structures are all in agreement with the morphologies of densely packed microspheres with specific numbers [61]. Figure 8.3b shows the phase diagram as a function of σ_{total} and R at $N_A = N_B = 20$ and $\sigma_B/\sigma_A = 2$. In general, the island number decreases with decreasing R. The symmetry of the islanded structures will change with decreasing the island number. This is because a smaller R makes it easier for more polymer chains of the same kind to aggregate and the island number thus decreases. The Z12 islanded structure that can be easily obtained at $R = 4$ is never observed at $R = 1$. Also, the Z2 islanded structure cannot be obtained at $R = 4$ either. Decreasing the particle size and thus the surface area will cause the overall space for phase separation to decrease and thus amplify the confinement effect that dramatically influences the island number of the structures.

In addition to the self-assembly of the mixed brushes in individual hairy NPs, MBNPs can also self-assemble into hierarchical superstructures (Figure 8.4a) [24, 60]. Figure 8.4b summarizes the hierarchical structures with a crystalline arrangement of MBNPs. The morphology of single NPs changes subtly, if they are forced to form simple cubic (SC), body-centered cubic (BCC), face-centered cubic (FCC), and hexagonal (HEX) lattices. This means that the single-particle morphology tends to retain the geometric symmetry of the corresponding lattice (e.g. the cube-like NP in the SC lattice), which is reminiscent of the microphase separation of block copolymers under complex confinement [62, 63]. Therefore, the microphase-separated polymer structures in different lattice types are also different. These two interacting hard- and soft-ordered structures will lead to unconventional superstructures.

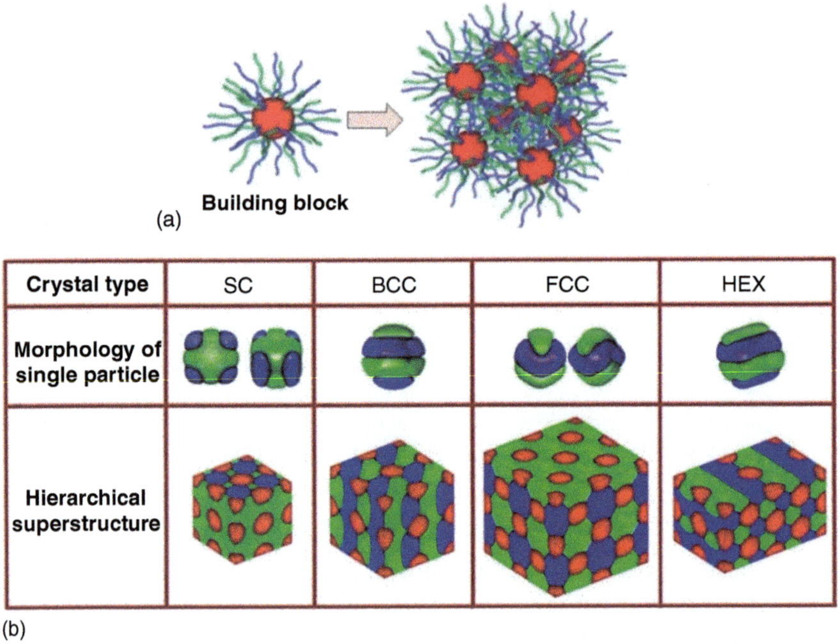

Figure 8.4 (a) Schematic representation of the assembly of MBNPs. (b) Self-assembled structures of single MBNPs and their hierarchical superstructures with fixed parameters: $R = 4$, $N_A = N_B = 32$, and $\sigma_A = \sigma_B = 0.1$. Polymer A is green and polymer B is blue, and the spherical core is red. SC: simple cubic, BCC: body-centered cubic, FCC: face-centered cubic, and HEX: hexagonal. Source: Adapted from Ref. [60]. Copyright American Chemical Society 2016.

8.3 Self-Assembled Morphologies of Well-Defined Binary Mixed Homopolymer Brushes Grafted on Silica NPs

8.3.1 Synthesis of Well-Defined Binary Mixed Homopolymer Brush-Grafted Silica NPs

To elucidate the morphologies of MBNPs under various conditions, we developed a synthetic method for the synthesis of binary mixed homopolymer brushes grafted on silica NPs with well-defined molecular weights, narrow dispersities, and controlled grafting densities by combining two distinct reversible deactivation radical polymerization techniques, atom transfer radical polymerization (ATRP) and nitroxide-mediated radical polymerization (NMRP) (Scheme 8.1) [64]. ATRP and NMRP are usually carried out under different conditions [39–43, 64–67]. Silica NPs are chosen in our studies because they are readily available either by the Stöber method [68, 69] or from a commercial source (e.g. Nissan Chemical Corporation), and the well-known robust silane chemistry can be employed to immobilize the initiators on the NP surface. Asymmetric difunctional initiators that contain one initiating group for ATRP and one initiating moiety for NMRP with a surface-immobilizable silane group (i.e. Y-Initiators [41–43, 64, 70, 71])

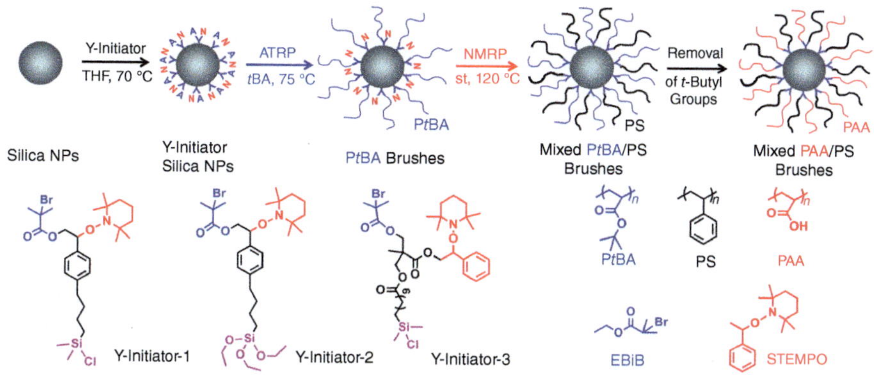

Scheme 8.1 Synthesis of well-defined binary mixed homopolymer bushes from asymmetric difunctional initiator-functionalized silica nanoparticles by sequential surface-initiated ATRP at 75 °C and surface-initiated NMRP at 120 °C with the addition of a corresponding free initiator in each step. Source: Adapted from Ref. [64]. Copyright 2005 American Chemical Society.

were prepared via multistep synthetic procedures and covalently immobilized onto the surface of silica NPs via the formation of Si—O—Si bonds with the NPs. Surface-initiated ATRP of *tert*-butyl acrylate (*t*BA) was then performed in the presence of a free ATRP initiator, ethyl 2-bromoisobutyrate (EBiB). The addition of a sacrificial initiator not only allows for better control of surface-initiated ATRP but also produces a free polymer analog [72–74]. Using hydrofluoric acid to etch the silica core, we confirmed that the molecular weights of the grafted polymer and the free polymer were essentially the same [64]. The NMRP initiating group on the surface of silica NPs was verified to be stable under the chosen ATRP conditions. Subsequently, surface-initiated NMRP of styrene was carried out at 120 °C, with the addition of a free NMRP initiator, typically 1-phenyl-1-(2′,2′,6′,6′-tetramethyl-1′-piperidinyloxy)ethane (STEMPO). Similarly, the molecular weight and dispersity of the grafted polymer were virtually the same as that of the free polymer formed in the solution [64]. One advantage of this method is the individual control of the molecular characteristics of two grafted polymers in the mixed brushes. Using this synthetic strategy, we prepared a series of well-defined MBNPs to systematically investigate the effects of various parameters on the morphology of binary mixed homopolymer brushes under various conditions.

8.3.2 Lateral Microphase Separation of Nearly Symmetric P*t*BA/PS MBNPs

Two P*t*BA/PS MBNP samples with a silica core diameter of 180 nm were synthesized (see the top panel of Figure 8.5 for characterization data). Here, the symmetric mixed brushes mean that the two polymers have similar molecular weights. TEM has been shown to be a powerful characterization technique to study the morphology of mixed brushes on silica particles [72, 75]. The P*t*BA polymer in **MBNP-I-1** had a number-average molecular weight (M_n) of 24.2 kDa with a grafting density

No.	M_n, Đ, and DP of P*t*BA	σ_{PtBA} (chains nm^{-2})	M_n, Đ, and DP of PS	σ_{PS} (chains nm^{-2})	Brush thickness
MBNP-I-1	24200, 1.09, 187	0.40	23000, 1.10, 221	0.37	~20 nm
MBNP-I-2	10400, 1.15, 87	0.32	11900, 1.19, 114	0.34	~10 nm

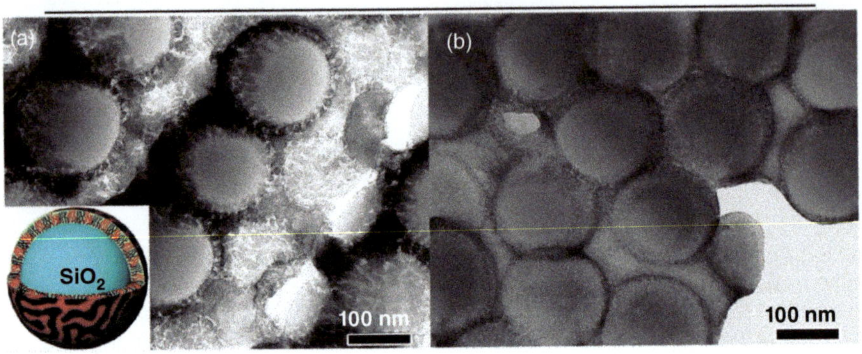

Figure 8.5 Bright-field TEM micrographs of self-assembled (a) **MBNP-I-1** and (b) **MBNP-I-2** after thermal annealing in the melt at 120 °C for 12 hours. The molecular characterization data of **MBNP-I-1** and **MBNP-I-2** are listed on the top panel. Source: Adapted from Ref. [75]. Copyright American Chemical Society 2006.

(σ_{PtBA}) of 0.40 chains nm^{-2}, and the PS in this sample had a M_n of 23.0 kDa with a σ_{PS} of 0.37 chains nm^{-2}. The average brush thickness was ~20 nm. As shown in Figure 8.5a, lateral microphase separation between the grafted P*t*BA and PS chains was observed with a rippled nanostructure on the surface of silica NPs [75]. For the first time, the laterally phase-separated morphology of symmetric mixed homopolymer brushes in melt predicted by Marko and Witten was experimentally directly visualized [26]. The P*t*BA in **MBNP-I-2** had an M_n of 10.4 kDa with a σ_{PtBA} of 0.32 chains nm^{-2}. The PS in **MBNP-I-2** had a M_n of 11.9 kDa with a σ_{PS} of 0.34 chains nm^{-2}. The average brush thickness was ~10 nm. As shown in Figure 8.5b, no pronounced microphase separation was observed in this sample. Therefore, the critical M_n for microphase separation of symmetric mixed P*t*BA/PS brushes grafted on 180 nm silica NPs in melt should be between 10 and 20 kDa when the individual polymer grafting density is around 0.3–0.4 chains nm^{-2}.

8.3.3 Effect of Chain Length Disparity on the Self-Assembled Morphology of P*t*BA/PS MBNPs

The effect of chain length disparity on the morphology of mixed homopolymer brushes has been theoretically studied [32, 56, 57]. Similar to what we have discussed in Figures 8.1 and 8.2, Roan used computer simulations to study the microphase separation of spherical MBNPs with a radius comparable to the root-mean-square end-to-end distance (<R_{rms}>) of the grafted polymer chains [56]. The equilibrium phase- separated nanostructure varies from rippled to 12-islanded and then layered when the chain length disparity of the two grafted polymers increases under the condition of equal grafting densities for the two polymers.

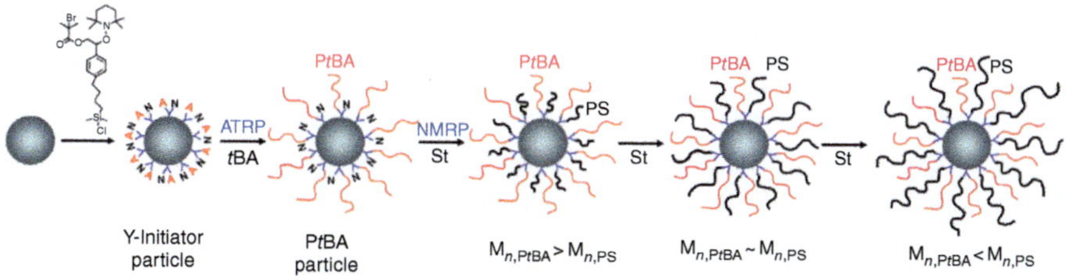

Scheme 8.2 Synthesis of P*t*BA/PS **MBNP-II** with a fixed P*t*BA M_n and various PS molecular weights by sequential ATRP and NMRP from Y-initiator-functionalized silica particles. Source: Reproduced from Ref. [76]. Copyright American Chemical Society 2010.

Using single-chain-in-mean-field simulations, Wang and Müller investigated the behavior of asymmetric binary mixed homopolymer brushes grafted on flat substrates [32]. They found that two layers could be distinguished for the mixed brushes with a large chain length disparity in a solvent that was marginally good for both grafted polymers with a slight preference for one, where the bottom layer was laterally phase separated and the top layer contained only the longer polymer species. Although asymmetric mixed brushes with different chain lengths for the two grafted polymers had been experimentally investigated [43], the morphologies had not been directly visualized.

Silica particles with a diameter of 160 nm were synthesized by the Stöber process and functionalized with a monochlorosilane-terminated Y-initiator (Y-Initiator-1 in Scheme 8.1) as illustrated in Scheme 8.2. Surface-initiated ATRP of *t*-BA was then conducted at 75 °C in the presence of a free initiator, EBiB, yielding P*t*BA brush-grafted silica particles with a P*t*BA $M_{n,SEC}$ of 24.5 kDa, a $Đ$ of 1.11, and a σ_{PtBA} of 0.36 chains nm^{-2} [76]. Mixed P*t*BA/PS brush-grafted silica particles (**MBNP-II**) were then synthesized by surface-initiated NMRP of styrene from the P*t*BA brush particles at 120 °C with the addition of STEMPO as the free initiator. NMRP is a reversible deactivation radical polymerization method [66], allowing us to synthesize a series of mixed P*t*BA/PS brush particles with different PS molecular weights in one pot (Scheme 8.2). Four mixed-brush NP samples with the same P*t*BA $M_{n,SEC}$ of 24.5 kDa but different PS M_n values (14.8, 18.7, 24.9, and 30.4 kDa with $Đ < 1.25$) were obtained at different polymerization times. The hairy particles were isolated by centrifugation and repeatedly washed with THF. Thermogravimetric analysis (TGA) showed that the total mass of the two grafted polymers relative to the silica residue increased with the increase of PS M_n in a nearly linear fashion. The grafting densities of PS in these four samples were calculated to be 0.21, 0.26, 0.27, and 0.32 chains nm^{-2}, and thus the total grafting densities were 0.57, 0.62, 0.63, and 0.68 chains nm^{-2}.

DSC showed that all four MBNP samples displayed two distinct glass transitions at ~47 and ~90 °C, which corresponded to the glass transition temperatures (T_gs) of P*t*BA and PS, respectively, indicating that the two grafted polymers microphase separated into separate domains in the brush layer [76]. In addition, the T_g of the PS increased from 88 to 90, 93, and 94 °C with the increase of PS M_n from 14.8, to 18.7, to 24.9, and 30.4 kDa. TEM was then used to visualize the morphologies

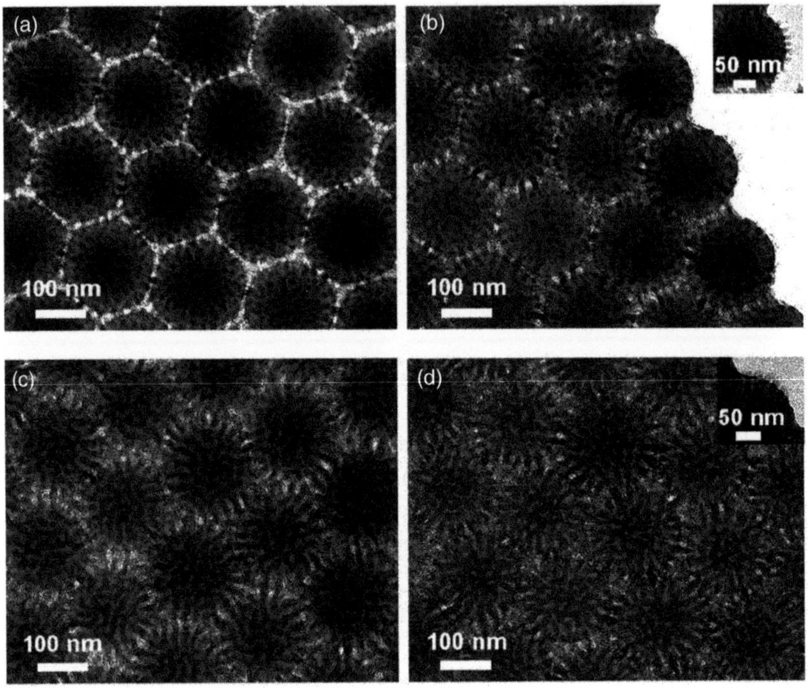

Figure 8.6 Bright-field TEM micrographs of (a) **MBNP-II-1** (P*t*BA $M_n = 24.5$ kDa, $\sigma_{PtBA} = 0.36$ chains nm^{-2}; PS $M_n = 14.8$ kDa, $\sigma_{PS} = 0.21$ chains nm^{-2}; $DP_{PS}/DP_{PtBA} = 0.74$), (b) **MBNP-II-2** (P*t*BA $M_n = 24.5$ kDa, $\sigma_{PtBA} = 0.36$ chains nm^{-2}; PS $M_n = 18.7$ kDa, $\sigma_{PS} = 0.26$ chains nm^{-2}; $DP_{PS}/DP_{PtBA} = 0.94$), (c) **MBNP-II-3** (P*t*BA $M_n = 24.5$ kDa, $\sigma_{PtBA} = 0.36$ chains nm^{-2}; PS $M_n = 24.9$ kDa, $\sigma_{PS} = 0.27$ chains nm^{-2}; $DP_{PS}/DP_{PtBA} = 1.25$), and (d) **MBNP-II-4** (P*t*BA $M_n = 24.5$ kDa, $\sigma_{PtBA} = 0.36$ chains nm^{-2}; PS $M_n = 30.4$ kDa, $\sigma_{PS} = 0.32$ chains nm^{-2}; $DP_{PS}/DP_{PtBA} = 1.53$) after being cast from $CHCl_3$, a nonselective good solvent for both P*t*BA and PS, and thermally annealed at 120 °C in vacuum for 3 hours. The samples were stained with RuO_4 vapor. Reproduced from Ref. [76]. Copyright American Chemical Society 2010.

of **MBNP-II**. The hairy particles were dispersed in $CHCl_3$, a nonselective good solvent for both polymers, and drop-cast onto carbon film-coated copper TEM grids, thermally annealed at 120 °C in vacuum for 3 h, and then stained with RuO_4 vapor at room temperature for 30 minutes. Clearly, all **MBNP-II** samples underwent microphase separation (Figure 8.6a), and the morphology changed with increasing PS molecular weight. For **MBNP-II-1** with a PS M_n of 14.8 kDa, much lower than that of P*t*BA (24.5 kDa), the grafted PS chains segregated into isolated, nearly spherical microdomains (Figure 8.6a). Interestingly, the silica core NPs in Figure 8.6a were separated from each other by a bright gap, which is believed to be filled with "invisible" P*t*BA chains, though occasionally dark PS domains formed bridges between neighboring particles. Thus, it was very likely that PS microdomains were buried inside the P*t*BA matrix (Scheme 8.3a). The average equivalent diameter of PS domains in Figure 8.6a from image analysis was 11.5 nm, comparable to the $\langle R_{rms} \rangle$ (8.0 nm) of PS with a molecular weight of 14.8 kDa in the unperturbed state.

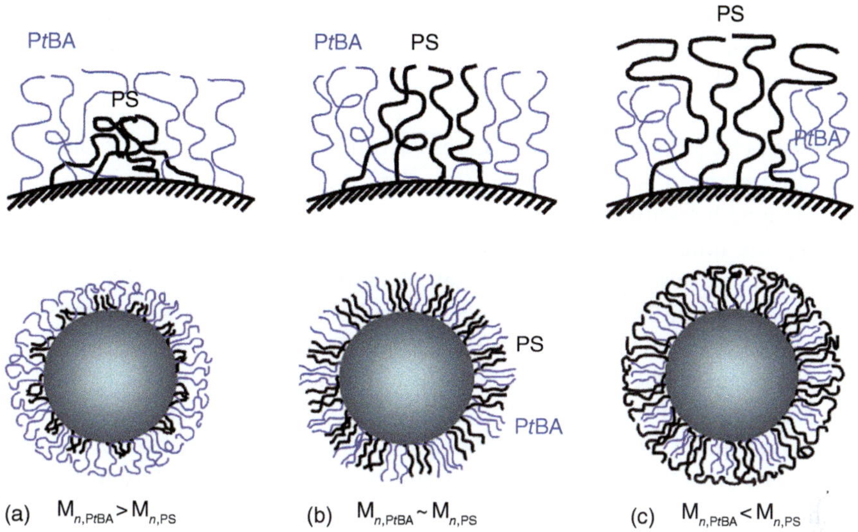

Scheme 8.3 Schematic illustration of morphologies of PtBA/PS **MBNP-II** with a fixed PtBA molecular weight and varying PS molecular weights. Source: Adapted from refs. [70, 76]. Copyright American Chemical Society 2010.

Upon increasing the PS M_n to 18.7 kDa (**MBNP-II-2**), the isolated PS domains began to merge into short cylinders but were not fully connected (Figure 8.6b), though the PtBA and PS chain lengths (DP_{PtBA} = 191 and DP_{PS} = 180) were very similar. This is probably caused by the slightly lower PS grafting density (0.26 chains nm^{-2}) compared with PtBA (0.36 chains nm^{-2}). The surface-tethered polymer chains spread out and covered the interstitials of particles, and some dark PS domains formed bridges among neighboring particles. From the edge of the particle in the inset of Figure 8.6b, it appeared that both dark PS and bright PtBA domains were present at the brush surface. For the **MBNP-II-3** sample with PtBA $M_{n,SEC}$ of 24.5 kDa and the PS M_n of 24.9 kDa (σ_{PtBA} = 0.36 chains nm^{-2} and σ_{PS} = 0.27 chains nm^{-2}), a nearly bicontinuous, random worm-like nanostructure was observed (Figure 8.6c), similar to the computer simulation results from Wang and Müller [32], with the cross-section of the brush layer schematically illustrated in Scheme 8.3b. Image analysis showed that the widths of PS and PtBA stripes were 14.0 and 12.3 nm, respectively, similar to the values of $\langle R_{rms} \rangle$ of the corresponding free PS and PtBA in the unperturbed states (10.4 and 9.3 nm, respectively). This result confirmed the theoretical prediction that the ripple wavelength was about twice the $\langle R_{rms} \rangle$ of polymers in the symmetric case. For the **MBNP-II-4** with PtBA $M_{n,SEC}$ of 24.5 kDa) and PS M_n of 30.4 kDa (σ_{PtBA} = 0.36 chains nm^{-2} and σ_{PS} = 0.32 chains nm^{-2}), a morphological transition was observed, where the PS dark domains were more connected and the bright PtBA domains became more isolated (Figure 8.6d). However, the shape of isolated PtBA domains was more irregular than the dark PS domains in Figure 8.6a. An inspection of the interstitials between core silica particles and the edge of the MBNP (see the inset in Figure 8.6d) implied that the isolated P*t*BA domains were buried in the PS matrix (Scheme 8.3c). This is likely because the PS chain length (DP = 292) was significantly longer than that of P*t*BA (DP = 191) while the grafting densities of the two polymers were similar. This result is consistent with the theoretical prediction for asymmetric mixed brushes [32] that a two-layered nanostructure is formed in which the laterally microphase-separated bottom layer is covered by the longer polymer brushes at a large chain length asymmetry.

By using the initiator particles made from 172 nm silica NPs and Y-Initiator-2 (Scheme 8.1), we also synthesized a set of asymmetric MBNPs with a fixed P*t*BA $M_{n,SEC}$ of 18.6 kDa and various PS molecular weights from 8.7 to 28.0 kDa (**MBNP-III**). The use of the triethoxysilane-terminated Y-initiator with a mass ratio of 1 : 1 with respect to the silica particles in the initiator immobilization step allowed for the preparation of high-density mixed brushes with the σ_{total} in the range of 0.9–1.2 chains nm^{-2} [70]. Similar morphological evolution was observed for this set of MBNPs (Figure 8.7). In addition, a dark ring was observed around each particle when the DP_{PS} was larger than DP_{PtBA}, which is not very visible in Figure 8.6c,d. This is likely because at lower grafting densities the PS domains can accommodate more PS segments, while at higher grafting densities the polymer chains were more stretched, leading to easier formation of a thin PS top layer. This further confirmed that the two grafted polymers microphase separated into a two-layered nanostructure with the laterally microphase-separated bottom layer covered by a thin PS top layer [32]. However, the feature size was significantly smaller. To confirm this observation, we synthesized a high grafting density MBNP sample with P*t*BA M_n of 23.7 kDa (σ_{PtBA} = 0.48 chains nm^{-2}) and PS M_n of 25.7 kDa (σ_{PS} = 0.51 chains nm^{-2}) [70], similar to that shown in Figure 8.6c (**MBNP-II-3**: P*t*BA M_n = 24.5 kDa, σ_{PtBA} = 0.36 chains nm^{-2}; PS M_n = 24.9 kDa, σ_{PS} = 0.27 chains nm^{-2}). The typical periodicity in the microphase-separated bottom layer was ~12 nm for this sample, in contrast to ~20 nm for the sample in Figure 8.6c. The much smaller feature size is likely a result of the competition between entropy and enthalpy. At high grafting densities, polymer chains are highly stretched; the entropy of polymer chains is thus much lower than that of intermediate-density mixed brushes. As a result, the lateral microphase separation of mixed brushes is restricted to a smaller lateral length scale – any further stretching of polymer chains in the lateral direction would mean a further decrease in entropy, which could not be compensated by the energy gain from the phase separation. This observation prompted us to systematically investigate the effect of grafting density on the morphology of MBNPs, as described in the next section.

8.3.4 Effect of Overall Grafting Density on Morphology of P*t*BA/PS MBNPs

The effects of overall and relative grafting densities of two grafted polymers on the morphology of MBNPs have been theoretically studied [30, 32, 56, 57]. In general, if the grafting density decreases, the microphase separation of the two grafted polymers in the brush layer weakens. This is because the grafting density

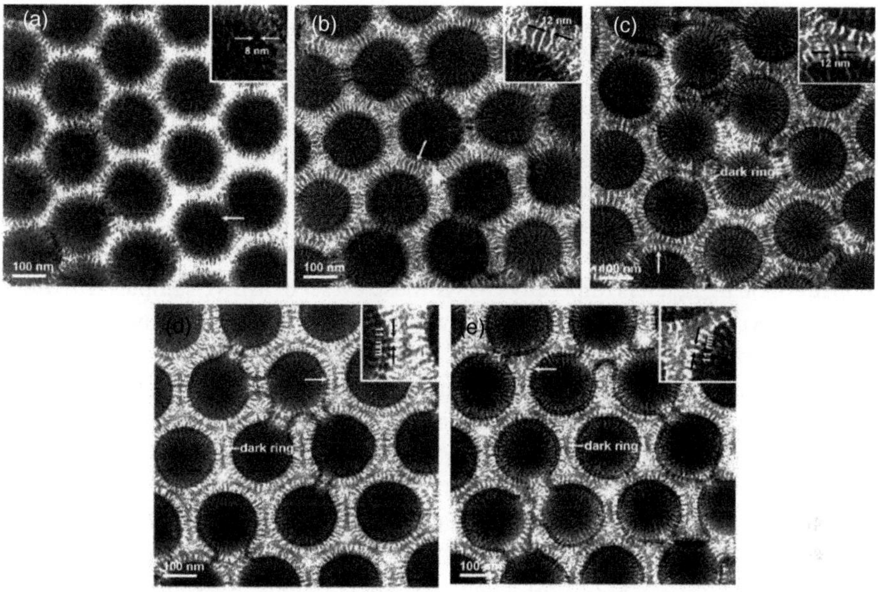

Figure 8.7 Bright-field TEM micrographs of (a) **MBNP-III-1** (P*t*BA M_n = 18.6 kDa, σ_{PtBA} = 0.63 chains nm^{-2}; PS M_n = 8.7 kDa, σ_{PS} = 0.26 chains nm^{-2}; DP_{PS}/DP_{PtBA} = 0.59), (b) **MBNP-III-2** (P*t*BA M_n = 18.6 kDa, σ_{PtBA} = 0.63 chains nm^{-2}; PS M_n = 13.4 kDa, σ_{PS} = 0.33 chains nm^{-2}; DP_{PS}/DP_{PtBA} = 0.90), (c) **MBNP-III-3** (P*t*BA M_n = 18.6 kDa, σ_{PtBA} = 0.63 chains nm^{-2}; PS M_n = 19.4 kDa, σ_{PS} = 0.43 chains nm^{-2}; DP_{PS}/DP_{PtBA} = 1.31), (d) **MBNP-III-4** (P*t*BA M_n = 18.6 kDa, σ_{PtBA} = 0.63 chains nm^{-2}; PS M_n = 25.3 kDa, σ_{PS} = 0.56 chains nm^{-2}; DP_{PS}/DP_{PtBA} = 1.70), and (e) **MBNP-III-5** (P*t*BA M_n = 18.6 kDa, σ_{PtBA} = 0.63 chains nm^{-2}; PS M_n = 28.0 kDa, σ_{PS} = 0.58 chains nm^{-2}; DP_{PS}/DP_{PtBA} = 1.88) after being cast from a $CHCl_3$ dispersion and annealed with $CHCl_3$ vapor. The samples were stained with RuO_4 vapor at room temperature for 20 minutes. The inset in each TEM micrograph shows the enlarged area marked by the white arrows. Reproduced from Ref. [70]. Copyright American Chemical Society 2010.

determines the degree of chain stretching, and one can imagine that the energy gain from microphase separation may not exceed the entropy loss caused by the additional chain stretching from the phase separation if the grafting densities of the two polymers are too low. The theoretical study by Zhulina and Balazs showed that the periodicity of the ripple pattern formed by symmetric Y-shaped mixed homopolymer brushes on planar substrates in nonselective poor solvents scaled inversely with the one-sixth power of the grafting density (i.e. $D \propto \sigma^{-1/6}$) in the limit of high immiscibility between two polymers and high grafting densities [30]. Computer simulations have revealed a profound effect of relative grafting densities of two polymers on the morphology of mixed brushes. Wang and Müller showed that with the increase of the ratio of individual grafting densities of polymer A to polymer B in the mixed-brush layer on a flat substrate, the morphology changed from isolated domains of polymer A in the matrix of polymer B to a rippled phase and then isolated domains of B in the matrix of A [32]. For the mixed homopolymer brushes grafted on nanospheres, the morphology of mixed brushes evolved from

AB-layered to ordered B-islanded, rippled, ordered A-islanded, and finally to BA-layered nanostructures with the change of relative grafting densities of two polymers [56, 57].

Inspired by the theoretical predictions, we sought to experimentally elucidate the effect of overall grafting density ($\sigma_{overall}$) on the periodicity of the nanopattern formed from the lateral microphase separation of mixed brushes on 172 nm silica particles [77]. To systematically study the effect of $\sigma_{overall}$ on microphase separation of mixed brushes and to quantify the dependence of the ripple wavelength D of the nanopattern on $\sigma_{overall}$, it is imperative to synthesize a set of MBNP samples with different $\sigma_{overall}$ values but similar molecular weights and comparable individual grafting densities for the two grafted polymers. To achieve this, we prepared a series of Y-initiator-functionalized, 172 nm silica particles with different Y-initiator densities by gradually changing the mass ratio of the precursor of Y-Initiator-2 (Scheme 8.1) to bare silica particles, ranging from ~109% to ~8%, in the initiator immobilization step while keeping other conditions the same (the left panel of Figure 8.8) [77]. The Y-initiator particles were repeatedly washed and used to synthesize mixed P*t*BA/PS brushes with molecular weights of ~23 kDa for both polymers (**MBNP-IV**) using the established two-step procedure. The right panel of Figure 8.8 shows the plots of $\sigma_{overall}$ of mixed brushes and individual grafting densities of P*t*BA and PS (σ_{PtBA} and σ_{PS}) versus mass ratio of the precursor of Y-Initiator-2 to silica particles used in the preparation of initiator particles. There was a clear correlation between the $\sigma_{overall}$ and the mass ratio of Y-initiator to silica particles; the higher the mass ratio, the higher the overall grafting density. Moreover, in most of the **MBNP-IV** samples, the σ_{PtBA} and σ_{PS} values were similar; the small difference in the individual grafting densities of P*t*BA and PS should not affect the ripple wavelength much, as many

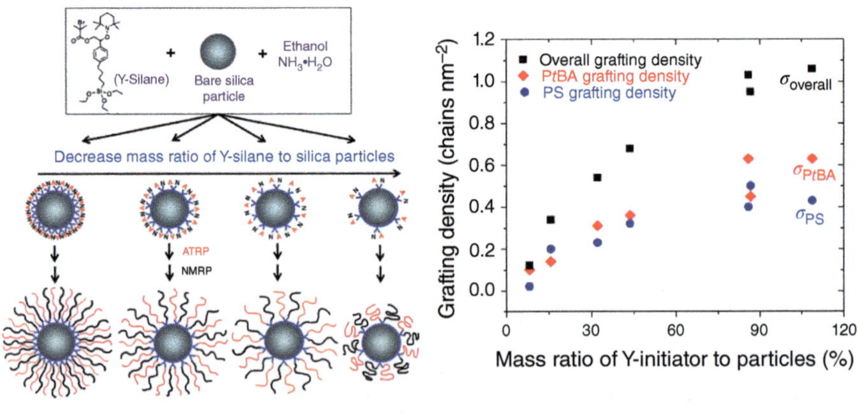

Figure 8.8 (left) Synthesis of mixed P*t*BA/PS brushes with varying overall grafting densities by changing the mass ratio of Y-initiator to silica particles in the initiator immobilization step. (right) Plots of overall grafting density of mixed P*t*BA/PS brushes ($\sigma_{overall}$, black solid square), grafting density of P*t*BA (σ_{PtBA}, red solid diamond), and grafting density of PS (σ_{PS}, blue solid circle) on silica particles versus mass ratio of Y-initiator to silica particles used in the process of immobilizing the Y-initiator onto the surface of bare silica particles. Source: Adapted from Ref. [77]. Copyright American Chemical Society 2012.

simulation studies showed that the lateral microphase separation can tolerate small variations in grafting densities and molecular weights of two grafted homopolymers.

Figure 8.9 shows bright-field TEM micrographs of seven P*t*BA/PS **MBNP-IV** samples arranged in the order of decreasing the overall grafting density of mixed brushes. Clearly, all the **MBNP-IV** samples, except for the lowest grafting density sample, were microphase separated, showing a rippled morphology with dark PS and bright P*t*BA worm-like domains. The feature size gradually increased with the decrease of $\sigma_{overall}$, which was qualitatively in agreement with the theoretical prediction [30]. For the sample with the lowest $\sigma_{overall}$, the grafted polymer chains can be clearly seen between neighboring hairy silica particles, but there was no microphase separation, indicating that the grafted P*t*BA and PS were in the miscible state. Evidently, the microphase separation was unfavorable at such a low $\sigma_{overall}$, and the lowest Gibbs free energy state is the miscible state. Image analysis of the TEM micrographs in Figure 8.9 was conducted to measure the average ripple wavelengths by selecting the features in the area between the half radius and the periphery (R) from the center of the particles, where the features were the clearest. The molecular weights of the two polymers in the MBNP samples were not exactly the same. To obtain the scaling relationship between ripple wavelength D and $\sigma_{overall}$, the average ripple wavelengths from the image analysis were normalized against the sample with a P*t*BA M_n of 22.9 kDa and a PS M_n of 22.2 kDa.

Figure 8.10 shows the plot of normalized ripple wavelength (D_N) versus $\sigma_{overall}$ on a double logarithmic scale. A linear dependence of $\log D_N$ on $\log \sigma_{overall}$ was observed, with a slope of −0.47, i.e. $D_N \sim \sigma_{overall}^{-0.47}$ in the range of $\sigma_{overall}$ of 0.54 – 1.06 chains nm^{-2}. Thus, our experimentally determined D_N showed a stronger dependence on $\sigma_{overall}$ than that predicted theoretically ($D \sim \sigma_{overall}^{-1/6}$) [30], although the trend was the same. The discrepancy observed here could be caused by several factors. In our experiments, the mixed brushes, which were not perfect Y-brushes, were cast from $CHCl_3$ (a good solvent for both P*t*BA and PS) and annealed with $CHCl_3$ vapor, while the theoretical study considered perfect Y-shaped symmetric mixed brushes in nonselective poor solvents. Nevertheless, our study revealed how the basic feature of the nanopattern from lateral microphase separation of mixed brushes changes with $\sigma_{overall}$.

8.3.5 Effect of Molecular Weight on Morphology of Symmetric MBNPs

The periodicity or the ripple wavelength (D) of the nanopattern formed by lateral microphase separation of symmetric mixed homopolymer brushes on flat substrates has been predicted to scale with the square root of chain length in the melt and in nonselective poor solvents [26, 30]. For example, the theoretical study by Marko and Witten predicted that the feature size of the "rippled" nanostructure in the equilibrium melt state was 1.97 times the chain root-mean-square end-to-end distance [26], $\langle R_{rms} \rangle = (C_\infty n l^2)^{1/2}$, where C_∞ is the Flory characteristic ratio, n is the number of chemical bonds, and l is the bond length, in the unperturbed state [78]. To experimentally elucidate the effect of chain length on microphase separation of mixed brushes, it is necessary to synthesize a series of mixed brush

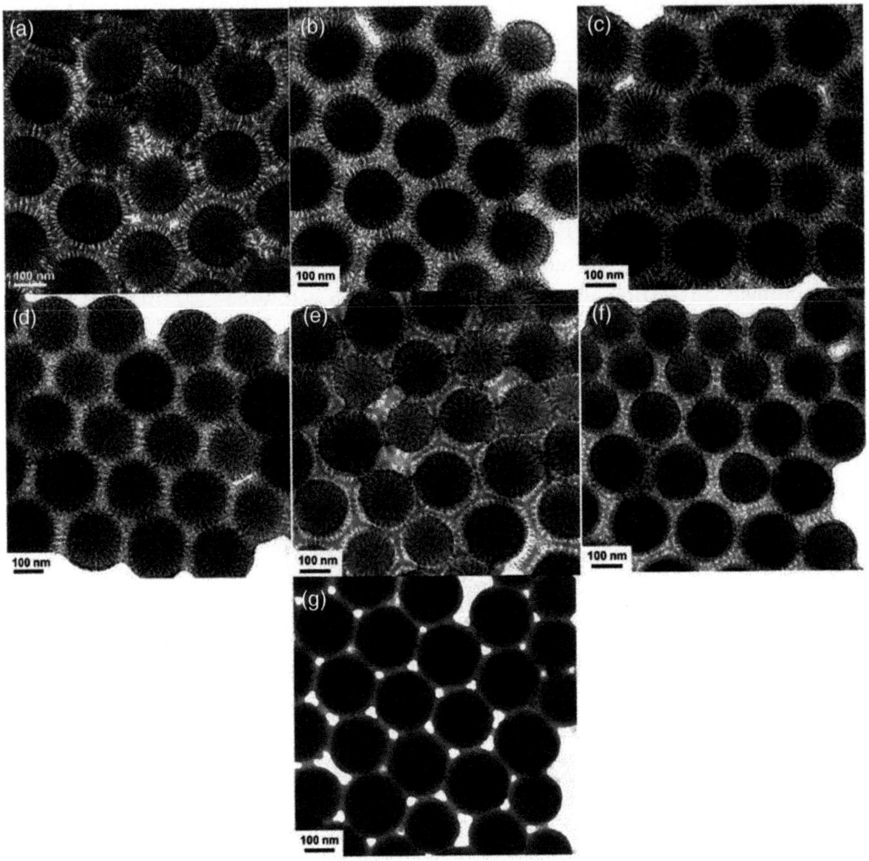

Figure 8.9 Bright-field TEM micrographs of mixed P*t*BA/PS brush-grafted 172 nm silica particles: (a) **MBNP-IV-1**: P*t*BA M_n = 18.6 kDa, σ_{PtBA} = 0.63 chains nm^{-2}, PS M_n = 19.4 kDa, σ_{PS} = 0.43 chains nm^{-2}, and $\sigma_{overall}$ = 1.06 chains nm^{-2}, (b) **MBNP-IV-2**: P*t*BA M_n = 23.6 kDa, σ_{PtBA} = 0.63 chains nm^{-2}, PS M_n = 23.5 kDa, σ_{PS} = 0.40 chains nm^{-2}, and $\sigma_{overall}$ = 1.03 chains nm^{-2}, (c) **MBNP-IV-3**: P*t*BA M_n = 25.2 kDa, σ_{PtBA} = 0.45 chains nm^{-2}, PS M_n = 21.9 kDa, σ_{PS} = 0.50 chains nm^{-2}, $\sigma_{overall}$ = 0.95 chains nm^{-2}, (d) **MBNP-IV-4**: P*t*BA M_n = 22.9 kDa, σ_{PtBA} = 0.36 chains nm^{-2}, PS M_n = 22.2 kDa, σ_{PS} = 0.32 chains nm^{-2}, and $\sigma_{overall}$ = 0.68 chains nm^{-2}, (e) **MBNP-IV-5**: P*t*BA M_n = 21.3 kDa, σ_{PtBA} = 0.31 chains nm– [2], PS M_n = 20.7 kDa, σ_{PS} = 0.23 chains nm^{-2}, and $\sigma_{overall}$ = 0.54 chains nm^{-2}, (f) **MBNP-IV-6**: P*t*BA M_n = 23.0 kDa, σ_{PtBA} = 0.14 chains nm^{-2}, PS M_n = 21.9 kDa, σ_{PS} = 0.20 chains nm^{-2}, and $\sigma_{overall}$ = 0.34 chains nm^{-2}, and (g) **MBNP-IV-7**: P*t*BA M_n = 22.1 kDa, σ_{PtBA} = 0.10 chains nm^{-2}, PS M_n = 23.5 kDa, σ_{PS} = 0.022 chains nm^{-2}, and $\sigma_{overall}$ = 0.122 chains nm^{-2} after being cast from $CHCl_3$ dispersions and annealed with $CHCl_3$ vapor for at least 3 hours. The samples were stained with RuO_4 vapor at room temperature for 20 minutes. Source: Reproduced from Ref. [77]. Copyright American Chemical Society 2012.

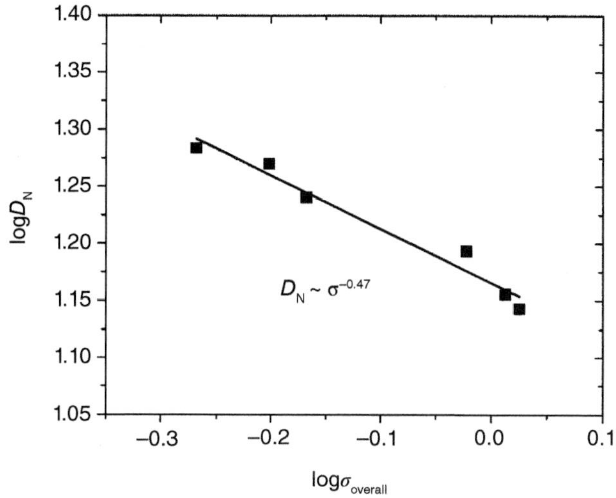

Figure 8.10 Plot of $\log D_N$ versus $\log \sigma_{overall}$, where D_N is the normalized ripple wavelength (normalized ripple wavelength = average ripple wavelength/([($M_{n,SEC\text{-}PtBA} + M_{n,SEC\text{-}PS}$)/($M_{n,SEC\text{-}PtBA\text{-}22.9k} + M_{n,SEC\text{-}PS\text{-}22.2k}$)]$^{1/2}$)) and $\sigma_{overall}$ is the overall grafting density of the mixed brushes. The straight solid line is a linear fit with $R = 0.983$ and a slope of −0.47. Source: Reproduced from Ref. [77]. Copyright American Chemical Society 2012.

samples with controlled molecular weights and similar individual and overall grafting densities. Using the same batch of Y-initiator-1-functionalized, 172 nm silica particles and taking advantage of the "living" nature of both ATRP and NMRP, we prepared a set of six mixed P*t*BA/PS brush-grafted particle samples with various average M_n of P*t*BA and PS, ranging from 13.8 to 33.1 kDa **(MBNP-V)** [79]. In each hairy particle sample, the molecular weights of the two grafted polymers were controlled to be as close to each other as possible so that nearly bicontinuous worm-like nanostructures were obtained from nonselective solvents. The average diameter of silica core particles (172 nm) was significantly larger than the polymers' unperturbed sizes, and thus the brushes in these hairy particle samples can be viewed as quasi-planar brushes. The overall grafting densities for the six samples were similar, in the range of 0.78–0.92 chains nm^{-2}, and the individual grafting densities of P*t*BA and PS were comparable, though σ_{PtBA} appeared to be slightly larger than σ_{PS} in most of the samples. The small differences in the $\sigma_{overall}$ of these particle samples should not have much influence on the ripple wavelength, as theoretical and simulation studies showed that the lateral microphase separation can tolerate small variations in grafting densities and molecular weights of two polymers in the brushes.

TEM was then used to study the nanopatterns formed from these MBNP samples drop-cast from chloroform, a nonselective good solvent for the two polymers, and surfactant-stabilized aqueous dispersions (water is a nonselective poor solvent for P*t*BA and PS). The TEM samples prepared from $CHCl_3$ were annealed by $CHCl_3$ vapor at room temperature in a closed container for ≥5 hours before staining with RuO_4, while the hairy particles drop-cast on carbon-coated TEM grids from surfactant-stabilized aqueous dispersions were stained with RuO_4 directly at room temperature for 20 minutes. Figure 8.11 shows bright-field TEM images of six **MBNP-V** samples arranged in the order of increasing average molecular weight, which showed that all of the hairy samples underwent microphase separation, forming a rippled morphology composed of dark PS and bright P*t*BA worm-like domains. The ripple wavelength increased gradually with the increase in molecular weights of the two polymers, as expected from the theoretical prediction [26, 30]. In contrast to the higher molecular weight sample, the microphase separation of the lowest molecular weight sample with an average M_n of 13.8 kDa (**MBNP-V-1**) was only observed in the interstitial area (Figure 8.11a). The dark and bright domains "on" the particles were hardly discernible. This is likely because the M_n values of P*t*BA and PS were rather low, perhaps just above the critical molecular weight for microphase separation of the brushes at this particular grafting density, resulting in weak phase separation.

Image analysis was performed on TEM micrographs in Figure 8.11b–f using the features in the region between the half radius ($R/2$) and the periphery (R) from the center of the particles. For **MBNP-V-1** in Figure 8.11a, the domains in the interstitial area were chosen for image analysis. The results are shown in Figure 8.12a as a double logarithmic plot of average periodicity D versus average M_n of the two polymers. A linear dependence of $\log D$ on $\log M_n$ was obtained, with a slope of 0.70. This result is different from the theoretically predicted relationship between D and degree of polymerization, N, of symmetric Y-shaped mixed homopolymer brushes, $D \sim N^{1/2}$, in nonselective poor solvents [30]. This discrepancy is likely caused by the effect of the good solvent in the preparation of TEM samples, which resulted in a nonuniform collapse of the mixed brushes on the particles during the drop-casting process/solvent evaporation. To eliminate this effect, it is imperative to prepare samples with uniformly collapsed mixed brushes on silica particles. To prepare uniformly collapsed MBNPs, the dispersion of the brush particle in $CHCl_3$ was added into water in the presence of a surfactant, cetyltrimethylammonium bromide (CTAB), under the vigorous stirring condition, followed by gradual evaporation of $CHCl_3$. CTAB, presumably neutral to P*t*BA and PS, stabilized small chloroform droplets. After evaporation of $CHCl_3$, the mixed brushes collapsed uniformly on the surface of silica core. The particles were drop-cast onto the carbon-coated TEM grids and stained with RuO_4 for the TEM study [79]. The results are shown in Figure 8.12b, with a double logarithmic plot of D versus average M_n of P*t*BA and PS. A linear dependence of $\log D$ on $\log M_n$ was obtained with a slope of 0.56. This result is close to the theoretical prediction ($D \sim MW^{0.5}$) [30], although there is still a small difference. The discrepancy probably comes from the fact that the mixed P*t*BA/PS brushes are not perfectly symmetric Y-brushes with identical DPs and grafting densities for the two grafted polymers.

The phase separation behavior of mixed brushes and P*t*BA-*b*-PS diblock copolymers was then compared. Marko and Witten predicted that the microphase separation of symmetric mixed homopolymer brushes grafted on a flat substrate was expected to occur at a critical point with $(\chi N)_c = 9.08$, where χ is Flory–Huggins

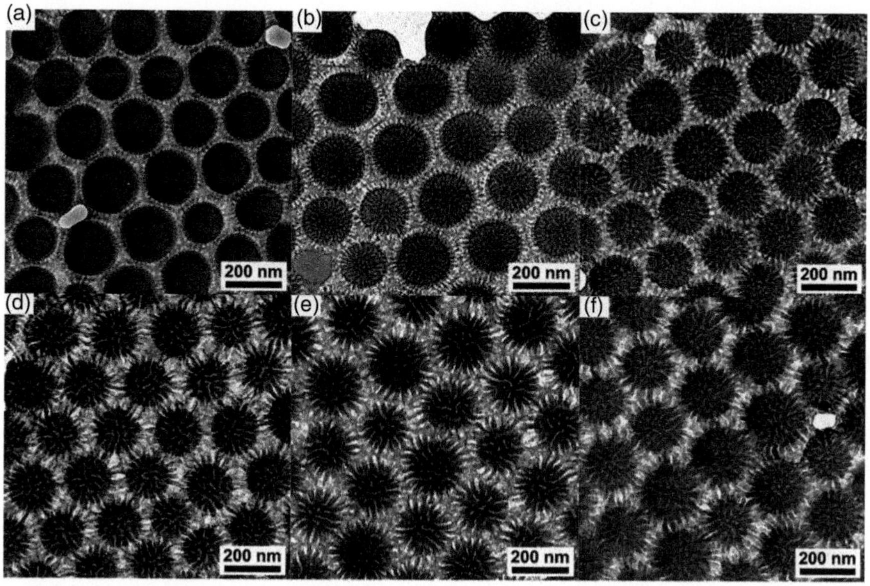

Figure 8.11 Bright-field TEM micrographs of (a) **MBNP-V-1**: P*t*BA M_n = 13.3 kDa, σ_{PtBA} = 0.47 chains nm^{-2}; PS M_n = 14.3 kDa, σ_{PS} = 0.31 chains nm^{-2}; $\sigma_{overall}$ = 0.78 chains nm^{-2}, (b) **MBNP-V-2**: P*t*BA M_n = 18.3 kDa, σ_{PtBA} = 0.42 chains nm^{-2}; PS M_n = 17.5 kDa, σ_{PS} = 0.33 chains nm^{-2}; $\sigma_{overall}$ = 0.75 chains nm^{-2}, (c) **MBNP-V-3**: P*t*BA M_n = 22.0 kDa, σ_{PtBA} = 0.37 chains nm^{-2}; PS M_n = 22.3 kDa, σ_{PS} = 0.41 chains nm^{-2}; $\sigma_{overall}$ = 0.78 chains nm^{-2}, (d) **MBNP-V-4**: P*t*BA M_n = 26.7 kDa, σ_{PtBA} = 0.51 chains nm^{-2}; PS M_n = 26.3 kDa, σ_{PS} = 0.29 chains nm^{-2}; $\sigma_{overall}$ = 0.80 chains nm^{-2}, (e) **MBNP-V-5**: P*t*BA M_n = 32.1 kDa, σ_{PtBA} = 0.51 chains nm^{-2}; PS M_n = 31.6 kDa, σ_{PS} = 0.41 chains nm^{-2}; $\sigma_{overall}$ = 0.92 chains nm^{-2}, and (f) **MBNP-V-6**: P*t*BA M_n = 34.1 kDa, σ_{PtBA} = 0.50 chains nm^{-2}; PS M_n = 32.1 kDa, σ_{PS} = 0.35 chains nm^{-2}; $\sigma_{overall}$ = 0.85 chains nm^{-2} after being cast from $CHCl_3$ dispersions and annealed with $CHCl_3$ vapor. The samples were stained with RuO_4 at room temperature for 20 minutes [46]. Source: Reproduced from Ref. [79]. Copyright American Chemical Society 2012.

interaction parameter and $N = N_A + N_B$ (N_A and N_B are the DPs of the grafted polymers A and B) [26]. It is known that for symmetric diblock copolymers, microphase separation occurs at a critical point of $(\chi N)_c = 10.5$ [80]. Thus, mixed brushes should undergo microphase separation at a lower molecular weight than the corresponding diblock copolymers. Our previous study showed that **MBNP-I-2** (180 nm silica particles, P*t*BA $M_{n,SEC}$ = 10.4 kDa, σ_{PtBA} = 0.32 chains nm^{-2}, PS $M_{n,SEC}$ = 11.9 kDa, and σ_{PS} of 0.35 chains nm^{-2}) did not exhibit microphase separation (Figure 8.5b). Combining the observations for **MBNP-I-2** and the **MBNP-V-1** sample in Figure 8.11a, it can be concluded that the critical average molecular weight for microphase separation of mixed P*t*BA/PS brushes at an overall grafting density of ~0.7–0.8 chains nm^{-2} should be between 11.2 and 13.3 kDa. Using RAFT polymerization, we synthesized a diblock copolymer with a P*t*BA M_n of 11.2 kDa and a PS M_n of 10.9 kDa; the molecular weight of this diblock copolymer is essentially the same as the aforementioned non-microphase-separated **MBNP-I-2**

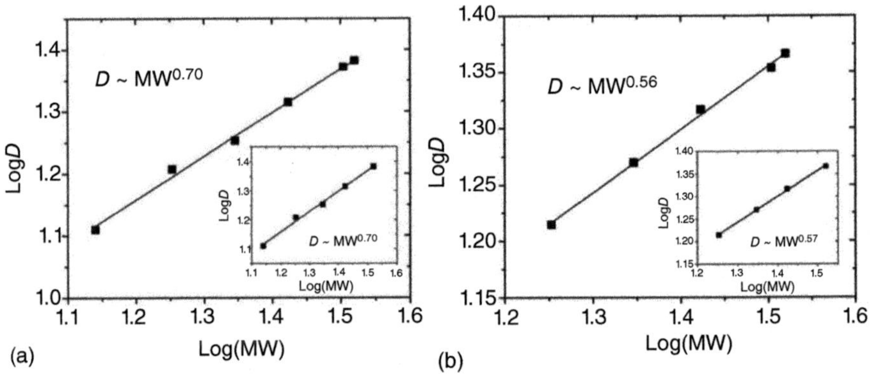

Figure 8.12 Plots of log$D$ versus log(MW) for various **MBNP-V** samples cast from (a) $CHCl_3$ and (b) water (with the help of surfactant). D is the average ripple wavelength, obtained from TEM image analysis, and MW is the average molecular weight of PtBA and PS in the unit of kDa. The insets show the plots of log$D$ versus log(MW) excluding **MBNP-V-5**. Source: Adapted from Ref. [79]. Copyright American Chemical Society 2014.

sample. DSC, SAXS, and TEM showed that this diblock copolymer strongly phase separated and formed a lamellar structure. Thus, our results showed that it is more difficult for mixed brushes to undergo microphase separation at an overall grafting density of ~0.7–0.8 chains nm^{-2} than PtBA-b-PS diblock copolymers. This is inconsistent with the theoretical study results of Marko and Witten [26]. However, in their theoretical work, the grafting density of mixed homopolymer brushes, an important molecular parameter, was not taken into account. Our results show that the grafting density plays a critical role in the microphase separation of mixed brushes. Moreover, unlike diblock copolymers in the melt state, when mixed brushes phase separate laterally, a crossover zone is formed in the bottom part of the brush layer, which undoubtedly increases the free energy of the system and thus decreases the tendency of mixed brushes to phase separate.

8.3.6 Effect of Core Particle Size on Morphology of PtBA/PS MBNPs

Mixed brushes grafted on particles are not simply a curved version of the brushes on flat substrates. The substrate curvature is expected to play an important role in the morphology of the mixed homopolymer brushes grafted on sub-100 nm NPs [56, 57]; with the increase of the distance from the grafting sites, polymer segments are less stretched because more space is available. For phase-separated mixed brushes grafted on small NPs, the formation of truncated wedge-shaped nanodomains instead of nanostructures with a uniform width from the core surface to the exterior of the brushes has been recognized in the computer simulation studies. As the first step toward the fabrication of mixed homopolymer brushes with controlled molecular weights and narrow dispersities on sub-100 nm silica NPs and the study of the effect of substrate curvature on the morphology of mixed brushes, we immobilized a triethoxysilane-terminated Y-initiator (Y-Initiator-2

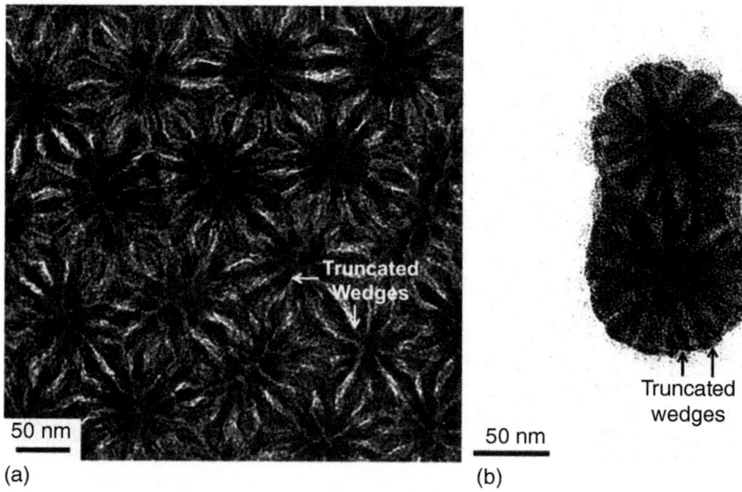

Figure 8.13 Bright-field TEM micrographs (A and B) of mixed P*t*BA/PS brush-grafted 67 nm **MBNP-VI-2** with P*t*BA M_n of 22.2 kDa and PS M_n of 23.4 kDa after being cast from $CHCl_3$ and annealed with $CHCl_3$ vapor for at least 3 hours. The sample was stained with RuO_4 vapor for 20 minutes. Source: Reproduced from Ref. [81]. Copyright American Chemical Society 2012.

in Scheme 8.1) onto the surface of 67 nm silica NPs and synthesized a series of P*t*BA/PS **MBNP-VI** by using the established method [81].

Y-Initiator-2 was immobilized onto the surface of silica NPs with an average diameter of 67 nm via an ammonia-catalyzed hydrolysis and condensation process in ethanol, and a series of **MBNP-VI** samples with the same P*t*BA M_n of 22.2 kDa but different PS molecular weights [18.7 kDa (**MBNP-VI-1**), 23.4 kDa (**MBNP-VI-2**), and 29.5 kDa (**MBNP-VI-3**)] were prepared and purified. The σ_{PtBA} in these three particle samples was 0.54 chains nm^{-2}, and the σ_{PS} values were 0.28, 0.31, and 0.31 chains nm^{-2}, respectively, which were appreciably smaller than σ_{PtBA}. For the TEM study, the hairy NPs were dispersed in $CHCl_3$, drop-cast on carbon-coated TEM grids, annealed with $CHCl_3$ vapor for at least three hours, and stained with RuO_4 vapor. For all three stained particle samples, truncated wedge-shaped dark PS domains were observed, and an example for **MBNP-VI-2** is shown in Figure 8.13. These unusual nanostructures were especially pronounced at the edges of the hairy NPs, likely due to the spreading of the brushes on the carbon film. In contrast, the alternating dark and bright nanodomains at the edge of 160 nm P*t*BA/PS **MBNP-II-2** (P*t*BA $M_n = 24.5$ kDa, $\sigma_{PtBA} = 0.36$ chains nm^{-2}; PS $M_n = 18.7$ kDa, $\sigma_{PS} = 0.26$ chains nm^{-2}, $DP_{PS}/DP_{PtBA} = 0.94$) were nearly uniform in the width (see Figure 8.6c). The observed distinct shapes of nanodomains formed by microphase-separated mixed brushes grafted on 67 and 160 nm silica particles demonstrated the effect of substrate curvature on the morphology of mixed brushes. This also hinted that even more interesting nanostructures might be observed for smaller NPs, which will be discussed next.

Computer simulation studies showed that for binary mixed homopolymer brushes on spherical NPs with a radius similar to the unperturbed R_{rms} values of the two grafted polymers, various islanded nanostructures were formed from lateral microphase separation of mixed brushes, yielding multivalent NPs with a small and defined number of nanodomains (from 2 to 12, except 11) as discussed earlier (Figures 8.1–8.3) [56, 57]. The valency (i.e. the number of islanded nanodomains) is governed by core size, σ, molecular weights, etc., among which core size plays a significant role. Using Y-Initiator-3-functionalized, 20.4 nm silica NPs, we synthesized a set of P*t*BA/PS MBNPs with the same P*t*BA M_n of 19.6 kDa ($Đ = 1.20$) and PS M_n values ranging from 6.1 to 31.5 kDa ($Đ = 1.17–1.28$) (**MBNP-VII)** [71]. To visualize their morphologies, TEM was used again. The **MBNP-VII** samples were drop-cast (from chloroform) on carbon-coated TEM grids, followed by chloroform vapor annealing for >3 hours and staining with RuO_4 for 7 minutes. Figure 8.14a shows a P*t*BA NP and Figure 8.14b–j shows a single NP of each of the nine different P*t*BA/PS **MBNP-VII** samples. Because of the microphase separation between P*t*BA and PS, spoke-like multivalent NPs were obtained with nanodomains extending out from the core. The most popular valency was 8, except for **MBNP-VII-6.1k**

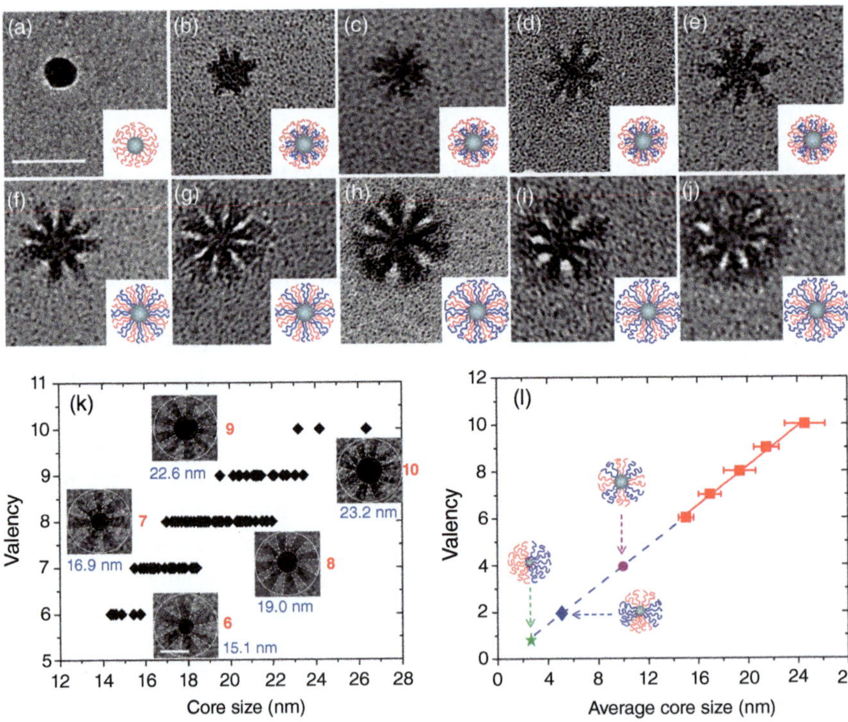

Figure 8.14 Bright-field TEM micrographs of one (a) P*t*BA-grafted silica NP, (b) **MBNP-VII-6.1k**, (c) **-9.7k**, (d) **-10.9k**, (e) **-12.9k**, (f) **-15.2k**, (g) **-20.0k**, (h) **-23.0k**, (i) **-27.6k**, and (j) **-31.5k**. Scale bar: 50 nm. The MBNPs were stained with RuO_4 for 7 minutes. (k) Plot of valency versus core size for **MBNP-VII-15.2k** after stained with RuO_4 for 2 min. Scale bar: 25 nm. (l) Plot of valency versus average core diameter for the corresponding valency with the standard deviation. The star, diamond, and circle indicate mono-, di, and tetra-valent NPs from extrapolation. Source: Adapted with permission from Ref. [71]. Copyright American Chemical Society 2021.

(Figure 8.14b). The MBNPs with PS M_n values from 10.9 to 23.0 kDa showed the best morphology (Figure 8.14d–h). It is interesting to note that the PS nanodomains were uniform in width for PS $M_n \leq 15.2$ kDa (~7 nm for **MBNP-VII-12.9k**). However, they became truncated wedge-like at PS M_n = 20.0 kDa [81]. With PS M_n increased, dark PS nanodomains gradually covered the bright P*t*BA nanodomains. For **MBNP-VII-27.6k** and **-31.5k**, the long PS chains fully covered the bright P*t*BA domains. Consistent with the computer simulation result, the laterally phase-separated bottom layer is covered by the grafted longer polymer chains when the chain length asymmetry is large [32].

Under two-dimensional (2D) TEM projections, the valency is obtained by $V \sim r/w$ (where r is the radius of core NP and w is the width of nanodomains). Because the valency is a result of intrinsic lateral microphase separation of mixed brushes, it can be tuned by M_n, volume fractions, Flory–Huggins interaction parameter, and σ. Here, r plays an important role as predicted by simulations [24, 56, 57]. To illustrate how the core size can affect the valency, **MBNP-VII-15.2k** was stained for two minutes in order to make both core NP and PS nanodomains visible (Figure 8.14k). Over 100 NPs were analyzed for core size and valency, see Figure 8.14k. The valency gradually decreased with decreasing the core size, and each valency had a size range. From this result, we obtained a linear relationship between valency and average core size (Figure 8.14l). The slope was 0.42, qualitatively fitting with the simulation results [60]. By extrapolation, monovalent (i.e. Janus NPs), divalent, and tetravalent NPs could be obtained for MBNPs with a core diameter of 2.8, 5.2, and 10 nm. Using neutron scattering, Kim et al. studied mixed brush-grafted ~3 nm gold NPs in toluene. With increasing molecular weight, MBNPs can phase separate into Janus NPs [82]. Di- and tetravalent NPs can self-assemble into square lattices [83].

8.3.7 3D Morphologies of P*t*BA/PS MBNPs by Cryo-TEM and Electron Tomography

So far, the morphologies of P*t*BA/PS MBNPs obtained from TEM studies were their projections on 2D substrates. To examine the 3D morphologies, TEM tomography was used to study the self-assembled monolayer of 67 nm P*t*BA/PS **MBNP-VI-2** [84]. The results are shown in Figure 8.15. Similar to Figure 8.13, Figure 8.15a shows a typical bright-field, conventional TEM micrograph of 67 nm P*t*BA/PS **MBNP-VI-2**, stained by RuO_4 vapor at room temperature for 20 minutes. Figure 8.15b,c shows the top views of reconstructed 3D boxes with the exclusive P*t*BA and PS/SiO_2 phase, respectively. Both P*t*BA and PS phases exhibited worm-like bicontinuous microdomains interconnecting in three dimensions. Figure 8.15d shows the top-view (or surface) image of the 3D-reconstructed box with P*t*BA (in red) and PS (in green) nanodomains.

From the 3D-reconstructed TEM result in Figure 8.15c, *z*-scan images are presented in Figure 8.16. The interval from the top layer (Figure 8.16a) to the bottom layer (Figure 8.16i) was 12 nm. On the top layer (Figure 8.16a,b), isolated PS domains existed on the protruded large silica particles (inside the yellow circle). Figure 8.16c,d shows the *z*-scan images at the top layer of the continuous film.

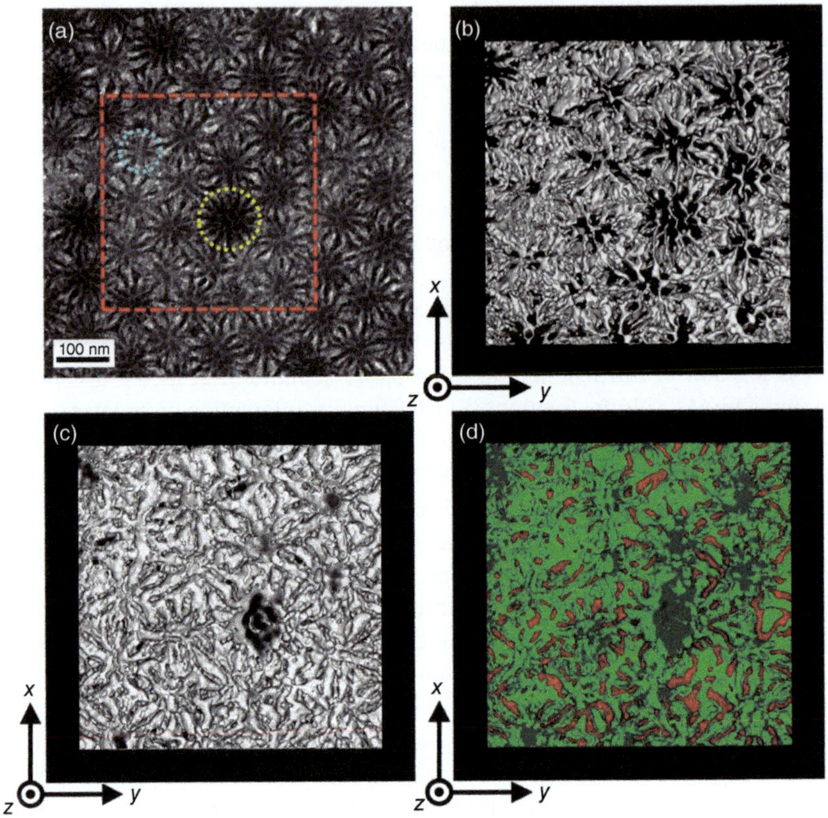

Figure 8.15 (a) Bright-field TEM micrograph of a dense monolayer film of 67 nm P*t*BA/PS **MBNP-VI-2** stained by RuO_4 vapor for 20 minutes. The red box indicates the area for reconstructed 3D image. The yellow and blue dotted circles highlight a large (90 nm) and a small (53 nm) nanoparticle, respectively. (b) Top-view 3D TEM image of the reconstructed box with visible P*t*BA and PS/SiO_2 are invisible. (c) Top-view 3D TEM image of the reconstructed box with visible PS/SiO_2 and P*t*BA is invisible. (d) Top surface image with combined P*t*BA (red) and PS (green) after 3D reconstruction. Source: Reproduced from Ref. [84]. Copyright American Chemical Society 2013.

Bright PS and dark P*t*BA microdomains radiated out from the silica core, connecting the corresponding domains from neighboring particles. In the middle region of the continuous film (Figure 8.16e–g), the microphase-separated P*t*BA and PS domains could still be seen in the interstitial regions. Figure 8.16h,i shows the *z*-scan images for the bottom region of the continuous film. The most microphase-separated morphology is seen because of the large space between silica NPs and the carbon film substrate. Both PS and P*t*BA microdomains radiated out from the projected center of the silica core, bridging neighboring NPs. In Figure 8.16h, the microphase separation was directed by the HEX packing of silica NPs. An example was shown among four closely packed MBNPs connected by blue dashed lines, which consisted of two triangles with 3-fold symmetry.

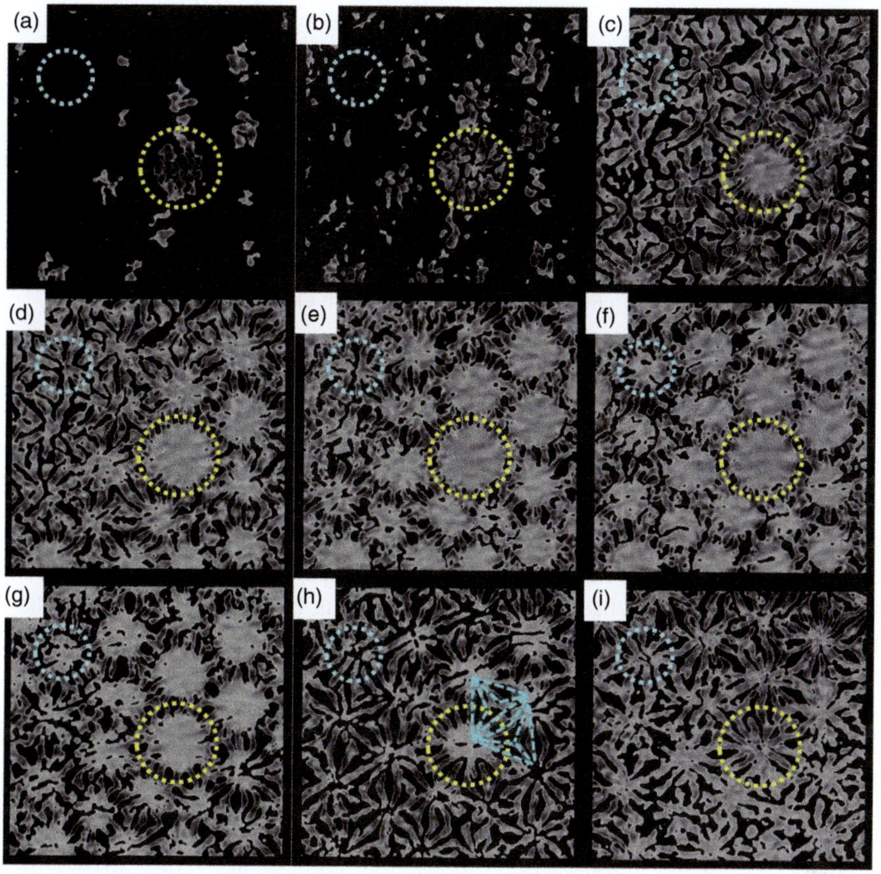

Figure 8.16 (a–i) Consecutive z-scan TEM images of the 3D-reconstructed box in Figure 8.15c. Each image is located at a 12 nm interval. The yellow and blue dotted circles indicate positions of the large (90 nm) and small (53 nm) particles, respectively. The blue dashed lines in (h) represent the 3-fold connections of PS microdomains among neighboring particles. Source: Reproduced from Ref. [84]. Copyright American Chemical Society 2013.

From the above 3D TEM results, we propose a layered nanostructure for the monolayer P*t*BA/PS **MBNP-VI-2**, as illustrated in Figure 8.17. In the protruded portion of large particles, PS chains self-assemble into isolated cylindrical domains in the P*t*BA matrix because the PS composition is 37 vol%. In the continuous layer, P*t*BA and PS chains formed a bicontinuous microphase-separated nanostructure because of the unfavorable interactions. Both the top and bottom layers had more well-defined phase separation due to the large interstitial spaces. In addition, the HEX packing of silica NPs directed the microphase separation of mixed P*t*BA/PS brushes. The truncated wedge-shaped microdomains generated a radiating structure from the projected centers, interconnecting neighboring NPs with a 3-fold symmetry. In the middle layer, microphase separation was poor because of limited interstitial space.

From the above 3D TEM results, the interparticle interactions via mixed brushes are important for the self-assembled morphology in a dense monolayer of MBNPs.

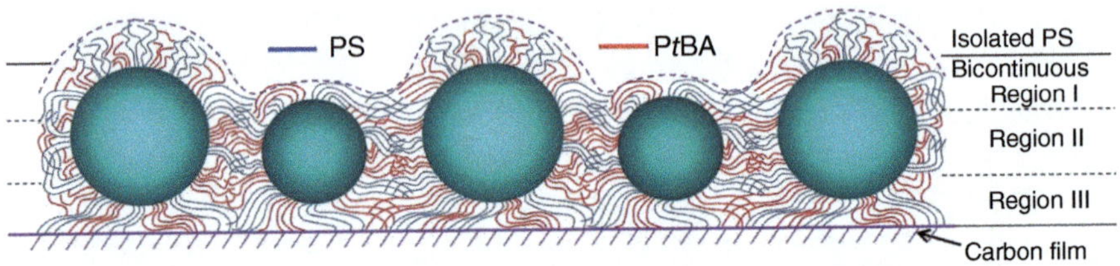

Figure 8.17 Schematic illustration of a dense monolayer of mixed PtBA/PS brush-grafted silica NPs. The PtBA and PS chains are in red and blue, respectively. Silica particles are in cyan. Source: Reproduced from Ref. [84]. Copyright American Chemical Society 2013.

It will be interesting to study the self-assembled morphology of individual MBNPs without any interparticle interaction. Figure 8.18a,d shows bright-field TEM images for a large and a small **MBNP-VI-2**, respectively. The silica core sizes were ~80 for the large particle and ~50 nm for the small particle. For the large isolated particle (Figure 8.18a), no isolated PS domains were seen in the particle center, as for the large particle in Figure 8.15a. Instead, truncated wedge-shaped PS domains were seen radiating out from the article center. For the small particle (Figure 8.18d), a similar morphology was observed with more obvious truncated PS wedges. The 3D morphologies are shown in Figures 8.18b,e (top views) and Figure 8.18c,f (side views), respectively. When viewing from the side (Figures 8.18c,f), bell-shaped particles were seen, which could be attributed to the solution-casting process. As chloroform was evaporating, mixed PtBA/PS brushes were pulled down on the carbon film substrate by the solvent, leading to less polymer on the top than at the bottom of the individual particle. The pulled-down polymer chains formed a skirt, which exhibited clear microphase separation between PtBA and PS.

From the above 3D TEM results, we propose a schematic representation of solution-cast individual particles on a carbon substrate (Figure 8.18g). In the top hemisphere, the mixed PtBA/PS chains are pulled down from solvent evaporation.

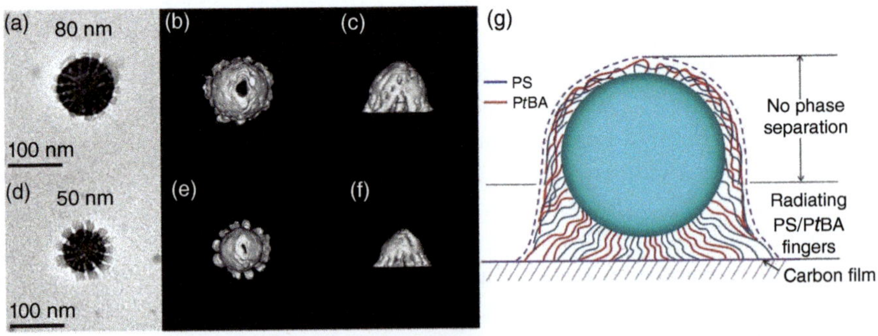

Figure 8.18 (a, d) Bright-field 2D TEM images and (b, e) top-view and (c, f) side-view reconstructed 3D TEM images of a large (~80 nm) and a small (~50 nm) PtBA/PS **MBNP-VI-2**, respectively. (g) Schematic illustration of an individual, mixed PtBA/PS brush-grafted silica NP. The PtBA and PS chains are in red and blue, respectively. Silica NPs are in cyan. Source: Adapted from Ref. [84]. Copyright American Chemical Society 2013.

Then, the mixed-brush layer in the top hemisphere was too thin to induce any microphase separation. In the bottom hemisphere, the interstitial space is large enough, and easy microphase separation leads to PS and P*t*BA microdomains radiating out from the projected center of the silica core.

8.4 Self-Assembled Morphology in Solvents and Homopolymer Matrices

8.4.1 Self-Assembly of MBNPs in Good and Selective Solvents

By fully hydrolyzing the P*t*BA in **MBNP-I-1**, amphiphilic poly(acrylic acid) (PAA)/PS MBNPs were obtained (**MBNP-I-3**) [64]. After the hydrolysis, the calculated M_n of the PAA from the DP of P*t*BA decreased to 13.6 kDa (note that the DP of PAA is the same as the P*t*BA precursor in **MBNP-I-1**), whereas the M_n of the PS remained the same (23.0 kDa). As shown in Figure 8.19a–d, bicontinuous rippled morphology was observed when **MBNP-I-3** was cast from a nonselective good solvent DMF [85]. However, when cast from water in the presence of uranyl acetate, isolated PS surface micelles were seen (Figure 8.19e–g). The average PS domain size was 14.4 ± 2.4 nm.

In the above study, the MBNP morphology was obtained after solution casting/solvent evaporation on the carbon film of the TEM grid. It is unclear whether the observed images reflected the genuine morphologies in the solution states. To investigate the in situ morphologies in different solvents, cryo-TEM combined with tomography was performed. By hydrolyzing the P*t*BA

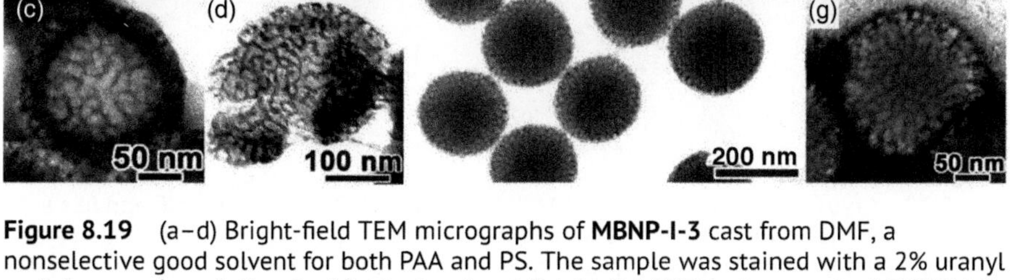

Figure 8.19 (a–d) Bright-field TEM micrographs of **MBNP-I-3** cast from DMF, a nonselective good solvent for both PAA and PS. The sample was stained with a 2% uranyl acetate aqueous solution for 15 minutes. (e–g) Bright-field TEM micrograph of **MBNP-I-3** cast from water, a selective solvent for PAA. Micrograph (f) shows a magnified view of a PAA/PS MBNP in (e) after subtracting the silica particle background. The sample was positively stained with uranyl acetate for 15 minutes. Micrograph (g) shows a negatively stained sample using uranyl acetate. Source: Adapted with permission from Ref. [85]. Copyright American Chemical Society 2008.

brushes in **MBNP-VI-1**, we obtained **MBNP-VI-4** with the calculated PAA M_n = 12.5 kDa ($\sigma_{PAA} = \sigma_{PtBA}$ = 0.54 chains nm^{-2}) and the PS M_n = 18.7 kDa (σ_{PS} = 0.28 chains nm^{-2}). Representative cryo-TEM micrographs of the PAA/PS **MBNP-VI-4** in DMF are shown in Figure 8.20a,b [86]. Prior to plunge-freezing the sample on a cryo-TEM grid, the PAA chains were selectively stained by uranyl acetate. Because DMF is a nonselective good solvent, both PAA and PS chains are highly stretched out in it. In Figure 8.20a,b, the mixed-brush layer formed a halo around the silica NPs with subtle striations of alternating lighter and darker gray colors, which corresponded to microphase-separated PS (lighter gray) and PAA (darker gray) microdomains. Some isolated bright PS domains were seen on the top of the silica NPs. Obviously, PS and PAA mixed brushes should form a rippled structure on the NP surface.

Representative cryo-TEM micrographs of the PAA/PS **MBNP-VI-4** in water are shown in Figure 8.20c,d. Because water is a good solvent for PAA but a poor solvent for PS, the self-assembled morphology of PAA/PS mixed brushes in water should be different from that in DMF. Based on the theoretical prediction [32] and a previous experimental study [85], mixed brushes in a selective solvent should adopt a vertical phase separation with the solvophilic chains stretching out to form the top layer and the solvophobic chains clustering in the bottom layer. Comparing the cryo-TEM micrographs in Figure 8.20c,d with those in Figure 8.20a,b, isolated bright PS nanodomains were seen on top of the silica NPs in water, similar to the situation in DMF. Around the edge of **MBNP-VI-4** in water, weak and irregular striations of lighter and darker gray were seen in the mixed-brush halo. The only difference was

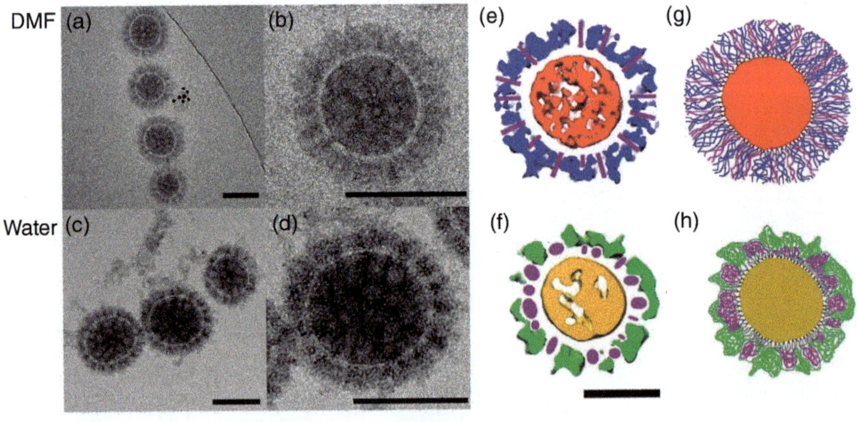

Figure 8.20 Cryo-TEM micrographs of PAA/PS **MBNP-VI-4** in frozen (a, b) DMF and (c, d) water. The PAA chains were stained by uranyl acetate to bear a darker contrast. Scale bars are 100 nm. Comparison of density slabs from extracted **MBNP-VI-4** in (e) DMF and (f) water, and schematic illustrations of the mixed-brush layers in (g) DMF and (h) water. The hypothesized volumes for the clustered PS chains are shown in magenta within the voids and cavities of the density for the PAA chains. Each slab is 5 nm thick. The scale bar is 50 nm. (g) and (h) Schematic illustrations of the mixed-brush layer in DMF and water, respectively. PAA chains and domains are blue in (e) and (g) and green in (f) and (h). Source: Adapted with permission from Ref. [86]. Copyright American Chemical Society 2015.

that the mixed-brush layer in water was 18 ± 3 nm thick and the mixed-brush layer in DMF was ~27 nm thick. The thinner brush layer in water suggested that the PS chains collapse in water with reduced overall brush-layer thickness. In this sense, 2D cryo-TEM was not powerful enough to differentiate the 3D morphology of surface micelle morphology from the bicontinuous rippled structure, and we needed to resort to 3D cryo-TEM.

Cryo-electron tomography was performed to visualize PAA/PS **MBNP-VI-4** particles in 3D. Complete tilt series (from −64 to +64°) was successfully obtained on a JEOL JEM2200FS cryo-TEM with the Serial EM tomography software package. 3D tomograms were generated from the tilt series of **MBNP-VI-4** in DMF and water with the IMOD software package [87]. From the cryo-electron tomography, we can see that the PS chains adopted a different aggregation state in DMF from that in water. For **MBNP-VI-4** in DMF, regular gaps were seen between the PAA nanodomains; therefore, PS domains also extended outward from the core (Figure 8.20e). Because of the lower grafting density of PS chains (0.28 chains nm^{-2}) than that of PAA (0.52 chains nm^{-2}) [81], the PS domains assembled into truncated cone-like structures. For **MBNP-VI-4** in water, larger voids were seen near the silica core. These voids must be constructed by the collapsed PS chains (Figure 8.20f). The density slab images for the particles in DMF showed that PS domains were interspersed between the PAA bundles in the brush layer (Figure 8.20g). In water, the PS chains collapsed near the silica core and were shielded by the swollen PAA shell (Figure 8.20h).

8.4.2 Self-Assembly of MBNPs in Homopolymer Matrices with Different Molecular Weights

In addition to the self-assembly of small molecule in solvents, it would be interesting to understand the self-assembly of MBNPs in homopolymer matrices [88]. For this purpose, we chose **MBNP-VI-2** for this study, where P*t*BA $M_n = 22.2$ kDa ($\sigma_{PtBA} = 0.54$ chains nm^{-2}) and PS $M_n = 23.4$ kDa ($\sigma_{PS} = 0.31$ chains nm^{-2}). The first set of polymer matrices is chosen to be the P*t*BA homopolymers with $M_n = 5$, 23, and 65 kDa, which are miscible with the grafted P*t*BA chains. The second set of polymer matrices is chosen to be the poly(cyclohexyl methacrylate) (PCHMA), which is miscible with the PS brushes. The M_n values of PCHMA are 3, 19, and 70 kDa, respectively. The molecular weights are selected to be below, similar to, and above the M_n values of the P*t*BA and PS brushes in **MBNP VI-2**.

First, **MBNP-VI-2** was blended into the 5 kDa P*t*BA homopolymer matrix. Figure 8.21a shows the bright-field TEM micrograph of **MBNP-VI-2** in the 5 kDa P*t*BA matrix. RuO_4 was used to stain the PS chains, which appeared dark in the micrograph. The MBNP particles were isolated with dark PS microdomains protruding out. The unstained P*t*BA brushes and the matrix must form the continuous "invisible" phase. The wet-brush theory for block copolymers can be used to explain this observation [89]. Namely, when the length of the P*t*BA homopolymer is lower than that of the grafted P*t*BA chains, the P*t*BA homopolymer can penetrate into the P*t*BA brush domains. As a result, the grafted PS chains are pushed into isolated

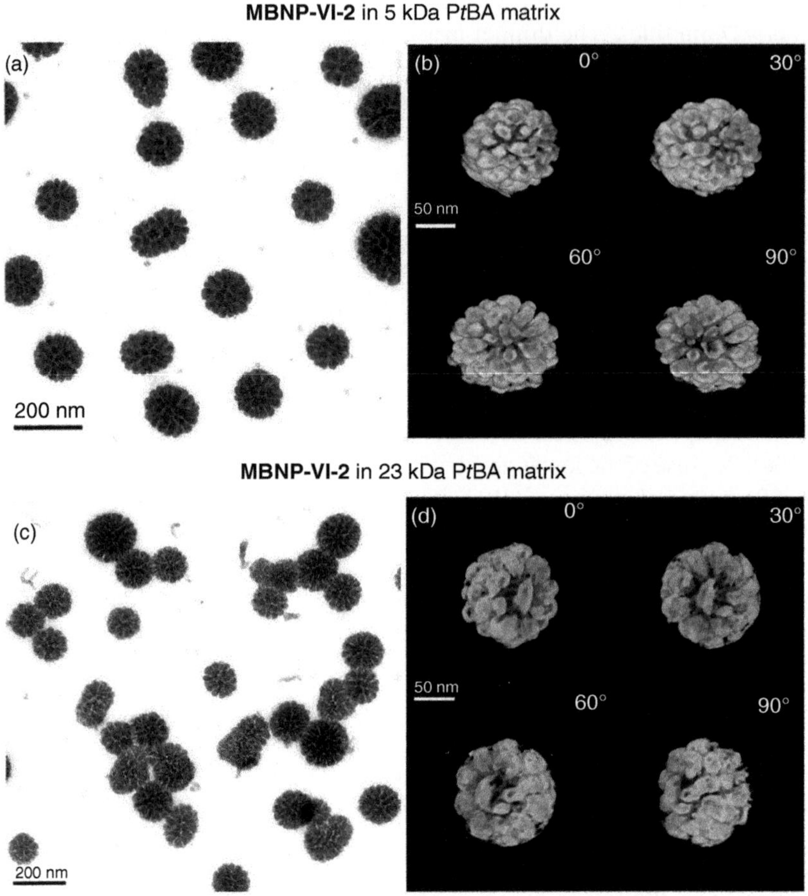

Figure 8.21 (a, c) Bright-field TEM micrographs of P*t*BA/PS **MBNP-VI-2** in the (a) 5 kDa and (c) 23 kDa P*t*BA matrix stained by RuO_4 for 20 min. (b, d) 3D TEM images (solid mode) of a reconstructed single **MBNP-VI-2** in the (b) 5 kDa and (d) 23 kDa P*t*BA matrix at different rotation angles. Source: Adapted with permission from Ref. [88]. Copyright Royal Society of Chemistry 2015.

domains. Figure 8.21b presents 3D tomographs of a single MBNP in the 5 kDa P*t*BA matrix at different viewing angles. From these images, isolated PS domains clearly protruded outward from the particle core.

Figure 8.21c shows P*t*BA/PS **MBNP-VI-2** in the 23 kDa P*t*BA matrix. Different from the situation of **MBNP-VI-2** in the 5 kDa P*t*BA matrix with isolated and protruding PS domains (Figure 8.21a), the dark PS domains were more or less connected among each other, suggesting the dry-brush regime. This is because the length of the P*t*BA matrix polymer was similar to that of the grafted P*t*BA chains. It became difficult for the matrix polymers to penetrate into the grafted mixed brushes. Figure 8.21d shows the 3D tomographs of a single **MBNP-VI-2** in the 23 kDa P*t*BA matrix at different viewing angles. From these images, short, worm-like PS domains were seen. The mixed brush particles in the 23 kDa P*t*BA matrix tended to aggregate together because of the dry-brush regime.

The morphology of P*t*BA/PS MBNPs in the 65 kDa P*t*BA matrix is shown in Figure 8.22 [88]. The MBNPs had a stronger tendency to aggregate, again because of the dry-brush regime due to a higher molecular weight of the P*t*BA matrix. The PS domains were more or less connected on average-sized particles. However, isolated PS domains were seen on small **MBNP-VI-2** (e.g. c. 90 nm overall diameter), similar to the wet-brush regime. This was clearly seen in two areas in Figure 8.22a, which are magnified in Figures 8.22b,c. Indeed, isolated and protruded PS domains were found (indicated by red arrows). The particle size effect could be better observed by 3D TEM. Figure 8.22d,e shows 3D TEM images for a larger (~143 nm diameter) and a smaller (~93 nm diameter) particle in the 65 kDa P*t*BA matrix, respectively. Connected PS domains were observed on the large particle, whereas isolated and protruding PS domains were seen on the smaller particle. We consider that the high particle curvature induced the wet brush-like behavior for the smaller particles.

The blends of P*t*BA/PS brush particles and PCHMA homopolymers were prepared using the same method as for the P*t*BA matrix systems [88]. Figure 8.23a,b shows 2D and 3D micrographs of **MBNP-VI-2** in the 3 kDa PCHMA matrix. Bright P*t*BA phase formed isolated domains in the dark PS matrix. The wet-brush situation was obtained because the shorter PCHMA matrix chains penetrated into the mixed-brush layer and swelled the PS domains. Consequently, the P*t*BA chains were squeezed into isolated microdomains. It is also observed that **MBNP-VI-2** particles disperse well in the 3 kDa PCHMA matrix. Figure 8.23b shows the 3D images of a single MBNP in the 3 kDa PCHMA matrix with isolated P*t*BA domains (in red) dispersed in the PS matrix (in cyan).When embedding the **MBNP-VI-2** in the 19 kDa PCHMA matrix (Figure 8.23c), P*t*BA domains formed worm-like structures. 3D tomographs are shown in Figure 8.23d. Worm-like P*t*BA domains existed on the particle surface. Because the length (19 kDa) of the PCHMA chains in the matrix was slightly lower than that (23.4 kDa) of the grafted PS chains, the PCHMA homopolymers partially penetrated into the mixed-brush layer, and the P*t*BA domains became less isolated to form short worm-like structures. By increasing the molecular weight of the PCHMA matrix to 70 kDa, significant particle aggregation was seen in Figure 8.23e. Because of the dry-brush regime, the bright P*t*BA phase formed the continuous phase. 3D tomographs of four aggregated particles are shown in Figure 8.23f. The PS phase (in cyan) formed either isolated or short worm-like domains in the P*t*BA phase (in red). Between three large particles in Figure 8.23f, PS chains connected together due to favorable particle–particle interactions.

As illustrated in Figure 8.24, we explain the microphase separation behavior of P*t*BA/PS MBNPs in selective P*t*BA and PCHMA matrices by the entropy-driven theories of wet- and dry-brushes [88]. On a large particle, the phase structure of mixed brushes is determined by the length of the matrix polymer chains compared with the grafted polymer chain length (Figure 8.24a,b). If the matrix polymer chains are shorter, they behave like a solvent and can penetrate into the brush layer, forming the wet-brush regime. As a result, the immiscible grafted chains are pushed into isolated domains (Figure 8.24a, the top panel). Meanwhile, the MBNPs are well dispersed (Figure 8.24a, the bottom panel). If the matrix polymer chains are longer, they cannot penetrate into the mixed-brush layer, forming the dry-brush regime. As a result, the morphology of the MBNPs is unperturbed (Figure 8.24b, the top panel),

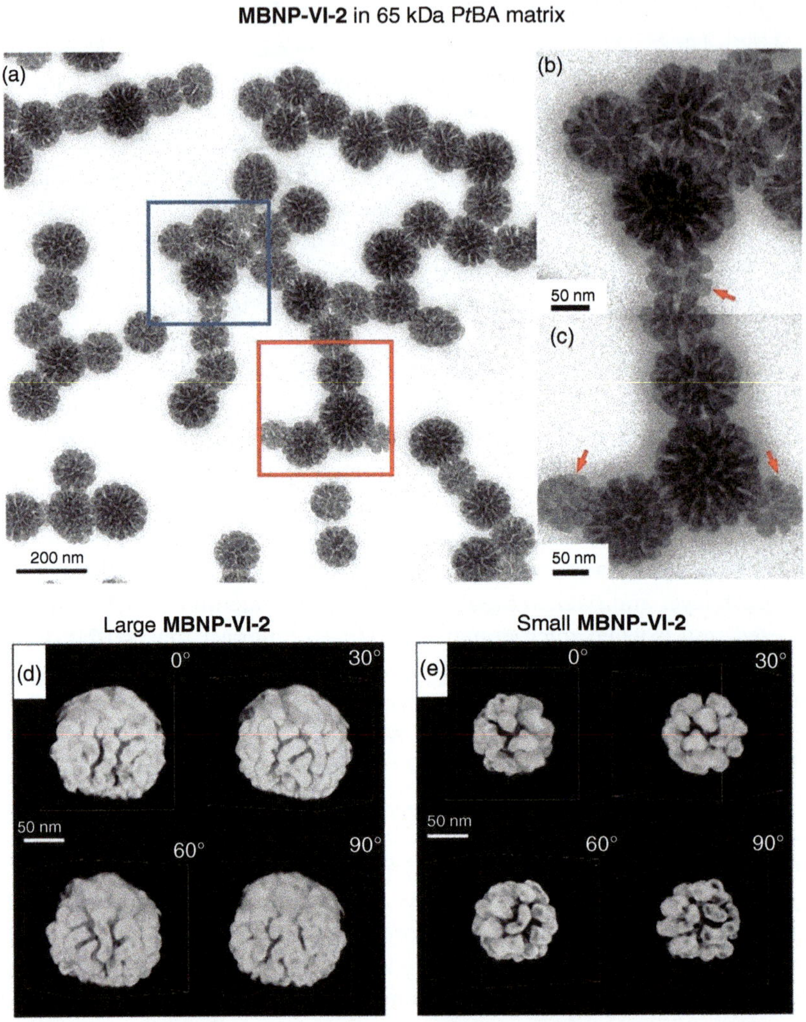

Figure 8.22 (a) Bright-field TEM micrograph of P*t*BA/PS **MBNP-VI-2** in the 65 kDa P*t*BA matrix stained by RuO_4 for 20 minutes. (b, c) Magnified images of the blue and red boxes in (a), respectively. 3D TEM images (solid mode) of reconstructed (d) large and (e) small P*t*BA/PS **MBNP-VI-2** in the 65 kDa P*t*BA matrix, viewed at various rotation angles. Source: Adapted with permission from Ref. [88]. Copyright Royal Society of Chemistry 2015.

and the MBNPs aggregate together (Figure 8.24b, the bottom panel). When the MBNPs are small (i.e. a high curvature), the matrix polymer chains can penetrate partially into the brush layer, pushing the immiscible grafted polymer chains to form isolated domains, as shown in the bottom panel of Figure 8.24c. In this sense, the self-assembly of MBNPs in a selective polymer matrix is entropy-driven. Taking advantage of this concept, NPs grafted with bimodal [90, 91] or heterogeneous [92, 93] length homopolymer brushes can enable the wet brush-like regime and disperse well in high molecular weight homopolymer matrices.

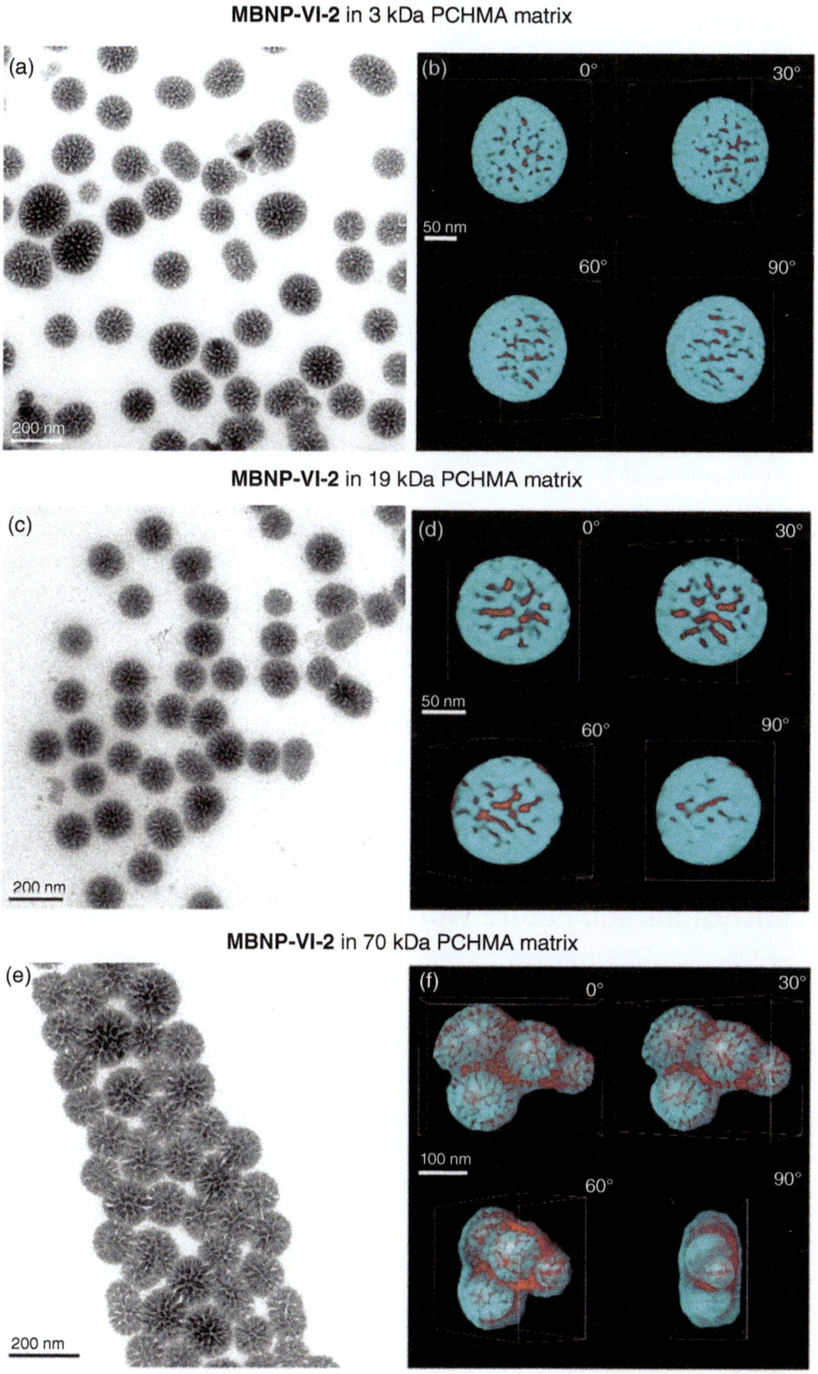

Figure 8.23 (a, c, e) Bright-field TEM micrographs of P*t*BA/PS **MBNP-VI-2** in (a) 3 kDa, (c) 19 kDa, (e) 70 kDa PCHMA matrix stained by RuO_4 for 20 min. (b, d, f) 3D TEM images (solid mode) of a reconstructed **MBNP-VI-2** in (b) 3 kDa, (d) 19 kDa, and (f) 70 kDa PCHMA matrix at different rotation angles. Source: Adapted with permission from Ref. [88]. Copyright Royal Society of Chemistry 2015.

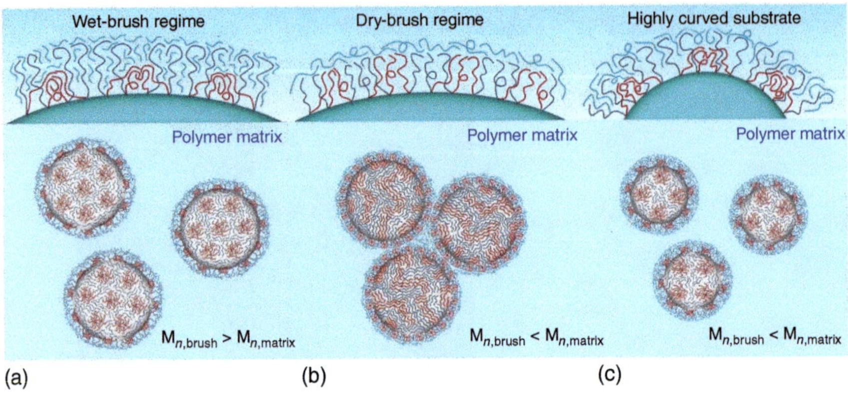

Figure 8.24 Schematic illustrations of (a) wet-brush regime, (b) dry-brush regime, and (c) highly curved substrate for MBNPs in selective homopolymer matrices. Source: Reproduced with permission from Ref. [88]. Copyright Royal Society of Chemistry 2015.

The criterion for the wet-to-dry brush transition needs to be further discussed here. Namely, the degree of penetration of the matrix chains into the grafted brush layer is the determining factor. The situation for MBNPs is more complicated [94–96], because the immiscible chains can lead to particle aggregation because of unfavorable enthalpic interactions. From the above results, the criterion for the wet-to-dry brush transition has an M_{brush}/M_{matrix} ratio between 5 and 1, where M_{brush} and M_{matrix} are the molecular weights of the miscible grafted polymer and the homopolymer matrix, respectively. Note, for single-component polymer brush-grafted particles in a homopolymer matrix, the criterion is $M_{brush}/M_{matrix} < 1$ [94–96].

8.5 Conclusions and Future Work

MBNPs represent a new class of environmentally responsive hybrid nanomaterials with hierarchical self-assembly at different length scales. At the molecular scale, microphase separation of the grafted chemically different polymers enables a suite of nanoscale morphologies, leading to patchy hairy NPs. At the mesoscale, colloidal packing of these MBNPs will provide a bottom-up approach to functional metamaterials. Currently, theory and simulation have predicted many intriguing nanoscale structures for MBNPs with the particle size similar to the brush dimensions, but experimental work lags behind these predictions. In the future, more research is needed to engineer these intriguing nanostructures for novel material science and biomedical applications.

Acknowledgment

This collaborative work was supported by the National Science Foundation [DMR-1007986 (B.Z.), DMR-1007918 (L.Z.), CHE-1709663 (B.Z.), and CHE-1709119

(C.Y.L)], the Cleveland Foundation (P.L.S.), and the National Natural Science Foundation of China [Grant Nos. 21320102005 and 21374023 (P.T.)]. We are also grateful for the fruitful collaboration with late Prof. Feng Qiu at Fudan University. B.Z. is indebted to his former graduate students, Dr. Dejin Li, Dr. Xiaoming Jiang, Dr. Xueguang Jiang, Dr. Jonathan M. Horton, Dr. Naixiong Jin, Dr. Chunhui Bao, Dr. Roger A. E. Wright, Mr. Andrew J. Chancellor (M.S.), and Dr. Caleb A. Bohannon, and his current graduate student Mr. Michael T. Kelly as well as former undergraduate student Ms. Tram T. Le for their contributions to the mixed brush project.

References

1 Moffitt, M.G. (2013). *J. Phys. Chem. Lett.* 4: 3654–2666.
2 (2009). *Adv. Mater.* 21: 1897–1905.
3 Nie, Z., Petukhova, A., and Kumacheva, E. (2010). *Nat. Nanotechnol.* 5: 15–25.
4 Pyun, J., Jia, S., Kowalewski, T. et al. (2003). *Macromolecules* 36: 5094–5104.
5 Wang, B., Li, B., Zhao, B. et al. (2008). *Chem. Soc.* 130: 11594–11595.
6 Zhao, B. and Zhu, L. (2009). *Macromolecules* 42: 9369–9383.
7 Wright, R.A.E., Hu, B., Henn, D.M., and Zhao, B. (2015). *Soft Matter* 11: 6808–6820.
8 Wright, R.A.E., Henn, D.M., and Zhao, B. (2016). *J. Phys. Chem. B* 120: 8036–8045.
9 Chmielarz, P., Yan, J., Krys, P. et al. (2017). *Macromolecules* 50: 4151–4159.
10 Rossner, C., Zhulina, E.B., and Kumacheva, E. (2019). *Angew. Chem. Int. Ed.* 58: 9269–9274.
11 Yi, C., Liu, H., Zhang, S. et al. (2020). *Science* 369: 1369–1374.
12 Stuart, M.A.C., Huck, W.T.S., Genzer, J. et al. (2010). *Nat. Mater.* 9: 101–113.
13 Li, M. and Pester, C.W. (2020). *Polymers* 12: 1553.
14 Kumar, S.K., Jouault, N., Benicewicz, B., and Neel, T. (2013). *Macromolecules* 46: 3199–3214.
15 Fu, Y., Yang, Y., Xiao, S. et al. (2019). *Prog. Org. Coat.* 130: 75–82.
16 Bratek-Skicki, A., Cristaudo, V., Savocco, J. et al. (2019). *Biomacromolecules* 20: 778–789.
17 Liu, Y., Liu, Y., Yin, J.J., and Nie, Z. (2015). *Macromol. Rapid Commun.* 36: 711–725.
18 Chen, L. and Klok, H.A. (2013). *Soft Matter* 9: 10678–10688.
19 Yi, C., Zhang, S., Webb, K.T., and Nie, Z. (2017). *Acc. Chem. Res.* 50: 12–21.
20 Motornov, M., Sheparovych, R., Lupitskyy, R. et al. (2007). *J. Colloid Interface Sci.* 310: 481–488.
21 Motornov, M., Sheparovych, R., Lupitskyy, R. et al. (2007). *Adv. Funct. Mater.* 17: 2307–2314.
22 He, J., Huang, X., Li, Y.C. et al. (2013). *J. Am. Chem. Soc.* 135: 7974–7984.
23 Song, J., Zhou, J., and Duan, H. (2012). *J. Am. Chem. Soc.* 134: 13458–13469.
24 Chen, C., Tang, P., and Qiu, F. (2014). *J. Polym. Sci., Part B: Polym. Phys.* 52: 1583–1599.
25 Wang, Z. and Li, B.H. (2016). *Chin. Phys. B* 25: 016402.

26 Marko, J.F. and Witten, T.A. (1991). *Phys. Rev. Lett.* 66: 1541–1544.
27 Dong, H. (1993). *J. Phys.* II (3): 999–1020.
28 Marko, J.F. and Witten, T.A. (1992). *Macromolecules* 25: 296–307.
29 Lai, P.Y. (1994). *J. Chem. Phys.* 100: 3351–3357.
30 Zhulina, E. and Balazs, A.C. (1996). *Macromolecules* 29: 2667–2673.
31 Müller, M. (2002). *Phys. Rev. E.* 65: 030802.
32 Wang, J. and Müller, M. (2009). *J. Phys. Chem. B* 113: 11384–11402.
33 Merlitz, H., He, G.L., Sommer, J.U., and Wu, C.X. (2009). *Macromolecules* 42: 445–451.
34 Wenning, L., Müller, M., and Binder, K. (2005). *Europhys. Lett.* 71: 639–645.
35 Sidorenko, A., Minko, S., Schenk-Meuser, K. et al. (1999). *Langmuir* 15: 8349–8355.
36 Wang, J., Kara, S., Long, T.E., and Ward, T.C. (2000). *J. Polym. Sci., Part A: Polym. Chem.* 38: 3742–3750.
37 Minko, S., Müller, M., Usov, D. et al. (2002). *Phys. Rev. Lett.* 88: 035502.
38 Minko, S., Usov, D., Goreshnik, E., and Stamm, M. (2001). *Macromol. Rapid Commun.* 22: 206–211.
39 Zhao, B. (2003). *Polymer* 44: 4079–4083.
40 Zhao, B. (2004). *Langmuir* 20: 11748–11755.
41 Zhao, B. and He, T. (2003). *Macromolecules* 36: 8599–8602.
42 Zhao, B., Haasch, R.T., and MacLaren, S. (2004). *Polymer* 45: 7979–7988.
43 Zhao, B., Haasch, R.T., and MacLaren, S. (2004). *J. Am. Chem. Soc.* 126: 6124–6134.
44 (2003). *Macromolecules* 36: 7268–7279.
45 Lemieux, M., Usov, D., Minko, S. et al. (2003). *Macromolecules* 36: 7244–7255.
46 Minko, S., Muller, M., Motornov, M. et al. (2003). *J. Am. Chem. Soc.* 125: 3896–3900.
47 Julthongpiput, D., Lin, Y.H., Teng, J. et al. (2003). *J. Am. Chem. Soc.* 125: 15912–15921.
48 Ionov, L., Sidorenko, A., Stamm, M. et al. (2004). *Macromolecules* 37: 7421–7423.
49 Shan, J., Nuopponen, M., Jiang, H. et al. (2005). *Macromolecules* 38: 2918–2926.
50 Santer, S., Kopyshev, A., Yang, H.K., and Ruhe, J. (2006). *Macromolecules* 39: 3056–3064.
51 Zubarev, E.R., Xu, J., Sayyad, A., and Gibson, J.D. (2006). *J. Am. Chem. Soc.* 128: 4958–4959.
52 Zubarev, E.R., Xu, J., Sayyad, A., and Gibson, J.D. (2006). *J. Am. Chem. Soc.* 128: 15098–15099.
53 Usov, D., Gruzdev, V., Nitschke, M. et al. (2007). *Macromolecules* 40: 8774–8783.
54 Santer, S., Kopyshev, A., Donges, J. et al. (2007). *Langmuir* 23: 279–285.
55 Kakkar, A. (2017). *Miktoarm Star Polymers: From Basics of Branched Architecture to Synthesis, Self-Assembly and Applications*. London: Royal Society of Chemistry.
56 Roan, J.R. (2006). *Phys. Rev. Lett.* 96: 248301.
57 Wang, Y., Yang, G., Tang, P. et al. (2011). *J. Chem. Phys.* 134: 134903.
58 Ma, X., Yang, Y., Zhu, L. et al. (2013). *J. Chem. Phys.* 139: 214902.
59 Ma, X., Chen, C., Yang, Y., and Qiu, F. (2014). *Soft Matter* 10: 6005–6013.

60 Chen, C., Zhang, T., Zhu, L. et al. (2016). *ACS Macro Lett.* 5: 718–723.
61 Manoharan, V.N., Elsesser, M.T., and Pine, D.J. (2003). *Science* 301: 483–487.
62 Chen, P., Liang, H., and Shi, A.C. (2007). *Macromolecules* 40: 7329–7335.
63 Zou, Z., He, X., and Wang, L. (2012). *J. Chem. Phys.* 136: 074902.
64 Li, D., Sheng, X., and Zhao, B. (2005). *J. Am. Chem. Soc.* 127: 6248–6256.
65 Matyjaszewski, K. and Xia, J. (2001). *Chem. Rev.* 101: 2921–2990.
66 Hawker, C.J., Bosman, A.W., and Harth, E. (2001). *Chem. Rev.* 101: 3661–3688.
67 Wright, R.A.E., Wang, K., Qu, J., and Zhao, B. (2016). *Angew. Chem. Int. Ed.* 55: 8656–8660.
68 Philipse, A.P. and Vrij, A. (1989). *J. Colloid Interface Sci.* 128: 121–136.
69 Bogush, G.H., Tracy, M.A., Zukoski, C.F., and Non-Cryst, J. (1988). *Solids* 104: 95–106.
70 Jiang, X., Zhao, B., Zhong, G. et al. (2010). *Macromolecules* 43: 8209–8217.
71 Bohannon, C., Chancellor, A., Kelly, M. et al. (2021). *J. Am. Chem. Soc.* 143: 16919–11624.
72 Chancellor, A.J., Seymour, B.T., and Zhao, B. (2019). *Anal. Chem.* 91: 6391–6402.
73 Li, D., Jones, G.L., Dunlap, J.R. et al. (2006). *Langmuir* 22: 3344–3351.
74 Husseman, M., Malmstrom, E.E., McNamara, M. et al. (1999). *Macromolecules* 32: 1424–1431.
75 Zhao, B. and Zhu, L. (2006). *J. Am. Chem. Soc.* 128: 4574–4575.
76 Jiang, X., Zhong, G., Horton, J.M. et al. (2010). *Macromolecules* 43: 5387–5395.
77 Bao, C., Tang, S., Horton, J.M. et al. (2012). *Macromolecules* 45: 8027–8036.
78 Hiemenz, P.C. and Lodge, T. (2007). *Polymer Chemistry*, 2e. Boca Raton, FL: CRC Press.
79 Bao, C., Tang, S., Wright, R.A.E. et al. (2014). *Macromolecules* 47: 6824–6835.
80 Leibler, L. (1980). *Macromolecules* 13: 1602–1617.
81 Horton, J.M., Tang, S., Bao, C. et al. (2012). *ACS Macro Lett.* 1: 1061–1065.
82 Kim, S., Kim, T.H., Huh, J. et al. (2015). *ACS Macro Lett.* 4: 417–421.
83 Tang, C., Lennon, E.M., Fredrickson, G.H. et al. (2008). *Science* 322: 429–432.
84 Tang, S., Lo, T.Y., Horton, J.M. et al. (2013). *Macromolecules* 46: 6575–6584.
85 Zhu, L. and Zhao, B. (2008). *J. Phys. Chem. B* 112: 11529–11536.
86 Fox, T.L., Tang, S., Horton, J.M. et al. (2015). *Langmuir* 31: 8680–8688.
87 Kremer, J.R., Mastronarde, D.N., and McIntosh, J.R. (1996). *J. Struct. Biol.* 116: 71–76.
88 Tang, S., Fox, T.L., Lo, T.Y. et al. (2015). *Soft Matter* 11: 5501–5512.
89 Jiang, M. and Xie, H. (1991). *Prog. Polym. Sci.* 16: 977–1026.
90 Rungta, A., Natarajan, B., Neely, T. et al. (2012). *Macromolecules* 45: 9303–9311.
91 Li, Y. and Tao, P. (2013). *Langmuir* 29: 1211–1220.
92 Martin, T.B. and Jayaraman, A. (2013). *Soft Matter* 9: 6876–6889.
93 Martin, T.B. and Jayaraman, A. (2013). *Macromolecules* 46: 9144–9150.
94 Green, P.F. (2011). *Soft Matter* 7: 7914–7926.
95 Ganesan, V. and Jayaraman, A. (2014). *Soft Matter* 10: 13–38.
96 Frischknecht, A.L., Hore, M.J.A., Ford, J., and Composto, R.J. (2013). *Macromolecules* 46: 2856–2869.

9

Hairy Plasmonic Nanoparticles

Christian Rossner[1], Tobias A.F. König[1,2], and Andreas Fery[1,3]

[1] *Leibniz-Institut für Polymerforschung Dresden e. V., Hohe Straße 6, 01069 Dresden, Germany*
[2] *Technische Universität Dresden, Center for Advancing Electronics Dresden (cfaed), 01062 Dresden, Germany*
[3] *Technische Universität Dresden, Physical Chemistry of Polymeric Materials, Bergstraße 66, 01069 Dresden, Germany*

9.1 Introduction

Why all the excitement about hairy plasmonic nanoparticles (NPs)? Basically, because the oscillation of the free electrons at the metallic NP surface creates a unique localized field enhancement, the full potential of which can be exploited by surface functionalization (see Figure 9.1). This light confinement at the particle surface makes them nanoscale probes that are highly sensitive to changes in size, shape, material, and environment [1]. Since the resonance is confined to the particle and cannot propagate, it is defined as localized surface plasmon resonance (LSPR). In order to obtain metallic NPs with high optical quality (narrow spectral width) and in large numbers (liter scale) in a reproducible form, wet-chemical synthesis is recommended. The crucial role of surfactants in nanocrystal growth is described in the review article by Xia and coworkers [2]. We will only tangent this topic to obtain specific shapes such as spheres, rods, and cubes as a starting product for further applications and device development. When plasmonic NPs are decorated with surface-immobilized macromolecular ligands, in addition, hairy plasmonic NPs are created. Two general strategies for this functionalization with polymer ligands exist, which are referred to as *in situ* and *ex situ* functionalization. In the former approach, a polymeric capping agent is present during the NP synthesis, resulting in hairy NPs from a simple one-pot reaction [3, 4]. The presence of macromolecular ligands may enable control over the process of NP growth, for example, in the synthesis of aluminum nanocrystals of different shapes [5]. But usually, better control is exerted by separating NP formation and polymer functionalization. In such an *ex situ* functionalization approach, small molecule ligands are (partly) replaced by macromolecular ligands, either by growing the polymers from the surface (grafting-from) [6] or by attaching presynthesized polymers to the surface (grafting-to) [7]. Functionalization of presynthesized NP surfaces with polymer ligands may lead to increased colloidal stability [8], which is critical for achieving

Hairy Nanoparticles: From Synthesis to Applications, First Edition. Edited by Zhiqun Lin and Yijiang Liu.

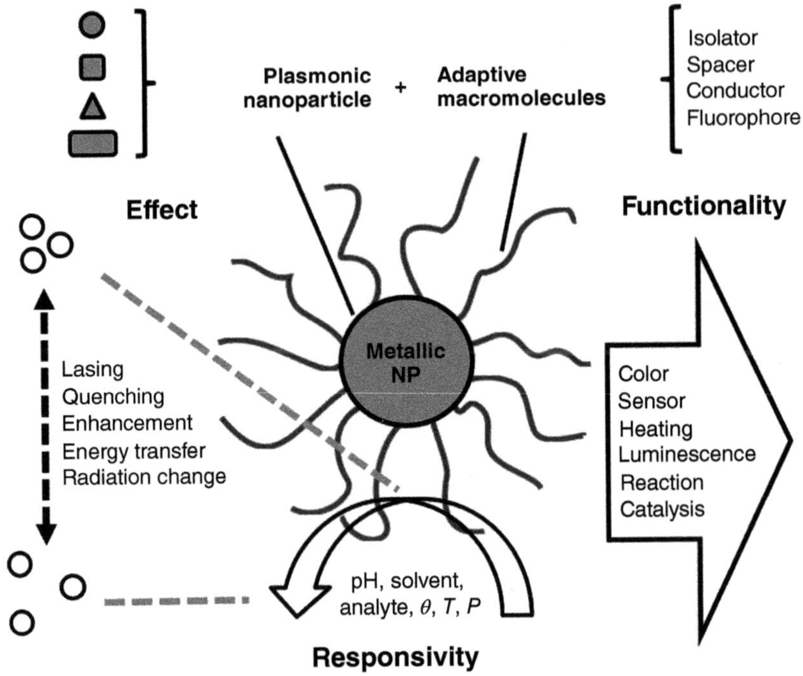

Figure 9.1 An overview about the potential of hairy plasmonic nanoparticles. A variety of distinct plasmonic nanoparticles can be merged with well-defined macromolecular ligands. The ligand's adaptive properties open up the possibility to dynamically tune secondary effects associated with the plasmon resonance excitation. These effects are exploited in a range of functional systems, which can thus be controlled by the polymeric coating layer.

consistent optical properties. Besides providing stability, polymer ligands can be used to adapt the interface of plasmonic NPs and their environment: Different properties, such as charge transport through the interface [9] or luminescence [10], can be controlled by tailoring the composition and architecture of the surface-immobilized macromolecules. Decoration with polymer ligands may also be used to equip plasmonic NPs with additional properties, such as functionality [11], surface charge of the polymer [12], conductivity [13], and adaptivity [14]. With regard to the latter, several polymer chemistries can be used for this purpose, thereby enabling adaptivity to a range of external stimuli, such as temperature [15], pH [16], solvent environment [17, 18], and light [19], or combinations of several of those stimuli. A comprehensive summary of application-oriented hairy NP's designs and the underlying switching mechanisms has been provided in a recent review [20]. Herein, we focus on plasmonic NPs and emphasize the targeted manipulation of the plasmonic properties and functionalities that can be harnessed through hairy ligands. This is summarized in Figure 9.1 and Table 9.1, which is not meant to be an exhaustive list, but rather intended to provide an overview of the rich variety of effects that can be exploited by functionalization of plasmonic NPs with hairy ligands.

Table 9.1 Isolated and coupled plasmonic NP systems and selected functionalities together with the underlying effect that can be achieved, enabled, and controlled by the respective hairy surface coating.

Functionality	Underlying effect	Contribution of the hairy surface functionalization
	Isolated	
Color displays [21]	(Orientation-dependent) absorption and scattering	Providing dispersibility in organic solvents [21]
Sensor capability [22, 23]	LSPR-sensitivity to the dielectric coating	Adjusting the thickness of the coating layer [23]
Boosted thermally activated reaction [24]	Plasmonic heating [25]	Linker and spacer between the plasmonic nanoparticle and catalytic particles [24]
Photothermal self-assembly [26, 27]	Plasmonic heating [25]	Thermoresponsive properties [26, 27]
Photoluminescence modulation [10, 28]	Fluorescence quenching [29]	Covalent linking of fluorescent entities to the plasmonic nanoparticle [10]
Photocatalysis [30, 31]	Near-field enhanced excitation of the photocatalyst [30, 31]	Linker and spacer between the plasmonic nanoparticle and photocatalysts [24]
	Coupled	
Enhanced spectroscopies [32–35]	Near-field enhancement	Nanoparticle linkers creating "hot spots" in colloidal dispersion [32]
Chiral sensing [36]	Chiroplasmonic activity [36, 37]	Achiral spacer [36], or chiral component in directed assembly [37]

Eventually, macromolecular linkers may also be used to direct the assembly of plasmonic NPs into structurally well-defined superstructures [38]. In fact, the electromagnetic field can be further amplified by coherent coupling and such arrangement can be tailored by directed assembly enabled by polymer ligands. Different coupling scenarios can be achieved via assembly of plasmonic NPs into 1D- [39–41], 2D- [42], or 3D-superstructures [43], which may strongly alter the plasmonic properties. In order to exploit these special properties and take full advantage of their potential, polymer ligands can be beneficially implemented in various ways (see also Table 9.1).

In this chapter, we will discuss recent achievements from the synergistic combination of plasmonic NPs with synthetic polymer ligands. For this purpose, a line of structure is followed that is oriented on the structural identity of isolated NPs on one hand and coupled systems on the other hand, as well as their surface functionalization achieved by hairy ligands. After introducing the basic concepts for understanding their plasmonic behavior, we discuss what effects may occur upon LSPR excitation, and how these may be harnessed depending on the nature of the surface coating.

9.2 Plasmonic Properties of Isolated NPs and Energy Transfer to Adjacent Hairy Environment

9.2.1 Plasmonic Principles of Hairy NPs

Because of their well-established synthesis [44, 45], stability, and ease of surface functionalization, gold nanospheres are by far the most commonly used plasmonic entities. Figure 9.2a shows the refractive index profile, electric field, and surface charges of a 45 nm diameter golden nanosphere in a homogeneous environment

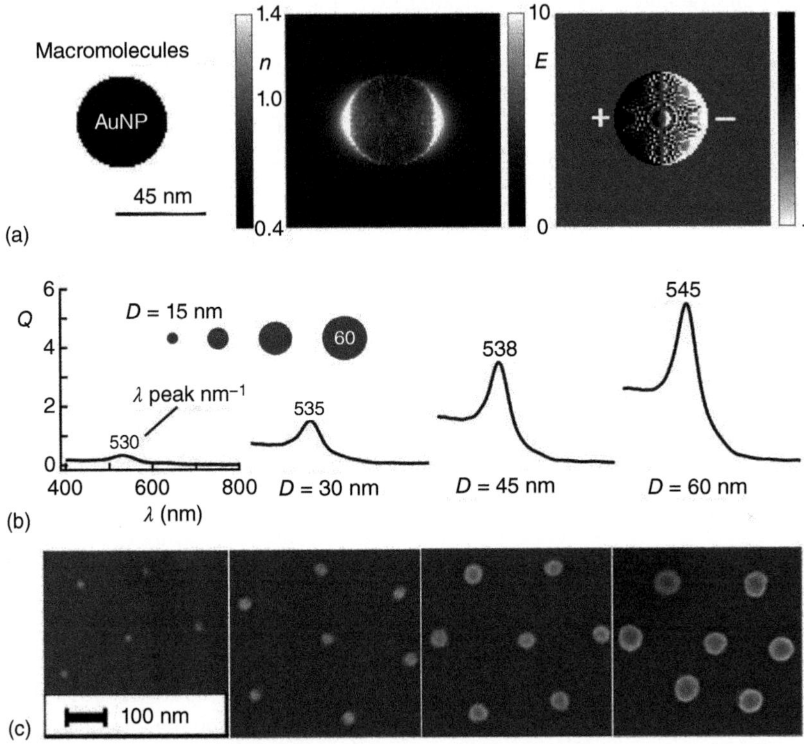

Figure 9.2 (a) Introduction to hairy plasmonic nanoparticles by a simplified setup ($n = 1.4$) to mimic the desired functional environment. Image of the refractive index (real[n]), electric-field enhancement (E), and surface charge (ρ) of a gold nanoparticle at the resonance wavelength of 538 nm. (b) Increase in extinction efficiency (Q) and redshift of plasmon resonance when the particle diameter is increased. Source: Reproduced from Rossner et al. [46], Creative Commons under CC BY-NC-ND 4.0. (c) SEM sections of a substrate that was dip-coated and then withdrawn in a controlled manner from a growth solution, leading to a gradient in particle diameters over a centimeter area. Source: Reproduced from Müller et al. [47], https://pubs.acs.org/doi/10.1021/nn503493c, further permissions related to the material excerpted should be directed to the ACS.

($n = 1.4$ to mimic the adaptive macromolecules). This selected case allows us to introduce the conditions and properties of a LSPR of an isolated NP, whose defining parameters (size, shape, material, environment, and self-assembly in higher-order structures) will be systematically modified in the following. Supported by simulation methods, the confinement and enhancement of the electric field at the particle surface can be demonstrated. Furthermore, this method allows us to identify the resonance as dipolar (see surface charge distribution). Characterization by extinction spectroscopy is common in the analysis of synthesized plasmonic NPs since it already shows quantitatively which plasmonic modes are present. Simulated data show (Figure 9.2b), if the particle diameter of a golden NP is increased, its extinction efficiency and line width increase, and the resonance shifts to the red (lower energy) [46]. The extinction efficiency (or extinction cross section per unit area) is unitless and defined by the ratio of the cross section to the scattering area. The observations in Figure 9.2b are explained by dispersion and the additional scattering of the conduction electrons at the particle surface [48]. The extinction is composed of scattering and absorption, which vary depending on the particle size. For small golden NPs (<20 nm), absorption dominates, at 80 nm the ratios are equal, and for even larger particles, scattering dominates [49]. This size-dependent dichroic effect was quantitatively described at the beginning of the twentieth century by Gustav Mie [50] and Richard Zsigmondy [51]. In solution, the size-dependent dichroic effect is clearly visible in the perceived color. To use this color change in laboratory courses and to obtain a match with the color from Mie's theory, the seed-mediated synthesis approach (high shape stability) must be used [52]. Today's advanced synthesis methods allow to adjust the gold NP dimension, as shown in Figure 9.2c, which highlights a work [47] in which a gradient of particle sizes over a large area could be achieved with reasonable effort. The particles are kept isolated at controlled interparticle spacings through a polymer coating layer on their surfaces [47]. Such surfaces with tailored gold NP spacings and density can form the basis of sensor applications [53] and also biophysical investigations of cell behavior on patterned substrates [54, 55]. Small particles are suitable for energy transfer, such as plasmonic solar cells because of their dominance in absorption [56]. Medium-sized (30~80 nm) particles are particularly suitable for plasmonic spectroscopy, as they exhibit new properties through coupling that go beyond the properties of the isolated particles [46]. How this case can be explicitly created and how it can be usefully combined with a functional polymer will be explained in more detail later in this chapter.

The plasmonic properties of NPs are also influenced by their macromolecular environment (see Figure 9.3) [22]. In modeling, it is usually sufficient to assume a homogeneous dielectric environment (or bulk environment) due to the light confinement at the particle surface. When this dependence is exploited for sensing purposes, it also means that the analyte should be brought as close as possible to the particle surface to ensure high sensitivity (defined as the quotient of resonance change and refractive index change). In this regard, the polymer coating can act to attract and accumulate analyte molecules. For example, polycyclic aromatic hydrocarbons- like pyrene were shown to be trapped from the gas phase (in which

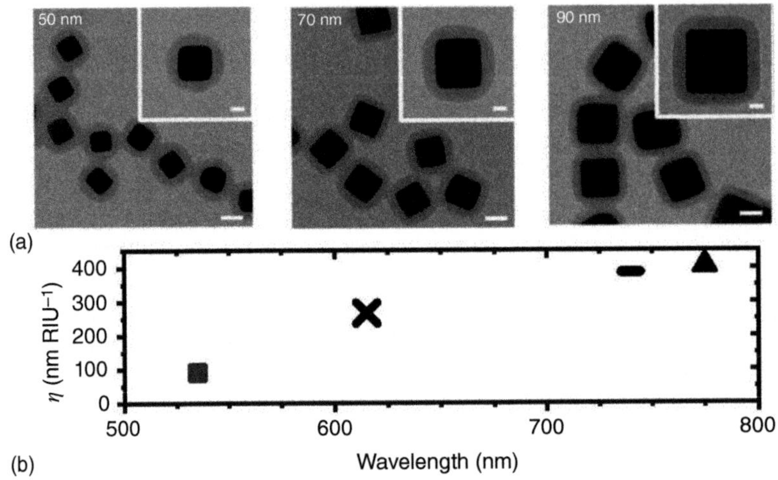

Figure 9.3 (a) TEM image of polystyrene-grafted gold nanocubes with different edge lengths. Source: Reproduced from Lee et al. [23], Creative Commons under CC BY 3.0. (b) Experimentally observed bulk sensitivity for gold nanoparticles with different shapes (from left to right: cubes, concave cubes, rods, trigonal prisms) in solution plotted against the corresponding resonance wavelength. Source: Replotted data taken from Martinsson et al. [22].

they were present together with solvent vapor) into poly(*N*-isopropylacrylamide) coating layers on gold nanostars [57]. In general, when the refractive index is increased, the plasmonic NP resonance shifts to the red (lower energy) and *vice versa* [48]. Branched, anisotropic, or polygonal plate structures show a particularly high sensitivity (Figure 9.3) [23]. Due to the larger geometric structure compared with the spherical particles discussed so far, the plasmonic NP modes appear in the red and infrared regions. In this wavelength range, the material damping for gold is significantly lower compared with the visible range [53]. And of course the LSPR red-shift scales with the thickness of the grafted hairy layer [23]. The refractive index sensitivity can be further enhanced upon forming supracolloidal plasmonic clusters (which will be discussed in Section 9.3 of this chapter) [58, 59], which enables applications in sensing, such as *in situ* reaction monitoring [60].

With this, in addition to spherical NPs, we have introduced geometries with fewer axes of symmetry (cubes and trigonal prisms) or axes of different lengths (rods).

At nonspherical NPs, significantly higher field enhancements occur at the corners and edges due to induced charge separation [61]. For plasmonic NPs with fewer symmetry axes, such as nanocubes (NCs), higher dipolar and quadrupolar modes can further arise (as exemplified for silver NCs) [62]. The higher dipolar modes have the property that field enhancement occurs specifically at vertices, edges, or faces. The quadrupolar modes cannot be observed directly (dark modes) and show their property in symmetry breaking. This case can be generated by a dielectric surface (plasmon hybridization), whereby field enhancements occur either at the surface (resulting in a nanocavity) or at the adjacent medium. With regard to their surface-chemical properties, the NC's faces, edges, and vertices may be discriminated upon functionalization with polymer ligands. In the solution state, constrained surface-dewetting of end-grafted polymer layers can lead to pinned micelle formation selectively of the edges and vertices of AuNCs, leading to polymer-deprived AuNC faces [63–65]. In a polymer matrix, polymer-functionalized NCs may segregate into chain assemblies, in which inter-NC contacts can be controlled by the properties of the hairy coating: Short-chain polymer ligands lead to inter-NC face-to-face assembly, whereas for NCs functionalized with longer polymer ligands, edge-to-edge assemblies are favored [66]. These examples show the possibility of addressing local hot spot sites of plasmonic NCs through grafting with polymer ligands. In addition, the adaptive property of the hairy layer can also be used to control the orientation of plasmonic NCs at interfaces [67]. The E-field enhancement observed for hairy plasmonic NCs can be further enhanced by introducing sharper apexes providing high local surface curvature in nanoprisms [68], nanostars [69], and nanostar-like structures [70].

Plasmonic nanorods support two plasmon modes, a transversal mode (arising from electron oscillations in the direction of the short geometric axis) and a longitudinal mode (oscillations along the long geometric axis). The latter mode occurs in the red and infrared range, a range in which the material attenuation is low for gold, thus this mode dominates the extinction spectrum [71]. Other higher modes occur in geometries with high aspect ratios [71]. In analogy to the particle-in-the-box model from quantum mechanics, the plasmonic modes can be described as integer multiples of the energetically lowest mode. This model was recently developed on particle chains [72]. For applications, it is interesting to distinguish between antisymmetric and symmetric modes. The latter has a dipole moment that does not couple to the far field, a mode also called dark mode, and therefore shows a much better energy transport [41, 73]. The dynamic alignment of gold NRs allows the step to tunable anisotropic material plasmonic properties. When the gold NRs are embedded in an elastomer matrix, their alignment can be strain-induced, and the particles can revert to the disordered configuration upon strain release [74]. In colloidal dispersion, external electric fields may be used for such alignment [75]. With such external operation, dynamic plasmonic pixels become possible, as has been demonstrated for polystyrene-coated gold NRs in the organic solution state [21].

So far, we have focused on plasmonic NPs made of gold, however, the optical properties of plasmonic NPs are also dependent on their chemical composition. Up to date, plasmonic NP made from other metals did not drive the field of plasmonics as much as their golden counterparts, which is, however, not in every case due to their inferior characteristic optical properties, but rather due to stability issues of the surface. For example, compared with gold, silver NPs feature LSPR at higher energies and a much higher extinction cross section, whereas exploitation of these features is hindered because the AgNP surface is prone to oxidation, thereby complicating the consistent control of their surface functionalization. One approach to circumvent these issues is the *in situ* formation of AgNPs with polymer ligands present during the NP growth, but this approach commonly leads to only small-sized (and hence only weakly plasmonically active) hairy AgNPs [76]. When operated with care, hairy

AgNPs with larger silver cores can also be obtained in "grafting-to" processes, for which the same anchor moieties as for gold (sulfur-containing groups, for example, thiols or trithiocarbonates) can be used [77]. Still, it is highly desirable to combine the extraordinary plasmonic properties of silver with the chemical stability of gold surfaces. This can be achieved in bimetallic gold/silver NPs with thin gold shells [61]. The study of bimetallic plasmonic NPs goes hand in hand with their in-depth structural characterization, which nowadays is possible with atomic resolution [78]. Silver NCs are accessible upon silver overgrowth of suited AuNS seed particles. Such silver NCs can also be coated with thin gold shells, providing the desired surface properties [61]. Also, plasmonic sensor systems with inverse sensitivity have been developed based on this process [79].

After introducing the size-, shape-, and composition-dependent optical properties of plasmonic NPs, we will now move on to discussing energy transfer to the hairy environment, which can be tailored using the toolbox of polymer chemistry. For example, this may lead to the incorporation of fluorophores that can create a strong coupling scenario. This strong light–matter interaction is necessary for future applications, such as in quantum information systems [80], high-contrast all-optical modulations [81], or one-atom laser [82]. Against this background, in the following, we will introduce distinct mechanisms of energy transfer in plasmonically excited NPs in the context of their hairy functionalization.

9.2.2 Energy Transfer to Adjacent Hairy Environment

Upon excitation of the plasmonic NP resonance, various secondary processes can be initiated. A suited polymer functionalization allows to guide the locally confined energy into specific applications.

9.2.2.1 Hairy NPs for Photothermal Heating

The excitation of the LSPR is a dissipative process and as such, the confined plasmonic energy is available as photothermal heating [25]. Figure 9.4a shows the relative temperature increase for a 100 nm gold NP in water. An appropriate macromolecular surface coating makes it possible to take advantage of the local temperature rise. Note, that depending on the nature of the hairy layer, the refractive index environment will also change during heating, thus altering the plasmon resonance. Critical phenomena occupy a prominent position in the context of photothermal heating, as they enable sharp phase transitions occurring at a certain critical temperature. For example, poly(*N*-isopropylacrylamide) solutions display a lower critical solution temperature (LCST), and heating above this LCST results in a coil-to-globule conformational transition [84]. In hairy plasmonic NPs, this phase transition gives rise to photothermally induced brush collapse [83, 85]. In such systems, brush collapse may be induced upon illumination, whereas under dark conditions these systems may quickly revert to their initial state since the rehydration (swelling) of the hairy layer is a fast (~100 ns) process [83]. This enables fast switching dynamics, which could hardly be achieved by heating and cooling the entire colloidal solution. Based on such heat-induced

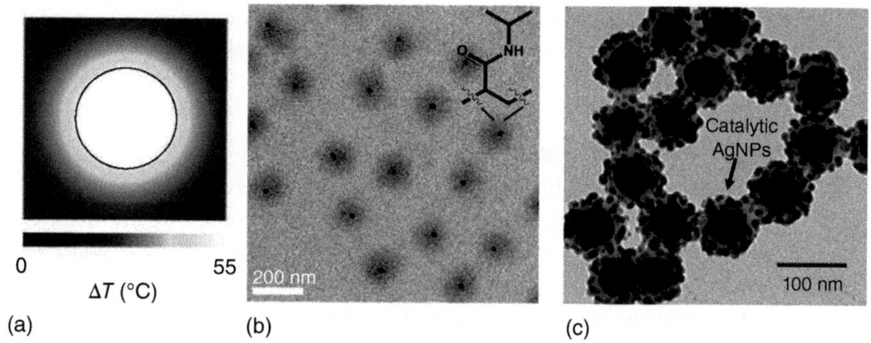

Figure 9.4 (a) Relative temperature increase calculated as a result of photothermal heating. Source: Reprinted with permission from Baffou et al. [25]. Copyright 2010 American Chemical Society. (b) Heating of the NP environment can induce the collapse of hairy LCST ligands, for example, poly(*N*-isopropylacrylamide). Source: Adapted with permission from Murphy et al. [83]. Copyright 2016 American Chemical Society. (c) By attaching catalytic NPs to a hairy layer surrounding gold NPs, plasmonic heating can be used to speed up chemical reactions. Source: Adapted with permission from Zakia et al. [24]. Copyright 2021, Royal Society of Chemistry.

volume phase transitions of a brush coating, thermoplasmonic self-assembly of NPs grafted with LCST-type hairs can be realized [26, 27]. This has been demonstrated for poly(*N*-isopropylacrylamide)-grafted gold nanorods, which bring in the additional advantage that their longitudinal LSPR is red-shifted into the NIR region. Their assembly can follow different scenarios depending on polymer load and localization: (uncontrolled) nanorod agglomeration may occur upon heating [27], but preferential polymer-functionalization of the two end-caps of the nanorods may also lead to nanorod end-to-end chain assembly [26]. In principle, any heat-induced process in brush-modified plasmonic NPs, for example, thermally induced surface patterning [86] or thermal activation of surface functions [87, 88], among many others, may be induced as well by illumination and LSPR excitation. This offers tempting opportunities in the realm of light-controlled nanosystems.

The localized heating can also be applied to drive a (thermally activated) chemical reaction. This can be realized using gold NPs surrounded by a polydopamine layer to which catalytically active silver NPs are attached, as illustrated in Figure 9.4c. Illumination at the plasmon resonance frequency of the gold core leads to higher reaction rates [24].

The local heating that can be achieved in these mentioned hairy plasmonic systems poses the challenge of determining temperatures locally in a nanometer-scale region surrounding the plasmonic NPs. Therefore required nanoscale thermometers can be realized by an adaptive polymer coating that shows a gradual swelling upon increasing the temperature (for moderately increased temperatures). Poly(ethylene glycol) shows this behavior and can therefore be used as a thermosensitive linker that alters the distance between the plasmonic NPs and surrounding fluorescent NPs. As we will see in Section 9.2.2.2, their fluorescence intensity is correlated with the (temperature-dependent) distance to the plasmonic NP, and can therefore serve as a measure for the local temperature in the close environment of the plasmonic NP [89].

9.2.2.2 Hairy NPs Conjugated with Photoactive Entities

The versatile combination possibilities of the polymer-modified particles, in particular, render them suitable building blocks for generating specific excitation states. The generated electronically excited states in the proximity of plasmonic NPs may be used to drive photocatalytic reactions. The involved so-called antenna–reactor complexes require that the absorption band of the photocatalyst overlaps with the LSPR band of the plasmonic NP [90]. In such a setup, there is an optimum size for the plasmonic NP (antenna), which should be large enough to lead to considerable near-field enhancement, but small enough to still have a strong absorption contribution (see discussion in Section 9.1) [30]. Also, the distance of the photocatalytic centers to the plasmonic particle surface is a critical parameter. Close to the surface, energy transfer to the metal dominates, and photocatalysis is attenuated; whereas upon increasing the distance the near-field enhancement may overcompensate this energy transfer, and photocatalytic rate enhancement may be observed [31, 91]. Hence, there may be an optimum spacing for maximum rate enhancement. So far, attempts to create suited spacings have relied on rigid silica coatings [30, 31] (see Figure 9.5a) of the plasmonic antenna particle, or several polyelectrolyte layers [92], but could as well be achieved by a hairy ligand layer, which may add responsive features to such systems. Instead of photocatalytic entities, a light-emitting material may be added. To implement existing concepts from nanophotonics here, it is important to distinguish between conservative and nonconservative systems [93]. So far, we have considered conservative plasmonic systems excited and dissipated at the same energy level. Photoluminescent materials enable nonconservative systems, where light absorption is brought into excited energy states and fluoresces or phosphoresces at lower energies. Such photoluminescent materials can consist

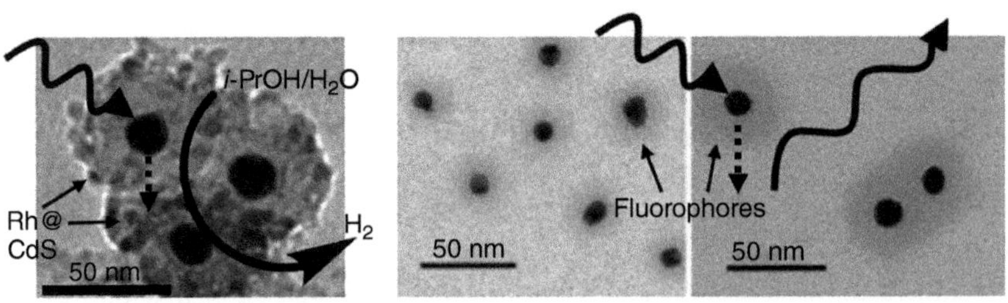

Figure 9.5 Near-field promoted energy transfer to the environment. (a) The energy of excited LSPR in Au nanospheres is transferred to Rhodium-coated quantum dots, which act as catalysts in a hydrogen evolution reaction. The catalytic quantum dots are separated from the gold core through a silica shell. Source: Adapted with permission from Torimoto et al. [31]. Copyright 2011 American Chemical Society. (b) Distance-dependent fluorescence of organic dyes in the exterior of a hairy layer (with defined chain length). Source: Adapted with permission from Cepraga et al. [10]. Copyright 2016, Royal Society of Chemistry.

of semiconductor nanocrystals [28] or fluorophore-containing polymers [10] that can be usefully combined with plasmonic NPs through surface functionalization. For example, fluorophores may be incorporated by covalently joining them with the plasmonic NP through polymer chain linkers (see Figure 9.5b) [10]. In such setup, the polymeric linker can act as a spacer to the metal surface, which is necessary to overcome the strong fluorescence quenching close to the metal surface [29]. In fact, successively increasing the chain length of a hairy linker leads to the observation of enhanced fluorescence intensity [10].

Fluorescence enhancement or high spectral energy density is a prerequisite to generating nonlinear processes. If the gain exceeds the losses (threshold condition), lasing effects can occur in plasmonic NPs. The lasing effect with colloidal NP was first demonstrated on plasmonic core-shell particles loaded with dye molecules [94]. This process requires the pumped excitation of the emitter material, which transfers its energy in a nonradiative decay to the plasmonic NP resonator. This colloidal approach to obtain lasing is defined as LSPR amplification by stimulated emission of radiation (spaser), i.e. with the confinement to the plasmonic NP and a localized plasmonic mode. Note, that the term plasmonic nanolaser is used to define semiconductor–metal nanostructures with propagating plasmonic modes. The introduced physical principles and challenges are the same for all structures. Current challenges are lowering the laser threshold, dynamic modulation, operation at room temperature, and efficient electrical/optical injection [95]. Also, the emitters should be brought close to the metallic surface but not too close. In this respect, using hairy particles with adjusted polymer chain length, this critical parameter could be investigated systematically.

9.2.2.3 Hairy NPs Conjugated with Acceptors

The "hot" electrons created in the process of the LSPR excitation may be donated to semiconductor particles or molecules by direct injection. The injection process is effective when the plasmonic NP and the electron acceptor are in direct contact with each other [96]. This mode of energy transfer from the excited plasmonic NP is promising because the transfer can occur in a resonant manner, which opens the possibility of selectivity control by tailoring the LSPR position through resonance engineering (for example, by controlling the plasmonic NP size, see Figure 9.2b) [97]. Transfer of hot electrons through conjugated polymer linkers to catalytic/reaction sites is yet to be developed, but may lead to useful features like enhanced reactivity through spatial separation of electron–hole pairs, as encountered in the natural paradigm of photosynthesis. It can be foreseen that such developments will occur in the near future, enabled by the very recent advances in the synthesis of fully conjugated hairy ligand coatings on plasmonic NPs [9, 98].

In this section, we have explored the current state and potential of hairy plasmonic NPs to transfer localized energy in a targeted and functional manner. In this discussion, we already had to rely on coupled plasmonic systems, which we will discuss in more detail in what follows. For achieving the coupling, it should be mentioned that besides the distance also the contact area between the particles plays an important role in energy transport. Particularly suitable for this purpose are core-satellite systems (which will be discussed in the following Section 9.3.1), which can be approximated as particle-to-film system (as will be discussed in Section 9.3.2). In these systems, a Fano resonance can be generated, which is also known as plasmon hybridization and shows a significantly higher sensitivity [62, 99].

9.3 Plasmonic Coupling Scenarios of Hairy Plasmonic NPs

From the assembly of hairy plasmonic NPs into supercolloidal structures, new properties will emerge, which range from the simple generation of localized hot spots to chiral plasmonic properties. In the following, we discuss the prospects of hairy ligands to create and tailor such assemblies, both in colloidal solution (Section 9.3.1) as well as on surfaces (Section 9.3.2).

9.3.1 Supercolloidal Structures in Solution

Recently, abiding effort was dedicated to devising solution-based strategies for the assembly of hairy plasmonic NPs into defined cluster architectures [43]. This is mostly motivated by the coherent energy transfer between the plasmonic NPs, which additionally enhance the localized electric field, also known as hot spot generation [46]. Hot spot engineering provides an approach to modulate and/or enhance light–matter interactions, for example, in surface-enhanced Raman spectroscopy (SERS) [32, 33], enabling drastically increased signal intensities, and thereby enabling signal collection from individual NP clusters [100]. For this as well as other reasons, controlling the gap regime in plasmonic NP clusters (i.e. the interparticle gap dimension, the number of interparticle gaps, and their localization in anisotropic structures) is of utmost importance.

Polymer ligands are ideally suited to guide the assembly of distinct individual NPs into hierarchical superstructures and for controlling their structure, functionality, and dynamic features. In particular, if the polymers are employed as macromolecular linker of NPs, adjusting the polymer's degree of polymerization can be used as a handle to achieve defined NP spacings. This is illustrated in Figure 9.6, which shows planet–satellite-type nanostructures in which star polymers of varying average molar mass connect the central "planet" particle with its surrounding satellites [38]. In these clusters, the planet–satellite distance is determined by the arm length of the linking star polymers [101]. By employing stimulus-responsive polymeric linkers, it is also possible to reversibly switch the planet–satellite distance between a contracted state (with shorter interparticle distances) and an expanded state (with larger interparticle distances) [102–104]. This has been realized by exploiting the coil-to-globule transition of poly(*N*-isopropylacrylamide) at its LCST, either by using poly(*N*-isopropylacrylamide) homopolymers [102, 103], or – if colloidal stability is would be compromised above the LCST – by using diblock copolymer architectures with a poly(*N*-isopropylacrylamide) block and an additional, non-responsive block providing colloidal stability [104]. By implementing a diblock

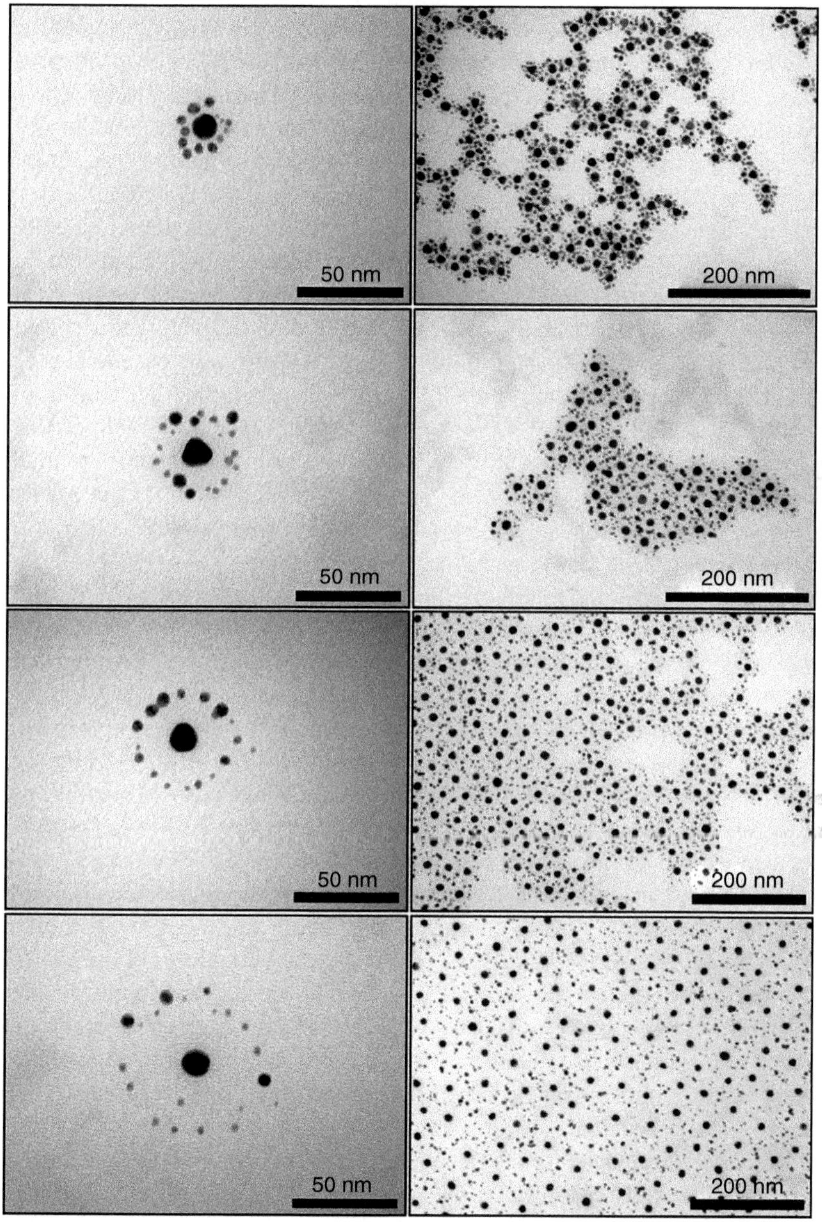

Figure 9.6 Left: Planet–satellite-type plasmonic clusters assembled via star polymer linkers (the degree of polymerization of the star polymer linker increases from top to bottom). Right: Close-packed 2D lattices assembled from the individual planet–satellite-type clusters. Source: Adapted with permission from Rossner and Vana [38], Copyright 2014, Wiley-VCH.

copolymer linker architecture comprising one LCST block and another responsive block that features an upper critical solutions temperature (UCST) such that UCST < LCST, it is also possible to fabricate planet–satellite clusters with expanded planet–satellite distances only in a temperature interval [UCST; LCST], but with collapsed particle distances at temperatures outside this interval [105]. If the planet and the satellite particles are sufficiently large, the switching of their gap distance can be associated with significant changes in their optical properties [102, 106]. When the single planet–satellite clusters are trapped in an optical tweezer setup, these optical properties can be studied on different individual planet–satellite clusters from an ensemble, which gives valuable insights: [106] Firstly, these laser tweezer experiments showed that satellites may "crash" upon laser-induced heating, due to the thermal instability of the thiol–gold surface anchorage involved [107]; and secondly, the results also highlight the typical dispersity of the cluster ensemble with regard to satellite number [106]. In fact, because the binding of satellite particles to a central hairy scaffold particle typically is a statistical process, only the average satellite number may be controlled in a straightforward manner, i.e. by adjusting the satellite stoichiometry during their attachment to the planet particle [103, 108]. We will discuss later how better control over the satellite number can be exerted in hairy systems.

Generally, to achieve strong coupling effects, short gap distances (<2 nm) and small gap volumes between the constituent plasmonic NPs of a cluster are sought [109]. In principle, the strong coupling can be described as a state in which the enhancement considerably exceeds the losses. This amplification differs significantly from the localized field enhancement discussed so far, in which a so-called hybrid system of light and matter is created. Spectrally, two energy levels are observed, which are also known as Rabi splitting [110]. While particle separations that small are not achievable in true hairy systems in the solvent-swollen solution state, they can be created in – to stay with the metaphor – comb-over hairy systems. This is realized in plasmonic NPs with a physisorbed polyelectrolyte layer, which may then be assembled with another type of oppositely charged plasmonic NPs in a modular manner [34, 111]. For example, strong coupling between gold nanorod cores and spherical satellite NPs can result in augmented plasmonic heating capability of nanorods, driven by NIR illumination [111]. There can be also situations in which a combination of both short and long planet–satellite distances, integrated into a single nanocluster architecture, may be useful. This combination may be beneficial for developing nanotheranostics, in which short particle spacings are employed for effective hot spot formation, whereas the longer spacings simply increase the overall size of the cluster, which affects its *in vivo* targeting [112]. Such cluster systems with two different particle spacings can be achieved by using reversible addition-fragmentation chain transfer polymerization (RAFT) star polymers of distinct degree of polymerization [113].

In addition to the (static and dynamic) control of particle spacings in planet–satellite-type clusters of plasmonic NPs discussed above, it is also possible to control the "valence" of the scaffold particle, giving rise to controlled satellite numbers [114]. These latter systems rely on electrostatic interactions between polyacids and polybases in the form of diblock terpolymers (comprising a stabilizing outer PEG block and a second block that is a statistical copolymer of styrene/acrylic acid or styrene/poly[dimethylaminoethyl methacrylate]). A neutralization reaction between these inner acidic and basic blocks drives the cluster formation. Satellite attachment ceases once all of those functionalities are consumed in their neutralization reaction; therefore, the satellite number can be controlled (up to moderate numbers) by adjusting the number of acid and base groups in the complementary terpolymers [114]. The satellite NPs also arrange uniformly around the planet NP core, because electrostatic repulsion interactions between the NP junctions lead to equilibrium angles of typical coordination polyhedral [114].

By deliberate control over the interface regime between the central scaffold particle and its hairy coating, it is also possible to impart directionality into plasmonic clusters, leading to anisotropic clusters [115]. A particularly potent way of achieving this is thermodynamically induced surface patterning in hairy NPs through constrained dewetting induced by reducing the solvent quality for the hairy layer [63]. This process may be driven by addition of a nonsolvent, or other external triggers such as electrooxidation [116], pH [63], or temperature [86]. With this approach, distinct surface patterns can be obtained, depending on the polymer's degree of polymerization and grafting density, and the NP's dimension and shape [63, 117, 118]. Spherical NPs desymmetrized in such a manner may undergo a triggered, secondary assembly step leading to finite-sized homogeneous clusters with small aggregation numbers [119, 120]. On anisotropic NPs, more complex surface patterns may emerge from the constrained surface dewetting, such as helicoidal polymer patches on gold NRs [121] as well as site selectively decorated gold NCs [64, 65]. Subsequent functionalization of the polymer-deprived surface sites may lead to the selective attachment of satellite particles onto those sites (Figure 9.7) [122]. This newly developed approach [122] of surface-encoding through constrained dewetting is more versatile with respect to possible particle shapes compared with previous strategies that relied on the direct site-specific functionalization of, for example, gold NRs [35]. With regard to such applications, adjusting the size of the central planet particle also

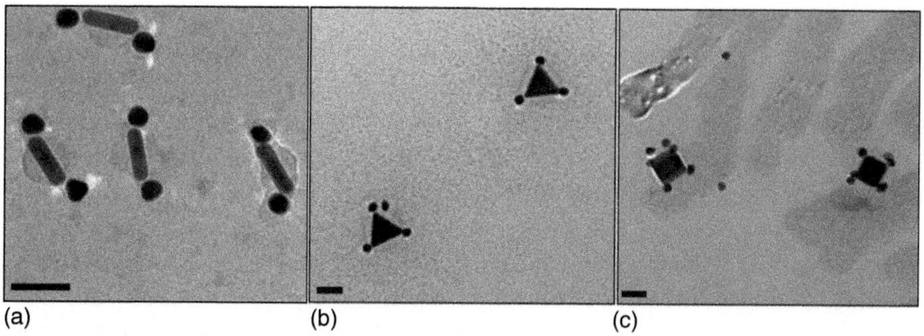

Figure 9.7 Selective attachment of satellite particles to the tips of NRs (a), and the vertices of triangular prisms (b) and NCs (c). Source: Adapted with permission from Chen et al. [122], copyright Springer Nature 2019.

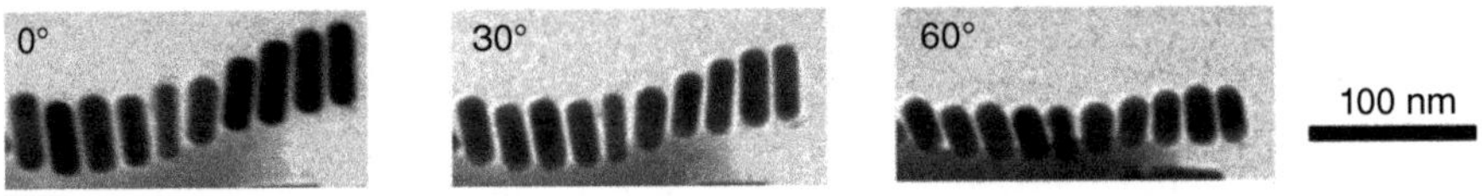

Figure 9.8 TEM tilt series emphasizing helical assemblies of AuNRs coated with RAFT polymers that direct the assembly. Adapted with permission from Cheng et al. [124]. Copyright 2019 American Chemical Society.

provides a handle to control the cluster optical features, such as resonance wavelength and the balance of scattering and absorption contributions, which in turn alter the SERS performance [34, 35]. Satellite attachment at the sharp vertices of anisotropic planet particle cores gives rise to electromagnetic hot spots with strong local field enhancement.

So far, we have discussed achiral hairy plasmonic nanocluster systems, whose optical properties do not depend on light polarization. In principle, the assembly of plasmonic NP building units into asymmetric cluster geometries can lead to chiroplasmonic clusters, even if the constituent NPs are nonchiral. To date, DNA-based approaches to achieve such chiral clusters are dominating [123], which makes the recently reported work [124] from Kun Liu's group reporting chiral assembly of hairy plasmonic NRs outstanding: [124]Those researchers established that such chiral assembly can be directed through different RAFT polymer ligands (Figure 9.8) [124]. They showed that such polymers – although made entirely from achiral monomer and chain transfer agent – may adopt solution conformations with a preferred handedness; and this chiral information can be transferred to the structure of helical AuNR assembly structures made therefrom [124]. It was shown that polymers with a higher degree of polymerization lead to a more pronounced CD of the AuNR superstructures.

9.3.2 Hairy NPs Linked to Surface and Self-assembly

Parallel to the solution-based strategies described in Section 9.3.1, also the surface-assembly of plasmonic NPs has been pursued. This is mostly motivated by the possibility of creating metasurfaces from such approach [125], which are expected to provide new material concepts that may eventually spark novel technological inventions. Metasurfaces feature properties that do not typically occur in nature, for example, negative refractive index [126] or strong circular dichroism [127]. Through the use of a responsive hairy NP coating, adaptive surface-assembled plasmonic NPs [128] may be established as well.

A straightforward way to create short-ranged self-assembled structures of plasmonic NPs on surfaces is to drop-cast the hairy NPs from their colloidal solution onto a substrate, as has been demonstrated, for example, for polystyrene-coated [129, 130] and also poly(*N*-isopropylacrylamide)-coated gold NPs [7]. Because the thickness of the hairy layer scales with the chain length of the hairy ligand [131, 132], distinct interparticle spacings will result from the casting of gold NPs grafted with polymers of various degree of polymerization (see Figure 9.9).

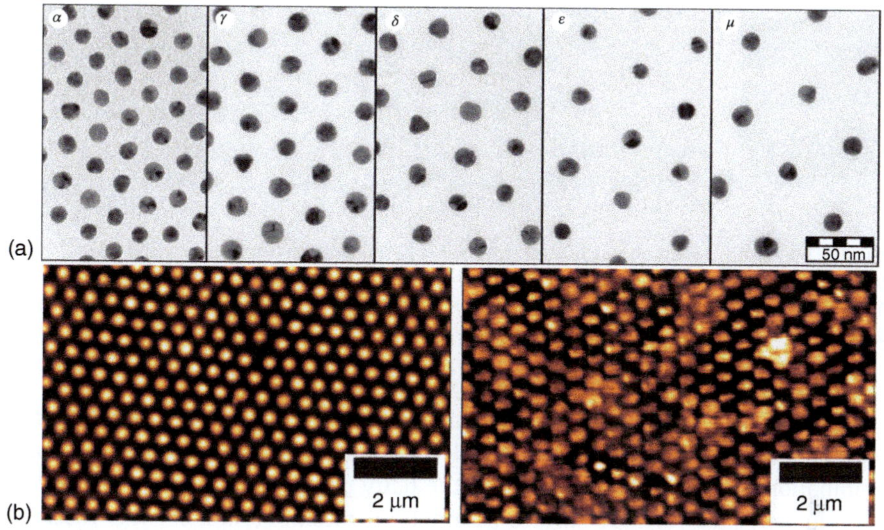

Figure 9.9 Top row: TEM images of self-assembled layers of poly(*N*-isopropylacrylamide)-coated spherical gold nanoparticles. The degree of polymerization of the polymer ligand increases from left to right. Source: Adapted with permission from Ebeling and Vana [7]. Copyright 2013 American Chemical Society. Bottom row: AFM images of metasurfaces from self-assembled silver@gold nanoparticles with a poly(*N*-isopropylacrylamide) coating in the collapsed state (against air, image on the left) and in the water-swollen state (immersed in water at room temperature, image on the right). Source: Adapted with permission from Volk et al. [133], Copyright 2017, Wiley-VCH.

And as a result, the extent of the (distance-dependent) plasmon coupling in these surface assemblies can be controlled by varying the length of the hairy ligands [129, 130]. Ordered self-assembled layers of hairy gold NPs on larger scales (up to square centimeters) can be obtained when using more controlled casting processes in comparison to drop-casting, such as controlled dip-coating [134] or Langmuir–Blodgett transfer [135].

Even better control of the surface self-assembly can be exerted through the use of a template surface that guides the plasmonic NPs deposition. The templated approach is highly versatile and as well allows to incorporate responsive assembly features. Such adaptivity can for example be used to create tunable antireflective surfaces: antireflective properties result from the efficient light extinction around the (narrow) LSPR band of gold NPs assembled on the surface, whereas the LSPR spectral position can be tuned owing to the responsive polymer coating, as has been demonstrated for pH- and redox-responsive polyaniline [136]. When the (templated) assembly of plasmonic NPs onto 2-dimensional substrates leads to highly regular NP deposition, different types of 2-dimensional plasmonic lattices can be created, which may feature surface lattice resonances (SLR): The SLR is a hybrid mode arising from LSPR and Bragg diffraction. Such coherent coupling leads to strongly reduced damping [133]. Such LSR constitutes a metasurface effect, and it can also be made dynamically tunable through the use of a responsive polymer coating [133].

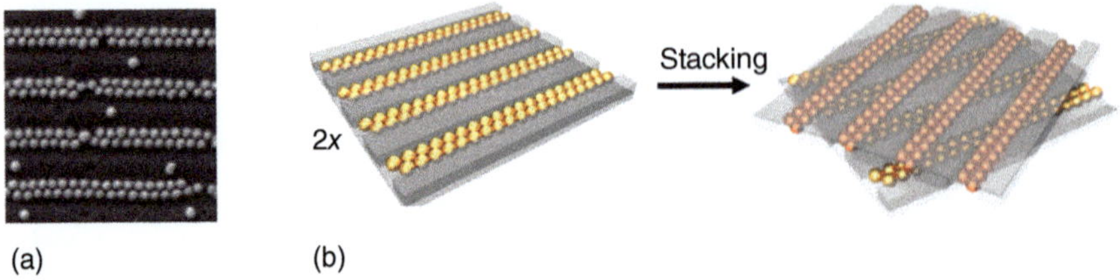

Figure 9.10 (a) SEM image of parallel gold NP (77 nm) double lines assembled on a template. (b) Process of stacking two surfaces created in such a manner at a certain tilt angle. Source: Adapted with permission from Probst et al. [36], copyright Springer Nature 2021.

Expanding from truly 2-dimensional plasmonic surfaces (with a monolayer of plasmonic NPs) into the third dimension further widens the range of plasmonic coupling phenomena, and new optical functionalities may emerge. A smart combination of both the plasmonic and the hairy properties to achieve controlled (laser-guided) and versatile surface assembly of plasmonic dimers was recently reported [137]. The strategy starts from surface-assembled, polymer-coated gold NP layers. A laser is then used to selectively expose the surface of the gold NPs in the irradiated area via photothermal polymer removal. This exposed surface can subsequently be used to attach another plasmonic NP, partly on top of the original NP, and thereby selectively form plasmonic dimers at the laser-irradiated regions of the former plasmonic NP monomer lattice, paving the way to plasmonic information inscription [137]. In another recent work [36], pronounced plasmonic circular dichroism in the vis–NIR wavelength range – the strongest of its kind – was reported in systems of surface-assembled, PEG-coated gold NP parallel (double) lines stacked on top of each other with a certain tilt angle (Figure 9.10). The PEG ligands, not containing any side chains, serve as an achiral AuNP surface coating to exclude contributions from molecular CD coupled to the surface plasmons [138]. The 2-dimensional templates required for the pursued approach could be obtained from (electron-beam free) laser-interference lithography. Adjusting the stacking angle provides a handle to tune the extent of the produced CD (up to 11°), whereas the spectral position of the CD band could be altered by mechanical compression [36]. The concept of generating circular dichroism from stacked structures can also be combined with photoluminescent gratings to reasonably extend chiroptical sensors [139].

Stacking of isolated NPs can also be pursued with the aid of a polymeric linker connecting the different layers. Adding several layers of distinct NPs (gold/silver plasmonic NPs and iron-oxide NPs) by covalent linking through a conjugated polymer establishes systems with potential nonreciprocal properties (such as transmission or Faraday rotation) [98].

9.4 Summary and Outlook Discussions

Hairy plasmonic NPs can enable or significantly enhance a range of functional material properties. Because their excitation by light of visible or NIR wavelengths is highly efficient, plasmonic NPs can be considered as nanoantennas that confine light and thus its energy at the particle surface only (evanescent field). This confinement constricts the range in which energy-transfer mechanisms can be effective, and, therefore, controlling the local chemistries in the plasmonic NP environment is of utmost importance for harnessing such energy transfer in functional material systems. The toolbox of nanomaterials science offers a range of distinct types of plasmonic NP coatings, each with its inherent benefits and limitations. Thus, concerning the role of a hairy coating, in particular, one may ask: what are the advantages of polymer ligands over other, alternative surface coatings? An obvious benefit inherent to polymer ligands is that their molecular dimensions (few nm to few tens of nm) naturally match the length scales at which the distance-dependent energy-transfer mechanisms are operative. By tailoring the polymeric microstructure, it is, therefore, possible to exert control over the secondary processes that can be initiated upon plasmonic NP excitation (see discussion in Section 9.2.2). Another important advantage of polymers is their high number of conformational degrees of freedom and the resulting strong dependency of their conformation on external parameters. This responsiveness can be exploited to manipulate the secondary processes mentioned above in a time-resolved manner. These characteristic polymer features provide the basis for the recent successful developments of individual hairy plasmonic NPs, for example, in sensing and catalysis. Similar reasons also constitute the basis for controlling hairy plasmonic supercolloidal assemblies. As discussed, the assembly structural properties (spacings between the constituent NPs, coordination number, and regioselectivity of inter-NP linking) as well as their dynamic properties can be adjusted using hairy ligands, which enables full control over LSPR coupling interactions. For those reasons, polymers can take an enabling role in the integration of plasmonic NPs into functional materials and devices. This brings us to the final paragraph of this chapter, in which we will discuss hairy plasmonic systems, which are not yet implemented but highly sought after as they may lead to new material/system developments.

Considering the coupled systems, especially the placing of the emitters into the optical cavity (hot spot) changes the coupling behavior so that the light can either be amplified or mixed states are created that are part light, part matter [140]. The mixed state case, also called strong coupling, is suitable to realize one-atom lasers [82], ultralow-current switches [141], or quantum information systems [80]. The state of strong coupling is achieved with coupled plasmonic systems, where the volume between the NPs must be as small as possible [140]. In principle, polymer linkers can be functionalized with the respective emitter functional unit with relative preparative ease, but because of their intrinsic dimensions, very small particle spacings usually do not result from hairy particle linkers. However, we have seen in Section 9.3.1 that polyelectrolyte linkers can create spacings smaller than 2 nm, because upon adsorbing onto the NP surface, they may adopt conformations in which they do not extend away from the surface, but rather the polymeric chains are wrapped around the NP surface. Hence, the particle separations created in such a manner are comparable to the dimension of one monomeric repeat unit, not the dimension of the overall macromolecule. Hence, incorporation of emitter entities into polyelectrolytes, for example via copolymerization, may establish the desired case of strong coupling with the aid of a polymeric linker. Weak coupling or lasing (spaser) is another possibility, where the photoluminescent material is brought into close contact with the NP. Herein, the plasmonic NP is used as a resonator to generate a population inversion by using the induced emission [95].

Another, more conceptual perspective on hairy NPs is connected to conformational switchability of functional supercolloidal structures. In analogy to proteins, where environmentally triggered conformational changes are essential for functionality, hairy polymer layers/linkers in supercolloidal structures could facilitate pronounced conformational changes on the supercolloidal scale that result in switching of functionalities mentioned above. Finally, colloidal enzymes or similar biomimetic systems could thus come into reach.

Acknowledgments

C.R. acknowledges receipt of a Liebig fellowship (Fonds der Chemischen Industrie). This project was financially supported by the Volkswagen Foundation through a Freigeist Fellowship to TK.

References

1 Garcia, M.A. (2011). *J. Phys. D: Appl. Phys.* 44: 283001.
2 Rodrigues, T.S., Zhao, M., Yang, T.H. et al. (2018). *Chem. A Eur. J.* 24: 16944–16963.
3 Lowe, A.B., Sumerlin, B.S., Donovan, M.S., and McCormick, C.L. (2002). *J. Am. Chem. Soc.* 124: 11562–11563.
4 Chen, Y., Yoon, Y.J., Pang, X. et al. (2016). *Small* 12: 6714–6723.
5 Lu, S., Yu, H., Gottheim, S. et al. (2018). *J. Am. Chem. Soc.* 140: 15412–15418.
6 Dong, H., Zhu, M., Jeong, A.Y. et al. (2008). *J. Am. Chem. Soc.* 130: 12852–12853.
7 Ebeling, B. and Vana, P. (2013). *Macromolecules* 46: 4862–4871.
8 Glaria, A., Beija, M., Bordes, R. et al. (2013). *Chem. Mater.* 25: 1868–1876.
9 Sun, N., Zhang, S.T., Simon, F. et al. (2021). *Angew. Chem. Int. Ed.* 60: 3912–3917.
10 Cepraga, C., Favier, A., Lerouge, F. et al. (2016). *Polym. Chem.* 7: 6812–6825.
11 Gibson, M.I., Danial, M., and Klok, H.-A. (2011). *ACS Comb. Sci.* 13: 286–297.
12 Boyer, C., Whittaker, M.R., Chuah, K. et al. (2010). *Langmuir* 26: 2721–2730.
13 Reiser, B., González-García, L., Kanelidis, I. et al. (2016). *Chem. Sci.* 7: 4190–4196.
14 Stuart, M.A.C., Huck, W.T.S., Genzer, J. et al. (2010). *Nat. Mater.* 9: 101–113.
15 Boyer, C., Whittaker, M.R., Luzon, M., and Davis, T.P. (2009). *Macromolecules* 42: 6917–6926.
16 Xia, H., Su, G., and Wang, D. (2013). *Angew. Chem. Int. Ed.* 52: 3726–3730.

17 Köhn Serrano, M.S., König, T.A.F., Haataja, J.S. et al. (2016). *Macromol. Rapid. Commun.* 37: 215–220.

18 Tao, H., Galati, E., and Kumacheva, E. (2018). *Macromolecules* 51: 6021–6027.

19 Huebner, D., Rossner, C., and Vana, P. (2016). *Polymer* 107: 503–508.

20 Talianov, P., Fatkhutdinova, L.I., Timin, A.S., and Milichko, V.A. (2021). *Laser Photonics Rev.* 2000421: 1–28.

21 Greybush, N.J., Charipar, K., Geldmeier, J.A. et al. (2019). *ACS. Nano.* 13: 3875–3883.

22 Martinsson, E., Shahjamali, M.M., Large, N. et al. (2016). *Small* 12: 330–342.

23 Lee, J., Bae, C., Ou, Z. et al. (2021). *Nanoscale. Adv.* 3: 1927–1933.

24 Zakia, M. and Yoo, S.I. (2020). *Soft. Matter.* 16: 10252–10259.

25 Baffou, G., Quidant, R., and García De Abajo, F.J. (2010). *ACS. Nano.* 4: 709–716.

26 Fava, D., Winnik, M.A., and Kumacheva, E. (2009). *Chem. Commun.* 2571–2573.

27 Rohleder, D. and Vana, P. (2021). *Biomacromolecules.* 22: 1614–1624.

28 Kulakovich, O., Strekal, N., Yaroshevich, A. et al. (2002). *Nano Lett.* 2: 1449–1452.

29 Anger, P., Bharadwaj, P., and Novotny, L. (2006). *Phys. Rev. Lett.* 96: 3–6.

30 Li, K., Hogan, N.J., Kale, M.J. et al. (2017). *Nano Lett.* 17: 3710–2717.

31 Torimoto, T., Horibe, H., Kameyama, T. et al. (2011). *J. Phys. Chem. Lett.* 2: 2057–2062.

32 Blanco-Formoso, M., Pazos-Perez, N., and Alvarez-Puebla, R.A. (2020). *Nanoscale* 12: 14948–14956.

33 Höller, R.P.M., Jahn, I.J., Cialla-May, D. et al. (2020). *ACS. Appl. Mater. Interfaces* 12: 57302–57313.

34 Höller, R.P.M., Kuttner, C., Mayer, M. et al. (2020). *ACS Photonics* 7: 1839–1848.

35 Xu, L., Kuang, H., Xu, C. et al. (2012). *J. Am. Chem. Soc.* 134: 1699–1709.

36 Probst, P.T., Mayer, M., Gupta, V. et al. (2021). *Nat. Mater.* 20: 1024–1028.

37 Lu, J., Chang, Y.X., Zhang, N.N. et al. (2017). *ACS. Nano.* 11: 3463–3475.

38 Rossner, C. and Vana, P. (2014). *Angew. Chem. Int. Ed.* 53: 12639–12642.

39 Nie, Z., Fava, D., Kumacheva, E. et al. (2007). *Nat. Mater.* 6: 609–614.

40 Vazirieh, L.S., Mayer, M., Wang, R. et al. (2022). *J. Phys. Chem. C* 126: 14017–14025.

41 Mayer, M., Potapov, P.L., Pohl, D. et al. (2019). *Nano Lett.* 19: 3854–3862.

42 Gupta, V., Probst, P.T., Goßler, F.R. et al. (2019). *ACS. Appl. Mater. Interfaces* 11: 28189–28196.

43 Rossner, C. and Fery, A. (2020). *MRS. Commun.* 10: 112–122.

44 Turkevich, J., Stevenson, P.C., and Hillier, J. (1951). *Discuss. Faraday. Soc.* 11: 55–75.

45 Xia, H., Xiahou, Y., Zhang, P. et al. (2016). *Langmuir* 32: 5870–5880.

46 Rossner, C., König, T.A.F., and Fery, A. (2021). *Adv. Opt. Mater.* 9: 2001869.

47 Müller, M.B., Kuttner, C., König, T.A.F. et al. (2014). *ACS. Nano* 8: 9410–9421.

48 Mulvaney, P. (1996). *Langmuir* 12: 788–800.

49 Jain, P.K., Lee, K.S., El-Sayed, I.H., and El-Sayed, M.A. (2006). *J. Phys. Chem. B* 110: 7238–7248.

50 Mie, G. (1908). *Ann. Phys.* 25: 377–445.

51 Zsigmondy, R. (1912). *Chemische Technologie in Einzeldarstellungen*, 22–38. Berlin, Heidelberg: Springer Verlag.

52 Aftenieva, O., Schletz, D., Meyer, A. et al. (2021). *J. Chem. Educ.* 98: 2566–2573.

53 Mayer, K.M. and Hafner, J.H. (2011). *Chem. Rev.* 111: 3828–3857.

54 Kruss, S., Erpenbeck, L., Amschler, K. et al. (2013). *ACS. Nano.* 7: 9984–9996.

55 Amschler, K., Erpenbeck, L., Kruss, S., and Schön, M.P. (2014). *ACS. Nano.* 8: 9113–9125.

56 Karg, M., König, T.A.F., Retsch, M. et al. (2015). *Mater. Today* 18: 185–205.

57 Mueller, M., Tebbe, M., Andreeva, D.V. et al. (2012). *Langmuir* 28: 9168–9173.

58 König, T., Kodiyath, R., Combs, Z.A. et al. (2014). *Part. Part. Syst. Charact.* 31: 274–283.

59 Prasad, J., Zins, I., Branscheid, R. et al. (2015). *J. Phys. Chem. C* 119: 5577–5582.

60 Li, K., Wang, K., Qin, W. et al. (2015). *J. Am. Chem. Soc.* 137: 4292–4295.

61 Mayer, M., Steiner, A.M., Röder, F. et al. (2017). *Angew. Chem. Int. Ed.* 56: 15866–15870.

62 Zhang, S., Bao, K., Halas, N.J. et al. (2011). *Nano Lett.* 11: 1657–1663.

63 Choueiri, R.M., Galati, E., Thérien-Aubin, H. et al. (2016). *Nature* 538: 79–83.

64 Galati, E., Tebbe, M., Querejeta-Fernández, A. et al. (2017). *ACS. Nano.* 11: 4995–5002.

65 Kim, J., Song, X., Kim, A. et al. (2018). *Macromol. Rapid Commun.* 39: 1800101.

66 Gao, B., Arya, G., and Tao, A.R. (2012). *Nat. Nanotechnol.* 7: 433–437.

67 Yang, Y., Lee, Y.H., Lay, C.L., and Ling, X.Y. (2017). *Chem. Mater.* 29: 6137–6144.

68 Elsner, C., Prager, A., Sobottka, A. et al. (2017). *Anal. Methods* 9: 4663–4672.

69 Pazos-Perez, N., Fitzgerald, J.M., Giannini, V. et al. (2019). *Nanoscale. Adv.* 1: 122–131.

70 Dey, P., Tabish, T.A., Mosca, S. et al. (2020). *Small* 16: 1906780.

71 Mayer, M., Scarabelli, L., March, K. et al. (2015). *Nano Lett.* 15: 5427–5437.

72 Willingham, B. and Link, S. (2011). *Opt. Express* 19: 6450.

73 Gür, F.N., McPolin, C.P.T., Raza, S. et al. (2018). *Nano Lett.* 18: 7323–7329.

74 (a) Pletsch, H., Tebbe, M., Dulle, M. et al. (2015). *Polymer* 66: 167–172. (b) Bokern, S., Getze, J., Agarwal, S., and Greiner, A. (2011). *Polymer* 52: 912–920.

75 Rizvi, M.H., Wang, R., Schubert, J. et al. (2022). *Adv. Mater.* 34: 2203366.

76 Bokern, S., Getze, J., Agarwal, S., and Greiner, A. (2011). *Polymer* 52: 912–920.

77 Wagner, J., Peng, W., and Vana, P. (2018). *Polymers* 10: 407.

78 Haberfehlner, G., Thaler, P., Knez, D. et al. (2015). *Nat. Commun.* 6: 8779.

79 Rodríguez-Lorenzo, L., De La Rica, R., Álvarez-Puebla, R.A. et al. (2012). *Nat. Mater.* 11: 604–607.

80 Akimov, V.A., Mukherjee, A., Yu, C.L. et al. (2007). *Nature* 450: 402–406.

81 Stern, L., Desiatov, B., Mazurski, N., and Levy, U. (2017). *Nat. Commun.* 8: 14461.

82 McKeever, J., Boca, A., Boozer, A.D. et al. (2003). *Nature* 425: 268–271.

83 Murphy, S., Jaber, S., Ritchie, C. et al. (2016). *Langmuir* 32: 12497–12503.

84 Ebeling, B., Eggers, S., Hendrich, M. et al. (2014). *Macromolecules* 47: 1462–1469.

85 Penelas, M.J., Contreras, C.B., Angelomé, P.C. et al. (2020). *Langmuir* 36: 1965–1974.

86 Rossner, C., Letofsky-Papst, I., Fery, A. et al. (2018). *Langmuir* 34: 8622–8628.

87 Zhang, K., Zhu, X., Jia, F. et al. (2013). *J. Am. Chem. Soc.* 135: 14102–14105.

88 Won, S., Richards, S.J., Walker, M., and Gibson, M.I. (2017). *Nanoscale. Horizons.* 2: 106–109.

89 Lee, J., Govorov, A.O., and Kotov, N.A. (2005). *Angew. Chem. Int. Ed.* 44: 7439–7442.

90 Ingram, D.B., Christopher, P., Bauer, J.L., and Linic, S. (2011). *ACS Catal.* 1: 1441–1447.

91 Linic, S., Christopher, P., and Ingram, D.B. (2011). *Nat. Mater.* 10: 911–921.

92 Asapu, R., Claes, N., Bals, S. et al. (2017). *Appl. Catal., B* 200: 31–38.

93 Miri, M.A. and Alù, A. (2019). *Science* 363.

94 Noginov, M.A., Zhu, G., Belgrave, A.M. et al. (2009). *Nature* 460: 1110–1112.

95 Azzam, S.I., Kildishev, V.A., Ma, R.M. et al. (2020). *Light. Sci. Appl.* 9: 90.

96 Primo, A., Corma, A., and García, H. (2011). *Phys. Chem. Chem. Phys.* 13: 886–910.

97 Christopher, P., Xin, H., and Linic, S. (2011). *Nat. Chem.* 3: 467–472.

98 Delabie, J., De Winter, J., Deschaume, O. et al. (2020). *Macromolecules* 53: 11098–11105.

99 König, T.A.F., Ledin, P.A., Kerszulis, J. et al. (2014). *ACS. Nano.* 8: 6182–6192.

100 Pazos-Perez, N., Fitzgerald, J.M., Giannini, V. et al. (2019). *Nanoscale. Adv.* 1: 122–131.

101 Rossner, C., Tang, Q., Glatter, O. et al. (2017). *Langmuir* 33: 2017–2026.

102 Han, F., Vivekchand, S.R.C., Soeriyadi, A.H. et al. (2018). *Nanoscale* 10: 4284–4290.

103 Wu, L., Glebe, U., and Böker, A. (2017). *Adv. Mater. Interfaces* 4: 1–10.

104 Rossner, C., Glatter, O., and Vana, P. (2017). *Macromolecules* 50: 7344–7350.

105 Han, F., Soeriyadi, A.H., and Gooding, J.J. (2018). *Macromol. Rapid Commun.* 39: 1800451.

106 Han, F., Armstrong, T., Andres-Arroyo, A. et al. (2020). *Nanoscale* 12: 1680–1687.

107 Galati, E., Tao, H., Rossner, C. et al. (2020). *ACS. Nano.* 14: 4577–4584.

108 Dey, P., Zhu, S., Thurecht, K.J. et al. (2014). *J. Mater. Chem. B* 2: 2827–2837.

109 Yang, Y., Gu, C., and Li, J. (2019). *Small* 15: 1–26.

110 Bitton, O., Gupta, S.N., and Haran, G. (2019). *Nanophotonics* 8: 559–575.

111 Kuttner, C., Höller, R.P.M., Quintanilla, M. et al. (2019). *Nanoscale* 11: 17655–17663.

112 Chou, L.Y.T., Zagorovsky, K., and Chan, W.C.W. (2014). *Nat. Nanotechnol.* 9: 148–155.

113 Tang, Q., Rossner, C., Vana, P., and Müller, M. (2020). *Biomacromolecules* 21: 5008–5020.

114 Yi, C., Liu, H., Zhang, S. et al. (2020). *Science* 369: 1369–1374.

115 Rossner, C., Tang, Q., Müller, M., and Kothleitner, G. (2018). *Soft Matter.* 14: 4551–4557.

116 Choueiri, R.M., Klinkova, A., Pearce, S. et al. (2018). *Macromol. Rapid Commun.* 39: 1700554.

117 Yu, L., Shi, R., Qian, H.J., and Lu, Z.Y. (2019). *Phys. Chem. Chem. Phys.* 21: 1417–1427.

118 Gong, M., Yu, Q., Wang, C., and Wang, R. (2019). *Langmuir* 35: 5534–5540.

119 Rossner, C., Zhulina, E.B., and Kumacheva, E. (2019). *Angew. Chem. Int. Ed.* 58: 9269–9274.

120 Chen, T., Yang, M., Wang, X. et al. (2008). *J. Am. Chem. Soc.* 130: 11858–11859.

121 Galati, E., Tao, H., Tebbe, M. et al. (2019). *Angew. Chem. Int. Ed.* 58: 3123–3127.

122 Chen, G., Gibson, K.J., Liu, D. et al. (2019). *Nat. Mater.* 18: 169–174.

123 Ha, M., Kim, J.H., You, M. et al. (2019). *Chem. Rev.* 119: 12208–12278.

124 Cheng, G., Xu, D., Lu, Z., and Liu, K. (2019). *ACS. Nano.* 13: 1479–1489.

125 Mayer, M., Schnepf, M.J., König, T.A.F., and Fery, A. (2019). *Adv. Opt. Mater.* 7: 1800564–1800580.

126 Soukoulis, C.M., Linden, S., and Wegener, M. (2007). *Science* 315: 47–49.

127 Kwon, D., Werner, P.L., and Werner, D.H. (2008). *Opt. Express* 16: 9267–9272.

128 Magnozzi, M., Brasse, Y., König, T.A.F. et al. (2020). *ACS. Appl. Nano. Mater.* 3: 1674–1682.

129 Yockell-Lelièvre, H., Desbiens, J., and Ritcey, A.M. (2007). *Langmuir* 23: 2843–2850.

130 Yockell-Lelièvre, H., Gingras, D., Vallée, R., and Ritcey, A.M. (2009). *J. Phys. Chem. C* 113: 21293–21302.

131 Dukes, D., Li, Y., Lewis, S. et al. (2010). *Macromolecules* 43: 1564–1570.

132 Wijmans, C.M. and Zhulina, E.B. (1993). *Macromolecules* 26: 7214–7224.

133 Volk, K., Fitzgerald, J.P.S., Ruckdeschel, P. et al. (2017). *Adv. Opt. Mater.* 5: 1600971.

134 Spatz, J.P., Mössmer, S., Hartmann, C. et al. (2000). *Langmuir* 16: 407–415.

135 Ohno, K., Koh, K., Tsujii, Y., and Fukuda, T. (2003). *Angew. Chem. Int. Ed.* 42: 2751–2754.

136 Brasse, Y., Ng, C., Magnozzi, M. et al. (2020). *Macromol. Rapid Commun.* 41: 1900415.

137 Yang, F., Ye, S., Dong, W. et al. (2021). *Adv. Mater.* 33: 2100325.

138 Zhang, Q., Hernandez, T., Smith, K.W. et al. (2019). *Science* 365: 1475–1478.

139 Aftenieva, O., Schnepf, M., Mehlhorn, B., and König, T.A.F. (2021). *Adv. Opt. Mater.* 9: 2001280.

140 Chikkaraddy, R., De Nijs, B., Benz, F. et al. (2016). *Nature* 535: 127–130.

141 Nozaki, K., Shinya, A., Matsuo, S. et al. (2012). *Nat. Photonics* 6: 248–252.

10

Hairy Metal Nanoparticles for Catalysis: Polymer Ligand-Mediated Catalysis

Zichao Wei and Jie He

Department of Chemistry, and Polymer Program, Institute of Materials Science, University of Connecticut, Storrs, CT 06269, United States

10.1 Nanocatalysis Mediated by Surface Ligands

10.1.1 Surface Ligands as an Important Component for Nanocatalysis

Metal nanoparticles (NPs) are a key component in many heterogeneous catalysts, e.g. Pt supported on carbon (Pt/C), widely used in organic synthesis for hydrogenation. Given their ultrasmall sizes in nanoscale, the intrinsic "nano effects" of metal NPs show significant enhancement in terms of catalytic activity compared to their bulk counterparts. Those unsaturated surface atoms, especially the ones on the edges and corners, are catalytically more active. The reactivity of NPs, also known as their catalytic activity and selectivity, is highly surface-dependent, i.e. surface area and surface energy. Surface energy is a key parameter that controls the binding strength of substrates, while the surface area is critical to determine the density of active sites for catalysts. Although surface area and surface energy closely impact each other, the surface energy of metal NPs is independently tunable through the control of surface facets. Surface properties are key to the catalytic properties of nanocatalysts in terms of their activity and selectivity. An interesting example is body-centered cubic copper (Cu) NPs having their selectivity for CO_2 reduction on different facets. Among those low-index facets, Cu (100) provides the most selectivity for the C–C formation to produce so-called deep reduction products as compared to (111) and (110) [1]. Numerous excellent reviews have covered the impact of that surface energy on the catalytic activity of various metal NPs [2, 3].

On the other hand, surface ligands that provide an alternative way to control the surface properties of metal NPs, have often been overlooked in the past. Taking inspiration from nature, the catalytic performance of metalloenzymes heavily relies on the coordination environment of metal sites in both the first and second coordination spheres. Despite not being directly involved in catalysis, the protein frameworks play an important role in tuning the activity and selectivity of metal centers. The question that has been long debated is whether surface ligands (particularly as polymers described below) of metal NPs promote their catalytic efficiency, in a similar fashion

Hairy Nanoparticles: From Synthesis to Applications, First Edition. Edited by Zhiqun Lin and Yijiang Liu.

as the protein frameworks in metalloenzymes [4–7]. Given the synthetic advances made in the last two decades, controlled synthesis of many metal NPs in terms of their sizes, shapes, and surface facets is available now. However, such synthesis usually relies on the use of capping ligands to maintain the colloidal stability and/or stabilize one or more specific crystalline facets. In many other cases, capping ligands can be replaced by other functionalized molecules (namely surface ligands here) purposely through ligand exchange to vary and modulate the local chemical environment of surface catalytic sites. As surface ligands, functionalized molecules regardless of organic or inorganic species can bind to metal NPs through mostly coordination with N, S, O, and P atoms [8]. The typical examples are alkyl thiols that are well-known to form self-assembly monolayers (SAMs) on a variety of metals [9, 10].

Surface ligands can control the catalytic efficiency of metal NPs in many reactions, although there are contradictory reports in literature [7, 11]. There are many examples with "positive" ligand effects where ligands improve the catalytic efficiency of metal NPs, despite the dynamic nature of those capping ligands during catalysis [12, 13]. At the same time, the "negative" ligand effects have also been reported, in particularly for electrochemical catalysis [14]. Those surface ligands like the protein frameworks of metalloenzymes that are not involved in catalysis directly can gate the catalytic surface potentially to favor the binding of specific substrates. They provide the proximity effect to tune the reaction selectivity and stabilize key intermediate states through hydrogen bonding or hydrophobic interaction. Lessons on ligand-controlled catalysis have also been taken from homogeneous catalysis, where the complexation of metals, mostly transition metals, with various ligands to tune the steric and electronic nature of metal sites.

Similarly, the role of surface ligands on metal NPs is threefold [6]. First, surface ligands can "gate" the NP surface by limiting the accessibility of undesired catalytic sites. Unlike discrete metal sites in homogeneous catalysis, exposed atoms on the surface of metal NPs show "heterogeneity" in terms of their reactivity. The atoms on the edges and corners with a lower coordination number are usually more active than those on the flat facets. The binding models of substrates on the edge and corner atoms are sometimes different from their binding on the flat facets. While most surface ligands form SAMs on the flat facets of metal NPs [15], they can create more barriers for substrates to bind directly on the flat facets. As compared to ligands on the flat facets, they are less ordered and defined on the edge and corner sites, which allows the diffusion and binding of substrates faster. This potentially shuts down some unwanted reaction pathways; and, consequently, the reaction selectivity becomes controllable in the presence of surface ligands [16, 17].

Second, the packing of ligands will crowd the surface of metal NPs [18]. The surface crowding can provide the proximity effect, e.g. to align the substrate with binding in a specific configuration. The formation of SAMs creates a non-reactive layer to pre-organize or pre-orient substrates. For substrates with multiple reactive bonds, molecular orientation plays a key role in controlling the reactivity of bonds. For example, when modifying Pd NPs supported on Al_2O_3 by 1-octadecanethiol (C18) and 1-adamantanethiol (AT), the two ligands pack on the surface of Pd very differently. With linear alkyl tails, C18 can crystallize and pack denser as compared to the bulky AT. The average ligand-to-ligand distance of C18 and atom Pd/Al_2O_3 is 4.7 and 6.9 Å, respectively, as measured by ambient scanning tunneling microscopy (STM) [19]. The two catalysts show very different selectivity for furfural hydrogenation. Pd/Al_2O_3 modified with C18 shows high selectivity to reduce the aldehyde group. The selectivity toward furfural alcohol and methylfuran is ~35% and 55%, respectively [19]. On the contrary, Pd/Al_2O_3 modified with AT has a high selectivity to reduce the furan ring, leading to the full hydrogenation to tetrahydrofuran (THF) and furan (selectivity ~95%) [19]. Using CO as a probe in diffuse reflectance infrared Fourier transform spectroscopy (DRIFTS), Pd/Al_2O_3 shows different binding sites for CO with surface ligands, where threefold hollow sites are seen on C18-modified Pd NPs and terrace bridging sites are present on AT-modified Pd NPs. When modifying the surface of metal NPs with alkyl thiols of different chain lengths, the surface crowding effect also varies the hydrogenation selectivity of substrates with multiple double reactive groups, like 2-hexen-1-ol [20] and cinnamic aldehyde [18].

Lastly, surface ligands change the electronic nature of metal sites of NPs in a very close fashion as in homogeneous catalysis [21–23]. Ligands with a similar bulkiness but a different binding motif, have been demonstrated to impact the catalytic reactivity of metal NPs. The nature of binding motifs varying the local electron density or even oxidation state of surface metal atoms potentially varies the catalytic efficiency of metal NPs. One example is the hydrogenation of alkynes using Pd nanocatalysts, where thiolate ligands can form PdS_x species to alter their hydrogenation selectivity, similar to that of the Lindlar catalyst. Using 3,4-difluorothiophenol ($HSPhF_2$) to modify Pd nanosheets, semihydrogenation of 1-phenyl-1-propyne can be achieved with a selectivity of 98.1% to 1-phenyl-1-propene [24]. As compared to other synthetic methods to vary the catalytic efficiency of metal NPs, e.g. control of surface facets or formation of alloys/intermetallic compounds [25], surface ligands can be added as a post-synthesis functionalization of metal NPs. Therefore, engineering surface ligands can provide a powerful tool to tune the catalytic efficiency of pre-synthesized metal NPs without re-synthesizing.

10.1.2 Polymers as Better Ligands for NPs

There are several advantages to using synthetic polymers as surface ligands for nanocatalysts. On one hand, polymer ligands provide much better colloidal stability for metal NPs as compared to small molecular ligands. Polymers usually have their size comparable to that of metal NPs to balance out the interparticle attraction. In the highly solvated state, polymer ligands have strong osmotic repulsion to overcome interparticle van der Waals interaction [26, 27]; therefore, polymer-grafted metal nanoparticles (PGNPs) are extremely stable in good solvents of polymer ligands [28]. Polymer ligands can be designed to have various binding motifs and ligand length/topologies as shown in Scheme 10.1. In the case of PGNPs-coated hydrophilic polymers like polyvinylpyrrolidone (PVP), those PGNPs are water-soluble even in the presence of electrolytes, acids, and bases. Under catalytic conditions (e.g. high temperature or corrosive atmosphere), the stability provided by polymer ligands is far superior as compared with small molecules. On the other hand, polymer ligands

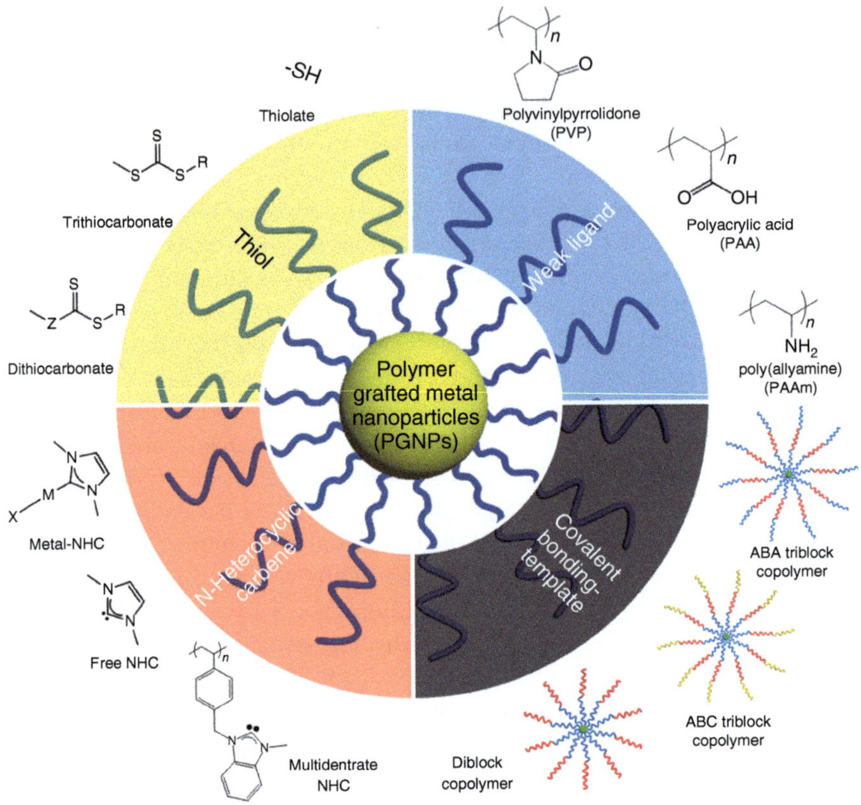

Scheme 10.1 Scheme to illustrate the PGNPs with different types of polymer ligands. (1) Yellow zoom: polymers ended with thiolate, trithiocarbonate, and dithiocarbonate. (2) Orange zoom: polymers ended with N-heterocyclic carbene (NHC). (3) Gray zoom: covalent polymer ligands with different topologies synthesized from di-BCP and tri-BCP (ABA, ABC) templates. (4) Blue zoom: polymers with the weak bonding strength of polyvinylpyrrolidone (PVP), poly(acrylic acid) (PAA), and poly(allyamine) (PAAm).

offer more precise control over surface accessibility to metal NPs as compared with small molecular ligands. The pioneering studies on metal-containing dendrimers demonstrate many excellent examples of how to control the surface crowdedness or hydrophobicity of metal sites (including metal ions and NPs) to improve their catalytic efficiency [29]. When loading metal NPs in dendrimers, the different generations of dendrimers can control the surface accessibility of metal NPs. For example, poly(amidoamine) (PAMAM) dendrimers can serve as nanoreactors to grow small Pd NPs with a diameter of 1.7 nm [30]. Using those Pd NPs as hydrogenation catalysts for allyl alcohols, there is a clear trend in the size selectivity of substrates. When the dendrimer frameworks have the same generation, maximum turnover frequency (TOF) decreases with larger substrates. With Pd loaded in the fourth generation (G4) dendrimer, the TOF is 480 mol H_2 h^{-1} per mol Pd for the simplest allyl alcohol, while the TOF dropped to 100 mol H_2 h^{-1} per mol Pd for 3-methyl-1-penten-3-ol. The increase of the dendrimer generation also increases the surface crowdedness that slows down the diffusion of large substrates significantly; while it has less impact on small substrates.

For PGNPs, control of the chain length, ligand density, and chemical composition of polymer ligands provides not only similar surface crowdedness to metal NPs but also creates a microenvironment around catalytically active sites to tune the accessibility of desired substrates. For example, when coating Cu catalysts with Nafion, a perfluorinated sulfonic acid (PFSA) ionomer containing the hydrophilic functional groups of $-SO_3^-$ and hydrophobic $-CF_2$ groups, there is a significant enhancement for CO_2 electroreduction efficiency with a gas-phase electrochemistry greater than $1\,A\,cm^{-2}$ and a high Faradaic efficiency (FE) of 65–75% to produce ethylene [31]. The coating shows a nearly 400-time enhancement for the diffusion of CO_2 relative to bulk electrolyte through hydrophobic channels built up by ionic Nafion. The chemical components of synthetic polymers are highly tunable; therefore, the microenvironment created through the combination of synthetic polymers and catalytically active metals can be designed on demand for a specific reaction.

Another level of control over surface accessibility of metal NPs is switchability; that is, polymer ligands can turn "ON" or "OFF" the surface accessibility of the catalytic metal NPs [5]. The switchability is usually built on stimuli-responsive polymers that can vary their physicochemical properties in response to the external stimulus, such as temperature, pH, and light. A very well-known example is poly(*N*-isopropylacrylamide) (PNIPAM) with a lower critical solution temperature (LCST). PNIPAM is water-soluble below its LCST, around 32 °C; while it becomes water-insoluble above its LCST. Using PNIPAM as surface ligands of metal NPs, those surface ligands can tune the surface accessibility in response to the change of temperature [32–34]. Above their LCST, the dehydrated polymer ligands collapse on the surface of metal NPs. The packed and dense layer of polymer ligands thus blocks the diffusion of substrates and thus turns "OFF" the activity of metal NPs. When polymer ligands are fully hydrated below their LCST, the layer of polymer ligands is permeable to substrates and thus turns "ON" the activity of metal NPs. The responsiveness of polymer ligands is highly tunable through their chemical composition that allows their switchability of catalysis in a broad range [35].

Grafting polymer ligands to metal NPs also provides additional tools to recycle and reuse expensive metal catalysts. Early studies on the synthesis of polymer-supported metal NPs in catalysis suggest that the use of polymer micro- or nanobead to support/load metal NPs facilitates the separation of metal catalysts from reaction mixtures [36]. Similar separation can be achieved for PGNPs by changing the solvation state of polymer ligands or transferring those PGNPs between different phases. For example, using pH-responsive poly(acrylic acid) (PAA) as surface ligands for AgNPs, the separation of AgNPs can simply be triggered by changing pH. Under acidic conditions, the protonation of PAA would result in the aggregation of AgNPs because of the solubility change for PAA ligands [37]. Simply increasing the pH of the solution will deprotonate PAA to increase the electrostatic repulsion of AgNPs, leading to the redispersion and reuse of AgNPs.

Polymer ligands can be added on metal NPs through "grafting-to" and "grafting-from" methods [38]. The "grafting-from" method is to grow polymer ligands from the surface of NPs. Surface-initiated atom transfer radical polymerization (ATRP) is commonly used where small molecular ATRP initiators are usually attached on metal NPs to allow the polymerization of monomers directly on the surface of metal NPs. This method has the advantage of adding polymer ligands with a high grafting density. The "grafting-to" method is to graft pre-synthesized polymers with end-functionalized groups (e.g. thiols) on metal NPs. It has gained popularity because of the synthetic availability of thiol-terminated polymers from reversible addition-fragmentation chain transfer (RAFT) polymerization [39]. Simple reduction of dithioester or trithioester end groups of polymers prepared through RAFT polymerization can produce thiol-terminated polymers [40]. The "grafting-to" method also allows the surface modification at a very small scale and the use of metal NPs with different shapes, sizes, and chemical compositions, which is not possible in the "grafting-from" method.

Surprisingly, the binding motif of polymer ligands is still of crucial importance for the catalytic performance of metal NPs, although it only represents a tiny fraction of the end group of polymers with tens to hundreds of repeating units. It is usually a result of electronic interaction between surface metal atoms and the binding motif, similar to that of small molecular ligands. A typical ligand comparison is thiol versus N-heterocyclic carbene (NHC). The NHC is a σ donor, where the lone pair on the NHC carbon can provide two electrons to coordinate transition metals. When bound with metal NPs, NHCs can enrich the electron density on the surface of metal NPs [22, 41]. Using an electrophile like CO_2 as a substrate, PGNPs with NHCs as the binding motif, therefore, show a higher reactivity as compared to other binding motifs [21].

This chapter will highlight the importance of polymeric ligands in metal NP-mediated chemical transformations. Using a few very recent examples, we focus on polymer ligand-promoted catalysis of metal NPs, i.e. the positive effect of surface ligands. Depending on the chemical identity of ligands that varies the surface properties of metal NPs in terms of surface accessibility and electronic states, we will discuss how the surface effects provided by different polymer ligands enhance the selectivity, activity, and durability for catalytic applications of metal NPs. Each section starts with the interaction between ligands and metal NPs to identify the different types of surface ligands. The polymer ligand-promoted catalysis is discussed in two main topics, including the impact of surface ligands on the catalytic activity and the cause of ligand-controlled reactivity of metal NPs.

10.2 Catalysis Mediated by PGNPs with Thiol-Terminated Polymers

The first concern we would like to address is whether polymer ligands will block the catalytic sites on metal NPs. Thiol is the most commonly used binding motif for PGNPs. It binds with a variety of metals as SAMs through metal–thiolate coordination. The binding energy of Au-thiolate as an example is around 126 kJ mol^{-1} [42, 43], comparable with the bond energy of the hydrogen bonding or even the

Au–Au interaction. Polymers terminated with thiols are strong ligands that can replace nearly any capping ligands of metal NPs. Polymer ligands can form a passive layer when bound with metal NPs and they stay at the interface of catalytic surface and substrates. With strong ligands like thiol-terminated polymers, will polymers still allow the diffusion, binding, and further catalysis of substrates? The early example from Huang and coworkers suggested that AuNPs modified by thiol-terminated polymers were catalytically active but the catalytic activity of AuNPs was impacted by polymer chain lengths [44]. Thiol-terminated block copolymers (BCPs) of poly(ethylene glycol)-*block*-polystyrene (PEG-*b*-PS) were synthesized using RAFT polymerization. Thiol anchoring groups were designed at the end of hydrophobic PS blocks having variable lengths, e.g. the average number of repeating units of 12–65. With a one-phase ligand exchange in the mixture of dimethylformamide (DMF)/THF/water, PEG-*b*-PS-SH was grafted to citrate-capped AuNPs (~12.9 nm in diameter) (Figure 10.1a). The PEG_{45}-*b*-PS_{12}-SH had a large grafting density (σ) of 4.2 chains nm^{-2}. The increase in the PS length resulted in a decrease in the ligand density. The σ of PEG_{45}-*b*-PS_{65}-SH was 1.8 chains nm^{-2}. All those densities were very high and thus they offered high colloidal stability for AuNPs. Those PGNPs with BCP ligands were stable in water against 2.0 M NaCl. From UV–vis spectroscopy, both PGNPs modified by PEG_{45}-SH and citrate-capped AuNPs showed a large redshift in a 2.0 M NaCl solution; while PGNPs with BCP ligands kept their LSPR at 525 nm without apparent changes in the presence of NaCl (Figure 10.1b).

The reduction of 4-nitrophenol (4-NP) was chosen as a model reaction to examine the catalytic performance of PGNPs with different chain lengths of polymer ligands. All PGNPs modified by PEG_{45}-*b*-PS_{12}-SH, PEG_{45}-*b*-PS_{30}-SH, and PEG_{45}-*b*-PS_{65}-SH were active to reduce 4-NP to 4-aminepheol (4-AP) in the presence of $NaBH_4$ (Figure 10.1c). The reactivity of those Au cores was dependent on the molecular weight of ligands. The first-order reaction rate constant (*k*) of PGNPs with PEG_{45}-*b*-PS_{12}-SH is $2.2\times10^{-3}\,s^{-1}$; while, PGNPs with PEG_{45}-*b*-PS_{65}-SH showed a *k* of $1.5\times10^{-3}\,s^{-1}$. As a control, citrate-AuNPs had the fastest reaction rate with a *k* of $4.6\times10^{-3}\,s^{-1}$. Those results suggested that polymer ligands became a diffusion barrier for the substrates and BH_4^- ions that formed a metal hydride on the surface of AuNPs. Hydrophobic PS with a longer chain length slowed down the diffusion more, possibly because of the high polarity of BH_4^- and hydride ions. Polymers PGNPs modified with those BCP ligands were recycled, and no loss in catalytic activity was observed for five cycles. Therefore, polymers as surface ligands provide high stability and recyclability to metal NPs.

The accessibility of metal NPs shows a strong dependence on the size of substrates and the molecular weight of polymer ligands. With grafted polymer brushes on metal NPs, the density of polymer ligands is still much lower than that of small molecular ligands, even at a concentrated polymer brush regime [28]. A larger interchain spacing or lower chain density of polymer ligands is highly important to avoid chain overlapping. This becomes more important for polymer ligands with high molecular weights, where a larger interchain spacing is expected. Those interchain spacings provide the diffusion channels for substrates. Therefore,

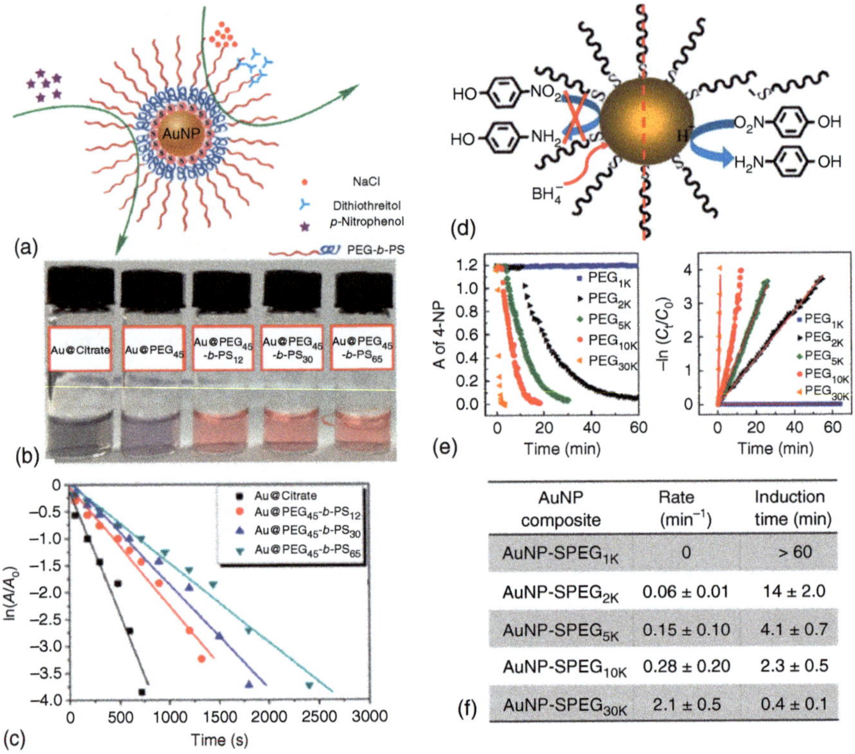

AuNP composite	Rate (min⁻¹)	Induction time (min)
AuNP-SPEG$_{1K}$	0	> 60
AuNP-SPEG$_{2K}$	0.06 ± 0.01	14 ± 2.0
AuNP-SPEG$_{5K}$	0.15 ± 0.10	4.1 ± 0.7
AuNP-SPEG$_{10K}$	0.28 ± 0.20	2.3 ± 0.5
AuNP-SPEG$_{30K}$	2.1 ± 0.5	0.4 ± 0.1

Figure 10.1 (a, d) Schemes to show the catalytic conversion of 4-NP to 4-AP catalyzed by AuNPs grafted by PEG-SH (d) and PEG-*b*-PS-SH (a), receptivity. (b) Images to confirm the dispersity of PGNPs coated by various PEG-*b*-PS-SH in an aqueous solution containing NaCl (2.0 M). (c, e) Kinetic plots of PGNPs modified by PEG-SH (e) and PEG-*b*-PS-SH (c) with different chain lengths. (f) Table summarizing the catalytic performance of citrate-capped AuNPs and AuNPs grafted by PEG-SH with different chain lengths. Reaction conditions: 0.05 M of 4-NP, 15 mM of $NaBH_4$, and 5.6 μM AuNPs at room temperature. The reduction kinetics were monitored by UV–vis spectroscopy using the absorption peak of 4-NP at 400 nm. Source: References [44, 45], Reproduced with permission from American Chemical Society.

polymer ligands will form a diffusion barrier likely for substrates, but they will not block the diffusion of small substrates even if fully collapsed on the surface of NPs. Polymer ligands with a lower molecular weight pack are denser simply because of less steric hindrance, which therefore could reduce the catalytic activity of metal NPs. Using AuNPs (13.7 nm in diameter) modified with PEG-SH as an example, the grafting density decreased with the increase of the molecular weight of PEG-SH, as reported by Kitchens and Ansar [45]. With PEG-SH of 1 kg mol^{-1} (or PEG$_{1K}$-SH), the grafting density was 2.61 chains nm^{-2}; while it decreased to 0.47 chains nm^{-2} for PEG$_{30K}$-SH. The high density of polymer ligands essentially took over the catalytic sites that became unavailable for adsorption and further catalysis. Using 2-mercaptobenzimidazole (MBI) as a molecular probe, PGNPs grafted with PEG$_{1K}$-SH had no adsorption for MBI. With increasing the chain length of PEG-SH, PGNPs showed much-improved adsorption of MBI. The adsorption amount of MBI was an indication of the "free" catalytic sites that were available for catalysis. The adsorption kinetic was independent of the molecular weight of ligands. That is, polymer ligands (PEG-SH) ultimately overcrowded the surface of AuNPs, but they had very little impact on the diffusion of MBI.

PGNPs with different chain lengths of PEG-SH showed dramatically different reactivity for the reduction of 4-NP, as shown in Figure 10.1d–f. First of all, PGNPs grafted with PEG_{1K}-SH having the highest grafting density were inactive. The catalytic activity of PGNPs increased with the molecular weight of PEG-SH. The k catalyzed by PGNPs with PEG_{2K}-SH was 0.06 min^{-1}; while it increased to 2.1 min^{-1} for PGNPs with PEG_{30K}-SH. It was interesting that the activity of PEG_{30K}-SH was close to that of citrate-capped AuNPs. Those results revealed that PGNPs had a molecular weight-dependent reactivity. Longer polymer ligands with a low grafting density would retain the accessibility of catalytic sites. Meanwhile, the induction time, defined as the initial slow stage of the reduction, was also highly dependent on the molecular weight and the ligand density of PEG-SH. Using PEG_{1K}-SH as an example, the grafting density of PEG_{1K}-SH could be controlled by the concentration of PEG_{1K}-SH. The reactivity of those PGNPs with identical polymer ligands but different grafting density showed a linear correlation with the ligand density. As for k, it was 0 min^{-1} with 150 μM of PEG_{1K}-SH; while it reached 1.5 min^{-1} with 2.5 μM of PEG_{1K}-SH. Therefore, PEG-SH likely took over some active sites on AuNPs, which decreased the free active sites at a high concentration of PEG_{1K}-SH. In addition, the induction time was attributed to the desorption and migration of PEG-SH. Using UV–vis spectroscopy, the catalytic reduction of 4-NP was monitored along with the LSPR peak of AuNPs. With citrate-capped AuNPs, the initial LSPR peak at 520 nm showed a fast blueshift to 513 nm along with the addition of $NaBH_4$. The peak shift was due to the hydride formation, where AuNPs became electron-rich. Under similar conditions, PGNPs with PEG_{5K}-SH had a two-stage LSRP shift. The first LSRP shift from 522 to 519 nm was likely assigned to the activation process, where the desorption of PEG_{5K}-SH occurred. The hydride ions that could reduce thiolate and bind with AuNPs as hydrides induced the surface mobility of PEG-SH to form more active sites for reduction.

A recent study from Xu and coworkers also suggests the dynamic nature of polymeric thiolate binding under mechanic perturbation, e.g. sonication, potentially providing another pathway to activate PGNPs (Figure 10.2a) [46]. PGNPs were prepared through co-reduction of disulfide-containing poly(methyl acrylate) (PMA), $HAuCl_4$, and $NaBH_4$. The Au core had an average diameter of 3 nm, while the molecular weight of PMA varied from 30 to 130 kg mol^{-1}. Under sonication, polymer ligands could dissociate from PGNPs with a critical PMA molecular weight of >60 kg mol^{-1}. By comparing the σ of PGNPs, the PMA ligands of 100 kg mol^{-1} continuously decreased along with sonication, while no change in the σ was observed for PMA of 30 and 60 kg mol^{-1}. High-resolution X-ray photoelectron spectroscopy (XPS) reveals the formation of new Au(I) species for PGNPs with PMA of 100 kg mol^{-1}. After separation of mechanically ruptured PMA chains, the Au(I) bound with PMA-thiolate species were found in solution. Those results

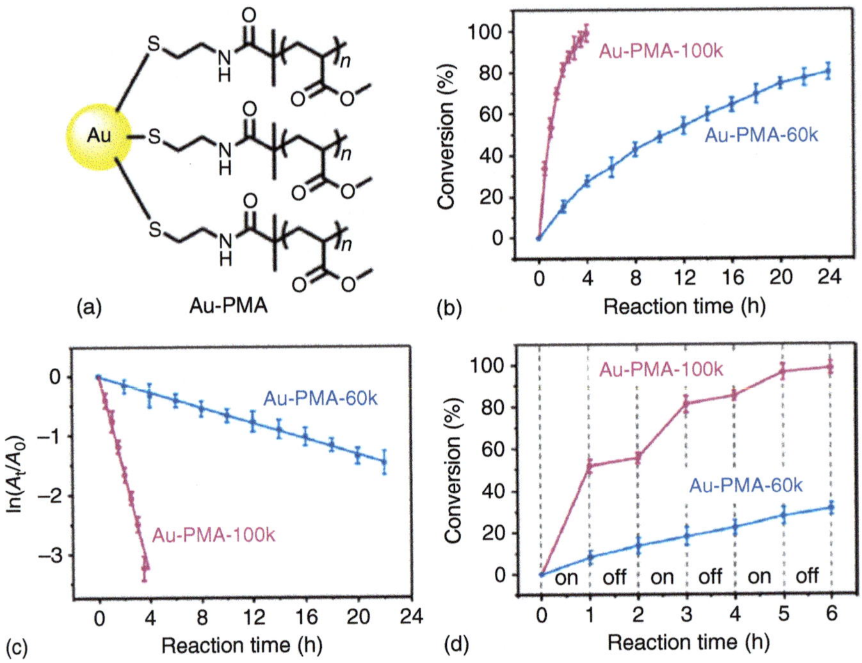

Figure 10.2 (a) Chemical structures of AuNPs grafted by PMA. (b) Plotting the reaction conversion, (c) kinetics plot, and (d) "ON-OFF" mechanical responsive kinetics catalyzed by AuNPs modified by PMA-60k and PMA-100k. Reaction conditions: 34.8 mg of 4-NP and 1.9 g of $NaBH_4$ were dissolved in 25 ml of CH_3CN/H_2O (20 : 1, vol) with different concentrations of AuNPs (10^{-4} M) under Ar and 0 °C. The kinetics was monitored by UV–vis spectroscopy using the absorption peak of 4-NP at 400 nm. Source: Reproduced with permission from Wu et al. [46]. Copyright 2020 American Chemistry Society.

suggested that the disruption of Au-thiolate bond would lead to the detachment of polymer-thiolate-Au species from AuNPs and the further formation of new interfaces between the Au core and solution. As a result, the dynamic polymer-NP interfaces would mechanically activate the reactivity of PGNPs.

The catalytic activity of those PGNPs showed an interesting dependence on the molecular weight of polymer ligands. Using 4-NP as a model reaction, PGNPs with PMA-100k showed a higher activity where the full conversion of 4-NP to 4-AP was in 4 hours and PMA-60k PGNPs only achieved 80% conversion for 24 hours under sonication. Since both PMA ligands had a high molecular weight, the surface accessibility of PGNPs was limited. While PGNPs with PMA-100k were mechanically responsive, the detachment of polymer ligands activated the Au cores and allowed the surface accessibility to substrates. Under sonication, the reactivity of PGNPs with PMA-100k was also switchable; i.e. higher activity was received under sonication and much less activity without sonication. As a control, the activity of PGNPs with PMA-60k was insensitive to sonication (Figures 10.2b–d). Similar to the chemical reduction, sonication as a strong mechanical force triggered the chain reorganization and detachment to activate metal NPs.

Polymers as surface ligands also control the dispersity of PGNPs. With stimuli-responsive polymer ligands, the changes in the physiochemical properties of polymer ligands also allow to vary the dispersity of PGNPs. Switching the hydrophobicity of polymer ligands, as an example, can tune the water solubility of PGNPs. In hydrated states, polymer ligands can stabilize metal NPs in water; while, in dehydrated states, polymer ligands precipitate out with metal NPs to recycle metal catalysts. AuNPs (~5 nm) grafted with a pH-responsive polymer, e.g. thiol-ended PDEAEMA, could vary their solubility in water. PDEAEMA with tertiary amines is only water-soluble under acid conditions when tertiary amines become protonated. When bubbled with CO_2, the produced carbonic acid as a weak acid could shift the solubility of PDEAEMA and thus AuNPs grafted by PDEAEMA [47]. With the gas flow cycles between N_2/CO_2, the reversible protonation and deprotonation of PDEAEMA switched the solubility of PGNPs (Figure 10.3). The transmittance of the aqueous solution containing AuNPs modified by PDEAEMA could switch from 100% to 20%. Those PGNPs were active to reduce 4-NP with the k of $1.9\times10^{-2}\,s^{-1}$ under room temperature, similar to that of ligand-free AuNPs prepared through pulsed laser ablation liquid (PLAL-AuNPs). PGNPs grafted by PDEAEMA could be separated from the reaction mixture by bubbling N_2, which triggered the deprotonation of PDEAEMA to recycle the Au catalyst. After four cycles, the activity of

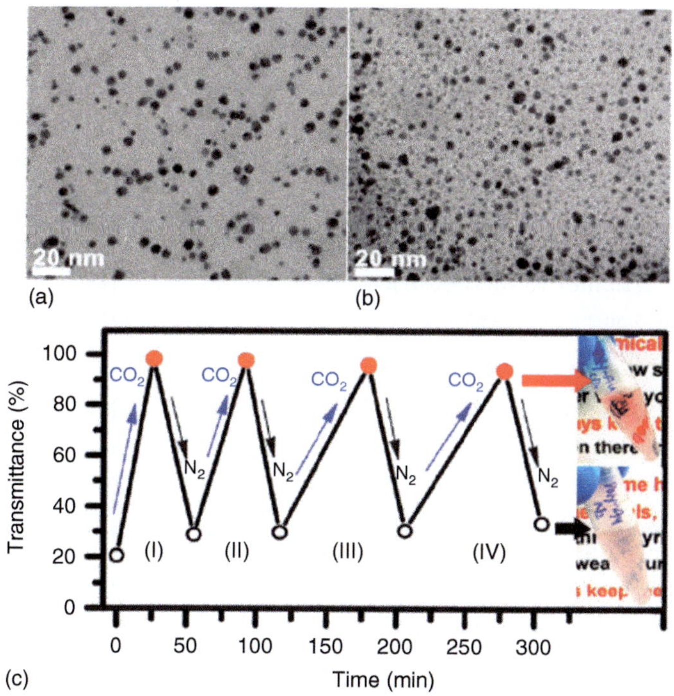

Figure 10.3 TEM images to show AuNPs (a) before and (b) after surface modification by PDEAEMA-SH. (c) Transmittance and images of the aqueous solution of AuNPs modified PDEAEMA-SH under cyclic bubbling of CO_2–N_2. Source: Chen et al. [84], Reproduced with permission from Royal Society of Chemistry.

AuNPs decreased by about 35%; while PLAL-AuNPs lost their activity completely. This principle is applicable for other responsive polymer ligands, e.g. PAA [48] and PNIPAM [49] to switch the solubility of PGNPs and reuse the expensive noble metal catalysts.

Other metal NPs can also be modified by thiol-terminated polymer ligands to improve the catalytic efficiency and increase the stability. For example, Pt NPs (1.8 nm in diameter) grafted by PNIPAM-SH were able to catalyze the asymmetric hydrogenation of ethyl pyruvate to R-(+)-ethyl lactate in the presence of cinchonidine (chiral ligands) [50]. With PNIPAM-SH of 16 kg mol^{-1}, Pt NPs show a 100% conversion and a high enantioselectivity (ee 76%) in 0.5 hours. A high TOF of 17 820 h^{-1} was received for Pt NPs grafted by PNIPAM-SH that outperformed other polymer ligands, like PVP and dendritic PEG. The molecular weight of PNIPAM-SH had a minimum impact on the activity of Pt NPs, but it varied the enantioselectivity. With the increase of the molecular weight of PNIPAM-SH from 10.2 to 28.6 kg mol^{-1}, the ee decreased from 77% to 70%. This was attributed to the higher steric hindrance of PNIPAM-SH with a longer chain length that inhibited the interaction between cinchonine and Pt NPs. As the mixture of acetic acid and water showed a cononsolvency for PNIPAM-SH, those Pt NPs grafted by PNIPAM-SH could be separated from the reaction mixture by adding water. Those PGNPs were recycled 11 times without losing their activity and enantioselectivity.

There are many other PGNPs with various metal NP cores that can catalyze similar reactions, like the reduction of 4-NP. We summarized their catalytic performance in Table 10.1. Although thiol is the most versatile ligand for surface

Table 10.1 Summary of 4-NP reduction catalyzed by other PGNPs.

Polymer ligands	NPs and size (diameter)	*k* (at 20 °C)	TOF (at 20 °C)	References
PAA-SH	Au (13.2 nm)	$0.32 \pm 0.07\ min^{-1}$	—	[48]
Dendronized triazolyl-containing ferrocenyl Polymers	Au (2 nm)	$2.63 \times 10^{-2}\ s^{-1}$	—	[51]
PNIPAM-*b*-poly(methyl-methacrylate-*co*-5-(2-methacryloylethyloxy-methyl)-8-quinolinol)-*b*-PS	Au (3 nm)	$0.87\ min^{-1}$	$927\ h^{-1}$	[52]
poly(3-hydroxybutyrate)–chitosan (PHB–chit)	Au (25 nm)	$0.57\ min^{-1}$	—	[53]
Fe_3O_4-poly(ionic liquid)	Au (20 nm)	—	$1176\ h^{-1}$	[54]
polyethyleneimine	Pt@Ag (186 nm)	$0.355\ min^{-1}$	—	[55]
PVP	Au (2.9 nm)	$0.68\ min^{-1}$	—	[56]
PNIPAM-*b*-poly(4-vinyl pyridine)	Au (3.3 nm)	$1.48 \times 10^{-3}\ s^{-1}$	$15.5\ h^{-1}$	[34]
PEG-*b*-PAA	Au (10 nm)	—	$800\ h^{-1}$	[57]
poly(ethyleneimine)	Au (7.6 nm)	—	$4.84 \times 10^{-16}\ min^{-1}$	[58]

medication of metal NPs, the stability of metal–thiolate binding has not been satisfactory under catalytic conditions. For example, metal–thiolate binding is vulnerable to oxidation. After thiolate is oxidized to sulfone or sulfoxide, the detachment of those thiolate ligands is usually seen. The reactivity of those PGNPs for many oxidation reactions has not been assessed. We noted that metal–thiolate is not stable under reductive conditions either. Other than the example reported by Kitchens et al., the removal of metal–thiolate has been reported in small molecular ligands. For example, AuNPs (1.8 nm) modified with *n*-dodecanethiol (DT) showed a dramatic size increase to 4.5 nm under excess $NaBH_4$ [56]. The likely reason was due to the labile metal–thiolate binding under reductive conditions. The dynamic character of metal–thiolate binding under elevated temperature also brings difficulties in the thermal stability of PGNPs [59], thus limiting their potentials in many catalytic reactions.

10.3 Catalysis Mediated by PGNPs with NHC-Terminated Polymers

More recently, NHC, as a stronger ligand to metal NPs, has gained increasing interest to tune the catalytic activity of metal NPs [21, 60–63]. NHC binds with transition metals through strong metal–carbon bonds where the long pair on carbon serves as a σ donor to metal atoms. The bond energy of Au–C of NHCs on Au (111) is measured to be 150 kJ mol^{-1} [64]. The NHC therefore is as strong as thiols, if not more so. Free NHCs synthesized from the deprotonation of imidazolium salts by a strong base can be directly used to graft metal NPs, as demonstrated by Fairlamb and Chechik in 2009 [65]. Although free NHCs are extremely reactive with oxygen and moisture, they can modify metal NPs prepared in oil phases. Mattoussi and coworkers recently developed the synthetic strategy of polymeric-free NHCs as surface ligands for AuNPs (~9 nm) capped with oleylamine (OA) [66]. The monodentate PEG NHCs were prepared through a simple tosylation of hydroxyl end groups to synthesize imidazolium-ended PEG. The PEG-free carbenes synthesized with a strong base of potassium *tert*-butoxide could exchange the capping ligands of OA *in situ* to modify AuNPs. Similar methods could be extended to prepare multidentate NHCs on a copolymer of poly(isobutylene-*alt*-maleic anhydride) (PIMA) functionalized with imidazolium and PEG side chains. AuNPs grafted by monodentate or multidentate PEG NHCs were water-soluble in a broad range of pHs 3–12 in phosphate buffer or strong electrolyte (0.1 M NaCl).

Free NHCs are not as accessible as thiols, especially for pre-synthesized metal NPs in aqueous phases. Given those synthetic challenges, the benchtop NHC precursors can be prepared to replace the air-sensitive free NHC in the preparation of NHC-stabilized metal NPs. In molecular NHC ligands, the imidazolium bicarbonate as the single-source NHC compound can replace free NHCs to bind metals. The bicarbonate counterions as a weak base can deprotonate imidazolium *in situ* to produce NHCs [67]. Since no free NHCs are involved, the imidazolium bicarbonate can work as a precursor of NHCs to modify metal NPs. In addition, it was also known

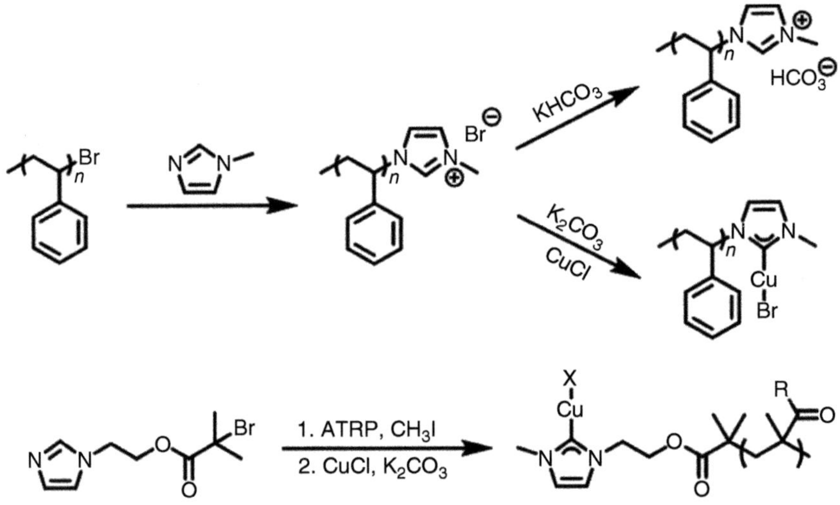

Figure 10.4 Synthesis of the two PS NHC precursors from bromide-terminated PS (top) and the imidazole-containing ATRP initiator for polymethacrylates (bottom).

that NHC-metal, e.g. Cu(I), Ag(I), and Au(I) complexes, can be prepared even in the presence of water [68, 69]. Those metal complexes can transmetalate with other metals under a mild condition [70, 71].

Aligning with those synthetic advances in molecular NHC ligands, the He group has developed the "grafting to" method for polymer NHC ligands using polymer imidazolium salts and polymer NHC-Cu(I) (Figure 10.4). Halogen-terminated polymers were first prepared through ATRP polymerization method. Then the prepared polymers are quaternized by 1-methylimidazole to synthesize imidazolium-terminated polymers. With those imidazolium-terminated polymers, there are two methods to prepare NHC-capped polymers. The first method is to carry out the counterion exchange of halogenated imidazolium with potassium bicarbonate ($KHCO_3$), as imidazolium-terminated polymers with bicarbonate. The imidazolium bicarbonate end group is an NHC precursor that binds metal NPs through the "*in situ*" formation of NHCs by deprotonation of bicarbonate imidazolium. The second method is to synthesize NHC-Cu(I)-terminated polymers. Imidazolium-terminated polymers, when treated with copper chloride (CuCl) and a weak base potassium carbonate (K_2CO_3), can be converted to polymer-NHC-Cu(I). Polymer-NHC-Cu(I) can bind to metal NPs through transmetalation or metal–metal bond formation. The preparation of those two polymer NHC precursors from quaternization of bromide-terminated polymer and 1-methylimidazole is very straightforward. This method can be applied to most of the polymers synthesized via ATRP. With polymethacrylates, the steric hindrance on the tertiary carbon may limit the conversion of the end group [72]. Alternatively, an imidazole-containing ATRP initiator can be used to synthesize imidazole-ended polymers (Figure 10.4), followed by methylation with iodomethane to afford a near 100% imidazolium yield.

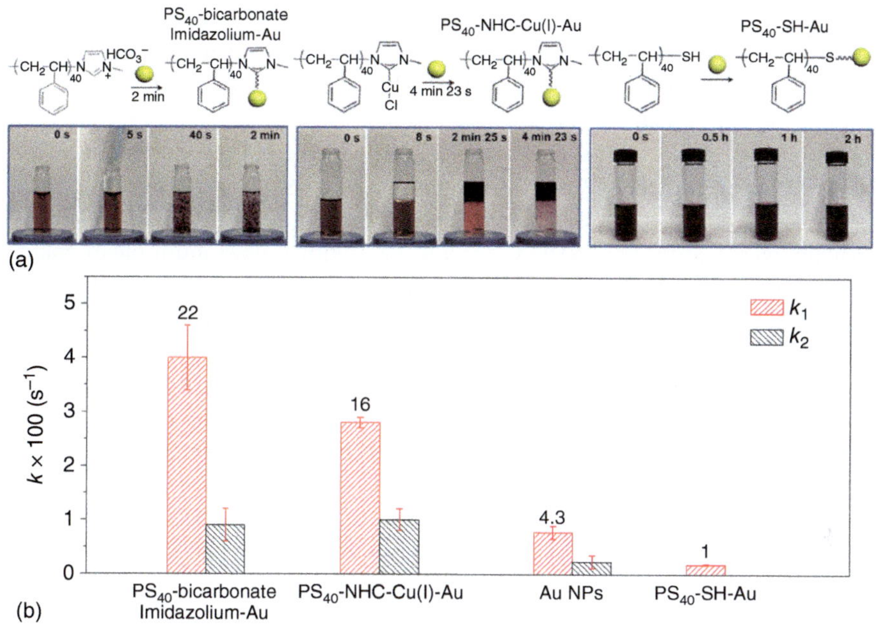

Figure 10.5 (a) Scheme and images to show the PGNPs prepared with PS_{40}-bicarbontae imidazolium, PS_{40}-NHC-Cu(I), and PS_{40}-SH. (b) The plot to show rate corresponding rate constant for the 4-NP reactions catalyzed by PGNPs with PS_{40}-bicarbontae imidazolium, PS_{40}-NHC-Cu(I), and PS_{40}-SH and citrate AuNPs. Source: (a) Lu et al. [73], Reproduced with permission from Royal Society of Chemistry.

The two precursors of two polymer NHCs can readily modify metal NPs through "grafting-to" method. For polymers ending with an imidazolium bicarbonate, a one-phase ligand exchange method has been developed [21]. Typically, polymer terminated with an imidazolium bicarbonate, e.g. hydrophobic PS with an imidazolium bicarbonate dissolved into THF, can be injected into the aqueous solution of metal NPs under vigorous stirring (Figure 10.5a). The bicarbonate works as a weak base; and, it deprotonates the imidazolium to produce NHCs *in situ*. Once polymer NHCs bind with metal NPs, PS-modified NPs would crush out of the solution. A simple redispersion of precipitates into a good solvent of PS, like DMF, can successfully transfer metal NPs into the organic phase. The one-phase method is extremely fast to complete the ligand exchange. Using AuNPs (14 nm in diameter) as an example, the grafting density of PS_{40} ended with an imidazolium bicarbonate can reach 0.35 chains nm^{-2} in two minutes, while PS_{40}-SH with the same molecular weight needs to take two hours to achieve a comparable grafting density of 0.2 chains nm^{-2} [48]. Polymer NHC-Cu(I), on the other hand, can modify metal NPs through a biphasic ligand exchange at the interface of water/oil (Figure 10.5a). Using citrate-capped AuNPs (14 nm) as an example, PS_{40}-NHC-Cu(I) dissolved in toluene (1 mg ml^{-1}) can be added on top of the aqueous solution of AuNPs. Under stirring, AuNPs can be transferred from water to toluene, as evidenced by the movement of the red color from the bottom (water) to the top layer (toluene).

Such biphasic ligand exchange takes only five minutes to reach a grafting density of 0.29 chains nm^{-2}. In addition, those two polymer NHC precursors can modify metal NPs through "grafting-to" method in a broad size range of NPs (3–40 nm) synthesized in either water or oil.

The stability of polymer ligands with thiols and NHCs has been compared using AuNPs (~14 nm) grafted by PS_{40}-SH (Figure 10.5a), PS_{40}-NHC-Cu(I), and PS_{40} with imidazolium bicarbonate. Upon thermal annealing at 110 °C in DMF under air, AuNPs modified by PS_{40}-NHC-Cu(I) and PS_{40} with imidazolium bicarbonate were extremely stable. There was no change in the LSPR peak of AuNPs grafted by those polymer NHC ligands. The LSPR band of AuNPs grafted by PS_{40}-SH showed an obvious broadening after 1.5 hours, and another peak at ~650 nm was seen shortly after. The plasmonic peak of AuNPs shifted, indicating that AuNPs grafted by PS_{40}-SH aggregated under thermal annealing likely, due to the detachment of PS_{40}-SH. The thermal stability of AuNPs modified by PS_{40}-NHC was attributed to the higher binding strength and high resistance to oxidation of Au-NHC binding, as compared to Au-thiolate. In terms of the chemical stability, AuNPs grafted by the two polymer NHC ligands were also stable against dithiothreitol (DTT) with a high concentration up to 1.5 M.

The binding motifs, despite a small fraction of polymer ligands, play a critical role in the catalytic activity of metal NPs (Figure 10.5b). He and coworkers investigated the difference in the catalytic activity of AuNPs (~14 nm) grafted by hydrophobic PS with various binding motifs [73], including PS_{40}-SH and the two NHC precursors prepared by PS_{40}-NHC-Cu(I) and PS_{40} terminated with imidazolium bicarbonate. Using the reduction of 4-NP as a model reaction, AuNPs grafted by the two PS_{40}-NHC showed a much higher activity as compared to those with PS_{40}-SH. The reduction rate constant k of PGNPs with PS_{40} imidazolium bicarbonate was $4 \times 10^{-2}\ s^{-1}$. It showed a similar activity to PGNPs with PS_{40}-NHC-Cu(I) ($k = 2.8 \times 10^{-2}\ s^{-1}$). Interestingly, AuNPs grafted by the two PS_{40}-NHC had an improved activity as compared to that of citrate-capped AuNPs with a k of $7.7 \times 10^{-3}\ s^{-1}$ and AuNPs grafted by PS_{40}-SH with a k of $1.8 \times 10^{-3}\ s^{-1}$. That is, AuNPs grafted by PS_{40}-NHC showed a 22-fold activity improvement than AuNPs grafted by PS_{40}-SH, without modifying the chemical composition of polymer ligands.

The impact of the binding motifs on the activity of metal NPs is applicable to other metal NPs. When modifying Pd NPs with those polymers, a similar trend was observed. Citrate-capped Pd NPs (18 nm) had a k of $4.3 \times 10^{-3}\ s^{-1}$ to reduce 4-NP with $NaBH_4$. PS_{40}-SH as surface ligands to modify Pd NPs led to the decrease of activity to $k = 1.0 \times 10^{-3}\ s^{-1}$. Pd NPs grafted by the two PS_{40}-NHC had a k of $1.4 \times 10^{-2}\ s^{-1}$ and $1.2 \times 10^{-2}\ s^{-1}$ for PGNPs with PS_{40} imidazolium bicarbonate and PS_{40}-NHC-Cu(I), respectively. Pd NPs with polymer NHCs, therefore, were >30 times more active, as compared to those grafted by thiol-terminated polymer ligands. The difference in reactivity of metal NPs was attributed to the electronic nature of the binding motifs. As a σ donor, the C atom of NHCs enriched the surface electron density of metal NPs, as compared to thiol (oxidizing metals instead). Both Au and Pd catalysts with a high electron density would favor the formation of metal hydride to improve the electron transport to 4-NP.

Combining synthetic polymers with catalytically active metal NPs also has an important implication in tuning the reaction selectivity. An interesting example is the electroreduction of CO_2 [74]. The reduction of CO_2 requires multi-electron and multi-proton transfer. However, the thermodynamic reduction potential of CO_2 is close to that of protons. When carrying the electroreduction in aqueous solution, there is a reduction competition between CO_2 and protons. Such competitive reduction often leads to a low FE, which describes the utilization efficiency of electrons to generate a specific product for CO_2 electroreduction. In PGNPs, polymer ligands can gate the surface of catalytic metal NPs to control the accessibility of metal NPs. For example, Au is one of the mostly used catalysts to reduce CO_2 to CO. Citrate-capped AuNPs (14 nm) showed a FE of 79.5% to produce CO at −0.9 V versus reversible hydrogen electrode (RHE, the same reference potential hereafter) [21]. After being grafted by hydrophobic PS_{65}-NHC, AuNPs became more selective to CO_2 electroreduction, and the CO FE increased to 90% under identical conditions. Likewise, commercial Pd supported on carbon (Pd/C, 5 wt% loading) was less selective to reduce CO_2. It had a CO FE of 40% and a H_2 FE of 60% at −1.26 V. Pd/C grafted by PS_{65}-NHC showed a much higher CO FE of 66%. Hydrophobic polymers as surface ligands, therefore, limited the diffusion of protons and consequently improved the selectivity to reduce CO_2. It was noteworthy that Pd/C modified with PS_{50}-SH did not provide a high CO FE (~48%) even with similar hydrophobicity as PS_{65}-NHC. The CO stripping test voltammograms suggested that polymer NHCs enriched the surface electron density of Pd surface to favor the binding of electrophiles like CO_2.

Polymer NHCs also provide high colloidal stability to metal NPs under redox conditions. Under reductive potentials for CO_2 electroreduction, as an example, fast sintering occurred through interparticle coalescence by the random walking of NPs. Alivisatos and coworkers demonstrated that 7 nm Cu spherical NPs grew into 23 nm after 10 minutes electrolysis at −1.25 V [75], and 4 nm AuNPs formed dendritic Au with a diameter of 13 nm after 10 minutes electrolysis at −1.2 V [76]. Note that, Au-thiolate in the latter example of AuNPs capped with DT did not slow down the sintering because those alkanethiolate desorbed at potentials close to the onset for CO_2 electroreduction [76]. On the other hand, metal-NHC binding is not redox-active [22, 41]. Polymeric NHCs can couple their hydrophobicity as polymers with the chemical stability of metal-NHC binding to mediate CO_2 electroreduction more efficiently.

He and coworkers recently demonstrated the use of monodentate and multidentate polymer NHC ligands to improve the NP stability during CO_2 electroreduction [21]. The multidentate poly(vinylbenzyl *N*-methylbenzyl N-heterocycliccarbene) ($PVBMB-NHC_{57}$, P1) was prepared through quaternization of *N*-methyl benzimidazole with poly(vinylbenzyl chloride) (PVBC). The monodentate NHC-terminated PS (PS_{65}-NHC, P2) was prepared through end-group functionalization of halogen-terminated polymers prepared through ATRP as described above (Figure 10.6a). After the counterion exchange with $KHCO_3$, both polymers with imidazolium bicarbonate could modify AuNPs (~14 nm) with a high grafting density of 1.3 and 0.9 chains nm^{-2} for P1 and P2, respectively. The stability of AuNPs was first examined by measuring the electrochemical active surface area (ECSA) of AuNPs.

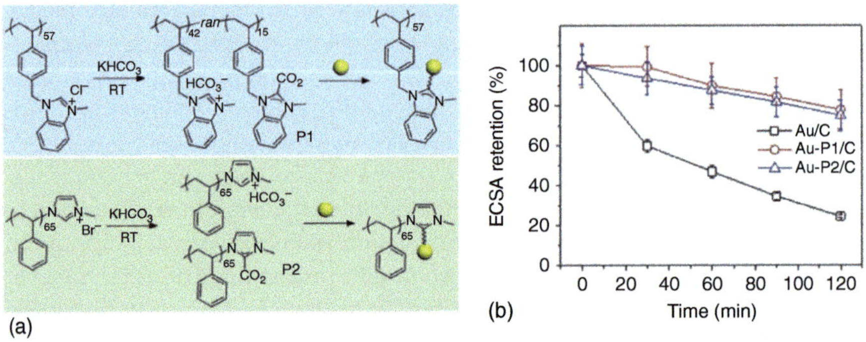

Figure 10.6 The retention of the ECSA of AuNPs under electrolysis at −0.9 V. Those three samples were supported on conductive carbon for the ECSA measurements. Source: Reproduced with permission from Zhang et al. [21]. Copyright 2019 Wiley-VCH.

Using cyclic voltammetry, the accessible surface of AuNPs could be estimated through the adsorbed oxygen monolayer. Under CO_2 electroreduction at −0.9 V, citrate-capped AuNPs showed a quick drop in their ECSA, where only 24.7% of ECSA remained after electrolysis for two hours. AuNPs modified by the two polymer NHCs had an ECSA retention of ~75% (Figure 10.6b). The TEM confirmed that citrate-capped AuNPs showed clustering and polymer NHC-grafted AuNPs remained their structural integrity without aggregation.

For commercial Pd/C, the two polymer NHCs also proved to be effective with stable Pd NPs (5 nm) (Figure 10.7). Without polymer NHCs, commercial Pd/C showed a 91% loss of its decay after electrolysis for two hours at −1.26 V. Pd/C after grafting by the two polymer NHCs showed an ECSA retention of ~70%. As controls, Pd/C modified by PS_{50}-SH was not stable for CO_2 electroreduction, and its ECSA retention was around 35% under identical conditions. Similar results were observed for 1-dodecanthiol and oleamide. Those results suggested that surface ligands with low binding strength to metal NPs, like thiolate and amine, would detach from metal NPs under CO_2 electroreduction conditions. Polymer NHCs that were resistant to

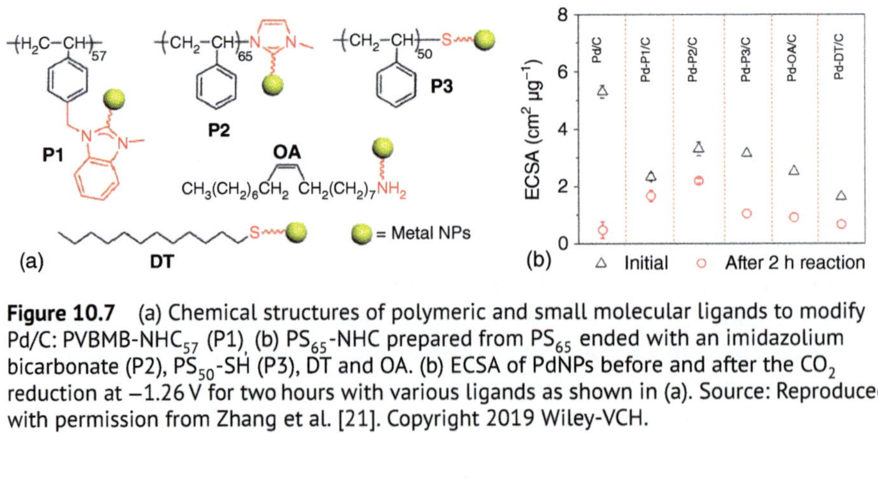

Figure 10.7 (a) Chemical structures of polymeric and small molecular ligands to modify Pd/C: PVBMB-NHC_{57} (P1), (b) PS_{65}-NHC prepared from PS_{65} ended with an imidazolium bicarbonate (P2), PS_{50}-SH (P3), DT and OA. (b) ECSA of PdNPs before and after the CO_2 reduction at −1.26 V for two hours with various ligands as shown in (a). Source: Reproduced with permission from Zhang et al. [21]. Copyright 2019 Wiley-VCH.

the redox reaction could provide the steric hindrance to prevent the interparticle sintering of metal NPs. Together with the hydrophobicity of those polymer ligands, metal NPs modified by polymer NHCs were ideal electrocatalysts to reduce CO_2 with a superior selectivity and stability.

Polymer NHCs provide excellent stability to metal NPs that are not redox-stable, e.g. Ni. Kobayashi and coworkers developed the synthesis of the polymer cage with NHCs to stabilize metallic Ni NPs [77]. Using a random copolymer of PS and poly[1-(4-vinylbenzyl) imidazole], the quaternization of imidazole to form imidazolium could stabilize Ni NPs with an average diameter of 1–4 nm through Ni—NHC bonds. Those Ni NPs were active for the Corriu–Kumada–Tamao (CKT) coupling with a high yield of 98%. The presence of polymer NHC ligands prevented the leaching of Ni during the reaction condition. The polymer NHCs, therefore, allowed the synthesis of organic products without metal contamination. Those Ni catalysts could be recycled three times without the loss of reactivity. When stored in air for two months, only 20% of its reactivity was lost.

10.4 Other PGNP Nanocatalysts

Although polymer ligands ending with thiols or NHCs bind strongly with a variety of metal NPs, non-covalent coordination is still dynamic. Under harsh reaction conditions, those ligands become labile. In previous examples, metal–thiolate has proven to be unstable under oxidative conditions due to the oxidation of thiolate to sulfone or sulfate. Likewise, metal-NHC is not stable in the presence of strong acids, where the protonation of amines in NHC destabilizes the metal–C coordination [62]. As such, those non-covalent interactions bring challenges to, (i) stabilize metal NPs under harsh reaction conditions; and (ii) more fundamentally, studying how polymer ligands impact the catalytic efficiency of metal NPs under those conditions.

One can design polymer ligands with neutral multidentate binding motifs to resolve the dynamic properties of those polymer ligands. For example, PVP binds with metal NPs through the weak coordination of carbonyl and metals. Despite being weak, PVP is multidentate with many repeating units; therefore, the multidentate PVP usually wraps around metal NPs, unlike those polymer ligands ending with strong binding motifs. Since each repeating unit of PVP can bind with metal NPs, the binding strength would increase exponentially with the number of repeating units [27]. The binding of PVP to metal NPs is stable in response to the change of pH and redox chemistry. The unwrapping of PVP is very unlikely because of the large energy input required to disrupt those interactions simultaneously. The carbonyl group of amides also shows strong electronic interaction with surface metal atoms to tune the activity of metal NPs. Tsunoyama et al. compared the activity of AuNPs grafted by PVP and poly(allyamine) (PAAm) for alcohol oxidation [23]. AuNPs (1.2 nm) coated by PVP were ~nine-times more active to convert *p*-hydroxybenzyl alcohol to *p*-hydroxybenzaldehyde as compared with Au (1.4 nm) grafted by PAAm. Using CO as a probe molecule, the electron density of AuNPs was examined through the vibrational frequency of adsorbed CO.

While unbound CO showed a typical C≡O stretching at 2137 cm^{-1}, the adsorbed CO on AuNPs coated by PVP had an obvious peak shift to a low wavenumber at 2100 cm^{-1}. The electron-donating C=O of PVP enriched the surface charge density of AuNPs to facilitate the activation of O_2 in the form of superoxo- or peroxide-like species, leading to a higher activity to oxidize. Similar studies have suggested that PVP as a multidentate polymer ligand can stabilize metal NPs for aerobic oxidation of phenols [78], alcohols [79], and amines [80], as well as the reduction of nitrobenzene [73] and hydrogenation of styrene oxide [81].

The other consideration to resolve the dynamic properties is the design of "covalent" bonds between polymer ligands and metal NPs, as recently proposed by Lin and coworkers [82, 83]. Metal NPs grafted by polymer ligands covalently are prepared through template synthesis [82]. In a typical synthesis, 21-armed star-like BCPs can be first synthesized on a rigid β-cyclodextrin initiator. The unimolecular micelles of those star-like BCPs as nanoreactors can template the growth of PGNPs with different architectures and chemical compositions, including spherical, core–shell, and hollow NPs. With amphiphilic star-like PAA-*b*-PS as an example, the inner PAA block can interact with metal precursors through electrostatic and coordination interaction. When reducing metal precursors within the inner core of the star-like BCPs, the hydrophilic PAA blocks will be encapsulated within NPs. Those hydrophobic PS blocks that have no interaction with metal precursors and their NPs will stay as permanent ligands. For example, star-like PAA-*b*-PS can template the growth of spherical AgNPs (6.1 nm) and AuNPs (5.8 nm) covalently grafted by PS ligands in the mixture of DMF and benzyl alcohol [82].

Those metal NPs with covalently grafted polymer ligands have excellent colloidal stability. Those PGNPs have unique polymer-regulated catalytic performance, which may be dismissed in the presence of labile polymer ligands. Lin and coworkers demonstrated that the catalytic activity of AuNPs grafted by PNIPAM covalently had very different optical and catalytic responses to temperature, i.e. a non-Arrhenius behavior with an increase of temperature [84]. Star-like BCPs of PAA-*b*-PNIPAM as nanoreactors could template the synthesis of AuNPs with a diameter of 14.5 nm and a PNIPAM shell of 10.6 nm. Since PNIPAM is temperature-responsive, the PNIPAM shell would collapse on annealing above its LCST. At 50 °C, the PNIPAM shell decreased to 4.8 nm. In the absence of free PNIPAM, AuNPs grafted by PNIPAM covalently were very stable in water, even with a fully dehydrated PNIPAM shell. The swollen-to-collapsed shell transition would result in the redshift of the LSPR peak of AuNPs from 522 to 532 nm by heating from 20 to 50 °C (Figure 10.8a). No aggregation of AuNPs was seen. In the meanwhile, the addition of a small amount of free PNIPAM to those PGNPs would lead to the aggregation of AuNPs above the LCST of PNIPAM. Free PNIPAM acted as cross-linkers to cluster AuNPs, where the plasmon coupling of AuNPs resulted in a new LSPR peak at 600–700 nm.

The reduction of 4-NP was used as a model reaction to evaluate temperature-dependent catalytic activity of AuNPs. At 25 °C, AuNPs grafted by PNIPAM covalently showed a k of $4.3 \times 10^{-3}\,s^{-1}$ Figure 10.8b. When the PNIPAM shell was fully hydrated below its LCST, the activity of AuNPs showed a conventional Arrhenius-type dependence on temperature. In the temperature range of 10–25 °C,

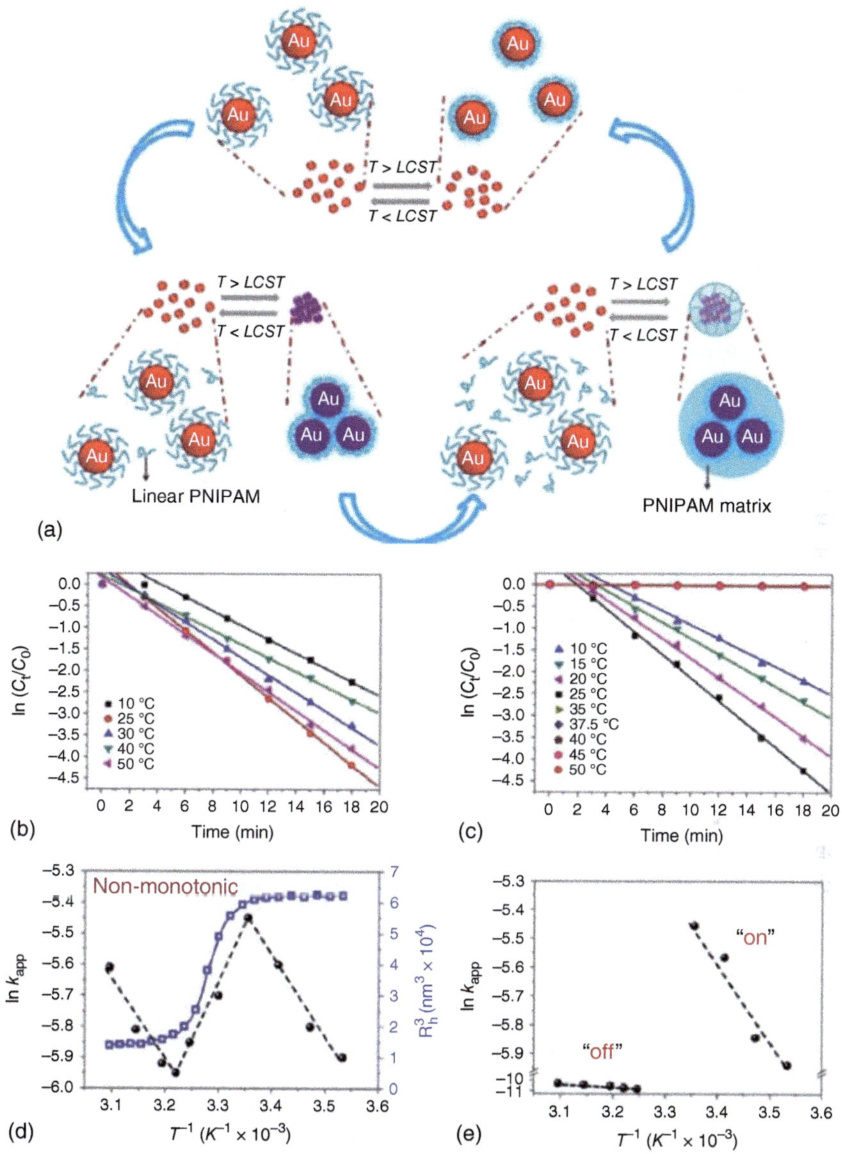

Figure 10.8 (a) Schematic illustration on the thermosresponsiveness of AuNPs grafted by PNIPAM prepared from templated synthesis. In the absence of free PNIPAM, PGNPs show a swollen-to-collapsed transition without interparticle clustering (a top) above the LCST of PNIPAM. By adding deliberately amount free PNIPAM (bottom left) and large free PNIPAM (bottom right), the formation of large clusters or encapsulated clusters within PNIPAM is seen. (b) Kinetic plot of AuNPs grafted by PNIPAM (without free PNIPAM) under different temperatures in the range of 10–50 °C and (d) its corresponding Arrhenius plot. (c) Kinetic plot of AuNPs grafted by PNIPAM in the presence of free PNIPAM and (e) its corresponding Arrhenius plot. Source: Reproduced with permission from Chen et al. [84]. Copyright 2019 Wiley-VCH.

ln$k$ had a linear dependence on $1/T$ (T for temperature) where the reduction rate of 4-NP increased with T (Figure 10.8d). Further increasing reaction temperature from 25 to 37.5 °C would result in the decrease of the activity as the non-Arrhenius response with temperature. This was attributed to the temperature-triggered dehydration of PNIPAM. The densely packed PNIPAM layer imposed a diffusion barrier, and the catalytic surface of AuNPs became inaccessible to the substrate. The catalytic reactivity of AuNPs set back to Arrhenius-type response in the temperature range of 37.5–50 °C where the PNIPAM layer was fully dehydrated and the diffusional barrier was overcome by the increase of temperature. On the contrary, the presence of free PNIAPM would turn OFF the activity of AuNP above the LCST of PNIPAM Figure 10.8c, where the reduction rate decreased by two orders of magnitude to $3.28 \times 10^{-5}\ s^{-1}$ at 50 °C (Figure 10.8e). Such a decrease in the reactivity of AuNPs was a result of thermal-induced aggregation where PNIPAM-grafted AuNPs precipitated out in the presence of free PNIPAM. The latter case was close to the previous reports on core-shell hydrogels of AuNPs encapsulated in PNIPAM [32, 85].

10.5 Conclusion and Outlook

We highlighted a few examples of hairy metal NPs with brush-like polymer ligands to engineer the catalytic surface of metal NPs as a post-synthesis tool, complementary to other synthetic methods to control the surface energy and catalytic activity of metal NPs. We first discussed whether polymers block the catalytic sites on metal NPs using polymer ligands with strong binding motifs like thiols. The reactivity of metal NPs grafted by polymer ligands showed a strong dependence on the grafting density and chain length of polymers with possible activation mechanisms through chemical or mechanical detachment. We addressed the significance of binding motifs in polymer ligands to vary the catalytic activity and selectivity of metal NPs, as well as the stability of metal NPs using NHC-terminated polymer ligands. We also elaborated the dynamic nature of polymer ligands. Using the examples of templated synthesis of metal NPs grafted by polymer ligands covalently, the synergies of polymers and metal NPs illustrated the intriguing applications to gate the catalytic sites.

The topic on the use of polymer ligands to control the catalytic performance of metal NPs is very exciting, but it is still in its early stages. The model reaction is very much limited to the reduction of nitroarenes in most cases. There is plenty of room to investigate the role of polymer ligands in complex reactions where the functionality of polymers is needed. For example, polymer ligands can shuttle protons to mediate multi-electron redox reactions like O_2 and CO_2 activation. Our understanding of how the polymer–metal interfaces contribute to the catalysis of metal NPs is limited, e.g. the interaction of reaction intermediates. There has been lots of knowledge in metalloenzymes and even metal with SAMs made of small molecular thiols that can be further applied in polymer-mediate catalysis. Studies on advanced spectroscopy and electron microscopy to characterize polymer–metal interfaces under reaction conditions are largely missing.

When seeking performance, ligands inevitably occupy catalytic sites on the surface of metal NPs, and the passive polymer layer will create a diffusional barrier to the substrate(s). In most of the examples discussed, polymer ligands are detrimental to the reactivity of metal NPs to a certain extend, although the microenvironment of polymer ligands can improve the selectivity for a few reactions. Existing studies have been primarily centered on the change of polymer compositions and binding motifs that may contribute the activity of metal NPs moderately. Another possible method is to borrow the concept from heterogeneous catalysts, e.g. using metal NPs supported on oxides. Polymer ligands instead of modifying metal NPs can be added to oxides. With polymer ligands physically present around metal NPs, similar microenvironment will be provided without any direct binding on metal NPs. The reported example of Pt supported on TiO_2 suggested that the ligands on TiO_2 vary the selectivity of Pt for CO_2 hydrogenation [86].

References

1 Kim, D., Kley, C.S., Li, Y., and Yang, P. (2017). *Proc. Natl. Acad. Sci. U.S.A.* 114: 10560–10565.

2 Liu, X., Iocozzia, J., Wang, Y. et al. (2017). *Energy Environ. Sci.* 10: 402–434.

3 Zhang, H., Jin, M., Xiong, Y. et al. (2013). *Acc. Chem. Res.* 46: 1783–1794.

4 Jin, L., Thanneeru, S., Cintron, D., and He, J. (2020). *ChemCatChem* 12: 5932–5937.

5 Wei, Z., Duan, H., Weng, G., and He, J. (2020). *J. Mater. Chem. C* 8: 15956–15980.

6 Jin, L., Liu, B., Duay, S.S., and He, J. (2017). *Catalysts* 7: 44.

7 Lu, L., Zou, S., and Fang, B. (2021). *ACS Catal.* 11: 6020–6058.

8 Kango, S., Kalia, S., Celli, A. et al. (2013). *Prog. Polym. Sci.* 38: 1232–1261.

9 Fenter, P., Eberhardt, A., and Eisenberger, P. (1994). *Science* 266: 1216–1218.

10 Badia, A., Singh, S., Demers, L. et al. (1996). *Chem. Eur. J.* 2: 359–363.

11 Shifrina, Z.B., Matveeva, V.G., and Bronstein, L.M. (2019). *Chem. Rev.* 120: 1350–1396.

12 Coan, P.D., Farberow, C.A., Griffin, M.B., and Medlin, J.W. (2021). *ACS Catal.* 11: 3730–3739.

13 Kahsar, K.R., Schwartz, D.K., and Medlin, J.W. (2013). *ACS Catal.* 3: 2041–2044.

14 Liu, B., Yao, H., Song, W. et al. (2016). *J. Am. Chem. Soc.* 138: 4718–4721.

15 Aizenberg, J., Black, A.J., and Whitesides, G.M. (1998). *Nature* 394: 868–871.

16 Chen, K., Wu, H., Hua, Q. et al. (2013). *Phys. Chem. Chem. Phys.* 15: 2273–2277.

17 Rossi, L.M., Fiorio, J.L., Garcia, M.A., and Ferraz, C.P. (2018). *Dalton Trans.* 47: 5889–5915.

18 Kahsar, K.R., Schwartz, D.K., and Medlin, J.W. (2014). *J. Am. Chem. Soc.* 136: 520–526.

19 Pang, S.H., Schoenbaum, C.A., Schwartz, D.K., and Medlin, J.W. (2013). *Nat. Commun.* 4: 1–6.

20 Hao, P., Pylypenko, S., Schwartz, D.K., and Medlin, J.W. (2016). *J. Catal.* 344: 722–728.

21 Zhang, L., Wei, Z., Thanneeru, S. et al. (2019). *Angew. Chem. Int. Ed.* 131: 15981–15987.

22 Cao, Z., Derrick, J.S., Xu, J. et al. (2018). *Angew. Chem. Int. Ed.* 130: 5075–5079.

23 Tsunoyama, H., Ichikuni, N., Sakurai, H., and Tsukuda, T. (2009). *J. Am. Chem. Soc.* 131: 7086–7093.

24 Zhao, X., Zhou, L., Zhang, W. et al. (2018). *Chemistry* 4: 1080–1091.

25 Liu, P. and Nørskov, J.K. (2001). *Phys. Chem. Chem. Phys.* 3: 3814–3818.

26 Boles, M.A., Engel, M., and Talapin, D.V. (2016). *Chem. Rev.* 116: 11220–11289.

27 Duan, H., Yang, Y., Zhan, Y. et al. (2020). *Giant* 4: 100033.

28 Yi, C., Yang, Y., Liu, B. et al. (2020). *Chem. Soc. Rev.* 49: 465–508.

29 Yamamot, K., Imaoka, T., Tanabe, M., and Kambe, T. (2020). *Chem. Rev.* 120: 1397–1437.

30 Niu, Y., Yeung, L.K., and Crooks, R.M. (2001). *J. Am. Chem. Soc.* 123: 6840–6846.

31 De Arquer, F.P.G., Dinh, C.-T., Ozden, A. et al. (2020). *Science* 367: 661–666.

32 Carregal-Romero, S., Buurma, N.J., Pérez-Juste, J. et al. (2010). *Chem. Mater.* 22: 3051–3059.

33 Feng, C., Shen, Z., Li, Y. et al. (2009). *J. Polym. Sci., Part A: Polym. Chem.* 47: 1811–1824.

34 Wang, Y., Wei, G., Zhang, W. et al. (2007). *J. Mol. Catal. A: Chem.* 266: 233–238.

35 Liu, X.-Y., Cheng, F., Liu, Y. et al. (2010). *J. Mater. Chem. A* 20: 360–368.

36 Bergbreiter, D.E., Tian, J., and Hongfa, C. (2009). *Chem. Rev.* 109: 530–582.

37 Liu, L., Gao, Z., Jiang, B. et al. (2018). *Nano Lett.* 18: 5312–5318.

38 Li, C.Y., Zhao, B., and Zhu, L. (2014). *J. Polym. Sci., Part B: Polym. Phys.* 52: 1581–1582.

39 Roth, P.J., Boyer, C., Lowe, A.B., and Davis, T.P. (2011). *Macromol. Rapid Commun.* 32: 1123–1143.

40 He, J., Liu, Y., Babu, T. et al. (2012). *J. Am. Chem. Soc.* 134: 11342–11345.

41 Cao, Z., Kim, D., Hong, D. et al. (2016). *J. Am. Chem. Soc.* 138: 8120–8125.

42 Engel, S., Fritz, E.-C., and Ravoo, B.J. (2017). *Chem. Soc. Rev.* 46: 2057–2075.

43 Lavrich, D.J., Wetterer, S.M., Bernasek, S.L., and Scoles, G. (1998). *J. Phys. Chem. B* 102: 3456–3465.

44 Que, Y., Feng, C., Zhang, S., and Huang, X. (2015). *J. Phys. Chem. C* 119: 1960–1970.

45 Ansar, S.M. and Kitchens, C.L. (2016). *ACS Catal.* 6: 5553–5560.

46 Wu, S., Wang, T., and Xu, H. (2020). *ACS Macro Lett.* 9: 1192–1197.

47 Zhang, J., Han, D., Zhang, H. et al. (2012). *Chem. Commun.* 48: 11510–11512.

48 Ansar, S.M., Fellows, B., Mispireta, P. et al. (2017). *Langmuir* 33: 7642–7648.

49 Tian, J., Huang, B., and Zhang, W. (2018). *Langmuir* 35: 266–275.

50 Yu, W., Lou, L.-L., Yu, K. et al. (2016). *RSC Adv.* 6: 52500–52508.

51 Liu, F., Liu, X., Astruc, D., and Gu, H. (2019). *J. Colloid Interface Sci.* 533: 161–170.

52 Lü, J., Yang, Y., Gao, J. et al. (2018). *Langmuir* 34: 8205–8214.
53 Silvestri, D., Wacławek, S., Sobel, B. et al. (2018). *Green Chem.* 20: 4975–4982.
54 Moghaddam, F.M., Ayati, S.E., Firouzi, H.R. et al. (2017). *Appl. Organomet. Chem.* 31: e3825.
55 Lv, Z.-S., Zhu, X.-Y., Meng, H.-B. et al. (2019). *J. Colloid Interface Sci.* 538: 349–356.
56 Dasog, M., Hou, W., and Scott, R.W. (2011). *Chem. Commun.* 47: 8569–8571.
57 Seo, E., Kim, J., Hong, Y. et al. (2013). *J. Phys. Chem. C* 117: 11686–11693.
58 Ortega-Muñoz, M., Blanco, V., Hernandez-Mateo, F. et al. (2017). *ChemCatChem* 9: 3965–3973.
59 Baira Donoeva, P. and de Jongh, P.E. (2018). *ChemCatChem* 10: 989.
60 Zhang, L., Wei, Z., Meng, M. et al. (2020). *J. Mater. Chem. A* 8: 15900–15908.
61 Chernyshev, V.M., Denisova, E.A., Eremin, D.B., and Ananikov, V.P. (2020). *Chem. Sci.* 11: 6957–6977.
62 Godbold, P., Johnson, G., Obi, A.D. et al. (2021). *J. Am. Chem. Soc.* 143: 2644–2648.
63 Koy, M., Bellotti, P., Das, M., and Glorius, F. (2021). *Nat. Catal.* 4: 352–363.
64 Crudden, C.M., Horton, J.H., Ebralidze, I.I. et al. (2014). *Nat. Chem.* 6: 409–414.
65 Hurst, E.C., Wilson, K., Fairlamb, I.J.S., and Chechik, V. (2009). *New J. Chem.* 33: 1837–1840.
66 Nosratabad, N.A., Jin, Z., Du, L. et al. (2021). *Chem. Mater.* 33: 921–933.
67 Crudden, C.M., Horton, J.H., Narouz, M.R. et al. (2016). *Nat. Commun.* 7: 1–7.
68 Saif, M.J. and Flower, K.R. (2013). *Transition Met. Chem.* 38: 113–118.
69 Schaper, L.A., Hock, S.J., Herrmann, W.A., and Kuehn, F.E. (2013). *Angew. Chem. Int. Ed.* 52: 270–289.
70 Wang, H.M. and Lin, I.J. (1998). *Organometallics* 17: 972–975.
71 Garrison, J.C. and Youngs, W.J. (2005). *Chem. Rev.* 105: 3978–4008.
72 Thanneeru, S., Ayers, K.M., Anuganti, M. et al. (2020). *J. Mater. Chem. C* 8: 2280–2288.
73 Huan, H.Y., Yang, Y.Q., Zhang, Y. et al. (2020). *Gaint* 4: 100033–100047.
74 Zhu, W., Michalsky, R., Metin, Ö. et al. (2013). *J. Am. Chem. Soc.* 135: 16833–16836.
75 Manthiram, K., Beberwyck, B.J., and Alivisatos, A.P. (2014). *J. Am. Chem. Soc.* 136: 13319–13325.
76 Manthiram, K., Surendranath, Y., and Alivisatos, A.P. (2014). *J. Am. Chem. Soc.* 136: 7237–7240.
77 Soule, J.-F., Miyamura, H., and Kobayashi, S. (2013). *J. Am. Chem. Soc.* 135: 10602–10605.
78 Tsunoyama, H., Ichikuni, N., and Tsukuda, T. (2008). *Langmuir* 24: 11327–11330.
79 Tsunoyama, H., Sakurai, H., Negishi, Y., and Tsukuda, T. (2005). *J. Am. Chem. Soc.* 127: 9374–9375.
80 Kitahara, H., Kamiya, I., and Sakurai, H. (2009). *Chem. Lett.* 38: 908–909.
81 Telkar, M., Rode, C., Chaudhari, R. et al. (2004). *Appl. Catal. A., Gen.* 273: 11–19.

82 Pang, X., Zhao, L., Han, W. et al. (2013). *Nat. Nanotechnol.* 8: 426–431.
83 Liu, Y., Wang, J., Zhang, M. et al. (2020). *ACS Nano* 14: 12491–12521.
84 Chen, Y., Wang, Z., Harn, Y.W. et al. (2019). *Angew. Chem. Int. Ed.* 58: 11910–11917.
85 Jiang, X., Xiong, D.A., An, Y. et al. (2007). *J. Polym. Sci., Part A: Polym. Chem.* 45: 2812–2819.
86 Zhang, J., Deo, S., Janik, M.J., and Medlin, J.W. (2020). *J. Am. Chem. Soc.* 142: 5184–5193.

11

Hairy Inorganic Nanoparticles for Oil Lubrication

Michael T. Kelly and Bin Zhao

University of Tennessee, Department of Chemistry, Knoxville, TN 37996, United States

11.1 Introduction

11.1.1 Oil Lubrication

Oil lubrication is critically important for the moving parts in automotive engines, wind turbines, and numerous other machines, for it not only improves the reliability of individual components and entire systems but also increases the energy conversion efficiency. For an internal combustion engine in a medium-size passenger car, about 15% of the energy derived from the combustion of fuel is lost to parasitic friction [1, 2]. Given the sheer number of vehicles on the road worldwide, even a minor reduction in the friction loss would save a huge amount of energy. This prospect, coupled with the increasing demands for environmentally benign lubricants driven by the requirements for environmental protection, has led to the continuous, vigorous development of advanced oil lubricants with increased lubrication performance and longer service intervals.

Oil lubrication in the engine can be divided into three general regimes: boundary, mixed, hydrodynamic/elastohydrodynamic lubrication regimes [1–5]. This is commonly represented by the Stribeck curve (Scheme 11.1a) [1–5], where the coefficient of friction (COF) is plotted against the Hersey number, a dimensionless quantity defined as the product of viscosity and speed divided by the load [5], or the ratio of the lubricant oil film thickness to the combined root-mean-square surface roughness of two interacting surfaces (i.e. the λ ratio) [1]. In the boundary regime, the oil lubricant film thickness is smaller than the surface roughness, and the asperities of the opposing rubbing surfaces come into direct contact, resulting in a high but fairly constant friction coefficient. In the mixed lubrication regime, the COF decreases significantly with the increase in speed (and hence the Hersey number) due to the increase in lubricant film thickness, and only the tallest asperities from the opposing interacting surfaces collide occasionally. For the hydrodynamic/elastohydrodynamic lubrication regime, the thickness of the load-carrying lubricant film is greater than the asperity heights, and the interacting surfaces are well separated from each other by the thick lubricant film. This is

Hairy Nanoparticles: From Synthesis to Applications, First Edition. Edited by Zhiqun Lin and Yijiang Liu.

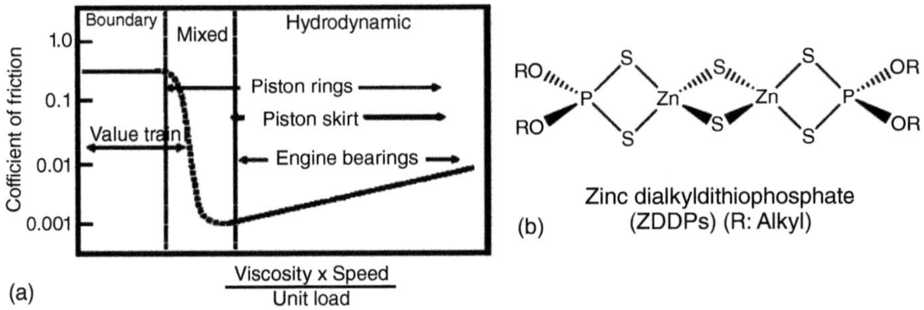

Scheme 11.1 (a) Schematic of the Stribeck curve for the lubrication regimes for engine components. Source: Reproduced with permission from Tung and McMillan [1]; © 2004, Elsevier. (b) General molecular structure of zinc dialkyldithiophosphates (ZDDPs).

a stable lubrication regime, and there is no contact between two surfaces during the steady-state operation conditions. After reaching the minimum, the friction coefficient increases gradually and steadily (Scheme 11.1a), mainly caused by the increase in the lubricant shear strain rate in the oil lubricant film.

Among the various sources of friction losses in automotive engines, hydrodynamic drag accounts for a significant portion [6, 7], and lubricant engineers have been continuously decreasing the viscosity of lubricating base oils to reduce friction from hydrodynamic drag. However, a low viscosity poses a significant challenge for wear protection because of the thinner lubricant film between the moving parts. Commercial lubricants are made up of a base lubricating oil, such as polyalphaolefin (PAO), and a variety of additives, including detergents, dispersants, antiwear reagents, friction reducers, antioxidants, and viscosity index modifiers, each with a distinct function [8]. For engine lubricants, currently, the most widely used antiwear reagent is zinc dialkyldithiophosphates (ZDDPs) (the general molecular structure is shown in Scheme 11.1b) [6, 7, 9–11], a class of coordination compounds discovered and developed in the 1940s due to their low cost and their demonstrated excellent ability to form protective tribofilms on the rubbing surfaces. Although widely used in the formulation of lubricants for wear protection and friction reduction, ZDDPs have been reported to poison the catalysts in catalytic converters in addition to posing a significant environmental issue [12]. Thus, there is an urgent need to develop next generation high performance and environmentally friendly friction reducers and antiwear additives to partially or entirely replace ZDDPs, allowing the use of lower-viscosity lubricating base oils to reduce friction loss from the hydrodynamic drag [12–14].

11.1.2 Nanoparticles as Oil Lubricant Additives for Friction and Wear Reduction

Nanoparticles (NPs) have shown promise as lubricant additives for friction and wear reduction [14–30]. Compared with traditional lubricant additives, NPs possess many advantages, including diverse chemical composition, tunable size, large surface area, tailorable morphology, good thermal conductivity, and rich surface chemistry. These properties make them particularly appealing as oil additives because they can be engineered to exhibit desired interactions with rubbing surfaces in order to reduce friction and wear. In general, there are four lubrication mechanisms when NPs are employed as lubricant additives [29]: (i) the rolling effect of spherical NPs, (ii) protective tribofilm formation as a result of tribochemical reactions on the rubbing surfaces, (iii) the mending effect afforded by the small sizes of NPs, and (iv) chemical–mechanical polishing effects of NPs to give increased λ ratios. In particular, NP-formulated lubricants, sometimes termed "nanolubricants" [14], have been shown to produce strong boundary films (i.e. tribofilms) on the rubbing surfaces from NP-involved mechano-chemical reactions, resulting in significant decreases in the interfacial frictions and wear volume losses. Note that the benefits generated from the improved lubrication performance of automotive engines include higher fuel efficiency, reduced oil consumption, increased engine power output and lifetime, and reduced harmful exhaust emissions [1].

Various NPs of different chemical compositions and morphologies have been evaluated for their effectiveness as lubricant additives for friction and wear reduction [14–30], including carbon and derivatives (e.g. carbon dots and graphene), metals (e.g. Cu, Fe, and Ni), metal oxides (e.g. Fe_2O_3, TiO_2, and SiO_2), metal sulfides (MoS_2, WS_2, and ZnS), rare earth compounds, and nanocomposites. Analysis showed that the effect of the chemical composition of NPs on friction reduction is subtle but important for wear resistance [29]. Wear protection is closely related to the properties of the tribofilms, which are formed on the rubbing surfaces in the boundary lubrication regime from either the thermomechanically induced melting of metallic NPs, the progressive delamination of layered materials such as graphite and MoS_2, or complex NP-involved tribochemical reactions (e.g. metal oxide NPs such as TiO_2, SiO_2, and ZnO). On the other hand, size optimization is mainly governed by the specific working conditions [29].

To achieve tribological benefits of NPs as oil lubricant additives, a critical requirement is that the NPs must be well dispersed in the lubricating base oils, forming homogeneous systems that exhibit long-term stabilities over a wide temperature range from the extreme temperature in winter (e.g. −20 °C) to typical automotive engine operating conditions (100 °C). However, inorganic and metallic NPs are known to have a high tendency toward aggregation in hydrophobic liquid media [30], induced by the high surface energy, large surface area, and van der Waals attractive interactions among NPs. Once aggregation occurs, the particle size increases and the diffusion to the rubbing interfacial zone decreases, resulting in the loss of tribological benefits. Therefore, the colloidal stability of individually dispersed NPs in lubricating base oils is of paramount importance for all nanolubricants. There are two general methods commonly employed to disperse or suspend NPs in lubricating base oils [30]: (i) use of surfactants or dispersants; and (ii) surface modification of NPs with appropriate organic compounds, e.g. alkoxyalkylsilanes for metal oxide NPs. Despite the progress made in recent years, it remains a grand challenge to obtain homogeneous dispersions of NPs in hydrophobic base oils with long-term stability in a wide temperature range and to achieve the full potential of NPs as friction reducers and antiwear reagents.

11.1.3 Polymer Brush-Grafted Nanoparticles: Definition and Synthesis

In general, there are two mechanisms for stabilizing NPs in a liquid medium: charge stabilization suitable for aqueous solutions, as described by the DLVO theory, and steric stabilization for all solvents, where polymer chains are attached to the surface of NPs [31]. Polymer brush-grafted NPs, also called hairy or brush NPs, are composed of a layer of polymer chains end grafted to the surface of NPs via a covalent bond with a sufficiently high grafting density [32–35]. These hybrid NPs display superior dispersibility and stability in good solvents compared with bare NPs or small molecule-modified NPs, thanks to the strong solvation forces from the favorable enthalpic interactions between the grafted polymer chains and the solvent and the entropic, steric repulsive forces between brush NPs [32]. Moreover, polymer brushes have been shown to exhibit excellent lubrication properties in their own right due to the stretched chain conformations in good solvents resisting compression [36–40]. However, despite tremendous potential, no study had been performed on brush NPs as oil lubricant additives for friction reduction and wear protection. In this chapter, we show that oil-soluble polymer brush-grafted silica and titania metal oxide NPs are highly effective lubricant additives for friction and wear reduction, with exceptional long-term stability in PAO over a wide range of temperature [41–43]. Note both SiO_2 and TiO_2 NPs were previously investigated as lubricant additives [15, 22, 23, 44].

Polymer brush-grafted metallic and inorganic NPs can be made by "grafting to" and "grafting from" methods [32–35]. While "grafting to" is straightforward [45], involving only the reaction between a reactive functional group of polymer chains and a complementary group on the surface of NPs, the obtained surface-tethered polymer layer can have a low grafting density due to the steric hindrance. Note that "grafting to" can also be implemented by using end-functionalized polymers as stabilizing ligands for the synthesis of inorganic or metallic core NPs, which generates hairy NPs with controlled sizes and shapes directly [46]. "Grafting from," also called surface-initiated polymerization, has become the predominant approach for the synthesis of brush NPs [32–35, 47–54], for it circumvents the steric hindrance problem in the "grafting to" process, allowing for the preparation of high grafting density polymer brushes on the NPs. In this method, an initiator or a chain transfer agent is first covalently fixed on the surface of the NPs, followed by surface-initiated polymerization to grow polymer brushes *in situ*. Monomer molecules can readily diffuse to the growing chain ends in the brush layer, and thus the steric hindrance presented by the grafted polymer chains for incoming macromolecules in the "grafting to" process is avoided, making it possible to generate high grafting density polymer brushes [32–35].

When a "living"/controlled polymerization technique, e.g. atom transfer radical polymerization (ATRP) [55], nitroxide-mediated radical polymerization [56], reversible addition-fragmentation chain transfer (RAFT) polymerization [57], or ring-opening metathesis polymerization [58], is employed in the surface-initiated polymerization, hairy NPs with predetermined molecular weights, narrow brush dispersities, high grafting densities, and well-defined architectures can be obtained [49–54]. For example, well-defined block copolymer brush- and binary mixed polymer brush-grafted inorganic particles have been successfully prepared via surface-initiated reversible deactivation radical polymerizations in a two-step process either using the same "living"/controlled polymerization method or combining two different controlled polymerization techniques [59–63]. Scheme 11.2 illustrates the preparation of well-defined homopolymer or random copolymer brush-grafted inorganic NPs by surface-initiated ATRP and RAFT polymerization with commonly used surface-immobilizable and free initiators or chain transfer agents, as well as the molecular structures of monomers that appear in this chapter [41, 42]. Despite all the advantages, one difficulty encountered in the characterization of hairy

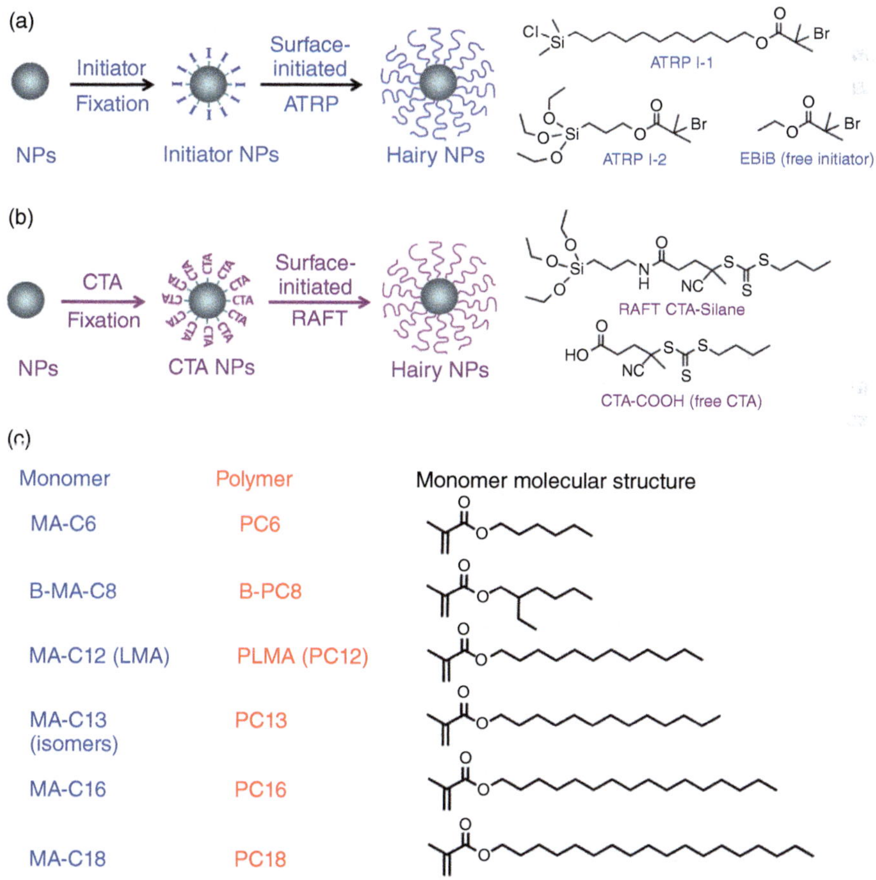

Scheme 11.2 Schematic illustration for the preparation of hairy nanoparticles by surface-initiated (a) ATRP and (b) RAFT polymerization with commonly used surface-immobilizable and free ATRP initiators and RAFT chain transfer agents as well as the molecular structures of monomers (c) used in the synthesis of polymer brush-grafted nanoparticles.

NPs made by “grafting from” is the determination of the chemical compositions, molecular weights, and dispersities (Đ) of the grafted polymers. Although one can cleave the grafted polymer chains off from the surface of NPs and collect them for characterization by ^{1}H NMR spectroscopy and size exclusion chromatography (SEC) analysis, the process is undoubtedly tedious. A widely used method is to add a sacrificial initiator or chain transfer agent into the polymerization mixture to simultaneously produce a free polymer analog for the characterization of molecular weights and compositions. By cleaving the surface-grafted polymers and comparing them with the free polymers, many research groups, including us, have found that the molecular weight and dispersity of the brushes grafted on the particles are virtually the same as those of the free polymer produced from the free initiator or CTA in the solution polymerization [50, 51, 60–62]. This method is more convenient than the degrafting of polymer brushes for the characterization of the brushes’ molecular weights and dispersities and is routinely practiced by us and others. Moreover, when the concentration of the surface-immobilized initiator/CTA is low, the addition of a free initiator/CTA can help achieve better control of surface-initiated reversible deactivation radical polymerization.

In the work presented in this chapter, we used surface-initiated ATRP (SI-ATRP) and RAFT (SI-RAFT) polymerizations (Scheme 11.2a,b) to prepare poly(alkyl methacrylate) brush-grafted inorganic NPs. Section 11.2 presents the synthesis, oil dispersibility, and lubrication property of poly(lauryl methacrylate) (PLMA or PC12) brush-grafted silica and titania NPs. Section 11.3 discusses the effects of alkyl pendant groups on the dispersibility in PAO and tribological properties of poly(alkyl methacrylate) brush-grafted silica NPs. A synergistic effect of combining oil-soluble PLMA-hairy silica NPs with a phosphonium–phosphate-based ionic liquid as friction reducers is described in Section 11.4. Section 11.5 presents a class of newly discovered upper critical solution temperature (UCST)-type thermoresponsive polymers in PAO and the use of ABA triblock copolymers to gel PAO. A summary is provided in Section 11.5.

11.2 Oil-Soluble Poly(lauryl methacrylate) Brush-Grafted Metal Oxide NPs as Lubricant Additives

11.2.1 Synthesis, Dispersibility, and Stability in PAO of Poly(lauryl methacrylate) Brush-Grafted Silica and Titania NPs

To demonstrate the excellent dispersibility and stability of oil-soluble brush NPs in synthetic oil PAO and the effectiveness of such hybrid NPs as friction and wear reducers, we used SI-ATRP to grow PLMA brushes from initiator-functionalized silica and titania NPs (Scheme 11.2a) [41]. Note that PLMA is known to be soluble in PAO, and lauryl methacrylate (LMA or MA-C12, Scheme 11.2c) is used in the synthesis of viscosity index modifiers [8, 40]. The ATRP initiator-functionalized, 23.8 nm silica NPs were prepared by immobilizing 11-(chlorodimethylsilyl)undecyl

2-bromoisobutyrate (ATRP I-1 in Scheme 11.2a), a monochlorosilane-terminated ATRP initiator, onto the surface of silica NPs. I-1 was prepared by hydrosilylation of 10-undecenyl 2-bromoisobutyrate with chlorodimethylsilane under a N_2 atmosphere in the presence of the Karstedt's catalyst and used immediately after the synthesis [41]. Because chlorosilanes are very sensitive to moisture, azeotropic distillation was performed with anhydrous toluene three times to remove as much water as possible from the dispersion of silica NPs in methyl isobutyl ketone (MIBK-ST from Nissan Chemical, Inc.). Subsequently, the surface immobilization reaction of the freshly synthesized I-1 was carried out at 90 °C under N_2 for 64 hours. The I-1 initiator-functionalized silica NPs were purified by multiple cycles of dispersion in *N*,*N*-dimethylformamide (DMF) and ultracentrifugation (Beckman Optima L-90K ultracentrifuge with type 60 Ti rotor, 35 000 rpm, 30 minutes). The initiator NPs were then dried under a stream of air flow, affording a slightly brown powder.

The surface-initiated ATRP of LMA from the initiator-functionalized silica NPs was performed in anisole at 50 °C using copper(I) bromide, copper(II) bromide, and 1,1,4,7,7-pentamethyldiethylenetriamine (PMDETA) as the catalyst system. The initiator NPs were dispersed in anisole via ultrasonication to form a homogeneous dispersion. To achieve better control of the SI-ATRP and to facilitate the characterization of the grafted polymer chains, a sacrificial initiator, ethyl 2-bromoisobutyrate (EBiB in Scheme 11.2a), was added into the polymerization mixture. The progress of the polymerization was monitored by ^{1}H NMR spectroscopy analysis using the ester peaks of the monomer and the polymer. After a desired monomer conversion was reached, the reaction was stopped, and the PLMA brush NPs were purified by multiple cycles of dispersion in tetrahydrofuran (THF) and ultracentrifugation, dried under vacuum, and characterized by thermogravimetric analysis (TGA) for the polymer content and transmission electron microscopy (TEM) for the morphology. The degree of polymerization (DP) was determined from the monomer conversion, calculated from ^{1}H NMR analysis, and the molar ratio of LMA to the sum of the free initiator and the effective surface initiator. The free polymer formed from the sacrificial initiator was analyzed by SEC for the number average molecular weight ($M_{n,SEC}$) and dispersity (Đ) with respect to polystyrene standards, which were taken as the $M_{n,SEC}$ and Đ, respectively, for the grafted polymer chains on NPs [50, 51, 60–62]. Four PLMA brush-grafted silica NP samples with $M_{n,SEC}$ values of 38.0, 21.7, 11.8, and 4.1 kDa and dispersities of 1.09, 1.10, 1.13, and 1.14, respectively, were made by using the same procedure and denoted as HNP-SiO_2-38.0k (DP of PLMA = 117), HNP-SiO_2-21.7k (DP = 66), HNP-SiO_2-11.8k (DP = 31), and HNP-SiO_2-4.1k (DP = 5), respectively. TGA analysis showed that the weight retention at 800 °C of these hairy silica NPs decreased with increasing brush molecular weight. TEM revealed that the brush NPs cast from their dispersions in chloroform, a good solvent for PLMA, self-assembled into close-packed patterns [64–66], and the interparticle distance increased with increasing brush molecular weight (Figure 11.1a–d). From the DPs, TGA data, and average core silica NP diameter, the grafting densities (σ) of the four brush NP samples were calculated to be 0.70, 0.72, 0.72, and 0.67 chains nm^{-2}, respectively. If the brushes formed a

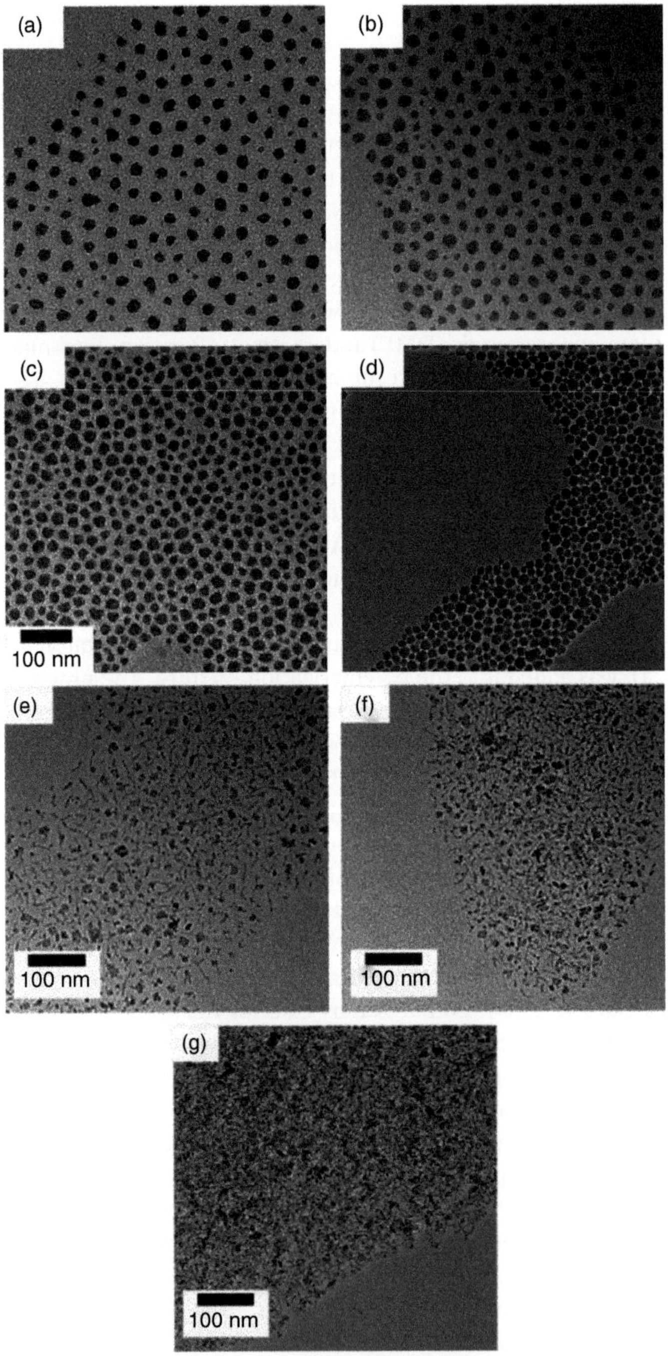

Figure 11.1 Bright field TEM micrographs of PLMA brush-grafted silica NPs with $M_{n,SEC}$ of 38.0 kDa (a), 21.7 kDa (b), 11.8 kDa (c), and 4.1 kDa (d), and PLMA brush-grafted titania NPs with $M_{n,SEC}$ of 21.5 kDa (e), 16.2 kDa (f), and 8.1 kDa (g). The scale bars for (a), (b), (c), and (d) are the same. The hairy NPs were cast onto carbon-coated, copper TEM grids from dispersions in a good solvent ($CHCl_3$ for hairy silica NPs and THF for hairy TiO_2 NPs) with a concentration of 4 mg ml^{-1}. Source: Wright et al. [41], Reproduced with permission from Wiley-VCH.

uniform film on the surface of silica NP with a diameter of 23.8 nm in the dry state, the film thickness of the grafted polymer brushes on silica NPs was 14.1, 10.2, 7.9, and 1.6 nm, respectively, for the four brush NP samples, calculated by using the TGA data and the density of PLMA (0.929 $g\,cm^{-3}$) [41].

Similarly, we synthesized PLMA brush-grafted TiO_2 NPs using SI-ATRP. Anatase TiO_2 NPs, with a size of 15 nm (according to the manufacturer, US Research Nanomaterials, Inc.), were functionalized with initiator I-2 (Scheme 11.2a), an ATRP initiator appended with a surface-immobilizable triethoxysilane group. Three PLMA brush-grafted TiO_2 NP samples with $M_{n,SEC}$ values of 21.5, 16.2, and 8.1 kDa and dispersities of 1.10, 1.12, and 1.16, respectively, were obtained and denoted as HNP-TiO_2-21.5k (DP = 66, which was the DP of HNP-SiO_2-21.7k considering the essentially same $M_{n,SEC}$ values for the two samples), HNP-TiO_2-16.2k (DP = 57), and HNP-TiO_2-8.1k (DP = 22), respectively. Figure 11.1e–g shows the TEM images of the three hairy TiO_2 NP samples cast from THF. In contrast to the nearly spherical SiO_2 NPs, the TiO_2 NPs used here were irregular, with various shapes, including plate- and rodlike. Similar to PLMA brush grafted, 23.8 nm silica NPs, the interparticle distance increased with increasing brush molecular weight. From TGA analysis, the weight retention of PLMA brush titania NPs at 800 °C was lower when the brush molecular weight was higher. Assuming that the TiO_2 NPs were spherical with a diameter of 15 nm as provided by the vendor, the grafting densities of these three NP samples were calculated to be 1.27, 0.83, and 0.80 chains nm^2, respectively. To confirm the presence of PLMA brushes in these hybrid NPs, we performed ^1H NMR spectroscopy analysis of a hairy TiO_2 NP sample, HNP-TiO_2-21.5k, and the corresponding free PLMA in $CDCl_3$, and found that the two ^1H NMR spectra were essentially identical, indicating the high mobility of the grafted PLMA brushes on the surface of titania NPs in $CDCl_3$ [41].

Upon ultrasonication, all four hairy silica NP samples can be readily dispersed in a PAO lubricating base oil and form completely transparent dispersions. The PAO used in our study is SpectraSyn™ 4 PAO oil (PAO-4) from ExxonMobil. The kinematic viscosities of PAO-4 are 19.0 cSt at 40 °C and 4.1 cSt at 100 °C, and the pour point of this base lubricating oil is−87 °C. To study the stability of these brush silica NPs in PAO-4 at different temperatures, we made three 1 wt% dispersions of HNP-SiO_2-4.1k, the NP sample with the lowest brush molecular weight, in PAO-4 and kept them at −20, 22, and 100 °C, respectively. The NP dispersions remained clear after 55 days, and there was no change in the transparency (Figure 11.2a,b). Similarly, all PLMA brush-grafted TiO_2 NP samples can be well dispersed in PAO-4 (Figure 11.2c). As an example to show the stability, three 1 wt% homogeneous dispersions of HNP-TiO_2-16.2k in PAO-4 were kept at −20, 22, and 100 °C, respectively, for 56 days (Figure 11.2d); there were no changes in transparency, except the appearance of a slightly yellow color for the sample kept at 100 °C, which was likely caused by the oxidation of the residual PMDETA ligand in the brush NPs. These experiments demonstrated the excellent dispersibility and stability of PLMA brush-grafted silica SiO_2 and TiO_2 NPs in PAO-4 over a wide temperature range.

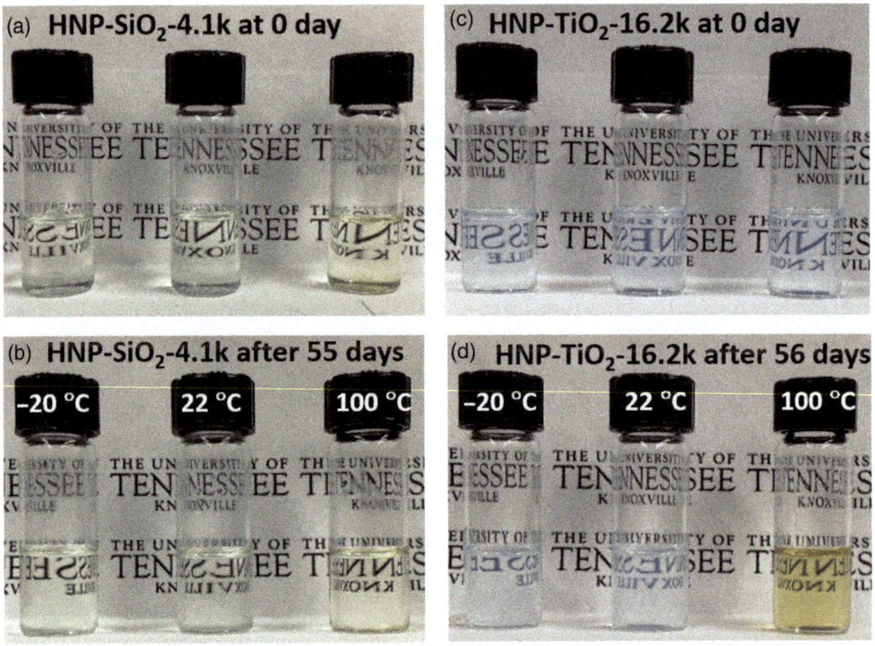

Figure 11.2 Photos of 1 wt% dispersions of HNP-SiO_2-4.1k in PAO in the initial state (a) and after being kept at −20, 22, and 100 °C for 55 days (b), and 1 wt% dispersions of HNP-TiO_2-16.2k in PAO in the initial state (c) and after being kept at −20, 22, and 100 °C for 56days (d). Source: Wright et al. [41], Reproduced with permission from Wiley-VCH.

11.2.2 Lubrication Properties of Poly(lauryl methacrylate) Brush-Grafted Silica and Titania NPs in PAO

The lubrication properties of PLMA brush-grafted silica and titania NPs in PAO-4 were then investigated at a concentration of 1.0 wt% in a high contact stress, ball-on-flat (52 100 steel ball against CL35 cast iron flat) reciprocating configuration at 100 °C under a normal load of 100 N using a Plint TE-77 tribotester. The oscillation frequency was 10 Hz with a stroke of 10 mm, and the total sliding distance was fixed at 1000 m. For each lubricant sample, two tribological tests were performed; the friction coefficients were averaged and are shown in Figure 11.3. For pure PAO-4, the COF exhibited an initial rapid increase, presumably caused by scuffing [67, 68], and then decreased to about 0.11 at a sliding distance of 115 m, followed by a gradual increase to 0.14 at the end of the sliding process (Figure 11.3A). The lubrication performance improved significantly with the addition of PLMA brush silica NPs to the base oil at a level of 1 wt%. For HNP-SiO_2-38.0k and HNP-SiO_2-21.7k with relatively long polymer chains (DP = 117 and 66, respectively), the friction coefficients were lower in the entire sliding experiment than those of PAO-4 (Figure 11.3C,D), and the COF values at the end of the test were 0.12 and 0.11, respectively. The two lower molecular weight samples, HNP-SiO_2-11.8k and -4.1k, showed even further improvement in reducing the friction compared with the higher molecular weight samples, and the final COFs were around 0.10 for both at the sliding distance of

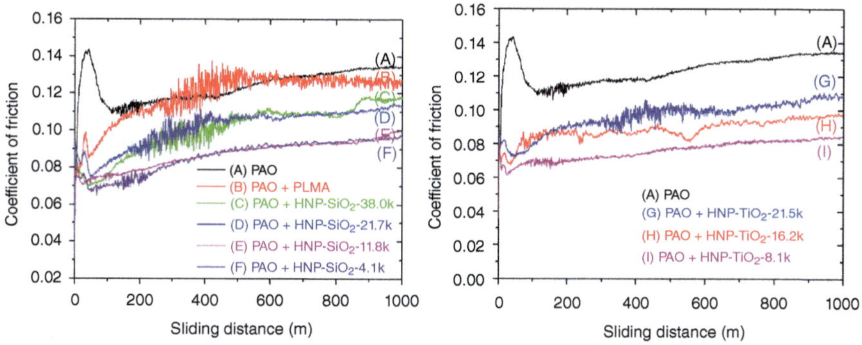

Figure 11.3 Friction curves for PAO SpectraSyn™ 4 (A), PAO containing 1 wt% of free PLMA with a $M_{n,SEC}$ of 38.0 kDa (B), HNP-SiO_2-38.0k (C), HNP-SiO_2-21.7k (D), HNP-SiO_2-11.8k (E), HNP-SiO_2-4.1k (F), HNP-TiO_2-21.5k (G), HNP-TiO_2-16.2k (H), and HNP-TiO_2-8.1k (I). The tribological tests were performed using a Plint TE-77 tribotester at 100 °C under a point contact load of 100 N for a sliding distance of 1000 m. Source: Reproduced with permission from Wright et al. [41]; © 2016, Wiley-VCH.

1000 m (Figure 11.3E,F). As a result, the friction at the end of the sliding process was reduced by 30% compared to neat PAO-based sliding. Overall, a general trend seemed to be present in Figure 11.3C-F that the friction reduction increased with decreasing brush molecular weight. The absence of scuffing with the addition of PLMA brush NPs likely stemmed from the participation of hairy silica NPs in the tribochemical reaction with the metal substrates under the rubbing conditions. We measured the wear volumes of both the cast iron flat and the steel ball after each tribological test using a Wyko NT9100 optical profilometer. The wear volumes for the iron flats and the steel balls decreased by ≥80% for the lubricants containing brush NPs compared with neat PAO-4, with 93% reduction for the iron flat from the best performer HNP-SiO_2-4.1k and 96% reduction for the steel ball from the best performer HNP-SiO_2-11.8k, which were likely due to the suppression of scuffing by hairy silica NPs.

To study the concentration effect of brush NPs on lubrication performance, three additional dispersions of HNP-SiO_2-21.7k in PAO with concentrations of 0.25, 2.0, and 4.0 wt% were prepared and tested. The friction coefficient decreased with increasing NP concentrations from 0.25 to 1, 2, and 4 wt%. However, there was a limit, as we found that the improvement was rather small when the concentration was raised from 2 to 4 wt%. To elucidate the roles of core silica NPs and grafted PLMA brushes in the lubrication performance of hairy NPs, we tested a free PLMA homopolymer with a $M_{n,SEC}$ of 38.0 kDa at a concentration of 1 wt% in PAO-4 under the same conditions. Compared with pure PAO-4, the PLMA solution performed slightly better in the beginning and at the end of the sliding experiment, but similarly for the majority of the sliding process (Figure 11.3B). Thus, a conclusion can be drawn that the major portion of the benefits of adding PLMA brush-grafted silica NPs into PAO-4 comes from the inorganic NPs. This was supported by the observation that lower molecular weight samples performed better because a lower brush molecular weight meant a higher number density of silica NPs in the oil at the same concentration of 1 wt% . However, it was unclear whether the polymer brush lubrication mechanism operated in the tribological test here due to the much harsher conditions adopted here compared with those typically used for brush lubrication studies [36–40].

The tribological properties of PLMA brush-grafted TiO_2 NPs were studied under the same conditions with the same tribotester as for hairy silica NPs. As shown in Figure 11.3G–I, all 1 wt% dispersions of hairy TiO_2 NPs in PAO-4 exhibited lower COFs in the entire sliding process than neat PAO-4, and there was a clear trend that the COF decreased with decreasing molecular weight of PLMA brushes. At the end of the sliding process, the COFs for 1 wt% dispersions of HNP-TiO_2-21.5k, HNP-TiO_2-16.2k, and HNP-TiO_2-8.1k in PAO-4 were 0.11, 0.10, and 0.08, respectively, which were significantly lower than the COF value of 0.14 for pure PAO-4. In particular, the lowest molecular weight sample, HNP-TiO_2-8.1k, showed a 40% reduction in friction, demonstrating the excellent lubrication property of hairy TiO_2 NPs. From Figure 11.3, hairy titania NPs appeared to perform better than hairy SiO_2 NPs at similar molecular weights. Although this could be caused by the difference in the chemical composition of the core NPs, many other factors could also come into play, such as size, shape, and the number of core NPs at the same concentration of 1.0 wt%. Like the PLMA brush silica NPs, the wears for hairy TiO_2 NPs were also reduced substantially, with >90% reductions in the wear volumes of iron flats for the two lower molecular weight hairy titania NP samples. It should be noted here that HNP-TiO_2-16.2k and HNP-SiO_2-4.1k performed comparably to the commercially used ZDDPs and amine-phosphate antiwear agents [69], and HNP-TiO_2-8.1k performed better than both commercial additives.

The friction and wear reductions observed for the lubricants containing PLMA brush-grafted NPs likely resulted from the formation of tribofilms on the rubbing metal substrates formed from the complex tribochemical reactions that involved metal oxide NPs. To characterize the tribofilm, the focused ion beam (FIB) technique was employed to cut a small, thin cross section from the wear scar on the iron flat that was tested with the lubricant containing 1 wt% HNP-SiO_2-4.1k for a TEM study. A 200–400 nm tribofilm was observed between the cast iron substrate and the carbon layer, which was used for protecting the surface during the FIB process, and the tribofilm appeared to be an amorphous matrix containing small nanocrystals. This tribofilm formed from the hairy silica NP-containing lubricant was similar, in terms of thickness and morphology, to those formed by low-molecular weight antiwear additives such as ZDDPs or ionic liquids [9, 10, 70, 71]. From energy-dispersive X-ray spectroscopy analysis (EDS), the tribofilm contained silicon, iron, and oxygen, indicating that hairy silica NPs were involved in the complex mechano-chemical reactions under the harsh tribological testing conditions with a point contact load of 100 N at 100 °C. The produced tribofilm protected the underneath material and hence reduced the friction and the wear loss. This study showed that oil-soluble PLMA brush-grafted SiO_2 and TiO_2 NPs, synthesized by SI-ATRP, were a promising class of friction- and wear-reduction additives for oil lubricants. These brush NPs exhibited exceptional stability in PAO-4 over a wide temperature range of −20 to 100 °C for ≥55 days with no changes in the transparency. Significant reductions in both COF (up to 40%) and wear (up to 90%) were achieved with the addition of 1 wt% hairy NPs in PAO.

11.3 Effects of Alkyl Pendant Groups on Oil Dispersibility, Stability, and Lubrication Property of Poly(alkyl methacrylate) Brush-Grafted Silica Nanoparticles

11.3.1 Synthesis of Poly(alkyl methacrylate) Brush-Grafted, 23-nm Silica NPs

Following the work described in Section 11.2, we conducted a systematic study of the effects of the alkyl pendant group of poly(alkyl methacrylate) on the dispersibility and stability of poly(alkyl methacrylate) brush-grafted silica NPs in PAO-4 and the tribological properties of hairy silica NPs as lubricant additives for friction and wear reduction [42]. Six methacrylate monomers with various alkyl pendant groups were used in this study (Scheme 11.2c), including *n*-hexyl methacrylate (MA-C6), 2-ethylhexyl methacrylate (B-MA-C8), LMA (MA-C12), tridecyl methacrylate (MA-C13, mixture of branched chain isomers), hexadecyl methacrylate (MA-C16), and stearyl methacrylate (MA-C18), and the corresponding polymers are denoted as PC6, B-PC8, PC12 (i.e. PLMA), PC13, PC16, and PC18, respectively (Scheme 11.2c). Here, SI-RAFT polymerization [72–75] was used, instead of SI-ATRP, to prepare a series of poly(alkyl methacrylate) brush grafted, 23-nm silica NP samples (Scheme 11.2b), because RAFT polymerization allows for readily achieving high monomer conversions and obtaining desired molecular weights as well as maintaining relatively low Đ values [76, 77]. A triethoxysilane-functionalized CTA, CTA-Silane (Scheme 11.2b), was designed, synthesized, and used to functionalize 23-nm silica NPs. A *n*-butyl group was used as the Z group for the CTA instead of the more popular *n*-dodecyl because we wanted to avoid possible steric hindrance from the larger *n*-dodecyl group during the SI-RAFT polymerization from the surface of silica NPs.

The CTA-functionalized silica NPs (CTA-NPs) were obtained by reacting CTA-Silane with the surface silanol groups of silica NPs (MIBK-ST, from Nissan Chemical) in a mixture of methyl isobutyl ketone and anhydrous THF at a mass ratio of 1 : 1 at 75 °C in a nitrogen atmosphere. The CTA-functionalized silica NPs were purified by repetitive ultracentrifugation and stored as dispersion in THF. Three batches of CTA-NPs, denoted as CTA-NPs-B1, -B2, and -B3, were prepared from MIBK-ST and used to synthesize various poly(alkyl methacrylate) brush-grafted silica NPs. The surface-initiated RAFT polymerizations of alkyl methacrylate monomers from CTA-NPs were performed at 70 °C using azobisisobutyronitrile (AIBN) as initiator and 4-(((butylthio)carbonothioyl)thio)-4-cyanopentanoic acid (CTA-COOH, Scheme 11.2b) as free CTA. CTA-COOH, synthesized using a procedure adapted from the literature [77], was added to achieve better control of SI-RAFT polymerization and to facilitate the characterization of the grafted polymer via the formation of a free polymer. As mentioned earlier, it has been well established by many research groups that the molecular weights and dispersities of the grafted polymer on the surface of NPs prepared by surface-initiated reversible deactivation radical polymerization are essentially the same as those of the corresponding free polymers formed from the free initiator/CTA [50, 51, 60–62, 75]. After the polymerization occurred for 18 hours, high monomer conversions of 92.1–98.5% were achieved. The brush NPs were isolated and purified by multiple cycles of ultracentrifugation/redispersion in THF, and the free polymers formed from the free CTA, CTA-COOH, were collected and analyzed by SEC for $M_{n,SEC}$ and Đ values with respect to polystyrene standards. The DP of the polymer was calculated from the final monomer conversion, determined from ^{1}H NMR spectroscopy analysis, and the molar ratio of the monomer to the sum of the sacrificial and effective surface CTA, as in the case of SI-ATRP.

A total of seven brush silica NP samples with similar DPs (around 30, in the range of 25–42) were prepared by SI-RAFT polymerization, one sample each of MA-C6, B-MA-C8, MA-C12, MA-C16, and MA-C18 and two samples from MA-C13. The DPs of these brush NP samples were purposely controlled to be similar so that we could compare their lubricating performance under the same conditions [41]. These hairy silica NP samples were named by the following method using PLMA (PC12) brush-grafted silica NPs with a $M_{n,SEC}$ of 9.5k synthesized from CTA-NPs-B1 as an example. This sample was denoted as PC12-NP1-9.5k, where PC12 is the grafted polymer (PLMA or PC12), NP1 refers to CTA-NPs-B1 used to make hairy NPs, and 9.5k is the $M_{n,SEC}$ of the corresponding free polymer, PC12. In addition, one higher molecular weight B-PC8 brush-grafted NP sample was made for studying the effect of molecular weight on the stability of B-PC8-NPs in PAO-4. SEC analysis showed that all polymerizations were controlled; yielding free polymers with Đ ≤ 1.25 with respect to narrow disperse PS standards. The organic polymer contents in hairy NP samples were determined by TGA; for all seven hairy NP samples, the weight retentions at 800 °C were around 50%, ranging from 44.2% to 51.6%, except for PC16-NP2-8.3k that showed slightly higher weight retention of 62.3%. Scanning TEM analysis of hairy NPs cast from THF showed that the average size of the core silica NPs was 23 nm. By using the TGA data, DPs, and the core silica NP size of 23 nm, the grafting densities for all hairy silica NP samples were calculated. The characterization data and the calculated grafting densities are summarized in Table 11.1. There appeared to be a general trend that for a larger alkyl pendant group, the grafting density of the brushes was lower, which was likely caused by the higher steric hindrance.

11.3.2 Dispersibility and Stability of 23-nm Silica NPs Grafted with Poly(alkyl methacrylate) Brushes with Various Pendant Groups in PAO-4

We then investigated how the alkyl pendant groups of poly(alkyl methacrylate)s affected the dispersibility and stability of poly(alkyl methacrylate) brush-grafted,

Table 11.1 Characterization data for hairy silica nanoparticles and corresponding free polymers.

Hairy Silica NPs	CTA-NPs[a]	Monomer[a]	$M_{n,SEC}$[b] (kDa)	Đ[b]	DP[c]	σ[d] (chains nm^{-2})
PC6-NP1-7.0k	CTA-NPs-B1	MA-C6	7.0	1.25	32	0.72
B-PC8-NP1-7.8k	CTA-NPs-B1	B-MA-C8	7.8	1.23	38	0.57
PC12-NP1-9.5k	CTA-NPs-B1	MA-C12	9.5	1.24	29	0.52
PC13-NP1-9.7k	CTA-NPs-B1	MA-C13	9.7	1.20	32	0.62
PC13-NP3-7.2k	CTA-NPs-B3	MA-C13	7.2	1.16	25	0.69
PC16-NP2-8.3k	CTA-NPs-B2	MA-C16	8.3	1.14	27	0.28
PC18-NP1-13.9k	CTA-NPs-B1	MA-C18	13.9	1.22	42	0.31
B-PC8-NP2-18.0k	CTA-NPs-B2	B-MA-C8	18.0	1.13	113	0.44

a) The batch of CTA-NPs used for the preparation of respective hairy NP sample.
b) Number average molecular weight ($M_{n,SEC}$) and dispersity (Đ) of the corresponding free polymer were determined by size exclusion chromatography (SEC) relative to polystyrene standards using THF as solvent.
c) Degree of polymerization (DP) was calculated from the monomer conversion and the molar ratio of monomer to the sum of free CTA and effective surface-grafted CTA.
d) Grafting density (σ) was calculated using the size of silica NPs (23 nm), DP, and TGA data.

23-nm silica NPs in PAO-4 by visual inspection and dynamic light scattering (DLS) [42]. To this end, we prepared 1.0 wt% dispersions of hairy NPs in PAO-4 by first diluting a dispersion of hairy NPs in THF with PAO-4 and then using a stream of nitrogen flow to remove THF, followed by applying a high vacuum to completely remove the volatile. All polymer brush-grafted SiO_2 NP samples, except PC6 and B-PC8 brush NPs, were fully dispersed in PAO-4 at room temperature and formed homogeneous, transparent dispersions. We note here that the use of THF is not necessary but convenient for preparing NP dispersions; readily dispersible hairy silica NPs, i.e. PC12, PC13, PC16, and PC18 brush NPs, can be dispersed in PAO-4 directly by vortex mixing or ultrasonication. We then examined if temperature variations would affect the stability of readily dispersible hairy NPs in PAO-4. Three dispersions of PC13-NP1-9.7k in PAO-4 with a concentration of 1.0 wt% were prepared and placed in a −15 °C freezer, at ambient conditions, and in a 100 °C oil bath. After two months, all three dispersions remained homogeneous and clear. DLS was then used to measure the hydrodynamic sizes of hairy NPs; the three samples were diluted to 0.1 mg ml^{-1} with PAO-4 at ambient conditions, and DLS measurements were taken. The average hydrodynamic diameters at 23 °C were found to be 56.2, 54.4, 57.5, and 58.9 nm for the freshly prepared sample and the three samples stored at −15, 18, and 100 °C for 60 days, respectively. These values were close to each other, indicating that the PC13-NP1-9.7k hairy NPs were stable in PAO-4 over a wide temperature range, similar to the PLMA hairy silica NPs made by surface-initiated ATRP described in Section 11.2 [41].

As mentioned earlier, PC6 and B-PC8 hairy NPs with shorter alkyl pendant groups in the grafted polymers could not be dispersed in PAO-4 at room temperature at a concentration of 1.0 wt% to form a clear, homogeneous dispersion. For PC6-NP1-7.0k, only an unstable, cloudy suspension was obtained after ultrasonication at room temperature; when left still, the hairy NPs completely precipitated out in <2 hours. Even prolonged heating and stirring at 80 °C did not lead to a stable dispersion. This is not surprising because PC6 is insoluble in PAO [40]. Quite unexpectedly, heating a 1.0 wt% suspension of B-PC8-NP1-7.8k in PAO-4 produced a clear, homogeneous dispersion similar to the dispersions of readily dispersible hairy NPs in PAO-4. Interestingly, the cloudy-to-clear transition was reversible; the clear dispersion became cloudy upon cooling from 80 °C to room temperature, and the cloud point, determined by visual inspection, was 54 °C. Because the free B-PC8 homopolymer is soluble in PAO-4 at ambient temperature, the poorer solubility of B-PC8 brush NPs apparently results from the presence of the silica NP core, which has unfavorable enthalpic interactions with PAO-4. To examine if increasing the brush molecular weight would improve the dispersibility of B-PC8 brush-grafted silica NPs, we synthesized a high-molecular weight B-PC8 brush NP sample, B-PC8-NP2-18.0k, with a $M_{n,SEC}$ of 18.0k. Indeed, the solubility of this sample in PAO-4 was better than B-PC8-NP1-7.8k, as indicated by a lower cloud point of 46 °C in PAO-4. This demonstrated that increasing the brush chain length could mitigate the unfavorable enthalpic interactions of the silica NP core with PAO-4, leading to improved dispersibility and stability of B-PC8 hairy silica NPs in the oil.

11.3.3 Effect of Alkyl Side Chains of Poly(alkyl methacrylate) Brushes on Lubrication Performance of 23-nm Hairy Silica NPs as Additives for PAO-4

Similar to the tribological testing of PLMA brush-grafted silica and titania NPs as friction- and wear-reduction additives for PAO-4 in Section 11.2, we used the Plint TE-77 tribotester with 9.525 mm AISI 52 100 steel balls and CL35 cast iron flats in a ball-on-flat reciprocating configuration at 100 °C to study the lubrication properties of seven hairy silica NP samples. We first tested a PAO-4 lubricant containing 1.0 wt% of a free polymer of MA-C13 with DP = 72, $M_{n,SEC}$ = 20.8 kDa, and Đ = 1.15 (PC13-20.8k). As shown in Figure 11.4a, the COF increased sharply at the beginning of the tribological test, which was a sign of scuffing failure. In the second half of the testing, there was some improvement, but not substantial. These observations indicated a minimal effect of adding free PC13 to PAO-4 for friction and wear reduction, which confirmed our previous result for the effect of the PLMA (PC12) free polymer in Figure 11.3 [41].

For the brush NPs that can be readily dispersed in PAO-4 at room temperature to form clear homogeneous dispersions (i.e. PC12-, PC13-, PC16-, and PC18 brush-grafted NPs), the COF values were significantly reduced compared with neat PAO-4, with much lower values observed throughout the sliding process (Figure 11.4a). Unlike pure PAO-4 and the 1.0 wt% solution of PC13-20.8k in PAO-4, there was no scuffing at the beginning of the test for all the NP lubricants. The COF values were between 0.08 and 0.10 at the end of the testing, and the largest friction reduction (38%), relative to PAO-4, was achieved with the addition of

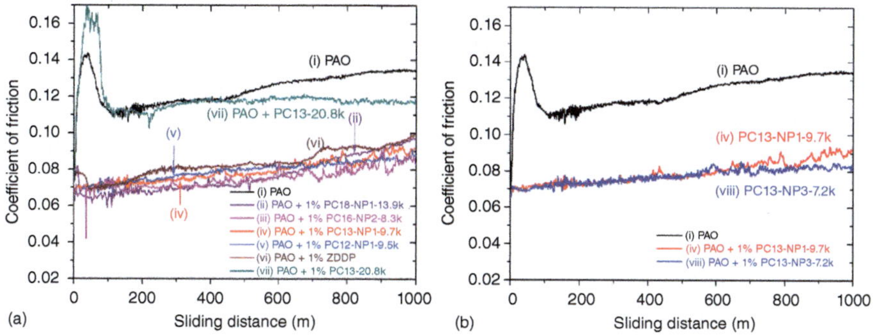

Figure 11.4 (a) Friction curves for PAO SpectraSyn™ 4 (i), PAO with 1.0 wt% PC18-NP1-13.9k (ii), PC16-NP2-8.3k (iii), PC13-NP1-9.7k (iv), PC12-NP1-9.5k (v), ZDDP (vi), and PC13-20.8k free polymer (vii). (b) Friction curves for PAO containing 1.0 wt% PC13-NP3-7.2k (viii) as well as neat PAO (i) and PAO containing 1.0 wt% PC13-NP1-9.7k (iv) for comparison. The tribological tests were performed using a Plint TE-77 tribotester at 100 °C under a point contact load of 100 N for a sliding distance of 1000 m. The curves for (i) PAO and (vi) PAO containing 1.0 wt% ZDDP were from a previous publication for comparison. Source: Reproduced with permission from Seymour et al. [42]; © 2017, American Chemical Society.

PC16-NP2-8.3k. Interestingly, these four brush NP samples all performed better than the commercially used ZDDP at the same concentration of 1.0 wt%, as can be seen from Figure 11.4a, where the friction curves of these hairy silica NPs were below that of the 1.0 wt% solution of ZDDP in PAO-4. Although the hairy NPs synthesized by SI-RAFT polymerization contained sulfur from the surface-immobilized CTA, their content was negligible compared with ZDDP. Two PC13 brush-grafted silica NP samples, PC13-NP1-9.7k and PC13-NP3-7.2k, performed rather similarly, although PC13-NP3-7.2k exhibited lower friction values after a sliding distance of 600 m (Figure 11.4b). Because the surface-grafted poly(alkyl methacrylate) chains had a relatively small effect on the friction reduction as shown in Figures 11.3 and 11.4a, the similar lubrication properties of five hairy silica NP samples with different alkyl side chains underscored the importance of the core silica NPs in the effectiveness of hairy NPs as friction reducers.

As discussed earlier, PC16-NP2-8.3k performed slightly better than other hairy NPs. While these hairy NPs had different chemical compositions, DPs, and grafting densities (Table 11.1), their lubrication performances could be explained by the observations from TGA analysis that PC16-NP2-8.3k had a noticeably higher weight retention at 800 °C (62.3%) than the other four samples (44.2% for PC13-NP1-9.7k, 48.2% for PC13-NP3-7.2k, 51.6% for PC12-NP1-9.5k, and 47.7% for PC18-NP1-13.9k). A higher weight retention at 800 °C for PC16-NP2-8.3k means more silica NPs in PAO-4 at the same concentration of 1.0 wt%, which would give a better friction reduction performance, while similar weight retentions at 800 °C mean similar numbers of silica NPs at the same concentration and thus comparable friction reductions. Note that we showed previously that the friction reduction increased with increasing concentrations of PC12 hairy NPs [41], although there was a limit. From these studies, we can conclude that the grafted polymer brushes mainly provide colloidal stability for the silica core NPs in the base oil PAO-4, and the lubrication property is predominantly determined by the number of silica NPs (or the silica content) in the lubricant. However, it remains unclear how the size of silica core NPs affects the tribological properties of hairy silica NPs. On the other hand, one could incorporate functional groups that contain triboactive elements, for example, phosphorous, into the brushes to increase the function of the grafted polymer brushes in the hairy NPs.

The material wear volumes of the wear scars formed on the cast iron flats and the steel balls from all tribological tests were measured by optical surface profilometry. Compared with neat PAO-4, the wear volumes of the iron flats for oil lubricants containing PC12-NP1-9.5k, PC13-NP1-9.7k, PC13-NP3-7.2k, PC16-NP2-8.3k, and PC18-NP1-13.9k were reduced by 80–90% (11.7 to 23×10^7 μm^3 versus 113.0×10^7 μm^3 for pure PAO-4). The material losses with the addition of 1.0 wt% hairy silica NPs, however, were still slightly higher than those for the 1.0 wt% ZDDP in PAO-4 (7.9×10^7 μm^3). The wear volumes of the steel balls for the tribological tests of these hairy NPs were in the range of 0.019 to 0.048×10^7 μm^3, which were similar to the value for the ZDDP but much smaller than pure PAO-4 (0.786×10^7 μm^3). In other words, they were reduced by >96% when compared to PAO-4.

Despite the fact that PC6 and B-PC8 brush-grafted silica NPs could not be dispersed in PAO-4 to form clear, homogeneous dispersions under ambient conditions, we studied their tribological properties under the same conditions. As noted earlier, the 1.0 wt% suspension of PC6-NP1-7.0k in PAO-4 remained cloudy upon heating, whereas B-PC8-NP1-7.8k became fully dispersed in PAO-4, forming a clear and stable system. We first measured the tribological property of the fully dispersed B-PC8-NP1-7.8k in PAO-4 with a concentration of 1.0 wt% obtained by heating to 80 °C and stirring for 30 minutes. The clear dispersion was added to the cast iron flats, whose temperature was preset at 100 °C and tested using the same tribological conditions. To compare, we prepared a cloudy mixture of 1.0 wt% PC6-NP1-7.0k in PAO-4 by heating and stirring at 80 °C to form a visually uniform cloudy suspension and then performed tribological tests on it. As can be seen from Figure 11.5, B-PC8-NP1-7.8k performed essentially the same as those more readily dispersed hairy NPs (Figure 11.5a), with the COF curve virtually overlapping with the curve for PC13-NP3-7.2k. It should be noted that B-PC8-NP1-7.8k and PC13-NP3-7.2k had very similar weight retentions, 49.1% and 48.2%, respectively, at 800 °C from TGA analysis, which highlighted again the critical role of the mass of silica core NPs in the lubricating performance. PC6-NP1-7.0k, even if not dispersed in PAO-4, also reduced the friction substantially compared with pure PAO-4, but the lubricating performance was worse than PC13-NP3-7.2k (Figure 11.5b). At the end of the sliding process, the COF for PC6-NP1-7.0k was higher by 0.02 than PC13-NP3-7.2k.

For comparison, we also evaluated the tribological properties of PC6-NP1-7.0k and B-PC8-NP1-7.8k in PAO-4 after sitting still at room temperature for seven days; the brush NPs appeared to completely precipitate out. The top, clear portion of the oil was used for tribological testing. Scuffing occurred early in the testing for the

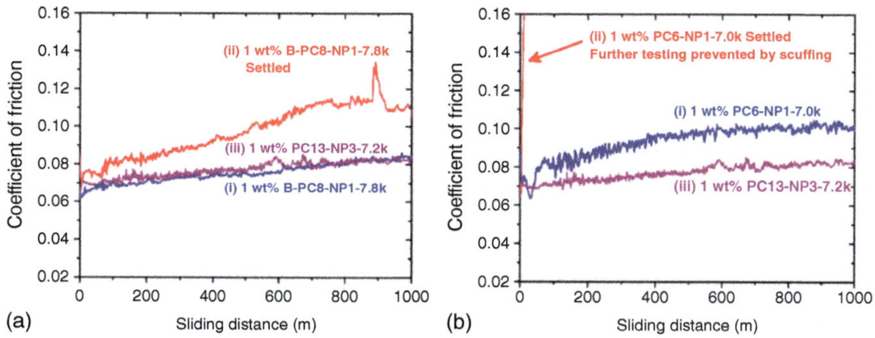

Figure 11.5 (a) Friction curves for PAO containing 1.0 wt% B-PC8-NP1-7.8k (i) preheated and stirred at 80 °C to achieve a uniform state and (ii) after sitting quiescently at room temperature for seven days. (b) Friction curves for PAO containing 1.0 wt% PC6-NP1-7.0k (i) preheated and stirred at 80 °C to achieve a uniform state and (ii) after sitting quiescently at room temperature for seven days. The curve for PAO containing 1.0 wt% PC13-NP3-7.2k (iii) was included for comparison in both (a) and (b). The tribological tests were performed using a Plint TE-77 tribotester at 100 °C under a load of 100 N. Source: Reproduced with permission from Seymour et al. [42]; © 2017, American Chemical Society.

PC6-NP1-7.0k sample, and the friction exceeded the testing limit, causing further testing to be halted (Figure 11.5b). In contrast, B-PC8-NP1-7.8k performed worse than that of the clear dispersion made at 80 °C but exhibited some level of friction reduction compared with neat PAO-4 (Figure 11.5a). Due to the higher affinity of B-PC8 for PAO-4 than PC6, it was possible that a small number of B-PC8-NP1-7.8k hairy NPs remained in the oil and provided some lubrication and protection effects. The higher molecular weight B-PC8 brush NPs were also evaluated tribologically after fully dispersing in PAO-4 by heating. The friction reduction of the fully dispersed B-PC8-NP2-18.0k was lower than that of the fully dispersed B-PC8-NP1-7.8k; at the sliding distance of 1000 m, the COF was 0.093 for B-PC8-NP2-18.0k and 0.084 for B-PC8-NP1-7.8k. This observation was consistent with our previous finding [41] and likely resulted from the different numbers of silica core NPs for the two samples at the same concentration of 1.0 wt%; higher molecular weight brushes meant less NPs in the dispersion and less friction reduction.

The improved lubrication performances of PAO-4 lubricants that contained fully dispersed hairy silica NPs were believed to stem from the protective tribofilms formed by the complex tribochemical reactions with the involvement of hairy silica NPs under the harsh conditions of local high temperatures and high pressures. As discussed earlier, protective tribofilms were observed for various friction reducers and antiwear additives, including ZDDPs and ionic liquids [9, 10, 13, 71, 78], and commonly contained elements from additives. The tribofilm on top of the wear track formed from the tribological testing of PLMA brush-grafted silica NPs in PAO-4 was enriched with element silicon from the silica core NP by TEM elemental mapping [41]. Here, EDS analysis coupled with scanning electron microscopy (SEM) was employed to confirm the presence of a tribofilm at the wear scar of the iron flat after the tribological test with 1.0 wt% PC13-NP3-7.2k in PAO-4. A significant increase in the contents of both silicon and oxygen inside the wear scar compared with the untested area of the iron flat was observed from the EDS spectra. Elemental mapping also showed that the contents of oxygen and silicon at the end of the wear track were clearly higher than outside of the wear scar. These observations confirmed the formation of a tribofilm on the wear track from the tribological test, which offered protection for the underneath metal substrate and prevented scuffing, leading to reduced friction and wear.

11.4 Improved Lubrication Performance by Combining Oil-Soluble Hairy Silica Nanoparticles and an Ionic Liquid as Additives for PAO-4

Besides NPs, ionic liquids are another promising class of friction- and wear-reducing additives for oil lubricants [13, 79–82]. These ionic compounds have relatively low melting points, often below ambient temperature, and exhibit many intriguing characteristics, including excellent thermal stability, high chemical stability, no or extremely low volatility, and a broad electrochemical window [80, 83]. Ye et al. were the first to explore the use of ionic liquids as lubricants in 2001 [79], and since then the tribological properties of various ionic liquids have been intensively investigated [13, 71, 79–82]. In particular, there has been an increasing interest in oil-miscible phosphonium–phosphate-based ionic liquids; one example, tetraoctylphosphonium bis(2-ethylhexyl) phosphate [P8888][DEHP] [13], is shown in Scheme 11.3. These ionic liquids have shown effective anti-scuffing, friction reducing, and anti-wear capabilities, which were ascribed to the adsorption of anions on the metal substrates and the tribofilm formation from the ionic liquid-involved tribochemical reactions on the rubbing substrates.

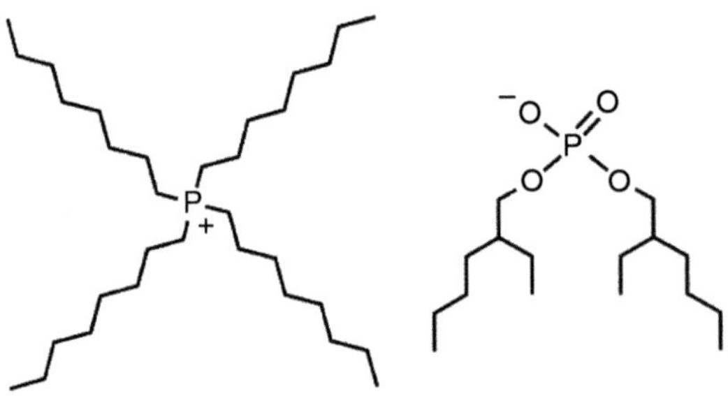

Scheme 11.3 Molecular structure of ionic liquid [P8888][DEHP] (denoted as IL).

Oil-soluble brush NPs and oil-miscible ionic liquids are two different classes of potential friction- and wear-reducing additives for oil lubricants, with vastly different physical and chemical characteristics. The tribofilms formed at the rubbing surfaces are from different tribochemical reactions under local high pressures and high temperatures for the two types of lubricant additives. The typical tribofilm thickness observed from TEM is 200–400 nm for hairy silica NPs [41] and 10–400 nm for oil-soluble ionic liquids [13]. For phosphonium–phosphate ionic liquids, the tribofilms were found to contain iron (poly)phosphates [13], which are believed to be from ionic liquids. On the other hand, phosphate anions have been reported to react with silica/silicate to form covalent linkages [84–86], which can be used to enhance the bonding strength. On the basis of all of these, we hypothesized that further improvement in the lubrication performance could be achieved with oil-soluble hairy silica NPs combined with a phosphonium–phosphate ionic liquid as additives for PAO-4. The tribochemical reactions that involve metal substrates, silica NPs, and phosphate anions in the rubbing process might produce mechanically more robust tribofilms with stronger bonding to the underlying metal substrates. To test this hypothesis, we prepared a series of PAO-4 lubricants containing both PLMA brush-grafted, 23-nm silica NPs and [P8888][DEHP] with a total concentration of 2.0 wt% but different individual concentrations for the two components and evaluated them in the boundary lubrication regime using a ball-on-flat reciprocating configuration at 100 °C under the same tribological testing conditions as in Section 11.3.3.

11.4.1 Preparation of PAO-4 Lubricants with Various Amounts of PLMA Hairy Silica NPs and [P8888][DEHP] and Stability of Hairy Silica NPs in the Presence of [P8888][DEHP]

The PLMA brush grafted, 23-nm silica NPs (denoted as HNP) were synthesized by SI-RAFT polymerization from the CTA-functionalized silica NPs in the presence of a free CTA, as described in Section 11.3 [42]. The DP of PLMA was calculated to be 25, and the $M_{n,SEC}$ and Đ of the free PLMA from SEC analysis were 6.5 kDa and 1.14, respectively, relative to polystyrene standards. The grafting density of PLMA brushes in this hairy NP sample was 0.51 chains nm^{-2} [43]. The ionic liquid used in this study was [P8888][DEHP] (denoted as IL, Scheme 11.3), synthesized by following the procedure in the literature [13].

To investigate the effects of mixing HNP and IL in PAO-4 on tribological properties, we prepared a set of PAO-4-based lubricants with various amounts of HNP and IL at a total concentration of 2%. The samples were: (a) 2% IL, (b) 2% HNP, (c) 0.34% HNP + 1.66% IL, (d) 0.66% HNP + 1.34% IL, (e) 1% HNP and 1% IL, (f) 1.34% HNP + 0.66% IL, (g) 1.66% HNP + 0.34% IL, and (h) 1.83% HNP + 0.17% IL. We previously showed that oil-soluble hairy NPs displayed excellent stability in PAO-4 at both low (−20 or −15 °C) and high temperatures (100 °C). [P8888][DEHP] is composed of charged species. To examine the stability of HNP in PAO-4 in the presence of [P8888][DEHP], we conducted DLS measurements at 23 °C using a Malvern Zetasizer Nano ZS instrument. A 1% HNP dispersion in PAO-4 and a dispersion of 1% HNP + 1% IL in PAO-4 were prepared and heated at 100 °C for 10 days, along with a 1% solution of IL in PAO-4. No aggregation or precipitation of NPs was observed for the HNP-containing dispersions; all three samples showed a slight yellow coloration compared with freshly prepared samples. Aliquots were subsequently withdrawn from the dispersions and diluted with pure PAO-4 to a concentration of 0.1 mg g^{-1} for the DLS study at 23 °C. DLS measurements revealed quite similar hydrodynamic diameters, in the range of 57.3–63.6 nm, for the hairy NPs from the freshly made dispersion of HNP in PAO-4, the dispersion of 1 wt% HNP + 1 wt% IL in PAO-4 after heating for 3 and 10 days at 100 °C, and the freshly prepared dispersion of 1 wt% HNP + 1 wt% IL in PAO-4. These DLS data demonstrated that the colloidal stability of HNP in PAO-4 was not affected by the addition of [P8888][DEHP] and they were compatible with each other in the lubricating base oil.

11.4.2 Lubrication Performances of PAO-4 Lubricants with the Addition of HNP, IL, and HNP + IL at Various Mass Ratios

To examine the possible further improvement in the lubrication performance by mixing HNP and IL in PAO-4 as additives, the aforementioned PAO-4-based lubricants with various amounts of HNP and IL at a total concentration of 2.0 wt% were evaluated using the identical tribological testing conditions detailed in the preceding sections. The results are summarized in Figure 11.6 for all the lubricant samples along with pure PAO-4 for comparison. Similar to the results described earlier or reported in the literature [13, 41–43], the PAO-4 lubricant with either 2 wt% HNP or 2 wt% IL exhibited a significant friction reduction with respect to neat PAO-4. At the end of the sliding process, the COF decreased by 16% for IL and 29% for HNP compared with the pure base oil. Unlike pure PAO-4, no scuffing peak was observed for all the lubricants with the addition of HNP, IL, or both, corroborating the previously reported anti-scuffing properties of both hairy silica NPs and IL [13, 41, 42]. Further friction reductions were found for the lubricants incorporated with both HNP and IL as additives at a total concentration of 2.0 wt%. When the concentration of one component was low, such as in (c) 0.34% HNP + 1.66% IL and (h) 1.83% HNP + 0.17% IL, the improvement in the friction reduction was small and the friction curves were just lower than the 2% HNP curve (b). Substantial friction decreases were found for all other lubricants with sufficient amounts of both HNP and IL: (d) 0.66% HNP + 1.34% IL, (e) 1.0% HNP + 1.0% IL, (f) 1.34% HNP + 0.66% IL, and (g) 1.66% HNP + 0.34%

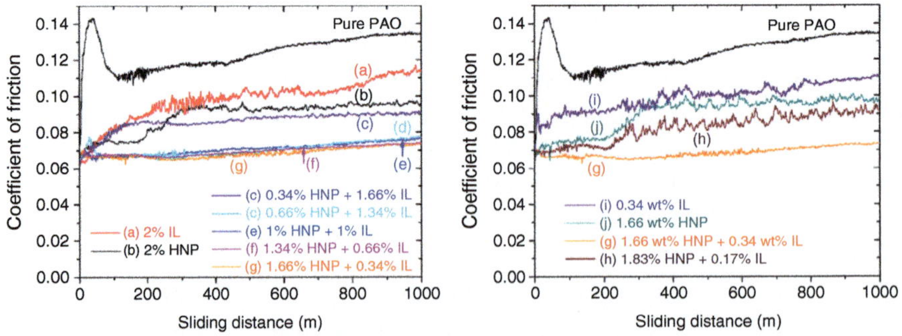

Figure 11.6 Friction curves for the PAO SpectraSyn™ 4 mixed with (a) 2% IL, (b) 2% HNP, (c) 0.34% HNP +1.66% IL, (d) 0.66% HNP + 1.34% IL, (e) 1% HNP and 1% IL, (f) 1.34% HNP + 0.66% IL, (g) 1.66% HNP + 0.34% IL, (h) 1.83% HNP + 0.17% IL, (i) 0.34% IL, and (j) 1.66% HNP. The tribological tests were performed using a Plint TE-77 tribotester at 100 °C under a point contact load of 100 N for a sliding distance of 1000 m. Source: Reproduced with permission from Seymour et al. [43]; © 2018, American Chemical Society.

IL. More precisely, at the end of the sliding process (i.e. at the sliding distance of 1000 m), the COF values decreased by 20–23% compared with the 2% HNP lubricant and by 32–35% with respect to the 2% IL. Moreover, as can be seen from Figure 11.6, curves (d), (e), (f), and (g) were noticeably smoother and flatter, indicating a positive effect of combining HNP and IL on the lubrication performance.

We then performed tribological tests for the lubricants containing only 0.34% IL and only 1.66% HNP to further confirm the synergistic effect of combining HNP and IL as additives for PAO. As can be seen from curves (i) and (j) in Figure 11.6, both lubricant samples showed higher COF values than those for (g) 1.66% HNP + 0.34% in PAO throughout the sliding process, demonstrating the positive effect of combining HNP and IL as additives on the lubrication performance. It should be noted here that throughout the sliding process the lubricant sample of (g) 1.66% HNP + 0.34% IL in PAO-4 performed noticeably better than 1 wt% ZDDP in PAO-4 in reducing the friction under the same testing conditions. Optical surface profilometry measurements of the wear scars on the iron flats and the steel balls after the tribological tests showed that the wear volumes were significantly reduced (Table 11.2). Total wear was reduced by 90% for all additive-containing lubricant samples when compared to neat PAO-4. The mixture of (c) 0.34% HNP + 1.66% IL resulted in the greatest wear reduction (93.9%), while (i) 0.34 wt% IL alone resulted in the smallest decrease (85.9%). Interestingly, the majority of the mixtures with a total concentration of 2% in PAO-4 resulted in lower wear volumes than either 2% HNP or 2% IL. For instance, the wear volume of the iron flat for (e) 1.0 wt% HNP + 1 wt% IL decreased by 33% and 17% relative to 2.0 wt% HNP and 2.0 wt% IL, respectively. There appeared to be a general trend that the wear decreased with increasing concentrations of IL from 0.17% to 1.66% in the mixture and simultaneously decreasing concentrations of HNP from 1.83% to 0.34%, suggesting that IL might have a better anti-wear ability than HNP.

Table 11.2 Wear volumes for iron flats and steel balls from tribological tests using various lubricants.

Lubricant sample	Flat wear volume ($\times 10^7$ μm^3)	Ball wear volume ($\times 10^7$ μm^3)
PAO	113.0 ± 35.9	0.786 ± 0.193
(a) PAO + 2% IL	9.90 ± 0.99	0.405 ± 0.120
(b) PAO + 2% HNP	12.26 ± 0.86	0.039 ± 0.026
(c) PAO + 0.34% HNP + 1.66% IL	6.91 ± 0.73	0.067 ± 0.007
(d) PAO + 0.66% HNP + 1.34% IL	7.46 ± 0.33	0.018 ± 0.006
(e) PAO + 1% HNP + 1% IL	8.21 ± 0.17	0.018 ± 0.006
(f) PAO + 1.34% HNP + 0.66% IL	9.77 ± 1.57	0.016 ± 0.010
(g) PAO + 1.66% HNP + 0.34% IL	13.32 ± 1.53	0.012 ± 0.013
(h) PAO + 1.83% HNP + 0.17% IL	12.90 ± 0.63	0.029 ± 0.020
(i) PAO + 0.34% IL	15.47 ± 1.21	0.571 ± 0.293
(j) PAO + 1.66% HNP	14.01 ± 0.44	0.057 ± 0.002

Source: Reproduced with permission from Seymour et al. [43]; © 2018, American Chemical Society.

11.4.3 SEM–EDS and XPS Analysis of Wear Scars Formed on Iron Flats from Tribological Tests

The further improvement in the friction reduction observed for PAO-4 lubricants containing both HNP and IL likely originated from the formation of stronger, more durable tribofilms on the surface of rubbing substrates. These complex tribofilms are usually enriched with elements from additives, as discussed earlier. To study and confirm the tribofilm formation on the iron flats from the tribological tests of PAO-4 lubricants containing both HNP and IL, we first used SEM–EDS. Figure 11.7 shows the SEM micrographs of the wear scars formed on the iron flats from the tribological testing of 2.0 wt% IL, 2.0 wt% HNP, and the mixture of 1.0 wt% HNP + 1.0 wt% IL in PAO-4, as well as the mapping of elements Fe, P, O, and Si. For all three lubricants, no difference was observed for elemental iron inside and outside the wear tracks. However, a noticeable increase in the oxygen content for all three samples was observed inside the wear scars compared with the region outside the wear scars, indicating that oxidation was involved in the tribochemical reactions. Element P was observed in the wear scar for the lubricant of 2.0 wt% IL but not for the one with 2.0 wt% HNP, whereas element Si was enriched in the wear track for the 2.0% HNP lubricant but not for the one with 2.0 wt% IL. For the sample of 1.0 wt% HNP and 1.0 wt% IL, a marked contrast was observed for both elements P and Si inside and outside the wear scar. These observations were further confirmed by EDS. A noticeable increase in the Si Kα peak intensity was found inside the wear scar relative to the pristine flat for the lubricants containing either 2.0 wt% HNP or the mixture of 1.0 wt% HNP + 1.0 wt% IL, but not for the 2.0% IL. On the other hand, the P Kα peak was found in the EDS spectra for the lubricants with 2.0 wt% IL and 1.0 wt% HNP + 1.0 wt% IL but not for the sample with 2.0 wt% HNP.

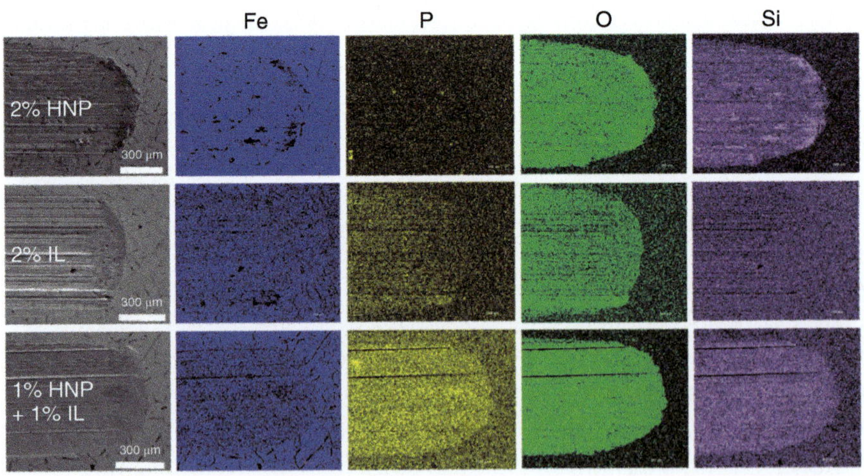

Figure 11.7 SEM micrographs and EDS elemental mapping of Fe, P, O, and Si of the wear scar at the end of wear track formed on the iron flat during the tribological test of the PAO mixed with 2% HNP (top row), 2% IL (middle row), and 1% HNP + 1% IL (bottom row). Source: Seymour et al. [43], Reproduced with permission from American Chemical Society.

Quantitative EDS analysis revealed the atomic compositions of the wear tracks. It should be pointed out that the chemical composition information underneath the tribofilm was also collected in EDS because the probing depth of EDS is a few micrometers [87] and the tribofilms are tens to a few hundreds of nanometers. For all lubricants with both hairy SiO_2 NPs and IL, a much higher P content was found compared with the lubricant containing only HNP, demonstrating that IL participated in the triboreaction. Understandably, the atomic ratio of P to Si increased with increasing concentrations of IL, from 0.080 for (h) 1.83% HNP + 0.17% IL to 0.486 for (c) 0.34% HNP + 1.66% IL. For the lubricants with both HNP and IL that showed a good improvement in friction reduction, the ratios of P to Si were found to be in the range of 0.16–0.27. These SEM–EDS data unequivocally showed that both HNP and IL, or their fragments, were involved in the tribochemical reactions to form the tribofilms for the lubricants containing both HNP and IL.

As discussed earlier, while the phosphate anions from [P8888][DEHP] can adsorb onto the iron substrate, PLMA brush-grafted silica NPs presumably do not interact with the metal surface before the tribological testing. Thus, we were intrigued to see if the IL participated in the tribochemical reaction in the early stage while hairy silica NPs joined in the later stage. We performed three additional tribological tests with the PAO-4 lubricant containing 1.0 wt% HNP + 1.0 wt% IL and halted the sliding at 100, 400, and 700 m. The three friction curves overlapped nicely with the friction curve for the full tribological testing process for the same lubricant. SEM–EDS were then employed to characterize the wear scars from these three shorter tests. The atomic ratio of P to Si appeared to increase slightly with increasing sliding distance, from 0.20 for 100 m to 0.26 for 1000 m, but all the values were in the range of 0.16–0.27 observed for those lubricants with both HNP and IL that showed significant friction reductions. These data suggested that both HNP and IL or their fragments were involved in the tribochemical reaction in the studied sliding distance range.

Because the probing depth of EDS (a few micrometers) is much greater than the typical thicknesses of tribofilms, X-ray photoelectron spectroscopy (XPS) was employed to further characterize the wear scars. Note that XPS has a much smaller probing depth of 0.1–10 nm [88], typically. The XPS spectrum of the wear scar formed from the 2.0 wt% HNP lubricant showed the peaks of Si 2s and 2p at 153 and 102 eV, respectively, but no peaks of P 2s and 2p, whereas the tribofilm produced from the 2.0 wt% IL lubricant was found to contain element P (P 2s at 190 eV and P 2p at 133 eV) but negligible Si. As expected, the XPS spectrum of the tribofilm from the lubricant with 1.0 wt% HNP and 1.0 wt% IL showed the 2s and 2p peaks of both Si and P, indicating that both HNP and IL were involved in the tribochemical reaction. These XPS results are consistent with the SEM–EDS data. The atomic compositions were obtained from quantitative XPS analysis, and the P/Si molar ratio for the lubricant containing 1.0% HNP and 1.0% IL was 0.393, larger than that from the EDS analysis (0.260), which was likely caused by the different probing depths of the two characterization techniques. Analysis of the core level spectra of C 1s, P 2p, O 1s, and Si 2p showed that there existed a possibility that a portion of the Si—O—Si bonds were replaced by new covalent bonds involving Si, O, and

P from the reaction between silica and phosphate anions. We further conducted depth profiling of the tribofilm from the tribological test of the lubricant with 1.0 wt% HNP and 1.0 wt% IL using argon-ion sputtering to etch the surface. The Fe content was found to increase with the increase of sputtering time, which was attributed to the gradual exposure of the underneath iron substrate. Despite the increasing iron signal, the atomic ratio of P to Si seemed to be similar from the top to the interior of the tribofilm. This indicated that both additives participated in the tribochemical reaction throughout the tribological testing process. Thus, both EDS and XPS analysis suggested that HNP and IL reacted with each other and also with the iron substrate during the rubbing process, which presumably enhanced the formation and the properties of the tribofilm and strengthened its bonding with the underlying metal, resulting in a further improvement in the lubricating performance of the PAO-4 lubricants with sufficient amounts of both HNP and IL.

11.5 Upper Critical Solution Temperature (UCST)-Type Thermoresponsive Poly(alkyl methacrylate)s in PAO-4

Thermoresponsive polymers exhibit reversible soluble–to–insoluble transitions in liquid media in response to temperature changes and can be classified into either lower critical solution temperature (LCST)-type or UCST-type [89, 90]. The LCST polymers are soluble in the liquid media at lower temperatures but become insoluble upon heating, whereas the UCST polymers are soluble at higher temperatures but turn insoluble upon cooling. These polymers have been intensively investigated in the past decades and have found (potential) applications in various areas, including injectable drug delivery systems, tissue engineering, and actuators [91–95]. While most thermoresponsive polymers are found in aqueous media, such polymers are also known in organic solvents [96, 97]. One notable example is that polystyrene exhibits a UCST in cyclohexane [96]. However, one major drawback of common organic solvents is their volatility, which has effectively limited the practical use of thermoresponsive polymers in such liquid media. In contrast, PAO is a class of non-volatile and non-crystallizable industrially important organic liquids [3], which are widely used as base lubricating oils for formulating motor lubricants, gear box oils, and hydraulic oils. These hydrocarbon organic liquids exhibit excellent thermal and chemical stability and superior lubrication properties, which largely originate from their branched and saturated hydrocarbon molecular structures.

We previously synthesized a series of poly(alkyl methacrylate) brush-grafted SiO_2 NPs with alkyl pendant groups of different sizes by SI-RAFT polymerization and used them as additives for PAO-4 for friction and wear reduction [42]. For poly(alkyl methacrylate) brushes with a sufficiently long side chain (>8 carbon atoms in the pendant group), the hairy NPs exhibited outstanding dispersibility and stability in PAO-4 in a wide temperature range from −20 to 100 °C and excellent lubricating properties. Intriguingly, we found that B-PC8-brush grafted, 23-nm silica NPs could not be dispersed in PAO-4 at room temperature but formed a clear homogeneous dispersion when heated at elevated temperatures. The cloudy-to-clear transition was reversible; upon cooling to room temperature, the clear dispersion of hairy NPs in PAO-4 turned cloudy, and eventually the brush silica NPs settled at the bottom of the container. Because B-PC8 is soluble in PAO-4 at room temperature, the observed UCST behavior was believed to be caused by the presence of silica in the NP core. The attractive interactions between silica NPs at lower temperatures outweigh the solvation of the B-PC8 brushes in PAO-4. These observations showed that the alkyl pendant length of poly(alkyl methacrylate) is crucial for the dispersibility and stability of brush NPs in PAO-4.

We hypothesized that poly(alkyl methacrylate)s with an appropriate number of carbon atoms in the side chain could exhibit UCST-type behavior in PAO-4 after being intrigued by the behavior of B-PC8-brush-grafted, 23-nm silica NPs in PAO-4. We used RAFT polymerization to prepare a large number of homopolymers and random copolymers from a set of alkyl methacrylate monomers (Scheme 11.4a) and investigated their solution behaviors in PAO-4 over a wide temperature range. As expected, poly(alkyl methacrylate)s with an appropriate (average) number of carbon atoms in the alkyl pendant group exhibited UCST transitions in PAO-4 [98]. We further synthesized linear ABA triblock copolymers with UCST-type thermoresponsive outer blocks and a PAO-philic middle block and used them as gelators for PAO-4 (Scheme 11.4b). The PAO-4 gels can be used as injectable gel lubricants for friction reduction.

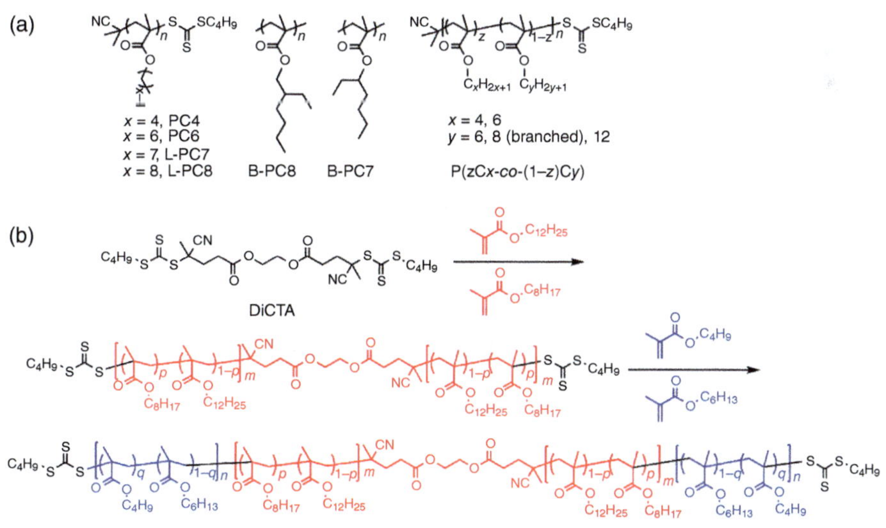

Scheme 11.4 (a) Molecular structures of poly(alkyl methacrylate)s with various alkyl pendants and random copolymers. (b) Synthesis of ABA triblock copolymers with a PAO-4-philic middle block and UCST thermoresponsive outer blocks by RAFT polymerization. Source: Reproduced with permission from Fu et al. [98]; © 2018, American Chemical Society.

11.5.1 Synthesis of Poly(alkyl methacrylate)s with Various Alkyl Pendant Groups by RAFT Polymerization and Their Thermoresponsive Properties in PAO-4

Poly(*n*-heptyl methacrylate) that contains seven carbon atoms in the linear alkyl side chain (L-PC7, Scheme 11.4a) was synthesized first by RAFT polymerization in anisole at 70 °C using *n*-butyl (2-cyano-2-propyl) trithiocarbonate as chain transfer agent and AIBN as initiator. The L-PC7 with a DP of 165 underwent a clear-to-cloudy transition upon cooling with a cloud point (CP) of 35.0 °C, determined from the plot of optical transmittance versus temperature (Figure 11.8). This UCST behavior was reversible, and a cloudy-to-clear transition was observed at 37.2 °C, with a small hysteresis of 2.2 °C between cooling and heating. The CP was found to increase from 28.2 °C at 0.5 wt% to 48.5 °C at 5 wt%, evidencing that the CP was sensitive to the polymer concentration. In addition, we found that the CP of L-PC7 in PAO-4 at a concentration of 1.0 wt% increased from 24.7 °C at a DP of 100–45.5 °C at a DP of 228. These observations are consistent with the behavior of typical UCST thermoresponsive polymers, whose CPs are heavily dependent on the polymer's concentration and molecular weight [90, 99–101].

To study how the alkyl pendant length affected the solution behavior of poly(alkyl methacrylate)s in PAO-4, we prepared by RAFT polymerization a set of homopolymers of various alkyl methacrylates with similar DPs that contained four to eight carbon atoms in the linear or branched alkyl pendant group (Scheme 11.4a). These polymers all underwent UCST transitions in PAO-4, with the CP at a concentration of 1.0 wt% ranging from 0.5 °C for L-PC8, to 3.8 °C for B-PC8, 36.1 °C for B-PC7,

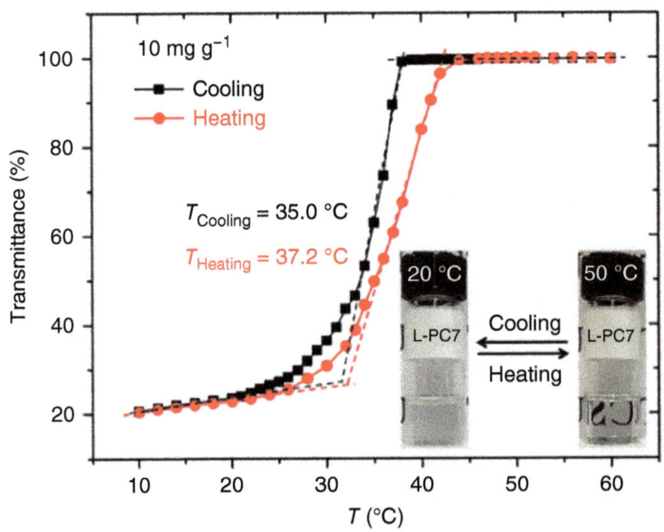

Figure 11.8 Optical transmittance as a function of temperature for a 10 mg g^{-1} solution of an L-PC7 in PAO in a cooling–heating cycle measured at 500 nm using a UV–vis spectrometer. Source: Fu et al. [98], Reproduced with permission from American Chemical Society.

71.5 °C for PC6, and 138 °C for PC4. As expected, the CP increased with the decreasing number of carbon atoms in the side chain, and PC12 is known to be completely soluble in PAO-4 [41]. One interesting observation is that the CPs of linear PC7 and PC8 were found to be lower than their branched counterparts by 1.1 and 3.3 °C, respectively, which can be attributed to the larger contact area of the linear alkyl pendant group than the branched side chain of the same number of carbon atoms and thus slightly stronger interactions with PAO-4 molecules. Linear fitting of the CPs of the four homopolymers with *n*-alkyl pendant groups showed CP (°C) = $276.4 - 34.4\,n_C$ with R^2 of 0.9995; in other words, decreasing the number of carbon atoms in the side chains by one increased the CP by 34.4 °C. Copolymerization of two or more monomers is widely practiced for achieving desired CPs [89–92, 102–104]. To see if the linear relationship holds between the CP and the average number of carbon atoms in the alkyl side chains, we synthesized a series of random copolymers of two distinct monomers by RAFT copolymerization. The monomers include 2-ethylhexyl methacrylate, *n*-hexyl methacrylate, *n*-butyl methacrylate, and LMA. The DPs of the random copolymers (around 155) were similar to those of the four homopolymers, and thus the effect of molecular weight on CP was minimized. The CPs of these copolymers in PAO-4 at a concentration of 1 wt% were determined and are plotted against the average number of carbon atoms in the side chain per repeat unit ($\langle n_C \rangle$), which was calculated using the mole fractions of two monomer units in the copolymer. There was a linear relationship between CP and $\langle n_C \rangle$; linear fitting showed CP = $275.4 - 34.5\langle n_C \rangle$ with R^2 of 0.9909, virtually the same as the dependence of CP on the number of carbon atoms for homopolymers of methacrylates with linear alkyl pendants. Thus, by varying the (average) number of carbon atoms in the side chains via homopolymerization or copolymerization, one can precisely control the CPs of thermoresponsive polymers in PAO over a large temperature range.

11.5.2 UCST-Type Thermoresponsive ABA Triblock Copolymers as Gelators for PAO-4

Linear ABA triblock copolymers composed of a water-soluble central block and two thermoresponsive outer blocks are known to reversibly self-assemble into a three-dimensional (3-D) network in water upon temperature changes when the polymer concentration is sufficiently high [105–108]. At temperatures above the LCST or below the UCST, the thermosensitive outer blocks become insoluble in water and associate with each other into micellar cores, which are bridged by the soluble middle blocks. PAO is a family of nonvolatile, saturated hydrocarbon organic liquids widely used in industry as lubricating base oils. The discovery of UCST-type thermoresponsive poly(alkyl methacrylate)s in PAO-4 provides an opportunity for using PAO in the gel form, e.g. as injectable gel lubricants.

Similar to the thermosensitive water-soluble ABA triblock copolymers used as gelators for water [105–108], it is possible to use UCST thermoresponsive ABA linear triblock copolymers to thermoreversibly gel PAO. We prepared a difunctional trithiocarbonate CTA (DiCTA) and used it to synthesize a series of ABA triblock copolymers with a PAO-4-philic central block and two UCST thermosensitive end blocks by two-step RAFT polymerizations (Scheme 11.4b). The first ABA triblock copolymer was composed of a PC12 middle block with a DP of 313 and PC6 outer blocks, each with a DP of 163, and was used to make a 15 wt% solution in PAO-4. Although the CP of PC6 in PAO-4 is quite high (71.5 °C), only a weak gel was observed by visual inspection for the 15 wt% PAO solution of this ABA copolymer even at 0 °C. Considering that the UCST of a thermoresponsive polymer in a liquid medium is very sensitive to the chain length and the linking to a soluble polymer, the poor gelation of this ABA triblock copolymer was likely caused by the high solubility of the PC12 central block. Therefore, we made copolymers of 2-ethylhexyl methacrylate and LMA with a molar ratio of about 1 : 1 as the central block and poly(*n*-butyl methacrylate)-*co*-(*n*-hexyl methacrylate) (P(C4-*co*-C6)) with a molar ratio of 4 : 1 as outer blocks; three ABA triblock copolymers were prepared: P($C4_{106}$-*co*-$C6_{27}$)-*b*-P($C8_{183}$-*co*-$C12_{193}$)-*b*-P($C4_{106}$-*co*-$C6_{27}$) with $M_{n,SEC}$ of 1.02×10^5 Da and Đ of 1.16 (ABA-a), P($C4_{104}$-*co*-$C6_{27}$)-*b*-P($C8_{240}$-*co*-$C12_{248}$)-*b*-P($C4_{104}$-*co*-$C6_{27}$) with $M_{n,SEC}$ of 1.09×10^5 Da and Đ of 1.22 (ABA-b), and P($C4_{104}$-*co*-$C6_{27}$)-*b*-P($C8_{240}$-*co*-$C12_{248}$)-*b*-P($C4_{104}$-*co*-$C6_{27}$) with $M_{n,SEC}$ of 1.19×10^5 Da and Đ of 1.23 (ABA-c). The CPs of the middle block and outer blocks were estimated to be −70 and 123 °C in PAO-4 when they were not linked together.

Solutions of these three ABA triblock copolymers in PAO-4 at a concentration of 15 wt% all showed a UCST-type transition from a clear sol to a clear gel upon decreasing temperature. To determine the sol–gel transition temperature ($T_{sol\text{-}gel}$), oscillatory shear rheological measurements were conducted in a temperature ramp at a cooling rate of 3 °C min^{-1}, a constant frequency of 1 Hz, a strain amplitude of 1.0%. The values of the dynamic storage modulus (G′) and loss modulus (G″) at elevated temperatures were small and fluctuating, evidencing a liquid state. Upon cooling, G′ increased faster than G″ and became larger than G″ below a crossover point, indicating a sol–gel transition. The crossover point of the G′ and G″ curves in the temperature ramp is commonly taken as the $T_{sol\text{-}gel}$ [106–109]. Upon heating, the gel was converted to a sol, and the gel–sol transition temperature ($T_{gel\text{-}sol}$) from the rheological measurement for the 15 wt% solution of ABA-a was 52.9 °C, which is essentially the same as $T_{sol\text{-}gel}$ (52.0 °C). In addition, the $T_{sol\text{-}gel}$ depended on the polymer concentration; it increased with increasing polymer concentration, from 44.6 °C at 10 wt% to 60.6 °C at 20 wt%, and we found that the hysteresis between $T_{sol\text{-}gel}$ and $T_{gel\text{-}sol}$ was quite small for all concentrations investigated.

For ABA-a and -b with similar end block lengths, at the same concentration of 15 wt%, the value of $T_{sol\text{-}gel}$ increased from 52.0 to 57.7 °C upon increasing the central block DP from 376 to 488. The middle blocks for ABA-b and -c were the same with a total DP of 488; rheological studies showed that the $T_{sol\text{-}gel}$ increased from 57.7 to 63.5 °C when the DP of the outer block was increased from 131 to 214, which was reasonable because a longer UCST thermoresponsive outer block would exhibit a higher cloud point, leading to a higher $T_{sol\text{-}gel}$ for the 15 wt% solution of the ABA triblock copolymer in PAO-4.

Because of the nonvolatility of PAO-4, these organic gels can be very stable and robust even in an open environment, which is in stark contrast to hydrogels from water and organogels from volatile organic solvents, where solvent evaporation is often a serious issue that leads to the drying out of gels. Furthermore, by adding a small amount of a thermoresponsive random copolymer and exploiting the clear-to-cloudy transition, the PAO-4 gels of ABA triblock copolymers could exhibit a reversible transparent-to-cloudy transition upon temperature changes. For example, we prepared a PAO-4 solution that contained 14.3 wt% ABA-a and 5 wt% of a random copolymer with a CP of 30.6 °C at a concentration of 1 wt%. Upon cooling from elevated temperatures, the sample showed clear sol-clear gel-cloudy gel transitions and the clouding occurred at 36 °C. The cloudy gels were observed to be very stable; even after being left open to air at ambient conditions for an extended period of time, no any changes were found, likely because the collapsed random copolymer was effectively trapped in the 3-D network without settling. The clear gel-to-cloudy gel transition can be varied by changing the thermoresponsive random copolymer. These stable PAO gels could be used to control optical transmittance.

The thermally induced reversible sol–gel transitions of ABA copolymer solutions in PAO-4 make it possible to use them as injectable gel lubricants, which can be applied as liquids at higher temperatures and used as gel lubricants at lower temperatures. The liquid state of the lubricants at elevated temperatures allows the gel lubricants to form conformal contact with (rough) solid surfaces. Note that gel lubricants have received increasing interest in recent years as they exhibit advantages under certain conditions of, e.g. high loads, over liquid lubricants [110, 111]. To demonstrate the lubrication properties of the gels of ABA triblock copolymers in PAO-4, we constructed a home-made setup in the same principle as Amonton's law (Figure 11.9A) [2, 112] to measure the static friction coefficient at the glass–glass

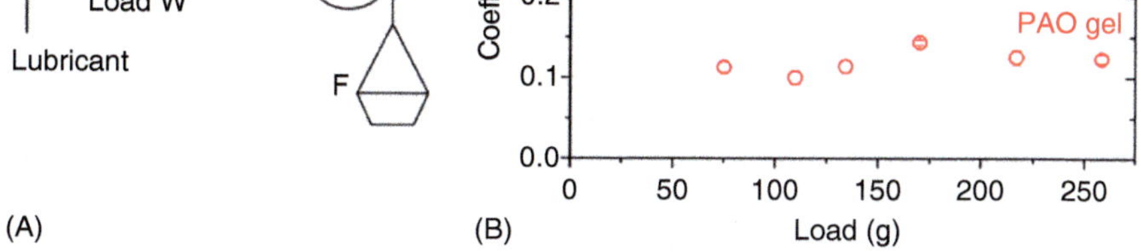

Figure 11.9 (A) Schematic illustration of the home-made friction test setup. Glass substrates (a) and (b) were precleaned microscope glass slides with a size of 7.5 × 5.0 × 0.1 cm; glass slide (a) was bonded firmly by glue to glass substrate (c) with dimensions of 3.0 × 3.0 cm to ensure no relative motion between (a) and (c). Note that the rotation friction of pulley was negligible compared with that between glasses. (B) Plots of static coefficients of friction of PAO- and 10 wt% ABA-2 PAO gel-lubricated glass–glass interface versus load. Source: Reproduced with permission from Fu et al. [98]; © 2018, American Chemical Society.

interface using the gel formed by 10 wt% ABA-a in PAO-4 as an example and pure PAO-4 for comparison. The PAO-4 gel or pure PAO-4 (9 mg) was applied to the glass substrate (b), and the bonded glass slides (a) and (c) with a string were then placed on top of the lubricant-covered substrate (b). We then applied a load (W) and gradually added water into the container suspended from the pulley until the bonds (a) and (c) began to move. The weight that resulted in the motion was measured (F), allowing for the calculation of the static COF (μ_s) for the glass–glass interface using $\mu_s = F/W$. The measurement for each load was repeated a total of three times, and the average μ_s value and the standard deviation were obtained. In the investigated load range of 50–250 g, the μ_s for the glass–glass interface lubricated by the gel decreased by four times compared with those lubricated by PAO-4 (Figure 11.9B). This is most likely due to the liquid nature of PAO-4 made it being squeezed out from the glass–glass interface much more easily, causing the surface asperities from the opposing glass slides to come into direct contact and thus a higher friction. Differently, the gel retained PAO-4 fluid in the 3-D polymer network even under higher loads and exhibited much small μ_s values. It should be noted that the μ_s values for the gel were too small to be measurable at low loads ($\leq$50 g).

11.6 Summary

In summary, this chapter shows that oil-soluble polymer brush-grafted inorganic NPs are a promising class of oil lubricant additives for friction reduction and wear protection. These organic–inorganic hybrid NPs can be synthesized by SI-ATRP or SI-RAFT polymerization with the addition of a free initiator or CTA, and the brush molecular weight can be readily tuned by either controlling the monomer-to-initiator/CTA molar ratio or the monomer conversion or both. When the length of the alkyl pendant group of poly(alkyl methacrylate) brushes is sufficiently large, the brush NPs exhibit excellent dispersibility in lubricating base oil PAO-4 and superior stability in the temperature range of −20 to 100 °C. High contact stress, ball-on-flat reciprocating tribological tests showed large friction decreases (up to 40%) and wear reductions (90%) for the lubricants containing 1 wt% of hairy NPs compared with neat PAO-4, and some of hairy NP samples performed even better than the commercially used ZDDP at the same concentration under the same tribological conditions. The analysis revealed that a substantial portion of the tribological benefits of hairy NPs came from the inorganic core NPs and the grafted polymer brushes mainly provide the colloidal stability to the NPs. It is possible to increase the function of polymer brushes by incorporating triboactive elements into the brushes. By combining oil-soluble hairy silica NPs and an oil-miscible phosphonium–phosphate ionic liquid as friction reducers for PAO-4 at a total concentration of 2 wt% with sufficient amounts for both components, further improvements in friction reduction were achieved, likely because both hairy silica NPs and the IL participated in the tribochemical reaction with the metal substrate, resulting in more robust tribofilms and stronger bonding with the metal substrate. Moreover, we discovered a class of UCST-type thermoresponsive poly(alkyl methacrylate)s with an appropriate alkyl pendant length and used them as building blocks to construct thermoresponsive ABA triblock copolymers for gelling PAO-4. At sufficiently high concentrations, the ABA copolymer solutions in PAO-4 were liquid-like at high temperatures but turned into gels upon cooling; the gels exhibited a much lower friction coefficient with a better load-bearing property compared with PAO-4. It is our belief that oil-dispersible, hairy inorganic NPs will be useful in real-world lubrication for friction reduction and wear protection, and the newly discovered UCST-type thermoresponsive polymers could be used for developing PAO-based injectable gel lubricants to further expand the applications of the already widely used synthetic lubricating base oils.

Acknowledgments

The authors would like to thank DOE-EERE for the support of the research presented in this chapter and NSF for the partial support of the work on UCST-type thermoresponsive polymers in PAO-4. We are indebted to our coworkers and collaborators, especially Dr. Jun Qu for tribological testing.

References

1 Tung, S.C. and McMillan, M.L. (2004). *Tribol. Int.* 37: 517–536.
2 Bhushan, B. (2013). *Introduction to Tribology*, 2e. New York: Wiley.
3 Rudnick, L.R. (2013). *Synthetics, Mineral Oils, and Bio-Based Lubricants*, 2e. Boca Raton: CRC Press.
4 Stachowiak, G.W. and Batchelor, A.W. (2006). *Engineering Tribology*. Amsterdam: Elsevier.
5 Zhu, D., Wang, J.X., and Wang, Q.J. (2015). *J. Tribol.* 137: 021501.
6 Barnes, A.M., Bartle, K.D., and Thibon, V.R.A. (2001). *Tribol. Int.* 34: 389–395.
7 Holmberg, K., Andersson, P., and Erdemir, A. (2012). *Tribol. Int.* 47: 221–234.
8 Rudnick, L.R. (2009). *Lubricant Additives: Chemistry and Applications*, 2e. Boca Raton: CRC Press.
9 Spikes, H. (2004). *Tribol. Lett.* 17: 469–489.
10 Nicholls, M.A., Do, T., Norton, P.R. et al. (2005). *Tribol. Int.* 38: 15–39.
11 Zhang, J. and Spikes, H. (2016). *Tribol. Lett.* 63: 24.
12 Demydov, D., Adhvaryu, A., McCluskey, P., and Malshe, A.P. (2010). *Nanoscale Materials in Chemistry: Environmental Applications*, ACS Symposium Series, Chapter 8, vol. 1045 (ed. L.E. Erickson, R.T. Koodali and R.M. Richards), 137–163. Washington, DC: American Chemical Society. Advanced Lubricant Additives of Dialkyldithiophosphate (DDP)-Functionalized Molybdenum Sulfide Nanoparticles and Their Tribological Performance for Boundary Lubrication.
13 Qu, J., Bansal, D.G., Yu, B. et al. (2012). *ACS Appl. Mater. Interfaces* 4: 997–1002.
14 Martin, J.M. and Ohmae, N. (2008). *Nanolubricants*. Chichester: Wiley.

15 Xue, Q.J., Liu, W.M., and Zhang, Z.J. (1997). *Wear* 213: 29–32.
16 Qiu, S., Zhou, Z., Dong, J., and Chen, G. (1999). *J. Tribol.* 123: 441–443.
17 Tarasov, S., Kolubaev, A., Belyaev, S. et al. (2002). *Wear* 252: 63–69.
18 Joly-Pottuz, L., Dassenoy, F., Belin, M. et al. (2005). *Tribol. Lett.* 18: 477–485.
19 Rapoport, L., Lvovsky, M., Lapsker, I. et al. (2001). *Nano Lett.* 1: 137–140.
20 Battez, A.H., Rico, J.E.F., Arias, A.N. et al. (2006). *Wear* 261: 256–263.
21 Li, B., Wang, X., Liu, W., and Xue, Q. (2006). *Tribol. Lett.* 22: 79–84.
22 Wu, Y.Y., Tsui, W.C., and Liu, T.C. (2007). *Wear* 262: 819–825.
23 Kim, D. and Archer, L.A. (2011). *Langmuir* 27: 3083–3094.
24 Padgurskas, J., Rukuiza, R., Prosyčevas, I., and Kreivaitis, R. (2013). *Tribol. Int.* 60: 224–232.
25 Rabaso, P., Ville, F., Dassenoy, F. et al. (2014). *Wear* 320: 161–178.
26 Spikes, H. (2015). *Tribol. Lett.* 60: 5.
27 Gulzar, M., Masjuki, H.H., Kalam, M.A. et al. (2016). *J. Nanopart. Res.* 18: 223.
28 Li, D.D., Xie, Y.C., Yong, H.S., and Sun, D.Z. (2017). *RSC Adv.* 7: 3727–3735.
29 Dai, W., Kheireddin, B., Gao, H., and Liang, H. (2016). *Tribol. Int.* 102: 88–98.
30 Chen, Y., Renner, P., and Liang, H. (2019). *Lubricants* 7: 7.
31 Hamley, I.W. (2007). *Introduction to Soft Matter: Synthetic and Biological Self-Assembling Materials*, Revised Ed. Chichester: Wiley.
32 Chancellor, A.J., Seymour, B.T., and Zhao, B. (2019). *Anal. Chem.* 91: 6391–6402.
33 Pyun, J. and Matyjaszewski, K. (2001). *Chem. Mater.* 13: 3436–3448.
34 Zhao, B. and Zhu, L. (2009). *Macromolecules* 42: 9369–9383.
35 Bao, C.H., Horton, J.M., Bai, Z.F. et al. (2014). *J. Polym. Sci. B Polym. Phys.* 52: 1600–1619.
36 Briscoe, W.H., Titmuss, S., Tiberg, F. et al. (2006). *Nature* 444: 191–194.
37 Nomura, A., Okayasu, K., Ohno, K. et al. (2011). *Macromolecules* 44: 5013–5019.
38 Liu, G.Q., Liu, Z.L., Li, N. et al. (2014). *ACS Appl. Mater. Interfaces* 6: 20452–20463.
39 Banquy, X., Burdynska, J., Lee, D.W. et al. (2014). *J. Am. Chem. Soc.* 136: 6199–6202.
40 Bielecki, R.M., Benetti, E.M., Kumar, D., and Spencer, N.D. (2012). *Tribol. Lett.* 45: 477–487.
41 Wright, R.A.E., Wang, K.W., Qu, J., and Zhao, B. (2016). *Angew. Chem. Int. Ed.* 55: 8656–8660.
42 Seymour, B.T., Wright, R.A.E., Parrott, A.C. et al. (2017). *ACS Appl. Mater. Interfaces* 9: 25038–25048.
43 Seymour, B.T., Fu, W., Wright, R.A.E. et al. (2018). *ACS Appl. Mater. Interfaces* 10: 15129–15139.
44 Ingole, S., Charanpahari, A., Kakade, A. et al. (2013). *Wear* 301: 776–785.
45 Zubarev, E.R., Xu, J., Sayyad, A., and Gibson, J.D. (2006). *J. Am. Chem. Soc.* 128: 15098–15099.
46 Shan, J., Nuopponen, M., Jiang, H. et al. (2005). *Macromolecules* 38: 2918–2926.
47 Prucker, O. and Rühe, J. (1998). *Macromolecules* 31: 592–601.
48 Prucker, O. and Rühe, J. (1998). *J. Macromolecules* 31: 602–613.

49 von Werne, T. and Patten, T.E. (1999). *J. Am. Chem. Soc.* 121: 7409–7410.

50 Husseman, M., Malmstrom, E.E., McNamara, M. et al. (1999). *Macromolecules* 32: 1424–1431.

51 Ohno, K., Morinaga, T., Koh, K. et al. (2005). *Macromolecules* 38: 2137–2142.

52 Bartholome, C., Beyou, E., Bourgeat-Lami, E. et al. (2003). *Macromolecules* 36: 7946–7952.

53 Li, C., Han, J., Ryu, C.Y., and Benicewicz, B.C. (2006). *Macromolecules* 39: 3175–3183.

54 Hotchkiss, J.W., Lowe, A.B., and Boyes, S.G. (2007). *Chem. Mater.* 19: 6–13.

55 Matyjaszewski, K. and Xia, J. (2001). *Chem. Rev.* 101: 2921–2990.

56 Hawker, C.J., Bosman, A.W., and Harth, E. (2001). *Chem. Rev.* 101: 3661–3688.

57 Moad, G., Rizzardo, E., and Thang, S.H. (2008). *Acc. Chem. Res.* 41: 1133–1142.

58 Xia, Y., Olsen, B.D., Kornfield, J.A., and Grubbs, R.H. (2009). *J. Am. Chem. Soc.* 131: 18525–18532.

59 Chmielarz, P., Yan, J., Krys, P. et al. (2017). *Macromolecules* 50: 4151–4159.

60 Wright, R.A.E., Hu, B., Henn, D.M., and Zhao, B. (2015). *Soft Matter* 11: 6808–6820.

61 Wright, R.A.E., Henn, D.M., and Zhao, B. (2016). *J. Phys. Chem. B* 120: 8036–8045.

62 Li, D.J., Sheng, X., and Zhao, B. (2005). *J. Am. Chem. Soc.* 127: 6248–6256.

63 Bao, C.H., Tang, S.D., Wright, R.A.E. et al. (2014). *Macromolecules* 47: 6824–6835.

64 Zhu, L. and Zhao, B. (2008). *J. Phys. Chem. B* 112: 11529–11536.

65 Jiang, X.M., Zhong, G.J., Horton, J.M. et al. (2010). *Macromolecules* 43: 5387–5395.

66 Jiang, X.M., Zhao, B., Zhong, G.J. et al. (2010). *Macromolecules* 43: 8209–8217.

67 Qu, J., Truhan, J.J., and Blau, P.J. (2005). *Int. J. Engine Res.* 6: 1–9.

68 Qu, J., Truhan, J.J., Blau, P.J., and Meyer, H.M. III, (2005). *Wear* 259: 1031–1040.

69 Zhou, Y., Dyck, J., Graham, T.W. et al. (2014). *Langmuir* 30: 13301–13311.

70 Qu, J., Chi, M., Meyer, H.M. III, et al. (2011). *Tribol. Lett.* 43: 205–211.

71 Qu, J., Barnhill, W.C., Luo, H. et al. (2015). *Adv. Mater.* 27: 4767–4774.

72 Li, C. and Benicewicz, B.C. (2005). *Macromolecules* 38: 5929–5936.

73 Peng, Q., Lai, D.M.Y., Kang, E.T., and Neoh, K.G. (2006). *Macromolecules* 39: 5577–5582.

74 Zhao, Y. and Perrier, S. (2007). *Macromolecules* 40: 9116–9124.

75 Ohno, K., Ma, Y., Huang, Y. et al. (2011). *Macromolecules* 44: 8944–8953.

76 Chiefari, J., Chong, Y.K., Ercole, F. et al. (1998). *Macromolecules* 31: 5559–5562.

77 Moad, G., Chong, Y.K., Postma, A. et al. (2005). *Polymer* 46: 8458–8468.

78 Spikes, H. (2008). *Lubr. Sci.* 20: 103–136.

79 Ye, C., Liu, W., Chen, Y., and Yu, L. (2001). *Chem. Commun.* 2244–2245.

80 Zhou, F., Liang, Y., and Liu, W. (2009). *Chem. Soc. Rev.* 38: 2590–2599.

81 Barnhill, W.C., Qu, J., Luo, H. et al. (2014). *ACS Appl. Mater. Interfaces* 6: 22585–22593.

82 Barnhill, W.C., Luo, H., Meyer, H.M. III, et al. (2016). *Tribol. Lett.* 63: 22.

83 Welton, T. (1999). *Chem. Rev.* 99: 2071–2084.

84 Sango, K., Sato, S., Naito, H. et al. (1984). *Ind. Eng. Chem. Prod. Res. Dev.* 23: 315–317.

85 Nurse, R.W., Welch, J.H., and Gutt, W. (1959). *J. Chem. Soc.* 1077–1083.

86 Iler, R.K. (1979). *The Chemistry of Silica: Solubility, Polymerization, Colloid, and Surface Properties, and Biochemistry*. New York: Wiley.

87 Goldstein, J., Newbury, D.E., Echlin, P. et al. (1981). *Scanning Electron Microscopy and X-Ray Microanalysis: A Text for Biologists, Materials Scientists, and Geologists*. New York and London: Plenum Press.

88 Briggs, D. (1998). *Surface Analysis of Polymers by XPS and Static SIMS*. Cambridge: Cambridge University Press.

89 Gil, E.S. and Hudson, S.M. (2004). *Prog. Polym. Sci.* 29: 1173–1222.

90 Seuring, J. and Agarwal, S. (2012). *Macromol. Rapid Commun.* 33: 1898–1920.

91 Jeong, B., Bae, Y.H., Lee, D.S., and Kim, S.W. (1997). *Nature* 388: 860–862.

92 He, C., Kim, S.W., and Lee, D.S. (2008). *J. Controlled Release* 127: 189–207.

93 Yamada, N., Okano, T., Sakai, H. et al. (1990). *Macromol. Rapid Commun.* 11: 571–576.

94 Lee, J., Panzer, M.J., He, Y. et al. (2007). *J. Am. Chem. Soc.* 129: 4532–4533.

95 Kim, J., Hanna, J.A., Byun, M. et al. (2012). *Science* 335: 1201–1205.

96 Hiemenz, P.C. and Lodge, T.P. (2007). *Polymer Chemistry*, 2e. Florida, USA: CRC Press.

97 Hou, L., Chen, Q., An, Z., and Wu, P. (2016). *Soft Matter* 12: 2473–2480.

98 Fu, W.X., Bai, W., Jiang, S.S. et al. (2018). *Macromolecules* 51: 1674–1680.

99 Seuring, J., Bayer, F.M., Huber, K., and Agarwal, S. (2012). *Macromolecules* 45: 374–384.

100 Shimada, N., Ino, H., Maie, K. et al. (2011). *Biomacromolecules* 12: 3418–3422.

101 Willcock, H., Lu, A., Hansell, C.F. et al. (2014). *Polym. Chem.* 5: 1023–1030.

102 Jiang, X.G., Lavender, C.A., Woodcock, J.W., and Zhao, B. (2008). *Macromolecules* 41: 2632–2643.

103 Jiang, X.G. and Zhao, B. (2008). *Macromolecules* 41: 9366–9375.

104 Jin, N.X., Woodcock, J.W., Xue, C.M. et al. (2011). *Macromolecules* 44: 3556–3566.

105 Li, C., Madsen, J., Armes, S.P., and Lewis, A.L. (2006). *Angew. Chem. Int. Ed.* 45: 3510–3513.

106 Woodcock, J.W., Jiang, X.G., Wright, R.A.E., and Zhao, B. (2011). *Macromolecules* 44: 5764–5775.

107 O'Lenick, T.G., Jin, N.X., Woodcock, J.W., and Zhao, B. (2011). *J. Phys. Chem. B* 115: 2870–2881.

108 Fu, W.X. and Zhao, B. (2016). *Polym. Chem.* 7: 6980–6991.

109 Noro, A., Matsushita, Y., and Lodge, T.P. (2009). *Macromolecules* 42: 5802–5810.

110 Huang, G.W., Yu, Q.L., Cai, M.R. et al. (2016). *Adv. Mater. Interfaces* 3: 1500489.

111 Yu, Q.L., Huang, G.W., Cai, M.R. et al. (2016). *Tribol. Int.* 95: 55–65.

112 Ikada, Y. and Uyama, Y. (1993). *Lubricating Polymer Surfaces*, 2–18. Lancaster: Technomic Publishing Company, Inc.

Index

Hairy Nanoparticles: From Synthesis to Applications, First Edition. Edited by Zhiqun Lin and Yijiang Liu.

f

g

h